Selected Titles in This Series

56 **Alexander Barg and Simon Litsyn, Editors,** Codes and Association Schemes

55 **Ding-Zhu Du, Panos M. Pardalos, and Jie Wang, Editors,** Discrete Mathematical Problems with Medical Applications

54 **Erik Winfree and David K. Gifford, Editors,** DNA Based Computers V

53 **Nathaniel Dean, D. Frank Hsu, and R. Ravi, Editors,** Robust Communication Networks: Interconnection and Survivability

52 **Sanguthevar Rajasekaran, Panos Pardalos, and D. Frank Hsu, Editors,** Mobile Networks and Computing

51 **Pierre Hansen, Patrick Fowler, and Maolin Zheng, Editors,** Discrete Mathematical Chemistry

50 **James M. Abello and Jeffrey Scott Vitter, Editors,** External Memory Algorithms

49 **Ronald L. Graham, Jan Kratochvíl, Jaroslav Nešetřil, and Fred S. Roberts, Editors,** Contemporary Trends in Discrete Mathematics

48 **Harvey Rubin and David Harlan Wood, Editors,** DNA Based Computers III

47 **Martin Farach-Colton, Fred S. Roberts, Martin Vingron, and Michael Waterman, Editors,** Mathematical Support for Molecular Biology

46 **Peng-Jun Wan, Ding-Zhu Du, and Panos M. Pardalos, Editors,** Multichannel Optical Networks: Theory and Practice

45 **Marios Mavronicolas, Michael Merritt, and Nir Shavit, Editors,** Networks in Distributed Computing

44 **Laura F. Landweber and Eric B. Baum, Editors,** DNA Based Computers II

43 **Panos Pardalos, Sanguthevar Rajasekaran, and José Rolim, Editors,** Randomization Methods in Algorithm Design

42 **Ding-Zhu Du and Frank K. Hwang, Editors,** Advances in Switching Networks

41 **David Aldous and James Propp, Editors,** Microsurveys in Discrete Probability

40 **Panos M. Pardalos and Dingzhu Du, Editors,** Network Design: Connectivity and Facilities Location

39 **Paul W. Beame and Samuel R Buss, Editors,** Proof Complexity and Feasible Arithmetics

38 **Rebecca N. Wright and Peter G. Neumann, Editors,** Network Threats

37 **Boris Mirkin, F. R. McMorris, Fred S. Roberts, and Andrey Rzhetsky, Editors,** Mathematical Hierarchies and Biology

36 **Joseph G. Rosenstein, Deborah S. Franzblau, and Fred S. Roberts, Editors,** Discrete Mathematics in the Schools

35 **Dingzhu Du, Jun Gu, and Panos M. Pardalos, Editors,** Satisfiability Problem: Theory and Applications

34 **Nathaniel Dean, Editor,** African Americans in Mathematics

33 **Ravi B. Boppana and James F. Lynch, Editors,** Logic and Random Structures

32 **Jean-Charles Grégoire, Gerard J. Holzmann, and Doron A. Peled, Editors,** The SPIN Verification System

31 **Neil Immerman and Phokion G. Kolaitis, Editors,** Descriptive Complexity and Finite Models

30 **Sandeep N. Bhatt, Editor,** Parallel Algorithms: Third DIMACS Implementation Challenge

29 **Doron A. Peled, Vaughan R. Pratt, and Gerard J. Holzmann, Editors,** Partial Order Methods in Verification

28 **Larry Finkelstein and William M. Kantor, Editors,** Groups and Computation II

27 **Richard J. Lipton and Eric B. Baum, Editors,** DNA Based Computers

(Continued in the back of this publication)

Codes and Association Schemes

DIMACS
Series in Discrete Mathematics
and Theoretical Computer Science

Volume 56

Codes and Association Schemes

DIMACS Workshop
Codes and Association Schemes
November 9–12, 1999
DIMACS Center

Alexander Barg
Simon Litsyn
Editors

NSF Science and Technology Center
in Discrete Mathematics and Theoretical Computer Science
A consortium of Rutgers University, Princeton University,
AT&T Labs–Research, Bell Labs (Lucent Technologies),
Telcordia Technologies, and NEC Research Institute

American Mathematical Society

This DIMACS volume contains papers from a DIMACS workshop on Codes and Association Schemes, held November 9–12, 1999.

2000 *Mathematics Subject Classification*. Primary 94Bxx, 05E30, 05E35.

Library of Congress Cataloging-in-Publication Data

DIMACS Workshop Codes and Association Schemes (1999 : DIMACS Center)
Codes and association schemes : DIMACS Workshop Codes and Association Schemes, November 9–12, 1999, DIMACS Center / Alexander Barg, Simon Litsyn, editors.
p. cm. — (DIMACS series in discrete mathematics and theoretical computer science ; v. 56)
Includes bibliographical references.
ISBN 0-8218-2074-5 (alk. paper)
1. Coding theory—Congresses. 2. Association schemes (Combinatorial analysis)—Congresses.
I. Barg, Alexander. II. Litsyn, Simon. III. Title.

QA268 .D55 1999
003′.54—dc21 00-053129

♾ The paper used in this book is acid-free and falls within the guidelines
established to ensure permanence and durability.
Visit the AMS home page at URL: http://www.ams.org/

10 9 8 7 6 5 4 3 2 1 06 05 04 03 02 01

Contents

Preface ix

Foreword xi

Harmonic weight enumerators of nonbinary codes and MacWilliams identities
CHRISTINE BACHOC 1

Numerical results on the asymptotic rate of binary codes
ALEXANDER BARG AND DAVID B. JAFFE 25

Weakly biased arrays, almost independent arrays and error-correcting codes
JÜRGEN BIERBRAUER AND HOLGER SCHELLWAT 33

Indexes of spherical codes
PETER BOYVALENKOV, DANYO DANEV, AND PETER KAZAKOV 47

Codes over $\mathbb{Z}_{p^n}$ and association schemes
PAUL CAMION 59

Bent, resilient functions and the Numerical Normal Form
CLAUDE CARLET AND PHILIPPE GUILLOT 87

On identifying codes
GÉRARD D. COHEN, IIRO HONKALA, ANTOINE LOBSTEIN, AND GILLES ZÉMOR 97

Ternary Hamming and binary perfect covering codes
ITALO J. DEJTER AND KEVIN T. PHELPS 111

A Riemann hypothesis analogue for self-dual codes
IWAN DUURSMA 115

On perfect codes in the Johnson scheme
TUVI ETZION 125

On perfect constant weight codes
TUVI ETZION AND JACK VAN LINT 131

On additive $GF(4)$ codes
PHILIPPE GABORIT, W. CARY HUFFMAN, JON-LARK KIM, AND VERA PLESS 135

Integral zeroes of Krawtchouk polynomials
LAURENT HABSIEGER 151

Some algebra related to P- and Q-polynomial association schemes
TATSURO ITO, KENICHIRO TANABE, AND PAUL TERWILLIGER 167

Bounds for the Christoffel-Darboux kernel of the binary Krawtchouk polynomials
ILIA KRASIKOV 193

Survey of binary Krawtchouk polynomials
ILIA KRASIKOV AND SIMON LITSYN 199

On an algebraic method for bounding the covering radius
TERO LAIHONEN 213

Design systems: Combinatorial characterizations of Delsarte $\mathcal{T}$-designs via partially ordered sets
WILLIAM J. MARTIN 223

The isomorphism problem for circulant graphs via Schur ring theory
MIKHAIL MUZYCHUK, MIKHAIL KLIN, AND REINHARD PÖSCHEL 241

New 1-generator quasi-twisted codes over $GF(5)$
IRFAN SIAP, NUH AYDIN, AND DIJEN RAY-CHAUDHURI 265

Communication complexity and orthogonal polynomials
ULRICH TAMM 277

Estimates of the distance distribution of nonbinary codes, with applications
ALEXEI ASHIKHMIN, ALEXANDER BARG, AND SIMON LITSYN 287

Preface

Though theory of association schemes first appeared in statistical applications, its development is mostly associated with a general view of error-correcting codes suggested by Delsarte in 1973. Since then it has grown into an independent branch of combinatorics, while maintaining a sustained connection with coding theory.

This volume collects papers presented at or closely related to the workshop "Codes and Association Schemes" held at DIMACS, Rutgers University, Piscataway, NJ in November 1999. The goal of the workshop was to demonstrate new applications of the theory of association schemes to coding and to give coding theorists an opportunity to familiarize themselves with new properties of association schemes, both polynomial and general.

The papers in the volume can be divided into the following categories: applications of association schemes to codes with a special subsection on the applications of the polynomial method, structural results for codes, structural results for schemes, and properties of orthogonal polynomials and their applications in combinatorics. In the first area the paper by C. Bachoc discusses a refined version of weight enumerators of nonbinary linear codes, called harmonic weight enumerators. As usual, they satisfy MacWilliams-type identities and provide a tool for computing the intersection numbers and coset weight distribution of self-dual codes. P. Camion in his paper characterizes Gray-map-like isometries between integer residue rings and their residue fields. T. Etzion rules out new sets of parameters of perfect codes in the Johnson space, making an advance in the difficult problem raised by Delsarte over 25 years ago. Applications of the polynomial method are discussed in the paper by T. Laihonen where he establishes a new upper bound on the covering radius of a linear code via its dual distance. Along the same lines, P. Boyvalenkov, D. Danev and P. Kazakov prove new results on indexes of spherical designs, A. Ashikhmin, A. Barg, and S. Litsyn give new bounds on the distance distribution of nonbinary codes and discuss their applications, and A. Barg and D. B. Jaffe collect some experimental results on the asymptotic rate of binary codes implied by the polynomial method and speculate on its potential limits.

The most prominent families of orthogonal polynomials appearing in coding theory are Krawtchouk polynomials, orthogonal on $\{0, 1, \dots, n-1\}$ with weight $q^{-n}\binom{n}{i}(q-1)^n$ and a particular family of discrete Hahn polynomials. A collection of properties of binary Krawtchouk polynomials appears in the paper by I. Krasikov and S. Litsyn, written with coding theory applications in mind. L. Habsieger makes advances in a difficult problem of bounding the number of integer zeros of Krawtchouk polynomials. I. Krasikov presents refined bounds for the Christoffel-Darboux kernel of binary Krawtchouk polynomials. U. Tamm suggests new applications of the Krawtchouk and Hahn polynomials in the rank method of bounding the communication complexity of functions related to the Hamming distance.

Structural properties of association schemes are studied in the paper by T. Ito, K. Tanabe, and P. Terwilliger which fits in the general program of classifying metric/cometric association schemes. The main results of the paper are related to classification of some irreducible modules in the subconstituent algebra of the scheme. W. J. Martin extends the notion of T-design to families of association schemes represented by maximal elements of partially ordered sets. The paper by M. Muzychuk, M. Klin and R. Pöschel surveys properties of Schur rings and applies them to obtain new necessary conditions for the isomorphism problem of circulant graphs.

Finally, papers in a large group devoted to properties of error-correcting codes, though they sometimes do not mention explicitly the context of association schemes, deal with problems motivated by the combinatorial view of codes. In this vein, T. Etzion and J. van Lint present a family of perfect single error correcting constant weight codes over the alphabets of size $2^k + 1$, and J. Bierbrauer and H. Schellwat investigate combinatorial applications of codes with a known dual distance. More diverse results in the remaining papers still follow the same general view of codes and fit well in the cultural paradigm of combinatorial coding theory. G. D. Cohen, I. Honkala, A. Lobstein and G. Zémor discuss parameters of codes in which every point is identified uniquely by the set of its closest neighbors. I. J. Dejter and K. T. Phelps give a simple characterization of a perfect dominating set in the 13-dimensional cube. C. Carlet and P. Guillot contribute to the program of classifying maximally nonlinear (bent) functions and P. Gaborit, W. C. Huffman, J.-L. Kim, and V. Pless classify short additive quaternary codes. New codes over $GF(5)$ are constructed in the paper by I. Siap, N. Aydin, and D. Ray-Chaudhuri. I. Duursma suggests an unexpected analogy between algebraic curves and linear codes based on a new form of the weight enumerator of the code and formulates a "Riemann hypothesis" for the zeros of the new enumerator.

We would like to use this opportunity to thank DIMACS Center for financial and organizational support of the workshop. The help provided by the Center turned the meeting into an enjoyable event for the organizers and hopefully for the other participants.

Alexander Barg
Math. Research
Bell Labs, Lucent Technologies
600 Mountain Ave., Rm. 2C-375
Murray Hill, NJ 07974, USA
abarg@research.bell-labs.com

Simon Litsyn
Dept. of EE-Systems
Tel-Aviv University
Ramat Aviv 69978, Israel
litsyn@eng.tau.ac.il

Foreword

The Workshop on Codes and Association Schemes was held November 9–12, 1999 at the DIMACS Center at Rutgers University in Piscataway, NJ. We would like to express our appreciation to Alexander Barg and Simon Litsyn for their efforts to organize and plan this successful conference. This volume consists of papers either presented at the workshop or closely related to it.

The theory of association schemes has taken its position as a branch of combinatorics, with close ties to coding theory. This volume collects papers on applications of association schemes to codes, structural results for codes, structural results for schemes, and properties of orthogonal polynomials (with applications in combinatorics).

DIMACS gratefully acknowledges the generous support that makes both its workshop and its publication programs possible. Special thanks go to the National Science Foundation, through its Science and Technology Centers program, the New Jersey Commission on Science and Technology, and DIMACS's partners at Rutgers, Princeton, AT&T Labs-Research, Bell Labs, NEC Research Institute, and Telcordia Technologies.

Fred S. Roberts
Director

Robert Sedgewick
Co-Director for Princeton

DIMACS Series in Discrete Mathematics
and Theoretical Computer Science
Volume **56**, 2001

Harmonic weight enumerators of nonbinary codes and MacWilliams identities

Christine Bachoc

ABSTRACT. We define harmonic weight enumerators asociated to codes defined over a group alphabet F of size q. They generalize the classical Hamming weight enumerator and are associated to the decomposition of the space $L(F^n)$ under the action of the group $S_{q-1}^n \rtimes S_n$. They satisfy a MacWilliams type identity, which leads to relative invariant polynomials in the case of self-dual codes. Applications to the study of additive quaternary codes are developed.

1. Introduction

This paper extends the ideas and results of [**1**] to the case of non-binary linear codes. Let C be a linear code of length n over an alphabet F which is an abelian group of size q. We introduce polynomials which generalize the Hamming weight enumerator of the code, using "harmonic functions", after the work of C. Dunkl on harmonic analysis for the group $S_{q-1}^n \rtimes S_n$ and of H. Tarnanen, J. Aaltonen, J.-M. Goethals on the non-binary Johnson scheme (in the setting of association schemes).

Let U_m be the set of words in F^n of Hamming weight equal to m. Let $L(U_m)$ be the space of complex valued functions on U_m; the group $S_{q-1}^n \rtimes S_n$ acts on this space and one can consider its irreducible decomposition, which is described by C. Dunkl in [**6**] (see Section 2.). If $\mathcal{B} \subset U_m$, the property that $\mathcal{B}$ is a "design of type $\lambda = q-1$" in the sense of Delsarte ([**5**]) is equivalent to the orthogonality of the characteristic function of $\mathcal{B}$ with some of these irreducible components (see Proposition 2.1).

If f belongs to one of these irreducible components, we define the weight enumerator associated to f and to an additive code C to be

$$W_{C,f}(X,Y) := \sum_{u \in C} f(u) X^{n-|u|} Y^{|u|} \tag{1}$$

where $|u|$ denotes the Hamming weight of u. Hence these polynomials are a nice tool to study the designs contained in the set of words of fixed Hamming weight in a given code.

We prove a MacWilliams type identity for these polynomials (more precisely for the polynomials obtained after division by some power of X and Y, see Definition

1991 *Mathematics Subject Classification.* Primary 94B05; Secondary 05B05, 05E35.

4.1 and Theorem 4.1). This identity involves an operator T acting on harmonic functions; in Section 5 we examine the case when T is homothetic. We prove that it corresponds to the binary, ternary and quaternary codes. In Section 6, we study more precisely the ternary and quaternary self-dual codes, because the harmonic weight enumerators are in these cases relative invariant polynomials. It allows us to extend, in Corollaries 6.1, 6.2, some results due to P. Delsarte on the existence of designs on the set of codewords of fixed Hamming weight of extremal codes. Section 7 is devoted to the application of this tool to the computation of intersection numbers. It makes use of the zonal functions associated to the subgroup $S_{q-1}^n \rtimes S_n$, which are expressed in terms of Krawtchouk and Hahn polynomials. We work out some examples in the case of even self-dual additive quaternary codes.

2. Group representation and the non-binary Johnson scheme

In this section we recall some notations and results on the Hamming and Johnson schemes previously settled by C. F. Dunkl ([**6**]) in the language of group representations and by Tarnanen, Aaltonen and Goethals ([**14**]) in terms of association schemes. We also discuss the connection with the notion of "design of type $\lambda = q-1$" introduced by P. Delsarte.

Let F be a finite alphabet of size $q \geq 2$. We denote by $F := \{a_0, \dots, a_{q-1}\}$ its elements. The group of permutations on q elements S_q acts on F by $a_i\sigma = a_{\sigma^{-1}(i)}$. We specify a series of subgroups of S_q denoted $\{1\} \subset \cdots \subset S_{q-1} \subset S_q$, where S_{q-i} is the subgroup of S_q fixing $a_0, \dots, a_{i-1}$ and is isomorphic to the group of permutations on $q-i$ elements.

Let $L(F)$ be the space of complex valued functions on F, equiped with the scalar product

$$< f, g > = \frac{1}{q} \sum_{x \in F} f(x)\overline{g(x)}.$$

It is endowed with the left action of S_q given by $(\sigma.f)(x) = f(x\sigma)$. The decomposition into irreducible S_q-modules of $L(F)$ is given by

$$L(F) = \mathbb{C}\mathbf{1} \perp L_1 \tag{2}$$

where $\mathbf{1}$ is the all-one function. We set $z_0 := \mathbf{1}$ and we define by induction an element z_i in L_i, and a subspace L_{i+1} of $L(F)$ such that the decomposition of L_i into S_{q-i}-irreducible modules, equals $\mathbb{C}z_i \perp L_{i+1}$, with trivial action on $\mathbb{C}z_i$. We normalize as in [**6**] the z_i by $(z_i, z_i) = 1/(q-1)$. Note that the z_i are determined from the above properties up to the multiplication by a complex number of module 1. The recursive computation of their values shows that they can be assumed to be real-valued, and hence are uniquely determined up to a sign. In particular, we fix $z_1(a_0) = 1$ and $z_1(a_i) = -1/(q-1)$ for $i \geq 1$. We shall make frequent use of the following properties: for $i \geq 2$, $z_i(a_0) = 0$ and, for $i \geq 1$, $\sum_{j=0}^{q-1} z_i(a_j) = 0$.

The functions z_i are explicitly realized in [**6**] as coordinate functions on F, seen as the regular simplex in $\mathbb{R}^{q-1}$.

Let $X := F^n$. The action of S_n on X given by

$$(x_1, \dots, x_n).\tau = (x_{\tau(1)}, \dots, x_{\tau(n)}),$$

combined with the action of S_q on F gives rise to a transitive action of the group $G := S_q^n \rtimes S_n$ on X. We denote by H the subgroup of G stabilizing $(a_0, \dots, a_0)$. It is isomorphic to $S_{q-1}^n \rtimes S_n$. Again we consider the space $L(X)$ together with the induced action of G (note that it is the representation of G induced by the trivial representation on H), and the hermitian product $< f, g >= \frac{1}{|X|} \sum_{x \in X} f(x)\overline{g(x)}$. Following [**6**], let, for all $u \in \{0, 1, \ldots, q-1\}^n$, $z_u \in L(X)$ be defined by

$$z_u(x) := \prod_{i=1}^{n} z_{u_i}(x_i).$$

If $|u|$ denotes the number of non-zero coordinates of u, then we have $< z_u, z_v >= \delta_{u,v}(q-1)^{-|u|}$. From (2), the subspace P_m spanned by $\{z_u \mid |u| = m\}$ is G-invariant and

$$L(X) = \oplus_{m=0}^{n} P_m \tag{3}$$

is the decomposition of $L(X)$ into G-irreducible subspaces.

We shall make use of the description given in [**6**] of the decomposition of each P_m as H-modules. Moreover, the spaces P_m are linked to the Johnson schemes via the following: for $b \in X$, let $|b|$ be the Hamming weight i.e. the number of coordinates of b not equal to a_0, and let $S(b)$ be its support, i.e. the set of coordinates i such that $b_i \neq a_0$. Let

$$U_m := \{b \in X \mid |b| = m\}.$$

Note that the set U_m is one orbit under the action of H. Then, as an H-module, the space $L(U_m)$ is isomorphic to P_m via the map: $\rho : P_m \to L(U_m)$ defined by

$$\rho z_u(b) = \begin{cases} 0 & \text{if } S(b) \neq S(u) \\ z_u(b) & \text{if } S(b) = S(u) \end{cases}.$$

Let, for $0 \leq l \leq m$, $P_{m,l}$ be the subspace of P_m spanned by the z_u such that $u_i = 1$ for exactly $m - l$ coordinates, and let $L(U_m)_l$ be the image of $P_{m,l}$ by ρ. In the special case $q = 2$, of course we only consider the case $l = 0$ and $P_{m,0} = P_m$. From the fact that S_{q-1} acts trivially on z_1, the space $P_{m,l}$ is H-invariant. It is worth noticing that $P_{m,0}$ is the space of functions f such that $f(b)$ only depends on $S(b)$, and hence can be identified with the corresponding space P_m over $\mathbb{F}_2^n$.

If x_i denotes the function defined by $x_i(b) = z_1(b_i)$, the operator

$$d := \sum_{i=1}^{n} \frac{\partial}{\partial x_i}$$

commutes with the action of H and maps $P_{m,l}$ to $P_{m-1,l}$. Its adjoint operator is denoted by d^*. Then, the decomposition of P_m into irreducible H-submodules is given by ([**6**]):

$$P_m = \oplus_{l=0}^{m} \oplus_{k=l}^{\min(m,n+l-m)} P_{m,l,k} \tag{4}$$

where

$$P_{m,l,k} := (d^*)^{m-k}(P_{k,l} \cap \ker d). \tag{5}$$

In the case $q = 2$, there is only the summand $l = 0$.

We define an incidence relation between the elements of X by:

$$x \leq y \text{ if } x_i \neq a_0 \Rightarrow x_i = y_i.$$

Following P. Delsarte ([**5**]), we define the notion of design in the context of a non-binary alphabet:

DEFINITION 2.1. *Let $\mathcal{B}$ be a set of elements of F^n, of equal Hamming weights m. $\mathcal{B}$ is called a t-design if, for all u in F^n, the number*

$$\operatorname{card}\{B \in \mathcal{B} \mid u \leq B\}$$

only depends on the weight of u.

REMARK 2.1. *This is called a "t-design of type $\lambda = q-1$" in [**5**]. It generalizes of course the notion of classical designs which corresponds to the case of a binary alphabet. It is worth noticing that, if $\mathcal{B}$ is a t-design in the sense of the previous definition then the supports of the elements of $\mathcal{B}$ hold a t-design in the classical sens with possibly repeated blocks.*

The t-design property has got a nice characterization in terms of orthogonality with some of the irreducible components in the decomposition of $L(U_m)$, just like in the binary case. This characterization is proved in [**13**] in the setting of association schemes. For sake of completness, we give a proof starting from the description of the irreducible decomposition of the spaces $L(U_m)$.

An easy computation shows that the operator d induces via the isomorphism ρ an operator again denoted d:

$$\begin{aligned} L(U_m) &\to L(U_{m-1}) \\ x &\to (df)(x) = -\sum_{\substack{y \in U_m \\ x \leq y}} f(y). \end{aligned} \tag{6}$$

We take the following notations:

$$L(U_m)_l := \rho(P_{m,l})$$

and

$$L(U_m)_{(k,l)} := \rho(P_{m,l,k}) = (d^*)^{m-k}(L(U_k)_l \cap \ker d)$$

so that

$$L(U_m) = \bigoplus_{L_m} L(U_m)_{(k,l)} \tag{7}$$

where

$$L_m := \{(k,l) \mid 0 \leq l \leq k \leq \min(m, n+l-m)\}. \tag{8}$$

PROPOSITION 2.1. *Let $\mathcal{B} \subset U_m$ and let $\beta := \sum_{B \in \mathcal{B}} B \in L(U_m)$ be the characteristic function of $\mathcal{B}$. The following conditions are equivalent:*

1. *$\mathcal{B}$ is a t-design.*
2. *β is orthogonal to $L(U_m)_{(k,l)}$ for all $(k,l) \in L_m$, $k \leq t$, $(l,k) \neq (0,0)$.*
3. *$\sum_{B \in \mathcal{B}} f(B) = 0$ for all $(k,l) \in L_m$, $k \leq t$, $(l,k) \neq (0,0)$.*

PROOF. Note first that, in the decomposition (7), $L(U_m)_{(0,0)} = \mathbb{C}.\mathbf{1}$ is the one-dimensional subspace spanned by the all-one function on U_m. Let τ be the characteristic function of an element $u \in U_t$. Clearly, from the above interpretation of $d : L(U_m) \to L(U_{m-1})$, the property required in the definition of a t-design is equivalent to ask that $< \beta, (d^*)^{m-t}\tau >_{L(U_m)}$ only depends on t. But $< (d^*)^{m-t}\tau, \mathbf{1} >_{L(U_m)} = \lambda(t) < \mathbf{1}, \mathbf{1} >_{L(U_m)}$ only depends on t since it counts the number of $x \in U_m$ such that $u \leq x$. From the decomposition (7),

$$\tau = \sum_{\substack{(k,l)\in L_m, k\leq t\\ (k,l)\neq(0,0)}} \tau_{k,l} + \lambda(t)\mathbf{1} \tag{9}$$

where $\tau_{k,l} \in L(U_m)_{(k,l)}$. This shows the equivalence of 1) and 2). Point 3) follows from $< \beta, f >_{L(U_m)} = 1/|U_m| \sum_{B\in\mathcal{B}} f(B)$. □

REMARK 2.2. *The weaker property that the supports of the blocks hold a t-design in the classical sense is equivalent to the orthogonality of β with the subspaces $L(U_m)_{(k,0)}$ for $k \leq t$ (since, as was already mentionned, $P_{k,0}$ is the space of functions f such that $f(B)$ only depends on $S(B)$).*

3. The alphabet F is a group

In this section we assume that the alphabet F has a structure of an abelian group $(F, +)$, for which $a_0 = 0$. We assume that F is endowed with a non degenerated symmetric bilinear map

$$\begin{aligned} F \times F &\to (\mathbb{C}^*, \times) \\ (x, y) &\to (x, y) \end{aligned} \tag{10}$$

The choice of such a duality is equivalent to a specification of an isomorphism between F and its group of characters, given by $x \to (., x)$. In particular, we shall make frequent use of the orthogonality relations between the characters of F.

The space $X = F^n$ is endowed with $(x, y) := \prod_{i=1}^n (x_i, y_i)$. If C is a linear code in X, i.e. a subgroup of X, $C^\perp$ denotes its orthogonal code with respect to $(,)$:

$$C^\perp := \{u \in X \mid (u, v) = 1 \ \forall\ v \in C\} \tag{11}$$

We define another operator on $L(U_m)$ which will be of major importance in the generalized MacWilliams formulas:

DEFINITION 3.1. *Let $f \in L(U_m)$. Let Tf be defined, for all $u \in U_m$, by*

$$Tf(u) := \sum_{\substack{b\in U_m\\ S(b)=S(u)}} (u, b) f(b) \tag{12}$$

Note that the operator T depends on the choice of the duality $(,)$. We denote by $\overline{T}$ the operator corresponding to the conjugate duality $\overline{(x, y)}$.

PROPOSITION 3.1. *The following properties hold for the operator T:*

1. *$T\mathbf{1} = (q-1)z_1$, $Tz_1 = -z_1$, and, for $i \geq 2$, Tz_i is a linear combination of $z_2, \dots, z_{q-1}$.*

2. *If* $f = \rho z_u$, $Tf(v) = \prod_{i\in S(u)} Tz_{u_i}(v_i)$.
3. *The operator* T *is linear, and maps* $L(U_m)_l$ *into itself injectively.*
4. *For all* $f, g \in L(U_m)$, $< Tf, g > = < f, \overline{T}g >$.
5. *For all* $f \in L(U_m)$, $dTf = -Tdf$ *and* $d^*Tf = -Td^*f$.

PROOF. We first prove 4.

$$
(13)\qquad
\begin{aligned}
< Tf, g > &= \frac{1}{|U_m|} \sum_{u\in U_m} (Tf)(u)\overline{g(u)} \\
&= \frac{1}{|U_m|} \sum_{u\in U_m} \Big(\sum_{\substack{b\in U_m \\ S(b)=S(u)}} (u,b) f(b)\Big)\overline{g(u)} \\
&= \frac{1}{|U_m|} \sum_{b\in U_m} f(b) \Big(\sum_{\substack{u\in U_m \\ S(u)=S(b)}} (u,b)\overline{g(u)}\Big) \\
&= < f, \overline{T}g > .
\end{aligned}
$$

In order to prove 1., we compute $Tz_1(u)$. From the orthogonality relations, $Tz_1(u) = \sum_{b\neq a_0} (u,b) z_1(b) = -1/(q-1)\sum_{b\neq a_0}(u,b) = -z_1(u)$. The computation of $T\mathbf{1}$ goes the same. The assertion on Tz_i for $i \geq 2$ follows from 4.

Let us prove 2.: we assume $f = \rho z_u$. Let $v \in U_m$ with $S(v) = S(u)$. Then $f(v) = \prod_{i\in S(u)} z_{u_i}(v_i)$ and

$$
(14)\qquad
\begin{aligned}
Tf(v) &= \sum_{\substack{b\in U_m \\ S(b)=S(u)}} (v,b) f(b) \\
&= \sum_{b_i\neq a_0} \prod_{i\in S(u)} (v_i, b_i) z_{u_i}(v_i) \\
&= \prod_{i\in S(u)} \Big(\sum_{\lambda\neq a_0} (v_i,\lambda) z_{u_i}(v_i)\Big) \\
&= \prod_{i\in S(u)} Tz_{u_i}(v_i)
\end{aligned}
$$

which proves 2. From 1. and 2., it follows that Tf is a linear combination of some ρz_w, where the number of coordinates of w which are not equal to 0, respectively which are equal to 1, is the same as for u. Hence T maps $L(U_m)_l$ into itself.

Assume $Tf = 0$. Let S be a fixed set of m coordinates. Since for all $u \in U_m$, such that $S(u) = S$, $Tf(u) = 0$, the following system of linear equations holds:

$$
(15)\qquad \sum_{\substack{b\in U_m \\ S(b)=S(u)}} (u,b) f(b) = 0 \text{ for all } u \in U_m
$$

We only have to prove that the matrix of this linear system is invertible. But this matrix is the $(q-1)^m \times (q-1)^m$-symetric matrix $((u,v))_{u,v\in U_m}$; it is a submatrix of the matrix $((u,v))_{u,v\in F^m}$ which is invertible since the characters $(.,v)$ span $L(F^m)$, and hence it has full rank.

We prove the equality $d^*Tf = Td^*f$; a similar proof can be given for d. Let $u \in U_{m+1}$.

$$Td^*f(u) = \sum_{\substack{c\in U_{m+1}\\ S(c)=S(u)}} (u,c)d^*f(c)$$

$$= - \sum_{\substack{c\in U_{m+1}\\ S(c)=S(u)}} (u,c) \sum_{\substack{b\in U_m\\ b\leq c}} f(b) \tag{16}$$

$$= - \sum_{\substack{b\in U_m\\ S(b)\subset S(u)}} \Big(\sum_{\substack{c\in U_{m+1}\\ S(c)=S(b)\\ b\leq c}} (u,c)\Big) f(b)$$

For a fixed $b \in U_m$ such that $S(b) \subset S(u)$, let i_b be the index in $S(u)$ but not in $S(b)$. Then, $\sum_{\substack{c\in U_{m+1}\\ S(c)=S(b)\\ b\leq c}} (u,c) = (u,b)\sum_{\substack{\lambda\in F\\ \lambda\neq a_0}} (u_{i_b},\lambda) = -(u,b)$, so

$$Td^*f(u) = \sum_{\substack{b\in U_m\\ S(b)\subset S(u)}} (u,b)f(b). \tag{17}$$

But, for a fixed $b \in U_m$ with $S(b) \subset S(u)$, we have $(u,b) = (v,b)$ where v is the only element in F^n such that $v \leq u$ and $S(v) = S(b)$. Hence

$$Td^*f(u) = \sum_{\substack{v\in U_m\\ v\leq u}} \Big(\sum_{\substack{b\in U_m\\ S(b)=S(v)}} (v,b)f(b)\Big) = -d^*Tf(u). \tag{18}$$

□

REMARK 3.1. *The operator T doesn't in general commute with the action of H. We discuss this possibility in section 5.*

In the case of the binary alphabet $F = \mathbb{F}_2$, and more generaly if $l = 0$, meaning that $f(u)$ only depends on $S(u)$, one easily sees that $Tf(u) = (-1)^{|u|}f(u)$.

4. A MacWilliams type identity

Let C be a linear code in F^n and let $C^\perp$ be its orthogonal code as defined in previous section. We are going to define harmonic weight enumerators associated to C and to prove a MacWilliams type identity for them.

Let $f \in L(U_k)_l \cap \ker d$. We set, for all $x \in X = F^n$,

$$D^*f(x) := \sum_{\substack{u\in U_k\\ u\leq x}} f(u) \tag{19}$$

Note that, if $|x| \geq k$, $D^*f(x) = (-1)^{|x|-k}\frac{(d^*)^{|x|-k}}{(|x|-k)!}f(x)$. If $|x| < k$, $D^*f(x) = 0$.

DEFINITION 4.1. *Let $f \in L(U_k)_l \cap \ker d$. The harmonic weight enumerator associated to C and f is*

$$Z_{C,f}(X,Y) := \sum_{u\in C} D^*f(u)X^{n-|u|-k+l}Y^{|u|-k}$$

REMARK 4.1. *Note that it is not clear yet that $Z_{C,f}$ is a polynomial. It will follow from Proposition 4.1. In the binary case, $l = 0$ is the only possibility, and $Z_{C,f}$ coincides with the notion introduced in* [**1**]. *The slightly more general case $l = 0$ over $\mathbb{F}_q$ is actually treated in* [**12**].

The following proposition is the key property needed to prove that $Z_{C,f}$ is a polynomial satisfying a MacWilliams type transformation formula. Its proof is postponed to Subsection 4.1.

PROPOSITION 4.1. *Let $f \in L(U_k)_l \cap \ker d$ and let $u \in X = F^n$. For all $i = 0, \dots, k$,*

$$\sum_{\substack{b \in U_k \\ |S(b) \cap S(u)| = k-i}} (u, b) f(b) = (q-1)^i \binom{k-l}{i} D^* T f(u) \tag{20}$$

The main property of the $Z_{C,f}$ is the following:

THEOREM 4.1. *Let $f \in L(U_k)_l \cap \ker d$ and let C be a linear subcode of F^n. Let $C^\perp$ be its orthogonal code. Then, $Z_{C,f}(X, Y)$ is a homogeneous polynomial of degree $n - 2k + l$, and*

$$Z_{C^\perp, f}(X, Y) = \frac{1}{|C|} q^{k-l} Z_{C, Tf}(X + (q-1)Y, X - Y) \tag{21}$$

4.1. Proof of Proposition 4.1. We proceed by induction on i. The case $i = 0$ is the identity

$$\sum_{\substack{b \in U_k \\ S(b) \subset S(u)}} (u, b) f(b) = D^* T f(u)$$

which follows from (17) and (18) by iteration of d^*. Let us now assume the identity for all $j \le i - 1$. From the fact that $f \in \ker d$ and from $Tdf = -dTf$, we know that $Tf \in \ker d$. Let $y \in U_{k-i}$. We have

$$\sum_{\substack{t \in U_k \\ y \le t}} Tf(t) = 0 \tag{22}$$

which means

$$\sum_{\substack{t \in U_k \\ y \le t}} \sum_{\substack{b \in U_k \\ S(b) = S(t)}} (t, b) f(b) = 0 \tag{23}$$

We sum up these equations over the set of $y \le u$, with $y \in U_{k-i}$. Hence all the $b \in U_k$ such that $|S(b) \cap S(u)| \ge k - i$ will contribute in the sum. We rearrange the sum over j such that $|S(b) \cap S(u)| = k - j$. We set

$$S(u, j) := \{b \in U_k \mid |S(b) \cap S(u)| = k - j\} \tag{24}$$

and obtain:

$$\sum_{j=0}^{i} \sum_{b \in S(u,j)} A_b f(b) = 0 \tag{25}$$

where

$$A_b := \sum_{\substack{y \in U_{k-i} \\ y \leq u \\ S(y) \subset S(u) \cap S(b)}} \sum_{\substack{t \in U_k \\ S(t)=S(b) \\ y \leq t}} (b, t). \tag{26}$$

If we denote by $b_{S(y)}$ the element of F^n which is equal to b over $S(y)$ and to a_0 outside, we have

$$\begin{aligned} \sum_{\substack{t \in U_k \\ S(t)=S(b) \\ y \leq t}} (b, t) &= (b_{S(y)}, u) \sum_{t_s \neq a_0} \prod_{s \in S(b) \setminus S(y)} (b_s, t_s) \\ &= (b_{S(y)}, u) \prod_{s \in S(b) \setminus S(y)} \left(\sum_{\lambda \neq a_o} (b_s, \lambda) \right) \\ &= (b_{S(y)}, u)(-1)^i \end{aligned} \tag{27}$$

and we are left with, if $S_{k-i} := S(y)$ runs over subsets of size $k-i$,

$$A_b = (-1)^i \sum_{S_{k-i} \subset S(b) \cap S(u)} (b_{S_{k-i}}, u). \tag{28}$$

LEMMA 4.1. *If $f \in L(U_k)_l$, for all $u \in F^n$,*

$$\sum_{b \in S(u,j)} A_b f(b) = (-1)^j \binom{k-l-j}{i-j} (q-1)^{i-j} \sum_{b \in S(u,j)} (b, u) f(b) \tag{29}$$

PROOF. The equality (29) is linear in f, so it is enough to verify it for $f = \rho z_v$. The fact that $f \in L(U_k)_l$ means that the number of non zero coordinates of v is k and that the number of coordinates of v equal to 1 is $k-l$. Let $S := S(v)$; from the definition of ρz_v, the only b having a non zero contribution in the left or right handsides are the ones with $S(b) = S$. Hence, if $|S(u) \cap S| \neq k-j$, both sides are equal to 0, so we assume that $|S(u) \cap S| = k-j$. Then, if L denotes the left handside of (29), and from the expression (28) for A_b,

$$\begin{aligned} L &= (-1)^i \sum_{\substack{b \in U_k \\ S(b)=S}} \sum_{S_{k-i} \subset S \cap S(u)} \prod_{s \in S_{k-i}} (b_s, u_s) \prod_{s \in S} z_{v_s}(b_s) \\ &= (-1)^i \sum_{S_{k-i} \subset S \cap S(u)} \sum_{\substack{b \in U_k \\ S(b)=S}} \prod_{s \in S_{k-i}} (b_s, u_s) z_{v_s}(b_s) \prod_{s \in S \setminus S_{k-i}} z_{v_s}(b_s) \\ &= (-1)^i \sum_{S_{k-i} \subset S \cap S(u)} \prod_{s \in S_{k-i}} \left(\sum_{\lambda \neq a_0} (\lambda, u_s) z_{v_s}(\lambda) \right) \prod_{s \in S \setminus S_{k-i}} \left(\sum_{\lambda \neq a_0} z_{v_s}(\lambda) \right). \end{aligned} \tag{30}$$

But $\sum_{\lambda\neq a_0} z_{v_s}(\lambda) = -z_{v_s}(a_0)$ since $s \in S$, and $z_{v_s}(a_0) = 1$ or 0, respectively if $v_s = 1$ or $v_s \geq 2$. Hence the only subsets S_{k-i} having a non zero contribution are the ones for which $S \setminus S_{k-i} \subset \{s \mid v_s = 1\}$. Such subsets exist only if $v_s = 1$ outside of $S(u)$, and if $l \leq k-i$ (since S_{k-i} must cover the coordinates s of v with $v_s \neq 1$). Moreover, in that case, since $v_s = 1$ outside of S_{k-i}, since $Tz_1 = -z_1$ and $z_1(\lambda) = -1/(q-1)$ if $\lambda \neq a_0$,

$$\prod_{s\in S_{k-i}} \Big(\sum_{\lambda\neq a_0} (\lambda, u_s) z_{v_s}(\lambda)\Big) = \prod_{s\in S_{k-i}} Tz_{v_s}(u_s) = (q-1)^{i-j} \prod_{s\in S\cap S(u)} Tz_{v_s}(u_s).$$

One sees easily that the number of S_{k-i} with such a contribution is equal to $\binom{k-l-j}{i-j}$. Finally, L equals 0 if $v_s \neq 1$ on $S \setminus S(u)$ and equals $\binom{k-l-j}{i}(q-1)^{i-j}\prod_{s\in S\cap S(u)} Tz_{v_s}(u_s)$ otherwise. Since $Tz_1(a_0) = -z_1(a_0) = -1$ while, for $s \geq 2$, $Tz_s(a_0) = 0$, in any case,

$$L = (-1)^j \binom{k-l-j}{i-j} (q-1)^{i-j} \prod_{s\in S} Tz_{v_s}(u_s).$$

The right handside of (29) is easily computed:

$$\begin{aligned}\sum_{b\in S(u,j)} (b,u) f(b) &= \sum_{\substack{b\in U_k\\ S(b)=S}} \prod_{s\in S} (b_s, u_s) z_{v_s}(b_s) \\ &= \prod_{s\in S} \Big(\sum_{\lambda\neq a_0} (\lambda, u_s) z_{v_s}(\lambda)\Big) \qquad (31)\\ &= \prod_{s\in S} Tz_{v_s}(u_s).\end{aligned}$$

□

We return to the proof of Proposition 4.1. Equation (25) becomes, applying Lemma 4.1,

$$\sum_{j=0}^{i} (-1)^j \binom{k-l-j}{i-j} (q-1)^{i-j} \sum_{b\in S(u,j)} (b,u) f(b) = 0 \qquad (32)$$

and, applying the induction hypothesis to $j \leq i-1$,

$$(-1)^i \sum_{b\in S(u,i)} (b,u) f(b) = -(q-1)^i \sum_{j=0}^{i-1} (-1)^j \binom{k-l-j}{i-j} \binom{k-l}{j} D^* Tf(u). \qquad (33)$$

From $\binom{k-l-j}{i-j}\binom{k-l}{j} = \binom{k-l}{i}\binom{i}{j}$, we get

$$\begin{aligned}(-1)^i \sum_{b\in S(u,i)} (b,u) f(b) &= -(q-1)^i \binom{k-l}{i} D^* Tf(u) \sum_{j=0}^{i-1} (-1)^j \binom{i}{j} \qquad (34)\\ &= (q-1)^i \binom{k-l}{i} D^* Tf(u) (-1)^i\end{aligned}$$

which is the expression of Proposition 4.1.

4.2. Proof of Theorem 4.1. We first prove that $Z_{C,f}(X,Y)$ is a polynomial. If $u \in C$ satisfies $n-|u|-k+l<0$, let i be an integer with $n-|u|<i\leq k-l$. Equation (20) of Proposition 4.1 proves that $D^*Tf(u)=0$, since no $b\in U_k$ with $|S(b)\cap S(u)|=k-i$ can exist. We conclude from the property $D^*Tf=-TD^*f$ and from the injectivity of T (see Proposition 3.1).

Now we prove the transformation formula (21). Therefore, as in [**1**], we compute the Fourier transform (over F^n) of

$$\Phi(u) := D^*f(u)X^{n-|u|-k+l}Y^{|u|-k} \tag{35}$$

which is

$$\hat{\Phi}(u) := \sum_{v\in F^n}(u,v)\Phi(v). \tag{36}$$

The formula will then follow directly from the Poisson summation formula:

$$\sum_{u\in C^\perp}\Phi(u) = \frac{1}{|C|}\sum_{v\in C}\hat{\Phi}(v). \tag{37}$$

Just like in the binary case [**1**], we first consider the case $f=\delta_b$ where δ_b is defined by $\delta_b(u)=1$ if $u=b$ and 0 if $u\neq b$. We denote by $u_{\overline{S}(b)}$ the element of F^n which is equal to u on the complementary set of $S(b)$ and to a_0 elsewhere.

LEMMA 4.2. *Let $f=\delta_b\in L(U_k)$. Then*

$$\hat{\Phi}(u) = (u,b)X^{-k+l}(X+(q-1)Y)^{n-k-|u_{\overline{S}(b)}|}(X-Y)^{|u_{\overline{S}(b)}|} \tag{38}$$

PROOF. Since $D^*\delta_b(u)=1$ if $b\leq u$ and 0 otherwise,

$$\hat{\Phi}(u) = \sum_{\substack{v\in F^n\\ b\leq v}}(u,v)X^{n-|v|-k+l}Y^{|v|-k}. \tag{39}$$

We can write $v=b+v'$, where $S(v')$ is the complementary set of $S(b)$. Hence $(u,v)=(u,b)(u,v')$ and v' can be considered to run over F^{n-k}.

$$\begin{aligned}\hat{\Phi}(u) &= (u,b)\sum_{v'\in F^{n-k}}(u,v')X^{n-2k+l-|v'|}Y^{|v'|}\\ &= (u,b)X^{-k+l}(X+(q-1)Y)^{n-k-|u_{\overline{S}(b)}|}(X-Y)^{|u_{\overline{S}(b)}|}\end{aligned} \tag{40}$$

where the last equality is the usual computation of the Fourier transform of the function $x\to X^{n-k-|x|}Y^{|x|}$ over F^{n-k}.

□

LEMMA 4.3. *Let $f\in L(U_k)_l\cap\ker d$.*

$$\hat{\Phi}(u) = q^{k-l}D^*Tf(u)(X+(q-1)Y)^{n-|u|-k+l}(X-Y)^{|u|-k} \tag{41}$$

PROOF. Since $f = \sum_{b\in U_k} f(b)\delta_b$ and from (38),

$$\hat{\Phi}(u) = X^{-k+l} \sum_{b\in U_k} (u,b)f(b)(X+(q-1)Y)^{n-k-|u_{\overline{S}(b)}|}(X-Y)^{|u_{\overline{S}(b)}|}. \tag{42}$$

Then $|u_{\overline{S}(b)}| = |u| - |S(u)\cap S(b)|$; we set $|S(u)\cap S(b)| = k-i$ and sum over $i \in \{0\dots k\}$. We get

$$\hat{\Phi}(u) = X^{-k+l}(X+(q-1)Y)^{n-k+l-|u|}(X-Y)^{|u|-k}\Psi(u) \tag{43}$$

where

$$\Psi(u) := \sum_{i=0}^{k}\Big(\sum_{\substack{b\in U_k\\ |S(b)\cap S(u)|=k-i}} (u,b)f(b)\Big)(X+(q-1)Y)^{k-l-i}(X-Y)^i \tag{44}$$

which becomes, by Proposition 4.1,

$$\begin{aligned}
\Psi(u) &= \sum_{i=0}^{k} (q-1)^i \binom{k-l}{i} D^*Tf(u)(X+(q-1)Y)^{k-l-i}(X-Y)^i \\
&= D^*Tf(u)\sum_{i=0}^{k-l}\binom{k-l}{i}(X+(q-1)Y)^{k-l-i}((q-1)(X-Y))^i \\
&= D^*Tf(u)(X+(q-1)Y+(q-1)(X-Y))^{k-l} \\
&= D^*Tf(u)(qX)^{k-l}.
\end{aligned} \tag{45}$$

Replacing (45) in (43), we obtain (41).

□

5. The operator T

We study in this section the possibility for the operator T to be homothetic on the spaces $L(U_k)_l \cap \ker d$. This case is especially interesting because Theorem 4.1 can in that case be read as a linear invariance property for a certain polynomial (see next section).

PROPOSITION 5.1. *The following statements are equivalent:*

- *For all k, l, there exists $\lambda_{k,l} \in \mathbb{C}$ such that, for all $f \in L(U_k)_l \cap \ker d$, $Tf = \lambda_{k,l} f$.*
- *$F = \mathbb{F}_2, \mathbb{F}_3$ and $(x,y) = \chi(xy)$, where χ is a non-trivial character of F, or $F = \mathbb{F}_4$ and $(x,y) = \chi(trace(xy^2))$ and χ is the non-trivial character of $\mathbb{F}_2$.*

If $F = \mathbb{F}_2, \mathbb{F}_4$, $\lambda_{k,l} = (-1)^{k-l}2^l$, and, if $F = \mathbb{F}_3$, $\lambda_{k,l} = (-1)^{k-l}(\sqrt{-3})^l$.

REMARK 5.1. *In terms of coding theory, the cases of the proposition are the binary, ternary, and quaternary additive codes, with the terminology of* [**11**]. *We have kept the usual notation for F, although the field structure has no importance here. In particular, the quaternary additive codes are the same as the Kleinian codes studied in* [**9**].

PROOF. From [**6**], we know that $L(U_k)_l \cap \ker d$ is a H-irreducible module. Hence, T is homothetic over $L(U_k)_l \cap \ker d$ if and only if T commutes with the action of H, i.e. if and only if $T(h.f) = h.T(f)$ for all $h \in H$, $f \in L(U_k)_l \cap \ker d$. If this is true for all k, l, since we have already seen that $Td^* = -d^*T$, and from the decomposition (7), it must be true for all $f \in L(U_k)$. The space $L(U_k)$ is generated by $\{\delta_b, b \in U_k\}$. We have $h.\delta_b = \delta_{bh^{-1}}$, and $T\delta_b(u) = (b,u)$ if $S(u) = S(b)$, and 0 otherwise. Hence it turns out that T commutes with H over $L(U_k)$ if and only if $(u,v) = (uh, vh)$ for all $h \in H$ and $u, v \in U_k$ with $S(u) = S(v)$. Since the permutation on the coordinates of the elements of F^n has no incidence on the duality (because we assume $(u,v) = \prod_i (u_i, v_i)$), we are left with the condition that $(,)$ must be constant on the orbits of $F \times F$ under the action of S_{q-1}, which means that $\alpha := (x,x)$ is independant of the choice of $x \neq 0$ in F, and $\beta := (x,y)$ is independant of the choice of $x \neq y \neq 0$ in F. Hence, $(,)$ takes at most three values: $\{1, \alpha, \beta\}$. Since they form a subgroup of $\mathbb{C}^*$ of order $d = 2, 3$, since $dF = 0$ and since, for each $x \neq 0$, the kernel of $(., x)$ has order q/d, one easily sees that the only possibilities left are the ones listed in the proposition.

We now compute $\lambda_{k,l}$. We have already seen that $Tz_1 = -z_1$. If $i \geq 2$, $Tz_i(x) = \sum_{y \neq a_0} (x,y) z_i(y) = \alpha z_i(x) + \beta \sum_{y \neq a_0, x} z_i(y)$. But $\sum_{y \neq a_0} z_i(y) = 0$, so $Tz_i(x) = (\alpha - \beta) z_i(x)$. Hence, if $f = \rho z_u \in L(U_k)_l$, from property 2. of Proposition 3.1, $Tf = (-1)^{k-l}(\alpha - \beta)^l f$. In the cases $F = \mathbb{F}_2, \mathbb{F}_4$, $\alpha - \beta = 1 - (-1) = 2$, and in the case $F = \mathbb{F}_3$, $\alpha - \beta = \pm(j - j^2) = \pm i\sqrt{3}$ where j is a cubique root of 1.

□

REMARK 5.2. *In the ternary case, there are two conjugate choices for the duality* $(,)$. *Since* $Tf(-u) = \overline{T}f(u)$, *we have* $Z_{C,Tf} = Z_{C,(Tf+\overline{T}f)/2}$. *The eigenvalue of* $\overline{T}$ *over* $L(U_k)_l$ *is the conjugate of the one associated to* T; *but, we have seen that* $\lambda_{k,l}$ *has a trivial real part in the case* $l \equiv 1 \mod 2$. *Hence, in this case,* $Z_{C,Tf} = 0$, *and so* $Z_{C,f} = 0$. *Note that it fits with the fact that the set* $C^\perp$ *only depends on the choice of the duality up to conjugation.*

6. Applications to self-dual codes

In this section, we consider the case of self-dual ternary and additive quaternary codes. The binary case was previously studied in [**1**]. In these cases, Theorem 4.1 shows that the polynomials $Z_{C,f}$ are relative invariants for the group of transformation acting on the Hamming weight enumerator.

6.1. Ternary self-dual codes. Let $M = \frac{1}{\sqrt{3}}\left(\begin{smallmatrix} 1 & 2 \\ 1 & -1 \end{smallmatrix}\right)$ and $D = \left(\begin{smallmatrix} 1 & 0 \\ 0 & j \end{smallmatrix}\right)$ where $j = e^{2i\pi/3}$. Let G_3 be the group generated by M and D. It is well-known that, if C is a self-dual ternary code, then its weight enumerator W_C is invariant under G_3, and hence belongs to its algebra of invariants $\mathcal{I}_{G_3}$, which is the polynomial ring in the polynomials $g_4 := x^4 + 8xy^3$, $g_{12} := y^3(x^3 - y^3)^3$ (see [**11**]).

We define the following characters $\psi_{u,v}$ of the group G_3:

$$\psi_{u,v}(M) = (-1)^u \quad \psi_{u,v}(D) = j^v \tag{46}$$

and we denote by $\mathcal{I}_{G_3,\psi_{u,v}}$ the space of relative invariants:

$$\mathcal{I}_{G_3,\psi_{u,v}} := \{P(x,y) \in \mathbb{C}[x,y] \mid P.M = \psi_{u,v}(M)P \text{ for all } M \in G_3\}. \tag{47}$$

The following polynomials are relative invariants, as one can check easily:

$$p_4 = y(x^3 - y^3) \in \mathcal{I}_{G_3,\psi_{0,1}} \tag{48}$$
$$p_6 = x^6 - 20x^3y^3 - 8y^6 \in \mathcal{I}_{G_3,\psi_{1,0}}$$

LEMMA 6.1. *The spaces of relative invariants* $\mathcal{I}_{G_3,\psi_{u,v}}$ *are free modules over* $\mathcal{I}_{G_3}$. *More precisely:*

- $\mathcal{I}_{G_3,\psi_{0,1}} = p_4\mathbb{C}[g_4, g_{12}]$
- $\mathcal{I}_{G_3,\psi_{0,2}} = p_4^2\mathbb{C}[g_4, g_{12}]$
- $\mathcal{I}_{G_3,\psi_{1,0}} = p_6\mathbb{C}[g_4, g_{12}]$
- $\mathcal{I}_{G_3,\psi_{1,1}} = p_4p_6\mathbb{C}[g_4, g_{12}]$
- $\mathcal{I}_{G_3,\psi_{1,2}} = p_4^2p_6\mathbb{C}[g_4, g_{12}]$

PROOF. It follows easily from the computation of the Molien series of the corresponding spaces (see [**11**] and [**1**] for examples of such computations). □

PROPOSITION 6.1. *Let C be a self-dual ternary code of length n. Let f belong to $L(U_k)_l \cap \ker d$. If $l \equiv 1 \mod 2$, $Z_{C,f} = 0$. If $l \equiv 0 \mod 2$, let $u \in \{0, 1\}$ be equal to $k + l/2$ modulo 2, and let $v \in \{0, 1, 2\}$ be equal to $-k$ modulo 3. Then*

$$Z_{C,f} \in \mathcal{I}_{G_3,\psi_{u,v}}.$$

PROOF. Straightforward from Theorem 4.1, Remark 5.2 and Proposition 5.1. □

In the special case of extremal codes, we can derive from previous proposition that $Z_{C,f} = 0$ for certain values of (k, l). We briefly recall what an extremal code is (see [**11**]): if C is a self-dual ternary code of length n, then n is a multiple of 4. Write $n = 12q + 4r$ with $r = 0, 1, 2$. Then the fact that W_C belongs to $\mathcal{I}_{G_3}$ shows that the minimum weight $w(C)$ of C satisfies

$$w(C) \leq 3q + 3. \tag{49}$$

A code meeting this bound is called extremal. Its weight enumerator is then uniquely determined.

COROLLARY 6.1. *Let C be an extremal self-dual ternary code of length $n = 12q + 4r$. Let $f \in L(U_k)_l \cap \ker d$. Then $Z_{C,f} = 0$ when $l \equiv 1 \mod 2$ from Remark 5.2, but also in the following cases:*

- *If $r = 0$: for $k = 1, 2, 3$ and all $l \leq k$, and for $(k, l) = (4, 0), (4, 2), (5, 0), (5, 4), (6, 2), (7, 0)$.*
- *If $r = 1$: for $k = 1, 2$ and all $l \leq k$, and for $(k, l) = (3, 0), (4, 2), (5, 0)$.*
- *If $r = 2$: for $(k, l) = (1, 0), (2, 2), (3, 0)$.*

REMARK 6.1. *In view of Proposition 2.1, the property $Z_{C,f} = 0$ is related to the fact that the set of codewords of fixed Hamming weight form a design. We recover here results already known from* [**5**] *and* [**3**], *namely that extremal ternary self-dual codes hold generalized 3-designs and classical $\{1, 2, 3, 4, 5, 7\}$-designs when $r = 0$ (respectively for the corresponding weaker results when $r = 1, 2$), which were derived from generalized Assmus-Mattson theorems. We obtain here some additional properties; in particular it is worth noticing that, for instance in the* *case $r = 0$, the generalized 3-designs are actually nearly 4-designs since the only property missing is orthogonality with $L(U_4)_4 \cap \ker d$.*

6.2. Even quaternary additive self-dual codes. Here we follow the same line as for ternary codes, so we omit the proofs and some comments since they are completely similar. Let $M = \frac{1}{2}\begin{pmatrix} 1 & 3 \\ 1 & -1 \end{pmatrix}$ and $D = \begin{pmatrix} 1 & 0 \\ 0 & -1 \end{pmatrix}$. Let G_4 be the group generated by M and D. If C is a self-dual even quaternary additive code, then its weight enumerator W_C is invariant under G_4, and hence belongs to its algebra of invariants $\mathcal{I}_{G_4}$, which is the polynomial ring in the polynomials $h_2 := x^2 + 3y^2$, $h_6 := y^2(x^2 - y^2)^2$ (see [**11**]).

We define the following characters $\psi_{u,v}$ of the group G_4:

$$\psi_{u,v}(M) = (-1)^u \quad \psi_{u,v}(D) = (-1)^v \tag{50}$$

and we denote by $\mathcal{I}_{G_4,\psi_{u,v}}$ the space of relative invariants. The following polynomials are relative invariants, as one can check easily:

$$\begin{aligned} q_3 &= y(x^2 - y^2) \in \mathcal{I}_{G_4,\psi_{0,1}} \\ r_3 &= x^3 - 9xy^2 \in \mathcal{I}_{G_4,\psi_{1,0}} \end{aligned} \tag{51}$$

LEMMA 6.2. *The spaces of relative invariants $\mathcal{I}_{G_4,\psi_{u,v}}$ are free modules over $\mathcal{I}_{G_4}$. More precisely:*

- $\mathcal{I}_{G_4,\psi_{0,1}} = q_3\mathbb{C}[h_2, h_6]$
- $\mathcal{I}_{G_4,\psi_{1,0}} = r_3\mathbb{C}[h_2, h_6]$
- $\mathcal{I}_{G_4,\psi_{1,1}} = q_3 r_3\mathbb{C}[h_2, h_6]$

PROPOSITION 6.2. *Let C be a self-dual even quaternary additive code of length n. Let $f \in L(U_k)_l \cap \ker d$. Let $u \in \{0,1\}$ be equal to $k - l$ modulo 2, and let $v \in \{0,1\}$ be equal to k modulo 2. Then*

$$Z_{C,f} \in \mathcal{I}_{G_4,\psi_{u,v}}.$$

Here the extremal codes have weight $w(C) = 2q + 2$, where $n = 6q + 2r$.

COROLLARY 6.2. *Let C be an extremal self-dual even quaternary additive code of length $n = 6q+2r$. Let $f \in L(U_k)_l \cap \ker d$. Then $Z_{C,f} = 0$ in the following cases:*

- *If $r = 0$: for $k = 1, 2$ and all $l \le k$, and for $(k,l) = (3,0), (3,1), (3,2), (4,0), (4,1), (4,3), (5,0), (5,2), (6,1), (7,0)$.*
- *If $r = 1$: for $(k,l) = (1,0), (1,1), (2,0), (2,1), (3,0), (3,2), (4,1), (5,0)$.*
- *If $r = 2$: for $(k,l) = (1,0), (2,1), (3,0)$.*

7. Intersection numbers

Just like in the binary case [**1**], the polynomials $Z_{C,f}$ and their invariance properties can be used to compute certain invariants associated to a fixed element $v \in F^n$ and to a code C. We fix $v \in U_t$.

DEFINITION 7.1. *Let $v \in U_t$ and $u \in U_w$. We define*

$$n(u, v) := \operatorname{card}(S(u) \cap S(v))$$

$$e(u, v) := \operatorname{card}\{s \in [1..n] \mid u_s = v_s \neq 0\}$$

and

Let $i = i(u,v) := t - e(u,v)$ and $j := j(u,v) = t - n(u,v)$. The group H acts on $U_t \times U_w$ and one can see easily that the orbits of this action are given by the values of (i,j). Equivalently, in the case $t = w$, (i,j) gives the relations of the non-binary Johnson scheme [**14**]. Clearly, the values of $(i(u,v), j(u,v))$ belong to the set

$$K_{t,w} := \{(i,j) \mid 0 \le j \le i \le t,\ t - w \le j \le n - w\}. \tag{52}$$

Let

$$K_t := K_{t,t} := \{(i,j) \mid 0 \le j \le i \le t,\ j \le n - t\} \tag{53}$$

and

$$L_t := \{(k,l) \mid 0 \le l \le k \le \min(t, n + l - t)\}. \tag{54}$$

The decomposition of the space $L(U_t)$ into irreducible H-subspaces given by $L(U_t) = \oplus_{(k,l)\in L_t} L(U_t)_{(k,l)}$ (see (7)) allows us to consider the zonal elements associated to this decomposition. These are functions $g_{k,l}$ having the characteristic property that $g_{k,l}(u)$ only depends on $i(u,v)$, $j(u,v)$. In our situation they are uniquely determined up to a scalar multiple because the subspaces $L(U_t)_{(k,l)}$ are pairwise non isomorphic. We recall that they can be constructed the following way: take any orthonormal basis (f_i) of $L(U_t)_{(k,l)}$, and set $E_{(k,l)}(u,v) := \sum_i f_i(u)\overline{f_i(v)}$. This definition is independant of the choice of the basis, and hence is invariant under the action of H on $(u,v) \in U_t^2$. Hence $E_{(k,l)}(u,v)$ is a function of $(i(u,v), j(u,v))$. In the setting of association schemes, the $E_{(k,l)}$ are the idempotents of the non-binary Johnson scheme and its expression in terms of $(i(u,v), j(u,v))$ is given by the second eigenvalues, which are computed in [**14**]. We recall in next proposition their expression in terms of Hahn and Krawtchouk polynomials.

PROPOSITION 7.1. [**14**]

$$E_{(k,l)}(u,v) = h_{(k,l),t}(i(u,v), j(u,v)), \tag{55}$$

where, for all $(k,l) \in L_t$ and $(i,j) \in K_t$,

$$h_{(k,l),t}(i,j) = \frac{\binom{n}{l}}{\binom{t}{l}} K_l(t-j, q-1, i-j) Q_{k-l}(n-l, t-l, j) \tag{56}$$

where K_k, Q_k are respectively Krawtchouk and Hahn polynomials, given by the following formulas:

$$\begin{aligned} K_k(n,q,x) &:= \sum_{i=0}^{k} (-1)^i (q-1)^{k-i} \binom{n-x}{k-i}\binom{x}{i} \\ E_k(n,t,x) &:= \sum_{r=0}^{k} (-1)^r \binom{x}{r}\binom{t-x}{k-r}\binom{n-t-x}{k-r} \\ Q_k(n,t,i) &= \frac{\binom{n}{k} - \binom{n}{k-1}}{\binom{t}{i}\binom{n-t}{i}} E_i(n,t,k) \end{aligned} \tag{57}$$

In view of applications to codes, we shall make use of the polynomials $Z_{C,f}$ for $f \in L(U_k)_l \cap \ker d$ such that $D^*f(u) = E_{(k,l)}(u,v)$ for $u \in U_t$. We denote by $H_{(k,l),v}$ such a function f. Therefore we need to compute $D^*f(u)$ for $u \in U_w$ for any $w \geq k$ (and not only for $w = t$). From last proposition, $D^*H_{(k,l),v}(u) = h_{(k,l),t}(i,j)$ for $u \in U_t$ and $i = i(u,v)$, $j = j(u,v)$. For the general case $u \in U_w$, we have a more complicated and rather ugly formula:

PROPOSITION 7.2. *For all $w \geq k$ and all $u \in U_w$, the value of $D^*H_{(k,l),v}(u)$ only depends on $(i(u,v), j(u,v))$. We set again*

$$D^*H_{(k,l),v}(u) = h^w_{(k,l),t}(i(u,v), j(u,v)), \tag{58}$$

where

$$\begin{aligned} h^w_{(k,l),t}(i,j) = c \sum_I (q-2)^{i_2+j_2+l_2}(q-3)^{k_3}(q-1)^{t-s}. \\ \binom{w+j-t}{i_1,i_2}\binom{t-i}{j_1,j_2}\binom{i-j}{k_1,k_2,k_3}\binom{j}{l_1,l_2}\binom{n-w-j}{t-s}\binom{i_1+j_1+k_1}{k}. \\ h_{(k,l),t}(t-j_1-k_2-l_1, t-j_1-j_2-k_1-k_2-k_3-l_1-l_2) \end{aligned} \tag{59}$$

where

$$c = \frac{1}{(q-1)^{t-k}\binom{n+l-2k}{t-k}} \tag{60}$$

and

$$\begin{aligned} I = \{ & (i_1, i_2, j_1, j_2, k_1, k_2, k_3, l_1, l_2) \mid \\ & i_1 \in [0..w+j-t], i_2 \in [0..w+j-t-i_1], \\ & j_1 \in [0..t-i], j_2 \in [0..t-i-j_1], \\ & k_1 \in [0..i-j], k_2 \in [0..i-j-k_1], k_3 \in [0..i-j-k_1-k_2], \\ & l_1 \in [0..j], l_2 \in [0..j-l_1], \\ & i_1+j_1+k_1 \geq k, t-(j_1+j_2+k_1+k_2+k_3+l_1+l_2) \leq n-t, \\ & s := i_1+i_2+j_1+j_2+k_1+k_2+k_3+l_1+l_2 \leq t \}. \end{aligned} \tag{61}$$

PROOF. We use the formula $(dd^* - d^*d)_{|L(U_k)_l} = (q-1)(n+l-2k)\mathrm{Id}$ ([**6**, Proposition 2.6]). Taking account of the fact that $H_{(k,l),v} \in \ker d$, and iterating it, we get

$$(q-1)^{t-k}\binom{n+l-2k}{t-k}H_{(k,l),v} = ((-1)^{t-k}\frac{d^{t-k}}{(t-k)!})((-1)^{t-k}\frac{{d^*}^{t-k}}{(t-k)!})H_{(k,l),v}. \tag{62}$$

Let $K := (-1)^{t-k}\frac{{d^*}^{t-k}}{(t-k)!}H_{(k,l),v} \in L(U_t)$. For all $u \in U_t$, $K(u) = D^*H_{(k,l),v}(u) = h_{(k,l),t}(i(u,v), j(u,v))$. We get, for all $u \in U_w$,

$$D^*H_{(k,l),v}(u) = \frac{1}{(q-1)^{t-k}\binom{n+l-2k}{t-k}} \sum_{\substack{u_k\in U_k\\ u_k\leq u}} \sum_{\substack{z\in U_t\\ u_k\leq z}} K(z)$$

(63)

$$= \frac{1}{(q-1)^{t-k}\binom{n+l-2k}{t-k}} \sum_{\substack{z\in U_t\\ e(z,u)\geq k}} \binom{e(z,u)}{k} K(z).$$

The parameters $e(z,u)$, $i(z,v)$, $j(z,v)$ express easily in terms of:
$i_1 := \mathrm{card}\{i \mid z_i = u_i \neq 0, v_i = 0\}$, $i_2 := \mathrm{card}\{i \mid z_i \neq u_i \neq 0, v_i = 0\}$,
$j_1 := \mathrm{card}\{i \mid z_i = u_i = v_i \neq 0\}$, $j_2 := \mathrm{card}\{i \mid z_i \neq u_i \neq 0, v_i = u_i\}$,
$k_1 := \mathrm{card}\{i \mid z_i = u_i \neq 0, v_i \neq u_i \neq 0\}$, $k_2 := \mathrm{card}\{i \mid z_i \neq u_i \neq v_i \neq 0\}$,
$k_3 := \mathrm{card}\{i \mid z_i \neq u_i \neq 0, z_i = v_i\}$, $l_1 := \mathrm{card}\{i \mid z_i = v_i \neq 0, u_i = 0\}$,
$l_2 := \mathrm{card}\{i \mid z_i \neq v_i \neq 0, u_i = 0\}$.
The formula then follows from the enumeration of all possibilities.

□

REMARK 7.1. *In the case when $t \leq w$, there is a more simple formula for $D^*H_{(k,l),v}(u)$ coming from: ${d^*}^{w-k}H_{(k,l),v} = {d^*}^{w-t}{d^*}^{t-k}H_{(k,l),v}$.*

Let C_w denote the set of words in C of Hamming weight w, i.e. $C_w = C \cap U_w$. Let $v \in U_t$. We set, for $(i,j) \in K_{t,w}$,

$$n_{w,(i,j)}(v) := \mathrm{card}\{u \in C_w \mid (i(u,v), j(u,v)) = (i,j)\}. \tag{64}$$

From Proposition 7.2, for each $(k,l) \in L_t$, the coefficient of $x^{n-w-k+l}y^{w-k}$ in $Z_{C,H_{(k,l),v}}$ is:

$$\sum_{(i,j)\in K_{t,w}} h^*_{(k,l),t}(i,j)\, n_{w,(i,j)}(v). \tag{65}$$

The claims in Propositions 6.1, 6.2, and the descriptions of the spaces of relative invariants involved, can then be turned out into linear relations between the unknowns $n_{w,(i,j)}(v)$. See [**1**] for examples in the binary case. We work out examples in the case of even quaternary additive self-dual codes.

7.1. The dodecacode. Let $n = 12$. It is known ([**11**]) that there is up to equivalence only one $[12, 6, 6]$ even self-dual quaternary additive code, the so-called dodecacode. We show what can be said *a priori* on the intersection numbers $n_{w,(i,j)}(v)$ defined in (64) of such a code C by use of the method described in section 7. We fix an element $v \in U_t$.

From Corollary 6.2, all the linear forms in (65) are equal to zero for $t \leq 2$ and $(k,l) \in L_t$. Moreover, the knowledge of the weight enumerator of such a code C

$$W_C(x,y) = x^{12} + 396x^6y^6 + 1485x^4y^8 + 1980x^2y^{10} + 234y^{12} \tag{66}$$

leads to more equations:

$$\sum_{(i,j)\in K_{t,w}} n_{w,(i,j)}(v) = \mathrm{card}(C_w). \tag{67}$$

TABLE 1. The $n_{w,(i,j)}(v)$ for $|v| = 1$

(i,j)	$(0,0)$	$(1,0)$	$(1,1)$
$w = 6$	66	132	198
$w = 8$	330	660	495
$w = 10$	550	1100	330
$w = 12$	78	156	0

TABLE 2. The $n_{w,(i,j)}(v)$ for $|v| = 2$

(i,j)	$(0,0)$	$(1,0)$	$(1,1)$	$(2,0)$	$(2,1)$	$(2,2)$
$w = 6$	10	40	72	40	144	90
$w = 8$	70	280	240	280	480	135
$w = 10$	150	600	200	600	400	30
$w = 12$	26	104	0	104	0	0

TABLE 3. The $n_{w,(i,j)}(v)$ for $|v| = 3$

(i,j)	$(0,0)$	$(1,0)$	$(1,1)$	$(2,0)$	$(2,1)$	$(2,2)$
$w = 6$	$x - 1$	$-3x + 15$	18	$3x + 9$	72	54
$w = 8$	$-3x + 21$	$9x + 63$	84	$-9x + 189$	336	108
$w = 10$	$3x + 133$	$-9x + 261$	90	$9x + 459$	360	30
$w = 12$	$-x + 11$	$3x + 45$	0	$-3x + 111$	0	0

(i,j)	$(3,0)$	$(3,1)$	$(3,2)$	$(3,3)$
$w = 6$	$-x + 13$	72	108	36
$w = 8$	$3x + 105$	336	216	27
$w = 10$	$-3x + 327$	360	60	0
$w = 12$	$x + 67$	0	0	0

We find for $t = 1, 2$ uniquely determined intersection numbers (see Table 1 and 2), in accordance with the fact that, for all w, C_w holds 2-generalized designs (see Corollary 6.2).

If $t = 3$, the intersection numbers are not uniquely determined (Table 3). They depend on one parameter $x(v) = x$ which is the number of weight 3 words in the coset $v + C$ (because clearly this number equals $1 + n_{6,(0,0)}(v)$). Clearly, x can take the values $1, 2, 3, 4$.

If $t = 4$, we moreover assume that v is a minimum weight word in its coset $v+C$. Again, the intersection numbers are computed from one of them. We don't give the full details of their expression but notice that the number of weight 6 codewords, the support of which contain the support of v, is a constant (because the supports hold classical 5-design from Assmus-Mattson theorem) and that this number equals $n_{6,(2,0)}+n_{6,(3,0)}+n_{6,(4,0)}$. The computation shows that the intersection numbers can all be expressed affinely in $x := n_{6,(4,0)}$ like $n_{6,(2,0)} = x + 4$ and $n_{6,(3,0)} = -2x + 8$.

TABLE 4. Coset distribution of a $[12, 6, 6]$ even self-dual code

1	**2**	**3**	**4**	**5**	**6**	**7**	**8**
1				66	132	528	660
	1		10	40	182	424	760
		x	$15-3x$	48	$8x+148$	$432-6x$	$810-6x$
			15	48	148	432	810

9	**10**	**11**	**12**
1045	1100	408	156
1080	961	504	134
$1040+8x$	948	$528-3x$	$127+x$
1040	948	528	127

The weight distribution of $v+C$ itself doesn't depend on x (see Table 4). Note that, in general, the weight distribution of $v+C$ follows from the intersection numbers because $|u+v| = |u| + i(u,v) + j(u,v) - t$.

We display in Table 4 the coset distribution of such a code. The weight of the cosets is at most 4. This can be proved directly by elementary counting arguments, or follows from [**5**]. Note that the coset distribution could also be computed using Delsarte method explained in [**5**]. If the coset has weight 3, it may contain $1, 2, 3$ or 4 coset leaders. For the dodecacode, one finds respectively 792, 1314, 756, 63 cosets of weight 3 with respectively 1, 2, 3, 4 leaders. Hence the dodecacode has got 540 cosets of weight 4.

7.2. Length 14. The extremal even self-dual codes of length 14 have weight 6 and weight enumerator

$$W_C(x,y) = x^{14} + 273x^8y^6 + 2457x^6y^8 + 7098x^4y^{10} + 6006x^2y^{12} + 549y^{14}. \tag{68}$$

It is known that there is a unique one which is $\mathbb{F}_4$-linear hermitian ([**11**]). The whole number of extremal additive codes seems huge, since we have found 490 such codes with the additional condition that at least two codewords of weight 6 have the same support ([**2**]). Information on the coset distribution of such codes could be computed in the same way as for $n = 12$. Here, we take v to be a codeword of weight 6. Because v is in the code C, additional constraints hold for $n_{w,(i,j)}(v)$: since $(u,v) = i(u,v) - j(u,v) \mod 2$, we have $n_{w,(i,j)}(v) = 0$ if $i + j \equiv 1 \mod 2$; also, since $|u+v| = |u| + i + j - t$, we have $n_{w,(i,j)}(v) = 0$ if $0 < w + i + j - t < 6$.

Altogether, the intersection numbers for $v \in C_6$ depend on three positive and integral parameters x, y, z. One of them is $2x := n_{6,(6,0)}(v)$ and counts the number of codewords with the same support as v. Clearly this number is either 0 or 2 (they come by pairs $(u, u+v)$ and they are minimal). The general expression shows that, if $x = 1$, i.e. if the support of v is also the support of some other word in C, then y and z are uniquely determined. Note that, if C is $\mathbb{F}_4$-linear, then it is the case for all $v \in C$ since wv and w^2v would provide codewords with the same support as v. If $x = 0$, then the fact that the intersection numbers are natural numbers show that $y \in [0..3]$ and $z \in [0..4]$. Among the 490 non equivalent codes found, all these possiblities for (y,z) occur. Table 5 gives the expression of the $n_{w,(i,j)}$ for the

TABLE 5. Computation of $n_{w,(i,j)}(v)$ for $v \in C_6$ and $w = 6, 8$

	w = 6	
(i,j)	$x=1$	$x=0$
$(0,0)$	1	1
$(3,3)$	0	$-4z+16$
$(4,2)$	24	$-4y+2z+24$
$(4,4)$	42	$2z+18$
$(5,1)$	0	$2y$
$(5,3)$	96	$8y+4z+72$
$(5,5)$	0	$-2y+24$
$(6,0)$	2	0
$(6,2)$	12	$-3y-2z+24$
$(6,4)$	84	$-2y-2z+90$
$(6,6)$	12	$y+4$

	w = 8	
(i,j)	$x=1$	$x=0$
$(2,2)$	42	$2z+18$
$(3,1)$	96	$8y+4z+72$
$(3,3)$	0	$4y+8z+48$
$(4,0)$	12	$-3y-2z+24$
$(4,2)$	528	$-16y-10z+564$
$(4,4)$	156	$-y-4z+84$
$(5,1)$	192	$2y-4z+192$
$(5,3)$	768	$-8z+816$
$(5,5)$	0	$2y+24$
$(6,0)$	0	$2z$
$(6,2)$	348	$5y+8z+312$
$(6,4)$	312	$4z+300$
$(6,6)$	3	$-y+3$

TABLE 6. Coset distribution for a $[18, 9, 8]$ even self-dual code

1	2	3	4	5	6	7	8	9
1						408	816	4930
	1				56	224	1304	3360
		1		x	$63-3x$	$303-2x$	$978+14x$	$3730-5x$
			y	z	$84-7y-3z$	$288-8y-2z$	$936+45y+14z$	$3680+8y-5z$
				t	$84-3t$	$288-2t$	$936+14t$	$3680-5t$
					84	288	936	3680

weights $w = 6, 8$. Note that the $n_{w,(i,j)}(v)$ for $j = 6$ give the weight distribution of the subcode

$$C_v := \{u \in C \mid S(u) \cap S(v) = \varnothing\}. \tag{69}$$

7.3. Length 18. The weight of the extremal codes of length 18 is 8. There is a unique $\mathbb{F}_4$-linear hermitian code of weight 8, named S_{18} ([**11**]). It is not known whether other additive self-dual codes meet this bound. We compute the coset distribution with the help of the intersection numbers. Table 6 gathers the results up to weight 9.

PROPOSITION 7.3. *Let C be an even self-dual additive quaternary code of weight 8. Then the covering radius R of C satisfies $5 \leq R \leq 6$.*

PROOF. For all i, we denote c_i a coset of C of weight i and by n_i the number of cosets of weight i. Let X_i denote the set of weight i words in F^{18}. Of course, $\operatorname{card}(X_i) = \binom{18}{i}3^i$. Note that, for $i = 1, 2, 3$, $n_i = \operatorname{card}(X_i)$. We start with the weight enumerators of the cosets of C of weight up to 6, computed with the help of the intersection numbers. With the notations of Table 6, we count the words of weight 4, 5, 6. We have:

$$\begin{aligned} \operatorname{card}(X_4) &= \sum_{c_4} y \\ \operatorname{card}(X_5) &= \sum_{c_3} x + \sum_{c_4} z + \sum_{c_5} t \\ \operatorname{card}(X_6) &= 56n_2 + \sum_{c_3}(63-3x) + \sum_{c_4}(84-7y-3z) + \sum_{c_5}(84-3z) + 84n_6 \end{aligned} \tag{70}$$

From these equations, we get

$$n_4 + n_5 + n_6 = 238680. \tag{71}$$

But we can check that $1 + n_1 + n_2 + n_3 + n_4 + n_5 + n_6 = 2^{18}$, so we have proved that $R \leq 6$ (it follows also from [**5**] since 6 is the number of different weights of $C^{\perp} = C$). Now we assume that $R \leq 4$, so that $n_5 = n_6 = 0$. For $y \in [1..4]$, we denote n_4^y the number of cosets of weight 4 containing exactly y coset leaders. Clearly, two coset leaders in a coset of weight 4 have disjoint supports because the code has minimal weight 8, so their number is upper bounded by 4. And the sum of two such words is a codeword of weight 8, so, if v is a coset leader of a fixed coset of weight 4,

$$y = 1 + \operatorname{card}\{u, u \in C_8 \mid v \leq u\} = n_{8,(0,0)}(v). \tag{72}$$

If C_8 was a 4-design, the cardinality of $\{u, u \in C_8 \mid v \leq u\}$ would be a constant. However, we can compute its average value

$$\mu := \frac{1}{\operatorname{card}(X_4)} \sum_{v \in X_4} \operatorname{card}\{u, u \in C_8 \mid v \leq u\} \tag{73}$$

by the usual formula:

$$\mu = \frac{\binom{6}{2}\lambda_2}{\binom{16}{2}3^2} \tag{74}$$

where, for $v \in U_2$, $\lambda_2 = \operatorname{card}\{u, u \in C_8 \mid v \leq u\} = 55$. This last value can be read in Table 6. Now, we have

$$\sum_{v \in X_4} y = \sum_{c_4} y^2 = \operatorname{card}(X_4)(1 + \mu). \tag{75}$$

Finally, we obtain the following system of equations:

$$\begin{aligned} n_4^1 + n_4^2 + n_4^3 + n_4^4 &= 238680 \\ n_4^1 + 2n_4^2 + 3n_4^3 + 4n_4^4 &= \operatorname{card}(X_4) \\ n_4^1 + 4n_4^2 + 9n_4^3 + 16n_4^4 &= \operatorname{card}(X_4)(1+\mu) \end{aligned} \tag{76}$$

One can parametrize the solutions by n_4^1 and see that it doesn't admit any positive integral solutions.

REMARK 7.2. *In the case of the code* S_{18}*, one can compute all these parameters. One finds:* $(n_4^1, n_4^2, n_4^3, n_4^4) = (122400, 36720, 12240, 3865)$*, and, with obvious notations,* $(n_5^{10}, n_5^{12}, n_5^{16}) = (24480, 38250, 765)$*. Hence the covering radius of* S_{18} *is equal to* 5*. It would be interesting to know if a non* $\mathbb{F}_4$*-linear code with covering radius* 6 *exists.*

□

References

[1] C. Bachoc, *On harmonic weight enumerators of binary codes*, Designs, Codes and Cryptography **18** (1999), 11-28

[2] C. Bachoc, P. Gaborit, *Extremal self-dual even quaternary additive codes*, to appear in Journal de Theorie des Nombres de Bordeaux.

[3] A.R. Calderbank, P. Delsarte, *On error-correcting codes and invariant linear forms* SIAM J. Disc. Math. **6.1** (1993), 1-23

[4] J. Conway, N.J.A. Sloane, "Sphere packings, Lattices and Groups", Springer-Verlag, 1988

[5] P. Delsarte, *Four fondamental parameters of a code and its combinatorial significance* Information and Control **23** (1973), 407-438

[6] C. F. Dunkl, *A Krawtchouk polynomial addition theorem and wreath products of symmetric groups* Ind. Univ. Math. Journal **25** 4 (1976), 335-358

[7] J. E. Fields, P. Gaborit, W. C. Huffman, V. Pless, *On the classification of extremal even formally self-dual codes*, Designs, Code and Cryptography **18** (1999), 125-148

[8] J. E. Fields, P. Gaborit, W. C. Huffman, V. Pless, *On the classification of extremal even formally self-dual codes of length* 20 *and* 22, to appear

[9] G. Höhn, *Self-dual codes over the Kleinian four-group*, preprint (1996)

[10] S. Karlin, J. McGregor, *The Hahn polynomials, formulas and an application* Scripta Math. **26** (1961), 33-46

[11] E. Rains, N.J.A. Sloane, *Self-dual codes*, Handbook of Coding Theory, V. Pless and W. C. Huffman editors, North Holland, Amsterdam, 1998

[12] K. Tanabe, *A new proof of Assmus-Mattson theorem for non-binary codes*, Codes, Designs and Cryptography, to appear.

[13] H. Tarnanen, *An approach to constant weight and Lee codes by using the methods of association schemes*, PHD thesis, Turku University 1982

[14] H. Tarnanen, M. J. Aaltonen, J.-M. Goethals, *On the nonbinary Johnson scheme*, Europ. J. Comb. **6** (1985), 279-285

LABORATOIRE A2X, UNIVERSITÉ DE BORDEAUX, 351, COURS DE LA LIBÉRATION, 33405 TALENCE

E-mail address: bachoc@math.u-bordeaux.fr

DIMACS Series in Discrete Mathematics
and Theoretical Computer Science
Volume **56**, 2001

Numerical results on the asymptotic rate of binary codes

Alexander Barg and David B. Jaffe

ABSTRACT. We compute upper bounds on the maximal size of a binary linear code of length $n = 1000$, dimension k, and distance d. For each value of d, the bound is found by solving the Delsarte linear programming problem. Relying on the results of the calculations, we discuss the known bounds on the size of codes and some recent conjectures made about them. The most important conclusion is that Delsarte's linear programming method is unlikely to yield major improvements of the known general upper bounds on the size of codes.

1. Introduction: bounds on codes

A code C is a subset of the binary Hamming space H_2^n. The minimum distance between a pair of distinct points in C is called the distance of C, denoted $d(C)$. One of the main problems of coding theory is to find the maximal size $A(n, d)$ of a code with given distance d. This problem is solved exactly only for some small values of n and d. The general results known are in the form of upper and lower bounds.

The aim of this paper is to explore the limits of Delsarte's linear programming (or polynomial) method of deriving upper bounds on $A(n, d)$. After explaining the method and citing the best known general bounds, we present the results of calculations for $n = 1000$. Based on them, we speculate that the bounds currently known are likely to be close to the limits of Delsarte's method.

The best known lower bound (the Varshamov-Gilbert or VG bound) can be stated in the following form: *Let M be the maximal number such that*

$$(M-1)\sum_{i=0}^{d-1}\binom{n}{i} < 2^n.$$

Then there exists a code C of length n with $d(C) = d$ and $|C| = M$.

1991 *Mathematics Subject Classification.* Primary 94B65, Secondary 05E30.
The second author was partially supported by NSF grant DMS-9801581.

On the other hand, for any code,

$$(1)\qquad |C| \le \begin{cases} 2^n / \sum_{i=0}^{(d-1)/2} \binom{n}{i} & d \text{ odd} \\ 2^{n-1} / \sum_{i=0}^{(d-2)/2} \binom{n-1}{i} & d \text{ even;} \end{cases}$$

this is the Hamming bound.

The best known upper bounds are derived by Delsarte's linear programming method [**4**], [**5**]. We have

$$(2)\qquad A(n,d) \le 1 + \max \sum_{i=d}^{n} A_i$$

$$A_i \ge 0 \quad (d \le i \le n); \qquad \sum_{i=1}^{n} A_i K_k(n,i) \ge -\binom{n}{k} \quad (0 \le k \le n).$$

Here $K_k(n,x)$ is a Krawtchouk polynomial,

$$K_k(n,x) = \sum_{i=0}^{n} (-1)^i \binom{x}{i} \binom{n-x}{k-i}.$$

Problem (2) in the dual form can be formulated as follows:

$$(3)\quad A(n,d) \le D(d) := \inf_f \{ f(0) : \sum_{i=0}^{n} f(i)\tbinom{n}{i} = 2^n;\ f(i) \le 0, i = d, d+1, \dots, n \},$$

where the infimum is taken over all polynomials $f \in \mathbb{R}[x]$ with nonnegative Fourier-Krawtchouk coefficients.

This method is often useful to derive bounds for specific values of n and d. General upper bounds on $A(n,d)$, i.e., those valid for any n, are derived by exposing a polynomial feasible with respect to the conditions in (3). The best known bounds are due to McEliece, Rodemich, Rumsey, and Welch (MRRW) [**11**] and Levenshtein [**9**]. Denote by $x_t(n)$ the first zero of $K_t(n,x)$. It is known that

$$x_t(n-1) \le x_t(n) \le x_{t-1}(n-1),$$

$t = 1, \dots, n-1$; we also put $x_0(n) := n+1$; $x_{n+1}(n) := 0$. The first of the two MRRW bounds has the form

$$(4)\qquad A(n,d) \le \binom{n}{t} \frac{(n+1)^2}{2x_{t+1}(n+1)(t+1)},$$

where $d \ge x_t(n)$.

Levenshtein's bound is of the following form (see also[**10**]):

$$(5)\qquad A(n,d) \le L_n(d),$$

where

$$L_n(d) = \begin{cases} L_{k,n}(d) & \text{if } x_k(n-1)+1 < d \le x_{k-1}(n-2)+1, \\ 2L_{k,n-1}(d) & \text{if } x_k(n-2)+1 < d \le x_k(n-1)+1 \end{cases}$$

and

$$L_{k,n}(z) = \sum_{i=1}^{k} \binom{n}{i} - \binom{n}{k} \frac{K_{k-1}(n-1, z-1)}{K_k(n,z)}.$$

This bound is better than (4) for all $d \le n/2$.

Both (4) and (5) admit improvements based on the inequality

$$A(n,d) \le \min_w \frac{2^n}{\binom{n}{w}} A(n,d,w), \tag{6}$$

where $A(n,d,w)$ is the maximal size of a constant weight code with distance d and weight of all vectors w, and the minimum is taken over all integer w between 0 and $n/2$. Bounds on $A(n,d,w)$ in [**11**] and [**9**] involve the extremal zero $y_t(n,w)$ of a dual Hahn polynomial of degree t (this family of polynomials is orthogonal on $(0,1,\dots,w)$ with weight $\binom{w}{i}\binom{n-w}{i}/\binom{n}{w}$). In particular, the result in [**11**] has the form

(7)

$$A(n,d,w) \le \binom{n}{t} \frac{(n^2-(2t-1)n-2t)^2(w-t)(n-w-t)}{y_{t+1}(n,w)(t+1)(n-t+1)(n-2t-1)(n-2t)(n-2t+1)}.$$

where t is such that $d \ge 2y_t(n,w)$. Inequalities (6)-(7) together form the second MRRW bound on $A(n,d)$. It is often better than (4), though notably more difficult to compute. Therefore, more specific comparison is easier for the asymptotic versions of these bounds. We do not cite Levenshtein's bounds on $A(n,d,w)$ since they are not used below.

Let us also cite the asymptotical behavior of these bounds. For this purpose, let

$$R(\delta) = \limsup_{n\to\infty} \frac{1}{n} \log_2 A(n,d),$$

where the limit is computed over all sequences of codes with $\lim(d/n) \ge \delta$. Then by the VG bound

$$R(\delta) \gtrsim R_{vg} := 1 - H(\delta), \tag{8}$$

where H is the binary entropy function.

The asymptotic form of the MRRW bounds is as follows:

$$R(\delta) \lesssim R_{m1} := H\Big(\frac{1}{2} - \sqrt{\delta(1-\delta)}\Big). \tag{9}$$

$$R(\delta) \lesssim R_{m2} := \min_{0\le u\le 1-2\delta} 1 + g(u^2) - g(u^2+2\delta u+2\delta), \tag{10}$$

where $g(x) = H((1-\sqrt{1-x})/2)$. It is known [**11**] that $R_{m2} \le R_{m1}$ with equality for all $\delta \in [0.273, 1/2]$. Levenshtein's bounds for large n also converge to R_{m1} and R_{m2}, respectively.

Tightening the gap between the lower and upper bounds is the main problem of asymptotic coding theory. Currently it is far from solution. Until recently it was not even known whether Delsarte's method can lead to the proof of the (asymptotic) tightness of the VG bound. In 1998 A. Samorodnitsky [**12**] proved that

$$D(d) \ge \frac{1}{n}\sqrt{\binom{n}{\lfloor x_d(n)\rfloor}2^n\Big/\binom{n}{d}}, \tag{11}$$

resolving the above question in the negative. For $n \to \infty$ this implies

$$(1/n)\log_2 D(\delta n) \gtrsim (R_{vg} + R_{m1})/2. \tag{12}$$

Samorodnitsky also conjectured that asymptotically this inequality is tight. This conjecture, if true, would imply that $R(\delta) \lesssim (R_{vg} + R_{m1})/2$. The present study is to some extent motivated by this result and conjecture.

2. n = 1000

2.1. Results. Here we present the results of the calculations. For $n = 1000$ and a given d we solved the linear programming problem (2). Two essential limitations of the calculations are as follows. Due to time constraints we were not able to deal with d odd or $d \leq 43$. Further, we only looked for the maximal k such that the existence of a linear $[n, 2^k, d]$ code does not contradict the inequalities in Delsarte's problem (2). This assumption simplified the calculations considerably; see also the next section. While we cannot make any claims on the maximum of (2) with linearity restriction lifted, the results for linear codes are so close to the upper bounds cited that this question seems irrelevant. The results are most easily visualized by plotting the curves in coordinates (δ, R), see Fig. 1, 2.

We also give a short table of the bounds. Some of the formulas computed in the table were stated above as bounds on the size of the code rather than on $R(\delta)$. In this case we compute the binary log of the answer and normalize it by n. The line typeset in boldface gives the results of our calculations of the maximum in (2).

d/n	0.05	0.01	0.15	0.2	0.25	0.3	0.35	0.4	0.45
(4)	0.884	0.752	0.627	0.505	0.392	0.292	0.196	0.121	0.057
(5)	0.878	0.748	0.621	0.501	0.387	0.283	0.191	0.115	0.0505
(1)	0.839	0.722	0.624	0.539	0.464	0.397	0.338	0.285	0.237
(2)	**0.839**	**0.712**	**0.597**	**0.488**	**0.380**	**0.280**	**0.188**	**0.109**	**0.047**
(11)	0.786	0.630	0.496	0.379	0.279	0.191	0.118	0.059	0.018
(9)	0.858	0.722	0.592	0.469	0.355	0.250	0.158	0.082	0.025
(10)	0.825	0.693	0.573	0.461	0.354	0.250	0.158	0.082	0.025
(12)	0.786	0.626	0.491	0.374	0.272	0.184	0.112	0.055	0.016
(8)	0.714	0.531	0.390	0.278	0.189	0.199	0.066	0.029	0.007

Here in the upper half we have collected the bounds for $n = 1000$: (top to bottom) the nonasymptotic MRRW-1, Levenshtein's, Hamming, a solution of the LP problem for $n = 1000$, and Samorodnistky's bound. In the lower half we show the asymptotic 1st and 2nd MRRW bounds, Samorodnistky's conjectured answer, and the VG bound. For small d/n the agreement of the computed solution for the LP problem and the Hamming bound (1) is very good:

d/n	0.044	0.046	0.048	0.05	0.052	.054	0.056
Hamming	0.8554	0.8500	0.8446	0.8393	0.8340	0.8287	0.8236
LP ($n = 1000$)	0.855	0.849	0.844	0.839	0.833	0.828	0.823

2.2. Comments. The results for $n = 1000$ follow the known bounds rather closely, both for large δ and in the range where (7) provides improvements over (4). Therefore, though the bounds most likely are not the best possible for finite (n, d) it is likely that the asymptotic MRRW bounds give the exact answer in the asymptotic Delsarte problem. The results do not, therefore, support Samorodnitsky's conjecture on the exponential behavior of $D(\delta n)$.

This leaves us in a rather uncomfortable situation with respect to finding $R(\delta)$. Namely, Delsarte's method is by far the most powerful method known in deriving upper bounds on this function; on the other hand it seems unlikely that the upper bounds (9)-(10) can be significantly improved within the frame of this method. So at present there seem to be no ideas around that would lead to tightening the gap between (8) and (10).

Another important parameter of a code is its distance distribution. Let us cite the values of an optimal assignment of variables $A_d, \dots, A_n$ in the LP problem for $d = 350$. Again the easiest way to present the results is to plot them. We take the set of binomial probabilities $\mathcal{A}_w = \binom{n}{w}2^{k-n}, d \le w \le n$, as a reference distance distribution. The reason for this is that $(\mathcal{A}_w)$ is the average distance distribution of a code chosen in H_2^n with uniform probability. Sequences of codes whose distance distribution is asymptotically binomial are known to exist and have a number of interesting properties. Therefore, we plot the numbers $(1/n)\log_2 A_w$ and $(1/n)\log_2 \mathcal{A}_w$ with w/n on the x-axis. This plot is shown in Fig. 3. The results are in good accord with a theorem in [**1**] that asserts that codes meeting the MRRW bounds, if they exist, must have distance spectrum that converges to the binomial distribution.

3. Details on computations

In this section we explain how the data for $n = 1000$ was derived. In short, the program `Split` was used to mechanize the computations. For example, in `Split`, the following commands

```
type [1000,502,194_2];
via lp [current] = ;
type [1000,501,194_2];
!via lp [current] = ;
```

instruct the program to *attempt* to show that:

- There is no even binary linear code with parameters $n = 1000$, $k = 502$, $d = 194$, and hence no binary linear code (even or not) with these parameters. This is accomplished by linear programming.
- By linear programming one *cannot* show that there is no even binary linear code with parameters $n = 1000$, $k = 501$, $d = 194$.

The program performs these linear programming calculations using the simplex method, with floating point arithmetic (at a sufficiently high level of precision), and then verifies the conclusions using exact arithmetic. Thus, barring a programming error, the calculations are guaranteed to be correct.[1]

`Split` uses its own implementation of the simplex method, because commercial linear programming solvers do not support the high level of precision needed for the calculations, which is roughly ten to twenty times double precision. While the simplex method is still an excellent method for solving linear programs, the implementation in `Split` is primitive by comparison with the implementations in the commercial solvers (such as `CPLEX`), apart from their inability to calculate at

[1]Owing to time and memory restrictions, not all the calculations of the second type were verified. But in all cases where they were verified, no inconsistency was observed.

high precision. Consequently, the calculations were far slower than they would be otherwise. As d decreased, the calculations became more and more time-consuming. For example, a single calculation for $d = 400$ took about an hour, whereas a single calculation for $d = 50$ took about 200 hours (using one 500 MhZ Alpha 21264 processor). This explains why the calculations terminate at $d = 44$.

In principle exactly the same method could be used to bound unrestricted binary codes, in which case we would fix $n = 1000$ and some d, and try to bound the number M of codewords. However, owing to the particular implementation of the simplex method in `Split`, it is easier to answer questions of the form "Is $M \geq C$?" or "Is $M \leq C$?" for fixed C, than to find M directly. Since in the linear case one need only use C values which are powers of 2, the calculations are easier.

More information about `Split` may be found in [**6**], [**7**], [**8**], [**2**], and [**3**].

Color plots of the data for $n = 1000$ may be found on the second author's webpage, `http://www.math.unl.edu/~djaffe`.

References

[1] A. Ashikhmin, A. Barg, and S. Litsyn, *Estimates of the distance distribution of codes and designs*, IEEE Transactions on Information Theory, to appear.

[2] I. Bouyoukliev, D. B. Jaffe, and V. Vavrek, *The smallest length of eight-dimensional binary linear codes with prescribed minimum distance*, IEEE Transactions on Information Theory, to appear.

[3] I. Bouyoukliev and D. B. Jaffe, *Optimal binary linear codes of dimension at most seven*, Discrete Mathematics, to appear.

[4] P. Delsarte, *Bounds for unrestricted codes, by linear programming*, Philips Res. Rep. **27** (1972), 272–289.

[5] P. Delsarte, *An algebraic approach to the association schemes of coding theory*, Philips Research Repts Suppl. **10** (1973), 1–97.

[6] D. B. Jaffe, *A brief tour of split linear programming* (Proc. AAECC 12, ed. T. Mora, H. Mattson), Lecture Notes in Computer Science **1255** (1997), 164–173.

[7] ———, *Binary linear codes: new results on nonexistence*, preprint (ongoing work), Version 0.6beta (3/21/2000), accessible over the World Wide Web via `http://www.math.unl.edu/~djaffe/codes/code.ps.gz` or `code.dvi.gz`; see `.../~djaffe/binary/codeform.html` for an online database, which is more frequently updated.

[8] ———, *Optimal binary linear codes of length* ≤ 30, Discrete Mathematics, to appear.

[9] V. I. Levenshtein, *On choosing polynomials to obtain bounds in packing problems*, Proc. 7th All-Union Conf. Coding Theory and Information Transmission, Part 2 (Moscow, Vilnius), 1978, (In Russian), pp. 103–108.

[10] ———, *Universal bounds for codes and designs*, Handbook of Coding Theory (V. Pless and W. C. Huffman, eds.), vol. 1, Elsevier Science, Amsterdam, 1998, pp. 499–648.

[11] R. J. McEliece, E. R. Rodemich, H. Rumsey, and L. R. Welch, *New upper bounds on the rate of a code via the Delsarte-MacWilliams inequalities*, IEEE Trans. Inform. Theory **23** (1977), no. 2, 157–166.

[12] A. Samorodnitsky, *On the optimum of Delsarte's linear program*, preprint (1998).

Bell Labs, Lucent Technologies, 600 Mountain Ave., Rm. 2C-375, Murray Hill, NJ 07974

Department of Mathematics and Statistics, University of Nebraska, Lincoln, NE 68588-0323

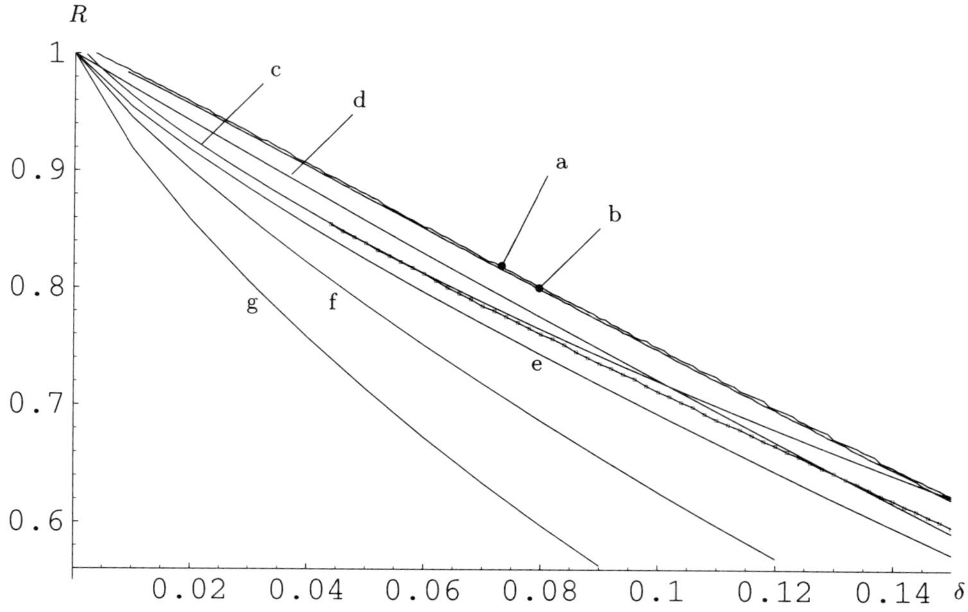

FIGURE 1. Bounds for low distances. The curve with marked dots represents the data for $n = 1000$ computed by (2). Nonasymptotic bounds ($n = 1000$): (a) MRRW bound (4), (b) Levenshtein's bound (5), (c) Hamming bound (1); Asymptotic upper bounds: (d) (9), (e) (10), (f) conjectured logarithmic asymptotics of $D(\delta n)$ (12) (g) the Varshamov-Gilbert bound.

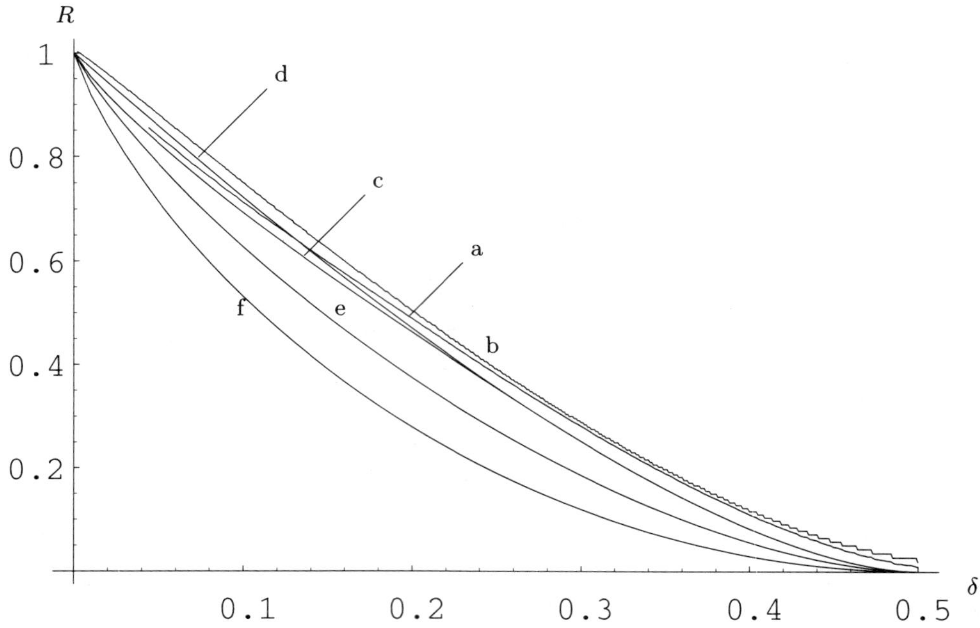

FIGURE 2. Bounds on codes: (a) data for $n = 1000$ computed by (2), (b) MRRW bound (4) for $n = 1000$, (c) (10), (d) (9), (e) (12), (f) the Varshamov-Gilbert bound.

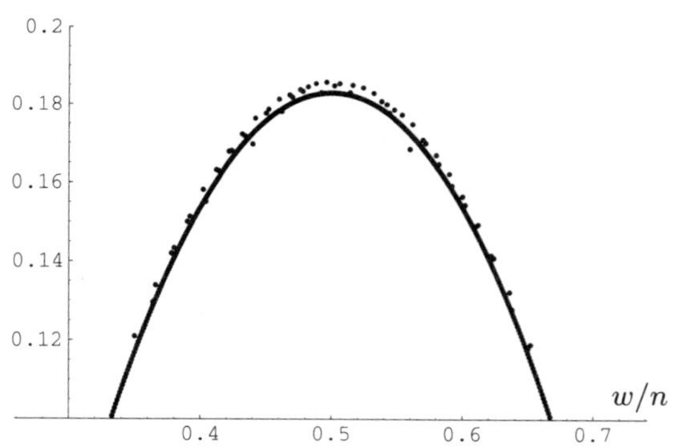

FIGURE 3. Optimal assignment of variables for $d = 320, k = 188$ (dots) and binomial distribution (solid curve).

DIMACS Series in Discrete Mathematics
and Theoretical Computer Science
Volume **56**, 2001

Weakly biased arrays, almost independent arrays and error-correcting codes

Jürgen Bierbrauer and Holger Schellwat

ABSTRACT. We exhibit the essentially coding-theoretic nature of the basic notions of limited bias and limited dependence. A new efficient construction sheds new light on the construction from [**4**] and leads to far-reaching improvements. The framework is Fourier analysis, ingredients are algebraic-geometric codes and Ramanujan graphs. The applications comprise universal hashing, authentication, resilient functions and pseudorandomness.

1. Introduction

The concepts of limited dependence and limited bias have manifold applications in cryptography and complexity theory. We mention universal hashing, authentication, resiliency against correlation attacks, pseudorandomness, block ciphers, derandomization, two-point based sampling, zero-knowledge, span programs, testing of combinatorial circuits, intersecting codes and oblivious transfer (see [**28, 24, 27, 4, 11, 30, 17, 7, 13, 8**]). A basic notion underlying these concepts are families of $\epsilon-$biased random variables. The heart of the influential papers by Naor and Naor [**27**] and by Alon, Goldreich, Håstad and Peralta [**4**] are constructions of families of binary $\epsilon-$biased random variables. In this paper we analyze and simplify these constructions, and we obtain far-reaching improvements. Our language is coding-theoretic and it will turn out that virtually all basic constructions are coding-theoretic.

DEFINITION 1. *An $(n,k)_q-$array $\mathcal{A}$ is an array with n rows and k columns, where the entries are taken from a set with q elements.*

Typically in complexity-theoretic applications one interprets the rows of the array as the points of a sample space with uniform distribution, where each column describes a random variable. One seeks to construct small sample spaces such that the random variables are not far from being independent. When used for the purposes of authentication of messages, the columns of the array are identified with the source, each row is a key and the entries are authenticators. In order to simplify description we chose to use neutral notation. Also, from the point of

---

2000 *Mathematics Subject Classification.* Primary 68P25, 94A60, 94B05, 94B27.

Key words and phrases. $\epsilon-$biased and $\epsilon-$dependent arrays (functions, random variables), (algebraic-geometric) codes, Hermitian codes, expander graphs, Fourier transform.

view of applications it is advisable not to restrict ourselves to the binary case. We consider the case when p is a prime.

DEFINITION 2. *Let p be a prime, $v = (v_1, v_2, \ldots, v_n) \in \mathbb{F}_p^n$. For every $x \in \mathbb{F}_p = \mathbb{Z}/p\mathbb{Z}$ let $\nu_v(x)$ be the number of times x appears as an entry of v and let $\delta_v(x)$ be defined by $\nu_v(x) = \frac{n}{p} + \delta_v(x)$. Let ζ be a primitive complex $p-$th root of unity. The* **bias** *of v is defined as*

$$bias(v) = \frac{1}{n}\left|\sum_{x\in\mathbb{F}_p} \delta_v(x)\zeta^x\right| = \frac{1}{n}\left|\sum_{x\in\mathbb{F}_p} \nu_v(x)\zeta^x\right|$$

Observe that $\sum_{x\in\mathbb{F}_p} \delta_v(x) = 0$. We see that $0 \leq bias(v) \leq 1$ for every vector v, and $bias(v) = 1$ if and only if v is a constant vector. The notion of $bias(v)$ depends on the choice of the root of unity ζ. However, we will always consider the bias of $\mathbb{F}_p-$linear spaces. As the change of root of unity corresponds to the consideration of a multiple of the vector, we see that the bias of an $\mathbb{F}_p-$vector space, defined as the maximum bias of its nonzero members, is defined with no ambiguity, and the result is the same independently of the choice of the p-th root of unity.

DEFINITION 3. *Let $0 \leq \epsilon < 1$. An $(n, k)_p-$array is $\epsilon-$***biased** *if every nontrivial linear combination of its columns has bias $\leq \epsilon$.*

Definition 3 shows that we should consider the linear code $\mathcal{C}$ generated by the columns of our $(n, k)_p-$array $\mathcal{A}$. If $dim(\mathcal{C}) < k$, then the 0-word is a nontrivial linear combination of the columns of $\mathcal{A}$. This shows that $\mathcal{A}$ cannot have a bias < 1 in this case. We will therefore assume that $\mathcal{C}$ is an $[n, k]_p-$code. Array $\mathcal{A}$ is the transpose of a generator matrix of $\mathcal{C}$ and the bias of $\mathcal{A}$ is the maximum bias of nonzero code-words of $\mathcal{C}$.

In the binary case it is easy to understand the relationship between the **bias** and the **minimum distance.** An $(n, k)_2$-array $\mathcal{A}$ of rank k is $\epsilon-$biased if and only if the nonzero words of the linear code $\mathcal{C}$ generated by the columns of $\mathcal{A}$ have weights w such that

$$\frac{1-\epsilon}{2} \leq \frac{w}{n} \leq \frac{1+\epsilon}{2}$$

In other words, $\mathcal{C}$ has to be a good code (large minimum distance) and also a good anticode (small maximum distance). It is easy to see that the theory of binary weakly biased arrays is equivalent to the theory of binary linear codes containing the all-1-word. When $p > 2$ the link between minimum distance and bias is less obvious. Here is a generalization of the binary construction for general p :

THEOREM 1. *Let p be a prime and $\mathcal{C}'$ a linear code $[n, k+1]_p$, which contains the all-1-word* $\mathbf{1}$. *Let D be the maximum weight among the words not contained in $\langle\mathbf{1}\rangle$. Let $\mathcal{C} \subset \mathcal{C}'$ be a hyperplane avoiding $\langle\mathbf{1}\rangle$ (and hence a code $[n, k]_p$ with maximum weight D). The transpose of a generator matrix of $\mathcal{C}$ is an $(n, k)_p-$ array with bias $\epsilon \leq p\frac{D}{n} - (p-1)$.*

PROOF. Let $0 \neq v \in \mathcal{C}$. The definition of D and the presence of $\mathbf{1}$ show that $\nu_v(x) \geq n - D$ for all $x \in \mathbb{F}_p$. Define $\mu_v(x) \geq 0$ by $\nu_v(x) = n - D + \mu_v(x)$. As the sum over all powers of ζ vanishes we have

$$\sum_x \delta_v(x)\zeta^x = \sum_x \nu_v(x)\zeta^x = \sum_x \mu_v(x)\zeta^x$$

and

$$\left|\sum_x \delta_v(x)\zeta^x\right| \leq \sum_x \mu_v(x) = (\sum_x \nu_v(x)) - p(n-D) = n - p(n-D) = pD - (p-1)n.$$

We conclude $bias(v) \leq p\frac{D}{n} - (p-1)$. □

Observe the trivial fact that the set of all k-tuples form the rows of an unbiased ($=0-$biased) $(p^k, k)_p-$array.

2. Weak bias and weak dependency

We define the notions that we are most interested in.

DEFINITION 4. *Let $0 \leq \epsilon < 1$. An $(n,k)_p-$array is **t-wise $\epsilon-$biased** if every nontrivial linear combination of at most t of its columns has bias $\leq \epsilon$.*

DEFINITION 5. *Let $0 \leq \epsilon < 1$. An $(n,k)_p-$array $\mathcal{A}$ is **t-wise $\epsilon-$dependent** if for every set U of $s \leq t$ columns and every $a \in \mathbb{F}_p^s$ the number $\nu_U(a)$ of rows of $\mathcal{A}$, whose projection onto U equals a satisfies*

$$\left|\frac{\nu_U(a)}{n} - 1/p^s\right| \leq \epsilon.$$

The notion of a t-wise $\epsilon-$dependent array generalizes the notion of an orthogonal array of strength t. An array is t-wise independent ($= t-$wise 0-dependent) if and only if it is an orthogonal array of strength t.

The standard method of deriving t-wise $\epsilon-$biased arrays from smaller $\epsilon-$biased arrays is essentially from [**27**].

THEOREM 2. *Let the following be given:*

- *An $(n,k)_p-$array B, which is $\epsilon-$biased.*
- *A linear code $[N, N-k, t+1]_p$.*

Then we can construct an $(n,N)_p-$array, which is t-wise $\epsilon-$biased.

PROOF. Let H be a check matrix of the code, hence an $(k,N)_p$-array each t columns of which are linearly independent. Our array is the product $\mathcal{A} = BH$. Denote by r_i the rows of B, by c_j the columns of $H, i = 1, \ldots, n; j = 1, \ldots, N$. The entries of BH are the dot products $r_i \cdot c_j$. Let $0 \neq \lambda = (\lambda_1, \ldots, \lambda_N)$ of weight $\leq t$. The corresponding linear combination of columns of $\mathcal{A}$ has entry $v_i = r_i \cdot (\sum_j \lambda_j c_j)$ in row i. The defining property of H shows that $\sum_j \lambda_j c_j \neq 0$. Vector $v = (v_i)$ is therefore a nontrivial linear combination of the columns of B, hence has bias $\leq \epsilon$. □

The most important structures from the point of view of applications are $t-$wise $\epsilon-$dependent arrays. The aim is to obtain efficient constructions. The main idea in [**27**] was that the notion of bias is easier to handle than dependency. In the next section we will introduce a very general notion of bias and use Fourier transform to show that $t-$wise $\epsilon-$biased arrays are also $t-$wise $\epsilon-$dependent. This shows that it suffices to construct $t-$wise $\epsilon-$biased arrays. Theorem 2 gives another reduction (via error-correcting codes): it suffices to construct $\epsilon-$biased arrays.

3. A general notion of bias

Let $G = \mathbb{F}_p^k$ be the elementary abelian group of order $g = p^k$. Consider the space $L^2(G)$ of complex-valued functions on G, with scalar product $\langle , \rangle$, where

$$\langle f, g \rangle = \frac{1}{g} \sum_{x \in G} f(x) \cdot \overline{g(x)}.$$

Here $x = (x_1, x_2, \ldots, x_k) \in G$. The characters of G form an orthonormal basis of $L^2(G)$. Let ζ be a primitive complex $p-$th root of unity. Every element $z = (z_1, z_2, \ldots, z_k)$ yields a character ϕ_z, where

$$\phi_z(x) = \zeta^{\sum x_i \cdot z_i} = \zeta^{x \cdot z}.$$

DEFINITION 6. *Let $f \in L^2(G)$. Express $f(x)$ in terms of the characters in the form*

$$f = \sum_{z \in G} \hat{f}(z) \phi_z.$$

The complex numbers $\hat{f}(z)$ are known as the **Fourier coefficients** *of f. The* **mass** *of f is defined as*

$$mass(f) = g\langle f, 1 \rangle = g\hat{f}(0) = \sum_{x \in G} f(x).$$

Finally we define the **bias** *of $z \in G$ with respect to function f by*

$$bias_f(z) = \frac{g}{mass(|f|)} |\hat{f}(z)|$$

Observe that because of the orthonormality of the characters the Fourier coefficients are given by

$$\hat{f}(z) = \langle f, \phi_z \rangle = \frac{1}{g} \sum_{x \in G} f(x) \zeta^{-x \cdot z}.$$

DEFINITION 7. *Let $\mathcal{A}$ be an $(n,k)_p-$array with entries in $\mathbb{F}_p$. The characteristic function $f(x) = \chi_{\mathcal{A}}(x)$ counts with which multiplicity x occurs as a row of $\mathcal{A}$. We define the Fourier transform $\hat{\mathcal{A}}(z)$, the mass and the bias of $\mathcal{A}$ by identifying $\mathcal{A}$ with its characteristic function and applying Definition 6. Observe $mass(\mathcal{A}) = n$.*

Let us compare the bias in Definition 7 with the combinatorial notion from Definition 3. So let $f = \chi_{\mathcal{A}}$ be the characteristic function of the $(n,k)_p-$array $\mathcal{A}$. As $f = |f|$ and $mass(f) = n$ we see that the bias of z in the sense of Definition 6 coincides with the bias of vector v, the linear combination of columns of $\mathcal{A}$ with z as vector of coefficients, in the sense of Definition 2.

LEMMA 1. *Let $f = \chi_{\mathcal{A}}$ be the characteristic function of an $(n,k)_p-$array $\mathcal{A}$ and $z \in G = \mathbb{F}_p^k$. Then $bias_f(z)$ in the sense of Definition 6 is the same as $bias(v)$ in the sense of Definition 3, where v is the linear combination of the columns of $\mathcal{A}$ with z as vector of coefficients.*

The following characterization of t-wise ϵ -biased arrays in terms of their Fourier transform follows directly from Definition 6 and Lemma 1.

THEOREM 3. *An $(n,k)_p-$array $\mathcal{A}$ is t-wise $\epsilon-$biased if and only if its Fourier coefficients $\hat{\mathcal{A}}(z)$ satisfy*

$$|\hat{\mathcal{A}}(z)| \leq \frac{n}{g}\epsilon \text{ for } 0 < wt(z) \leq t.$$

4. The relation between weak dependency and weak bias

DEFINITION 8. *Let $\mathcal{A}$ be an $(n,k)_p-$array, $f \in L^2(G)$. The* **expectation** *$E_{\mathcal{A}}(f)$ with respect to $\mathcal{A}$ is defined as the statistical expected value of f when the rows of $\mathcal{A}$ are used as a sample space with uniform distribution:*

$$E_{\mathcal{A}}(f) = \sum_{x \in G} \frac{\chi_{\mathcal{A}}(x)}{n} f(x)$$

The following lemma was inspired by [**12**]:

LEMMA 2. *Let $\mathcal{A}$ be an $(n,k)_p-$array. Then the following are equivalent:*

- *$\mathcal{A}$ is t-wise $\epsilon-$dependent.*
- *for every subset U of $s \leq t$ columns and every $f \in L^2((\mathbb{F}_p)^U)$ the following holds:*

$$|E_U(f) - mass(f)/p^s| \leq \epsilon \cdot mass(|f|).$$

Here the expectation $E_U(f)$ is with respect to the restriction of $\mathcal{A}$ to the columns in U.

PROOF. If f runs through the characteristic functions $\chi_a, a \in (\mathbb{F}_p)^U$, then the second condition expresses the definition of t-wise $\epsilon-$dependent arrays. So one direction of the equivalence is clear. Assume therefore $\mathcal{A}$ is t-wise $\epsilon-$dependent and fix U. Consider a general function $f = \sum_a c_a \chi_a$. Because of the linearity of the expectation and the mass, we get $|E_U(f) - \frac{mass(f)}{p^s}| = |\sum_a c_a(\frac{\chi_U(a)}{n} - \frac{1}{p^s})| \leq \sum_a |c_a| \cdot \epsilon = \epsilon \cdot mass(|f|)$. □

LEMMA 3. *Let $\mathcal{A}$ be an $(n,k)_p-$array. Then the following are equivalent:*

- *$\mathcal{A}$ is t-wise $\epsilon-$biased.*
- *For every set U of $s \leq t$ columns and every $0 \neq z \in G$ with support in U, the restriction g_z of the character ϕ_z to $\mathbb{F}_p^U$ satisfies*

$$|E_U(g_z)| \leq \epsilon$$

PROOF. We have $|E_U(g_z)| = |\sum_{x \in \mathbb{F}_p^U} \frac{\nu_U(x)}{n} \zeta^{x \cdot z}|$. Comparison with Definition 2 shows that $|E_U(g_z)|$ is the bias of vector $\mathcal{A}z$. □

After these preparations we can start comparing bias and dependency.

THEOREM 4. *Let $\mathcal{A}$ be a $p-$ary array, which is t-wise $\epsilon-$dependent. Then $\mathcal{A}$ is s-wise $p^s\epsilon$-biased for all $0 < s \leq t$*

PROOF. Lemma 2 gives a characterization of weak dependency in terms of a property, which all complex-valued functions defined on $(\mathbb{F}_p)^U$ have to satisfy, where U is a set of $s \leq t$ columns. Let U be given. Let g_z be the restriction to $(\mathbb{F}_p)^U$ of the character ϕ_z, where $0 \neq z$ has support in U. Apply Lemma 2, observing $mass(g_z) = 0, mass(|g_z|) = p^s$. We get $|E(g_z)| \leq p^s\epsilon$. Apply Lemma 3. □

THEOREM 5. *Let $\mathcal{A}$ be a $p-$ary array, which is s-wise δ_s-biased $(s = 1, 2, \ldots, t)$. Then $\mathcal{A}$ is t-wise ϵ -dependent, where $\epsilon = \left(\sum_{s=1}^t \binom{t}{s}(p-1)^s\delta_s\right)/p^t$.*

PROOF. Let $|U| = s \leq t, g_z$ as above, $h = p^s, a \in \mathbb{F}_p^U$. It is easy to check that $\chi_a = \frac{1}{h}\sum_z \zeta^{-a \cdot z} g_z$. By assumption $|E(g_z)| \leq \delta_{wt(z)}, z \neq 0$. Thus $|E(\chi_a) -$

$\frac{1}{h}| = \frac{1}{h}|\sum_{z\neq 0}\zeta^{-a\cdot z}E(g_z)| \leq \frac{1}{h}\sum_{i=1}^{s}\binom{s}{i}(p-1)^i\delta_i$. The δ_i form by definition a non-decreasing series. It is easily seen that the last expression increases with s. This leads to the statement of the theorem. □

Two special cases deserve being mentioned.

COROLLARY 1. *Let $\mathcal{A}$ be a p-ary array. If $\mathcal{A}$ is $(t-1)$-wise 0-biased (equivalently: is an orthogonal array of strength $t-1$) and t-wise $\epsilon-$biased, then $\mathcal{A}$ is t-wise $(\frac{p-1}{p})^t\epsilon-$dependent.*

THEOREM 6. *Let $\mathcal{A}$ be a p-ary array. If $\mathcal{A}$ is t-wise $\epsilon-$biased, $\epsilon > 0$, then $\mathcal{A}$ is t-wise $\epsilon'-$dependent for some $\epsilon' < \epsilon$.*

PROOF. If $\mathcal{A}$ is t-wise $\epsilon-$biased, then it is also s-wise $\epsilon-$biased for all $1 \leq s < t$. We can apply Theorem 5 with $\delta_s = \epsilon, s = 1, 2, \ldots, t$. It follows that $\mathcal{A}$ is t-wise ϵ'-dependent, where $\epsilon' = \epsilon\left(\sum_{s=1}^{t}\binom{t}{s}(p-1)^s\right)/p^t = \epsilon(p^t-1)/p^t < \epsilon$. □

5. The basic construction

We have seen that the construction problem of $t-$wise ϵ-dependent arrays can be reduced to ϵ-biased arrays. We concentrate on the following versions of the construction problem:

PROBLEM 1. *Given ϵ, for which $a > 0$ can we find asymptotic constructions for $\epsilon-$biased $(n, k)_p-$arrays such that $n \leq \left(\frac{k}{\epsilon}\right)^a$?*

PROBLEM 2. *Given ϵ, for which $b > 0$ can we find asymptotic constructions for $\epsilon-$biased $(n, k)_p-$arrays such that $n = O\left(\frac{k}{\epsilon^b}\right)$?*

We wish to minimize the exponents a, b. In the language of random variables as described in the introduction this amounts to construct families of random variables with the prescribed independence properties, which are defined on a small sample space. Moreover we are interested in efficient constructions, not just proofs of existence.

It has been proved in the final version of [**4**] that $b \geq 2$. A nonconstructive argument in [**6**] shows that $b = 2$. This yields the complete answer to Problem 2: $b = 2$. Our main result will be that there are efficient constructions for every $a > 1$ in Problem 1 and for every $b > 2$ in Problem 2.

Our basic construction is a generalization of what was called the **Powering construction** in [**4**]. It also generalizes and strengthens the **generalized powering construction** from [**9**].

THEOREM 7. *Let $\mathcal{C}$ be a code $[n, k, d]_q$, where $q = p^m$ and $\mathcal{B}$ an $(n_0, m)_p$-array of bias ϵ_0. We can construct an $(nn_0, km)_p-$array with bias $\epsilon = 1 - \delta + \delta\epsilon_0 < 1 - \delta + \epsilon_0$, where $\delta = d/n$ is the relative distance of code $\mathcal{C}$.*

PROOF. Fix a basis $\{a_r \mid r = 1, \ldots, m\}$ of $\mathbb{F}_q \mid \mathbb{F}_p$ and denote the columns of $\mathcal{B}$ by $b_r, r = 1, \ldots, m$. Construct an $(nn_0, km)_p$-array $A(\mathcal{C}, \mathcal{B})$ as follows: the columns $s(j, r)$ are indexed by pairs of basis elements $v(j)$ of the code and basis elements a_r of $\mathbb{F}_q \mid \mathbb{F}_p$. Let $v_i(j) = \sum_{r=1}^{m}\lambda_r^{(i,j)}a_r$. The rows come in sections of length n_0, the sections being indexed by the coordinates $i = 1, 2\ldots, n$. Section i of column $s(j, r)$ is $\lambda_r^{(i,j)}b_r$. We want to show that this array $A(\mathcal{C}, \mathcal{B})$ has bias $\leq 1 - \delta + \delta\epsilon_0$.

Let $s = \sum_{j,r} \mu(j,r)s(j,r)$ be a linear combination of the columns, where $\mu(j,r) \in \mathbb{F}_p$, not all zero. Every section of s is a linear combination of columns of $\mathcal{B}$, and hence either is 0 or has bias $\leq \epsilon_0$. It will suffice to show that there are at most $n-d$ indices i such that $s_i = 0$. We have

$$s_i = \sum_{j,r} \mu(j,r)\lambda_r^{(i,j)} b_r = \sum_r \left(\sum_j \mu(j,r)\lambda_r^{(i,j)} \right) b_r.$$

It follows that $s_i = 0$ if and only if $\sum_{j=1}^k \mu(j,r)\lambda_r^{(i,j)} = 0$ for all r.

Consider the codeword $w^{(r)} = \sum_{j=1}^k \mu(j,r)v(j) \in \mathcal{C}$. As not all the coefficients $\mu(j,r) \in \mathbb{F}_p$ vanish we must have $w^{(r)} \neq 0$ for some r. If we expand the entry $w_i^{(r)}$ with respect to the basis a_r we obtain

$$w_i^{(r)} = \sum_r \left(\sum_{j=1}^k \mu(j,r)\lambda_r^{(i,j)} \right) a_r.$$

Comparison shows that $s_i = 0$ if and only if $w_i^{(r)} = 0$. Choose r such that $w^{(r)} \neq 0$. It follows that at least d segments have $s_i \neq 0$. The bound on the bias follows from the following innocent Lemma 4, which is trivial. □

LEMMA 4. *Let $v = (v_1, v_2, \ldots, v_n)$, where $v_i \in \mathbb{F}_p^m$ and $bias(v_i) \leq \epsilon_i$. Then*

$$bias(v) \leq \left(\sum_{i=1}^n \epsilon_i \right) /n.$$

An application of Theorem 7 to Reed-Solomon codes $[p^m, Rp^m, (1-R)p^m]_{p^m}$ and inner unbiased arrays $(p^m, m)_p$ yields arrays $(p^{2m}, Rmp^m)_p$ with bias $\epsilon = R$. This shows that $a = 2$ can be reached in Problem 1. This reproduces an asymptotic result from [**4**]. There are therefore $(n_0, m)_p-$ arrays such that $n_0 \leq (m/\epsilon_0)^2$. Choose $R = \epsilon_0 = \epsilon/2$. Theorem 7 yields an $\epsilon-$biased $(N, K)_p-$array, where $N = n_0 p^m, K = km = Rmp^m$. For which exponent a' will it be true that

$$N = n_0 p^m = \frac{4m^2}{\epsilon^2} p^m \leq (K/\epsilon)^{a'} = (mp^m/2)^{a'}?$$

This will be true for every $a' > 1$ when m is large enough. We have a construction showing that every exponent $a > 1$ can be reached in Problem 1. We remark that this result is best possible in the sense, that it is impossible to replace the exponent a, which is arbitrarily close to 1, but fixed, by $a = 1$. In fact, consider the binary case. A family of binary codes such that $n \sim k/\epsilon$ would give a point in $(\delta, R)-$space, which is asymptotically reachable by binary linear codes and meets the asymptotic Plotkin bound with equality. This is impossible as it contradicts the JPL-bound (see [**26**]).

6. Block covering systems

An analysis of the constructions in [**27**] and [**2**] suggests the introduction of the following concept.

DEFINITION 9. *A* **block covering system** *$BCS(n, l, b, \gamma, \epsilon)$ is a family $\mathcal{B}$ of l-element subsets of a fixed $n-$set Ω, where $|\mathcal{B}| = b$ and the following holds:*

whenever $W \subset \Omega, \mid W \mid = w$ *such that* $1 - \frac{w}{n} \leq \gamma$, *then*

$$\frac{\textit{number of blocks avoiding } W}{b} \leq \epsilon.$$

Here it is permitted that a point occurs in a block with multiplicity more than 1. It is also permitted that different blocks are made of the same set of points, even with the same multiplicities. Clearly, we are interested in $BCS(n, l, b, \gamma, \epsilon)$ with small l, small b and small ϵ.

In the language of graphs we can describe a BCS as a bipartite graph with n vertices on the left, $b \geq n$ vertices on the right, where each right vertex has precisely l neighbours. The defining property says that every sufficiently large set W of vertices on the left has as set of neighbours almost all vertices on the right. Here the sizes are measured with respect to the size of the underlying set. Set W must satisfy that a random choice of a left vertex lands in W with probability at least $1 - \gamma$. Likewise a random choice among right vertices lands in $\Gamma(W)$ with probability $\geq 1-\epsilon$. This notion is clearly closely related to concepts like **expanders** and **dispersers,** which are often used in theoretical computer science, see the survey [**29**]. In fact, a BCS can be seen as some kind of a disperser. The main differences are the following: in a disperser the dispersion is from the large set of vertices to the small set. Typically dispersers are very unbalanced. In BCS the dispersion happens the other way round, from the small set to the large set. Also, the assumption of regularity is on the starting end for dispersers, on the receiving end for BCS. More importantly, in BCS the sizes of the subsets are measured relative to the size of the underlying set.

One source for BCS are orthogonal arrays. The result of Gopalakrishnan and Stinson [**20**] says in our terminology that the rows of an $OA_1(2, l, q)$ form a

$$BCS\left(q, l, q^2, \gamma, \frac{\gamma}{1 + (1 - \gamma)(l - 1)}\right).$$

Observe that $OA_1(2, l, q)$ can be obtained by projecting from 2-dimensional Reed-Solomon codes.

An often used construction idea for BCS (and for dispersers) are walks in expander graphs and neighbourhoods of expander graphs. A study of Eigenvalues (see [**22, 23**]) leads to the following:

THEOREM 8. *Let* Γ *be a graph on* n *vertices, which is connected and regular of valency* Δ. *Let* λ_1 *be the second-largest Eigenvalue of the adjacency operator of* Γ. *The family* $T_l(\Gamma)$ *of walks of length* l *leads after identification of inverse walks to a* $BCS(n, l, \frac{1}{2}n\Delta^{l-1}, \gamma, \gamma(\gamma + (1 - \gamma)\frac{\lambda_1}{\Delta})^{l-1})$.

Theorem 8 can be found in [**3**].

A different and in a way dual method of obtaining BCS from graphs is to use the family of neighbours of vertices.

Bounds from [**5**] and [**2**] lead to the following result:

THEOREM 9. *Let* Γ *be a graph on* n *vertices, which is connected and regular of valency* Δ. *Let* λ_1 *be the second-largest Eigenvalue of the adjacency operator of* Γ. *Denote by* $i^\perp$ *the set of neighbours of vertex* i. *The family* $\mathcal{B} = \{i^\perp \mid i \in V\}$ *is a* $BCS(n, \Delta, n, \gamma, Min\{\frac{\gamma}{1+4(1-\gamma)(1-\lambda_1/\Delta)}, \frac{\gamma}{1-\gamma}(\frac{\lambda_1}{\Delta})^2\})$.

6.1. BCS, codes and ϵ-biased arrays.

THEOREM 10. *Assume the following exist:*

- *A linear code* $\mathcal{C}$ *with parameters* $[n, k, d]_q$
- *A* $BCS(n, l, b, 1 - \frac{d}{n}, \epsilon)$ *for some* l.

Then we can construct a q^l- *ary* $\mathbb{F}_q-$*linear code* $\tilde{\mathcal{C}}$ *of length* b, *with just as many codewords as* $\mathcal{C}$ *and minimum distance (equal to minimum weight)* $\geq (1-\epsilon)b$.

PROOF. Let $v \in \mathcal{C}$. The corresponding codeword of $\tilde{\mathcal{C}}$ is $\tilde{v} = (\tilde{v}_B)$, where B varies through the blocks. Here $\tilde{v}_B = (v_i)_{i \in B}$. Observe that we interpret $\tilde{\mathcal{C}}$ as defined over $V(l, q)$, the vector space of dimension l over $\mathbb{F}_q$. Clearly $\tilde{\mathcal{C}}$ is linear over the ground field $\mathbb{F}_q$. This shows that minimum weight and minimum distance agree. Let $v \neq 0$. Then $\tilde{v}_B = 0$ if and only if $B \cap supp(v) = \emptyset$. The definition of a BCS shows that the number of such blocks is $\leq \epsilon b$. □

The output of Theorem 10 is a code, which is q^l-ary and $\mathbb{F}_q-$linear. Concatenation with any linear $q-$ary code of dimension l as inner code produces a q-ary linear code. This approach was used in [**2**] in an attempt to construct asymptotically nontrivial families of linear codes. We remark that codes, which are q^l-ary but only $q-$linear (also known as **additive codes** or **subgroup codes**)) have recently found renewed interest. In [**15, 21**] the mechanism of BCH-codes is extended to cover also additive codes, in [**16**] use is made of this generalization of BCH-codes to construct quantum codes.

Here we use concatenation for the construction of weakly biased arrays. All we need to know is the bias of the inner code.

THEOREM 11. *Assume the following exist:*

- *A linear code* $\mathcal{C}$ *with parameters* $[n, k, d]_p$.
- *A* $BCS(n, l, b, 1 - \frac{d}{n}, \epsilon)$ *for some* l.
- *An* ϵ_0-*biased* $(n_0, l)_p-$*array.*

Then we can construct an $(n_0 b, k)_p-$*array, which is* $\{\epsilon + \epsilon_0(1 - \epsilon)\}-$*biased.*

PROOF. Apply Theorem 10 and use concatenation in an obvious way. Denote the code generated by the columns of the resulting array by $\mathcal{C}'$. We need to control the bias of an arbitrary codeword v'. If $B \cap supp(v) = \emptyset$, then v'_B is a 0-tuple of length n_0. By definition of a BCS there are at least $(1 - \epsilon)b$ blocks B such that $B \cap supp(v) \neq \emptyset$. Each corresponding section v'_B is a nonzero codeword of the inner code, hence of bias $\leq \epsilon_0$. It follows that v'_B can have at most $\epsilon_0 n_0$ more zeroes than ones. The total surplus of zeroes in v' is therefore limited by $\epsilon b n_0 + (1 - \epsilon) b \epsilon_0 n_0$. Division by the length $b n_0$ yields the desired upper bound on the bias. □

The construction of weakly biased binary arrays as given in [**27**] is an application of Theorem 11 in the binary case, where the inner code is the Simplex code $[2^l, l, 2^{l-1}]$ of bias $\epsilon_0 = 0$ and the BCS is the family of walks (see Theorem 8). We remark that the family of codes constructed in [**27**] and used in the role of code $\mathcal{C}$ in Theorem 11 is not interesting. In fact, the use of 2-dimensional Reed-Solomon codes makes sure that the family has to be asymptotically trivial.

Apply Theorem 11 with the following ingredients: Code $\mathcal{C}$ belongs to a family of asymptotically nontrivial codes with relative minimum distance δ. The BCS is derived from Ramanujan graphs (see [**25**]) on n vertices with valency Δ via Theorem 9. It has parameters

$$BCS\left(n, \Delta, n, 1-\delta, \frac{1-\delta}{\delta}\cdot\frac{4}{\Delta}\right).$$

The weakly biased array are derived from Section 5 (Theorem 7), an ϵ_0-biased $(n_0, \Delta)_p-$ array, where $n_0 = (\Delta/\epsilon_0)^a$. Choose $\epsilon_0 = \epsilon/2$ and Δ such that $\frac{1-\delta}{\delta}\frac{4}{\Delta} \leq \epsilon/2$. We obtain an $\epsilon-$biased $(N, K)_p-$ array, where $N = O(K/\epsilon^{2a})$.

We reach the promised conclusion: every exponent $b > 2$ in Problem 2 is reachable with an effective construction.

7. Constructions of weakly dependent arrays

We use Theorem 2 with the weakly biased arrays constructed in Section 5 and in Subsection 6.1 as one ingredient. Primitive BCH-codes have parameters $[p^i-1, p^i-1-it, t+1]_p$. The following is obtained:

THEOREM 12. *Let p be a prime. For every natural number t and $a > 1$ there are efficient constructions for infinite families of t-wise $\epsilon-$biased $(n, N)_p-$arrays, where asymptotically*

$$n \leq \left(\frac{t\log_p(N)}{\epsilon}\right)^a.$$

For every natural number t and $b > 2$ there are efficient constructions for infinite families of t-wise $\epsilon-$biased $(n, N)_p-$arrays, where asymptotically

$$n = O\left(t\log_p(N)/\epsilon^b\right).$$

Theorem 6 shows that we can replace the term *t-wise $\epsilon-$biased* in Theorem 12 by *t-wise $\epsilon-$dependent.*

7.1. A problem in coding theory. As one of the ingredients for the derivation of Theorem 12 we used primitive BCH-codes. An attempt to optimize the code contribution leads to the following highly nontrivial problem in coding theory.

DEFINITION 10. *Fix q and t. Denote by $n_{t,q}(k)$ the maximal length n of a code $[n, n-k, t+1]_q$, equivalently the largest length of a linear orthogonal array of dimension k and strength t. Define*

$$l(t,q) = \liminf_{k\to\infty} \frac{k}{\log_q(n_{t,q}(k))}$$

The primitive $BCH-$codes show $l(t,q) \leq t$. This is what we used in the derivation of Theorem 12. The Gilbert-Varshamov bound seems to give the same bound $l(t,q) \leq t$ as the BCH-codes. The function $l(t,q)$ has been studied by Dumer in [**14**]. The sphere packing bound implies $l(t,q) \geq \lfloor t/2 \rfloor$. We have

$$\lfloor t/2 \rfloor \leq l(t,q) \leq t,$$

showing that asymptotically we cannot do better than replace the factors t in the statement of Theorem 12 by $\lfloor t/2 \rfloor$. Still, this is an interesting coding problem.

Let $t = 1$. Clearly $n_{1,q}(k) = q^k - 1$, hence $l(1,q) = 1$. Clearly we have $l(t,q) \geq l(t-1,q)$ in general. When $t = 2$ the maximum N such that a code $[N, N-k, 3]_p$ exists is of course $N = (p^k-1)/(p-1)$. This yields $l(2,q) = 1$.

THEOREM 13. *If there is an $\epsilon-$biased $(n,k)_p-$array, then there is a 2-wise $\epsilon-$biased $(n,\frac{p^k-1}{p-1})_p-$array. In particular for every $a>1$ there are efficient constructions for infinite families of 2-wise $\epsilon-$ biased $(n,N)_p-$arrays, where asymptotically*

$$n\leq\left(\frac{\log_p(N)+1}{\epsilon}\right)^a.$$

Also, in the binary case we have $l(t,2)=\lfloor t/2\rfloor$. To see this just use primitive $BCH-$codes with defining interval $[1,t/2]$ if t is even and $[0,(t-1)/2]$ if t is odd. In case $t=3$ we obtain a question on geometrical objects called caps. It is known that $l(3,q)\leq 3/2$, and this is the best known upper bound for general q. It is an open general question on caps if $l(3,q)=1$. The ternary case is special, too. It follows from a construction [**10**] based on the Hill cap that $l(3,3)<1/\log_3(2.21)$.

8. Authentication codes

In this application we interpret the rows of our array as keys, the columns as source states in an authentication scheme. An $(N,m)_p-$array is $(\delta,t)-$**almost strongly universal** (short $(\delta,t)-$ASU) if for every set $U=U_0\cup\{u\}$ of t columns and every $a'\in\mathbb{F}_p^{t-1}, x\in\mathbb{F}_p$ the frequencies $\nu_{U_0}(a')$ and $\nu_U(a',x)$ satisfy

$$\mid\nu_U(a',x)/\nu_{U_0}(a')\mid\leq\delta.$$

This means that when we authenticate t messages using the same key, the opponent's conditional probability of guessing the authenticator for message number t based on his knowledge of the authenticators of the preceding $t-1$ messages is still bounded by δ. The following obvious fact was proved in [**24**]:

THEOREM 14. *A t-wise $\epsilon-$ dependent array is $(\delta,t)-ASU$, where*

$$\delta=\frac{p^{-t}+\epsilon}{p^{-(t-1)}-\epsilon}.$$

Because of Theorem 5 we see that t-wise $\epsilon-$biased arrays yield authentication codes $(\delta,t)-$ASU. As an example start from $\epsilon-$biased arrays constructed out of Hermitian codes via Theorem 7. Consider the Hermitian code of dimension $k\sim\frac{1}{2}p^m$ defined over $\mathbb{F}_{p^m}$, where m is even. The parameters of the code are $[p^{3m/2},\frac{1}{2}p^m,p^{3m/2}-p^m]$. Theorem 7 yields an $\epsilon-$ biased $(p^{5m/2},\frac{1}{2}p^m)_p-$array, where $\epsilon=p^{-m/2}$. Apply Theorem 2 with codes $[p^{2i},p^{2i}-3i,4]_p$ (recall $l(3,p)\leq 3/2$ from Subsection 7.1) Here $k\sim\frac{1}{2}p^m\sim 3i$. The result is a

$$\text{3-wise } p^{-m/2} - \text{ biased } (p^{5m/2},p^{\frac{1}{3}p^m})_p - \text{array}.$$

In case $m=6$ we obtain a $(p^{15},p^{\frac{1}{3}p^6})_p-$array, which is $(\delta,3)-$ASU, where $\delta\sim 2/p$. This can be used to authenticate a source with $\frac{1}{3}p^6\log(p)$ bits with $15\log(p)$ key bits, where the same authenticator is used for 3 consecutive messages. Choose for instance $p\sim 2^{41}$. The opponent's probability of success is limited by 2^{-40} in this case. We can authenticate a source of up to 2^{249} bits, using 615 key bits. The method of [**24**] would require more than 750 bits of key in this situation. Further improvements can be achieved by using Theorem 7 and more sophisticated AG-codes.

9. Covering arrays

An $(n, N, t)_p$−**covering array** is an $(n, N)_p$−array with the property that $\nu_U(z) \neq 0$ for every set U of t columns and every t−tuple z. Covering arrays are also known as t−**independent sets** and as **universal sets.**

It follows from Theorem 6 that a t−wise p^{-t}−biased array is necessarily an $(n, N, t)_p$− covering array. Theorem 12 shows that there are efficient constructions for such covering arrays, where $n = O(t \log_p(N) \cdot p^{tb})$, $b > 2$.

DEFINITION 11. *Consider $(n, N, t)_p$−covering arrays. Define*

$$c_p(t) = \liminf_{N\to\infty} \frac{n}{\log_p(N)}.$$

We have seen that there are effective constructions of such covering arrays showing $c_p(t) = O(tp^{bt})$ for every $b > 2$, and non-constructively $c_p(t) = O(tp^{2t})$. A non-constructive upper bound is in [**19**]. In case of strength $t = 2$ the mechanism of **Sperner capacity** solves the asymptotic problem completely, for all q. The result in [**18**] is

$$c_q(2) = \frac{q \log_2 q}{2} \text{ for all } q.$$

The most general construction principle for covering arrays involves perfect hash families.

DEFINITION 12. *An $(N, n)_q$−array is* **perfectly** t−**hashing** *if for every set of t columns there is a row such that the entries in this set of columns are different. We use the notation $PHF(N; n, q, t)$. The most suggestive interpretation is obtained when we interpret the array as a family of N mappings from an n-set to a q−set. The defining property says then that for every set of t elements in the domain we can find a function whose restriction to these t points is injective.*

THEOREM 15. *If a set of n_1 functions from an N-set to an m−set is perfectly t−hashing and a set of n_2 functions from the m−set to a q-set is t-covering, then the composition is a set of n_1n_2 functions from an N-set to a q−set, which is again t-covering, hence an $(n_1n_2, N, t)_q$−covering array.*

This has been formulated in the language of mappings, where the statement is indeed obvious. In order to apply Theorem 15 we need perfect hash families and small covering arrays. The only effective construction of perfect hash families seems to follow from the following remark of Alon [**1**], which makes use of codes again.

LEMMA 5 (Alon). *Let C be a q-ary code of length N with minimum distance d and n codewords. Let the codewords be written as columns of an array. Then a $PHF(N; n, q, t)$ is obtained if t is chosen such that*

$$\frac{d}{N} > 1 - 1/\binom{t}{2}.$$

The main construction in [**13**] is an application of Theorem 15, families of algebraic-geometric codes furnishing the perfect hash families. **Intersecting codes** are used in order to construct small covering arrays to which to apply the construction. The main result of [**13**] is an efficient construction for families of binary covering arrays showing $c_2(t) = O(t^3 2^{2t})$. This is rather close to what our mechanism yields in the binary case.

References

1. N. Alon: *Explicit construction of exponential sized families of k-independent sets, Discrete Mathematics* **58** (1986), 191–193.
2. N. Alon, J. Bruck, J. Naor, M. Naor, R. Roth: *Construction of asymptotically good low-rate error-correcting codes through pseudo-random graphs, IEEE Transaction on Information Theory* **38** (1992), 509–516.
3. N. Alon, U. Feige, A. Wigderson, D. Zuckerman: *Derandomized graph products, Computational Complexity* (1995), 60–75.
4. N. Alon, O. Goldreich, J. Håstad, R. Peralta: *Simple constructions of almost k-wise independent random variables, Random Structures in Algorithms* **3** (1992), prelim. version Symposium 31st FOCS 1990, 544–553.
5. N. Alon and V. D. Milman: λ_1, *isoperimetric inequalities for graphs, and superconcentrators,* JCTB **38** (1985), 73–88.
6. Y. Azar, R. Motwani and J. Naor: *Approximating probability distributions using small sample spaces,* Combinatorics **18** (1998), 151–171.
7. J. Boyar, G. Brassard and R. Peralta: *Subquadratic zero-knowledge, JACM* **42** (1995), 1169–1193.
8. G. Brassard, C. Crépeau and M. Santha: *Oblivious transfers and intersecting codes, IEEE Transactions on Information Theory* **42** (1996), 1769–1780.
9. J. Bierbrauer and H. Schellwat: Constructive asymptotic lower bounds on linear codes and weakly biased arrays, submitted for publication in *IEEE Transactions on Information Theory.*
10. A. R. Calderbank and P. C. Fishburn: *Maximal three-independent subsets of* $\{0,1,2\}^n$, *Designs, Codes and Cryptography* **4** (1994), 203–211.
11. J. L. Carter and M. N. Wegman: *Universal Classes of Hash Functions,* J.Computer and System Sci. 18 (1979), 143–154.
12. B. Chor, O. Goldreich, J. Håstad, J. Friedman, S. Rudich and R. Smolensky: *The bit extraction problem or t-resilient functions,* Proc. 26th Symp. on Foundations of Computer Science (1985), 396–407.
13. G. D. Cohen and G. Zémor: *Intersecting codes and independent families, IEEE Transaction on Information Theory* **40** (1994), 1872–1881.
14. Ilya Dumer: *Nonbinary double-error-correcting codes designed by means of algebraic verieties, IEEE Transactions on Information Theory* **41** (1995), 1657–1666.
15. Y. Edel and J. Bierbrauer: *Twisted BCH-codes, Journal of Combinatorial Designs* **5** (1997), 377–389.
16. Yves Edel and J. Bierbrauer: *Quantum twisted codes, Journal of Combinatorial Designs* **8** (2000), 174–188.
17. A. Gal: *A characterization of span program size and improved lower bounds for monotone span programs, Proceedings* 13th *Symposium of the Theory of Computing* (1998), 429–437.
18. L. Gargano, J. Körner and U. Vaccaro: *Sperner Capacities, Graphs and Combinatorics* **9** (1993), 31–46.
19. A. Godbole, D. E. Skipper and R. A. Sunley: *t-Covering arrays: upper bounds and Poisson approximations, Combinatorics, Probability and Computing* **5** (1996), 105–118.
20. K. Gopalakrishnan and D. R. Stinson: *A simple analysis of the error probability of two-point based sampling, Information Processing Letters* **60** (1996), 91–96.
21. M. Hattori,R. J. McEliece and G. Solomon: *Subspace subcodes of Reed-Solomon codes, IEEE Transactions on Information Theory* **44** (1998), 1861-1880.
22. N. Kahale: *Better expansion for Ramanujan graphs, Proc. 33rd FOCS 1992,* 398–404.
23. N. Kahale : *Eigenvalues and expansion of regular graphs, Journal of the ACM* **42** (1995), 1091–1106.
24. K. Kurosawa, T. Johansson and D. Stinson: *Almost k-wise independent sample spaces and their cryptologic applications, Lecture Notes in Computer Science* **1233** (1997), 409–421 (Advances in Cryptology, Eurocrypt 97).
25. A. Lubotzky, R. Phillips and P. Sarnak: *Explicit expanders and the Ramanujan conjectures, Proceedings 18th ACM Symposium Theory of Computing 1986,* 240–246.
26. R. J. McEliece, E. R. Rodemich, H. Rumsey, L. R. Welch: *New upper bounds on the rate of a code via the Delsarte-MacWilliams inequalities, IEEE Transactions on Information Theory* **23** (1977), 157–166.

27. J. Naor and M. Naor: *Small-bias probability spaces: efficient constructions and applications, SIAM Journal on Computing* **22** (1993), 838–856, preliminary version: Proceedings STOC 1990, 213–223.
28. M. Naor and O. Reingold: *On the construction of pseudo-random permutations:Luby-Rackoff revisited,* Proceedings STOC 29 (1997), 189–199.
29. N. Nisan: *Extracting randomness: How and Why - A Survey, Complexity 96.*
30. M. N. Wegman and J. L. Carter: *New Hash Functions and Their Use in Authentication and Set Equality,* J.Computer and System Sci. 22 (1981), 265–279.

DEPARTMENT OF MATHEMATICAL SCIENCES, MICHIGAN TECHNOLOGICAL UNIVERSITY, HOUGHTON, MICHIGAN 49931
E-mail address: jbierbra@mtu.edu

DEPARTMENT OF NATURAL SCIENCES, UNIVERSITY OF ÖREBRO, 70182 ÖREBRO, SWEDEN
E-mail address: holger.schellwat@nat.oru.se

DIMACS Series in Discrete Mathematics
and Theoretical Computer Science
Volume **56**, 2001

Indexes of spherical codes

Peter Boyvalenkov, Danyo Danev, and Peter Kazakov

ABSTRACT. We characterize certain good spherical codes by finding their indexes. Three methods for finding indexes are proposed and used in several cases. Among the most interesting application, we show that the icosahedron has non-trivial indexes 8 and 14 and that both the 600-cell and the 120-cell have non-trivial indexes 14, 16, 18, 22, 26, 28, 34, 38, 46 and 58.

1. Introduction

Let $\mathbf{S}^{n-1} = \{x = (x_1, x_2, \dots, x_n) : x_1^2 + x_2^2 + \dots + x_n^2 = 1\}$ be the n-dimensional Euclidean sphere with the usual inner product $\langle x, y\rangle = \sum_{i=1}^n x_i y_i$, where $x = (x_1, x_2, \dots, x_n)$ and $y = (y_1, y_2, \dots, y_n)$.

A finite nonempty set $C \subset \mathbf{S^{n-1}}$ is said to be an (n, M, s) spherical code, $M = |C|$, if the maximal inner product of any two different vectors from C does not exceed s. A spherical code C is called antipodal if $C = -C$.

DEFINITION 1.1. A spherical code $C \subset \mathbf{S}^{n-1}$ is called a spherical τ-design if and only if the formula for numerical integration

$$\int_{\mathbf{S}^{n-1}} f(x)d\mu(x) = \frac{1}{|C|}\sum_{x\in C} f(x) \tag{1.1}$$

($\mu(.)$ is the Lebesgue measure, $\mu(\mathbf{S}^{n-1}) = 1$) is true for all real polynomials $f(x) = f(x_1, x_2, \dots, x_n)$ of total degrees $1, 2, \dots, \tau$. A spherical code $C \subset \mathbf{S}^{n-1}$ is said to have an index k if and only if the formula (1.1) holds for all real homogeneous harmonic polynomials $f(x) = f(x_1, x_2, \dots, x_n)$ of total degree k. Actually both sides of (1.1) are then equal to zero.

It is clear that any spherical τ-design has indexes $1, 2, \dots, \tau$, and that any antipodal spherical code has all odd indexes: 1,3,5, etc. We call such indexes trivial. In this paper we study nontrivial indexes of certain good (in most of the cases maximal) spherical codes. In Section 3 we describe three methods for finding indexes. Applications to some known optimal spherical codes are shown in Section 4. In Section 5 we make some further investigations on other interesting spherical codes. The most interesting results show that the icosahedron (which is a 5-design and an antipodal code) has non-trivial indexes 8 and 14 and both the 600-cell

1991 *Mathematics Subject Classification.* Primary 05B30; Secondary 52B10, 52B11.

and the 120-cell have non-trivial indexes 14, 16, 18, 22, 26, 28, 34, 38, 46 and 58. An elementary approach for upperbounding nontrivial indexes of concrete codes is explained in Section 6. The results are summarized in Table 1 (see also Theorem 6.3).

2. Some preliminaries

2.1. Gegenbauer polynomials. The theory of spherical codes and designs is naturally connected with the family of Gegenbauer polynomials. The Gegenbauer polynomials $\{P_i^{(n)}(t)\}_{i=0}^{\infty}$ [**1**, Chapter 22] (n is nothing but the relevant dimension) can be defined by the recurrence

$$(i+n-2)P_{i+1}^{(n)}(t) = (2i+n-2)tP_i^{(n)}(t) - iP_{i-1}^{(n)}(t),\ i \geq 1,$$

where $P_0^{(n)}(t) = 1$ and $P_1^{(n)}(t) = t$. It is well known that any real polynomial $f(t)$, $\deg(f) = m$, can be uniquely expanded in terms of Gegenbauer polynomials as

$$f(t) = \sum_{i=0}^{m} f_i P_i^{(n)}(t).$$

The following definition joins two properties of very different nature. It is convenient for explanation of the linear programming approach for bounding sizes of spherical codes and designs.

DEFINITION 2.1. Denote

$$\begin{aligned} A_{n,s} &= \{f(t) : f(t) \leq 0 \text{ for } t \in [-1,s], f_i \geq 0 \text{ for } i = 1,...,m = \deg f(t), f_0 > 0\}, \\ B_{n,\tau} &= \{f(t) : f(t) \geq 0 \text{ for } t \in [-1,1], f_i \leq 0 \text{ for } i = \tau+1,...,m = \deg f(t)\}. \end{aligned}$$

In our applications, we need polynomials in the good sets $A_{n,s}$ or $B_{n,\tau}$ for specified values of n and s, or n and τ, respectively.

2.2. The linear programming bounds. The most powerful tool for bounding the size of spherical codes and designs is the so called linear programming bound (LPB). It works for every polynomial metric space [**15**] and the Euclidean spheres appear to be among the most interesting examples of such spaces. We describe in brief the idea behind these bounds as it is exposed in [**15, 16**]. Another description of this approach can be found in [**11**].

The spaces $\mathbf{S}^{n-1}$ are under consideration with the usual Euclidean metric. The Hilbert space $\mathcal{L}_2(\mathbf{S}^{n-1}, \mu)$ of complex-valued functions defined on $\mathbf{S}^{n-1}$ with inner product

$$\langle u, v \rangle = \int_{\mathbf{S}^{n-1}} u(x)\overline{v(x)}d\mu(x)$$

can be decomposed into a countable direct sum of mutually orthogonal subspaces V_i, $i = 0, 1, \dots$, where $\dim(V_i) = r_i$. Actually the subspace V_i is the space of all functions, which are represented on $\mathbf{S}^{n-1}$ by harmonic homogeneous polynomials on n variables of total degree i. If $\{v_{ij}(x) : j = 1, 2, \dots, r_i\}$ is an orthonormal basis of V_i, one considers the so-called zonal spherical functions

$$P_i^{(n)}(\langle x, y\rangle) = \frac{1}{r_i}\sum_{j=1}^{r_i} v_{ij}(x)\overline{v_{ij}(y)}. \tag{2.1}$$

In this way one obtains nothing but the Gegenbauer polynomials exactly as defined in 2.1.

The next statement gives an equivalent definition of spherical designs and index of a spherical code.

LEMMA 2.2. *(i) A spherical code C is a τ-design iff $\sum_{x\in C} v(x) = 0$ for every function $v(x) \in \cup_{i=1}^{\tau} V_i$. (ii) A spherical code C has index k iff $\sum_{x\in C} v(x) = 0$ for every function $v(x) \in V_k$.*

PROOF. The integral in (1.1) vanishes for harmonic polynomials. This proves the "if" direction both in (i) and (ii). The other direction can be proved by the direct sum decomposition of the space of polynomials on $\mathbf{S}^{n-1}$. □

It is not difficult to see that, for any finite set $C \subset \mathbf{S}^{n-1}$ and any polynomial $f(t)$ of degree m, the following identity holds [**15, 16**] (see also [**11**, Corollary 3.8])

$$|C|f(1) + \sum_{x,y\in C, x\neq y} f(\langle x,y\rangle) = |C|^2 f_0 + \sum_{i=1}^{m} \frac{f_i}{r_i} \sum_{j=1}^{r_i} \left| \sum_{x\in C} v_{ij}(x) \right|^2 . \tag{2.2}$$

Using (2.2) and Lemma 2.2(i) one obtains the LPB for codes and designs [**11, 14**].

THEOREM 2.3. *(i) For any polynomial $f(t) \in A_{n,s}$ and any spherical (n, M, s) code we have $M \leq f(1)/f_0$.*
(ii) For any polynomial $f(t) \in B_{n,\tau}$ and any spherical τ-design C we have $|C| \geq f(1)/f_0$.

2.3. The Levenshtein bound. A general choice for good polynomials was proposed by Levenshtein [**14, 15, 16**] for Theorem 2.3(i) and by Delsarte, Goethals and Seidel [**11**] for Theorem 2.3(ii). The Levenshtein's polynomials give the best possible LPB among the polynomials of the same or lower degree. The bound can be easily described in terms of adjacent orthogonal polynomials and their zeroes.

In our case, the adjacent polynomials are simply Jacobi polynomials $P_i^{a+\frac{n-3}{2}, b+\frac{n-3}{2}}(t)$ [**1**, Chapter 22], where $a, b \in \{0, 1\}$. Set

$$T_k^{1,\varepsilon}(t,s) = \sum_{i=0}^{k} r_i^{1,\varepsilon} P_i^{\frac{n-1}{2}, \frac{n-3}{2}+\varepsilon}(t) P_i^{\frac{n-1}{2}, \frac{n-3}{2}+\varepsilon}(s), \tag{2.3}$$

where $r_i^{1,\varepsilon} = \left(\frac{n+2i-1+\varepsilon}{n+\varepsilon-1}\right)^{2-\varepsilon} \binom{n+i-2+\varepsilon}{i}$ and $\varepsilon \in \{0,1\}$ (i.e. $a = 1$ and $b = \varepsilon$). Let $t_k^{1,1}$ and $t_k^{1,0}$ be the greatest zeroes of the polynomials $P_i^{(n-1)/2,(n-1)/2}(t) = P_i^{(n+2)}(t)$ and $P_i^{(n-1)/2,(n-3)/2}(t)$, respectively. The Levenshtein's polynomials are defined to be

$$f_m^{(s)}(t) = (t+1)^{\varepsilon}(t-s)[T_{k-1}^{1,\varepsilon}(t,s)]^2$$

where $m = 2k - 1 + \varepsilon$ and $t_{k-1+\varepsilon}^{1,1-\varepsilon} \leq s < t_k^{1,\varepsilon}$.

THEOREM 2.4. *([**14, 15, 16**]) If C is an (n, M, s) code, then*

$$M \leq \begin{cases} L_{2k-1}(n,s) = \binom{k+n-3}{k-1}\left[\frac{2k+n-3}{n-1} - \frac{P_{k-1}^{(n)}(s) - P_k^{(n)}(s)}{(1-s)P_k^{(n)}(s)}\right], & \text{for } t_{k-1}^{1,1} \leq s < t_k^{1,0}, \\ L_{2k}(n,s) = \binom{k+n-2}{k}\left[\frac{2k+n-1}{n-1} - \frac{(1+s)(P_k^{(n)}(s) - P_{k+1}^{(n)}(s))}{(1-s)(P_k^{(n)}(s) + P_{k+1}^{(n)}(s))}\right], & \text{for } t_k^{1,0} \leq s < t_k^{1,1}. \end{cases}$$

Moreover, if $M = L_m(n,s)$*, then* C *is a spherical* m*-design and all inner products* $\langle x,y\rangle$, $x,y \in C$, $x \neq y$*, are zeros of* $f_m(t)$.

2.4. The test functions. In [**6**], test functions $Q_j(n,s)$ have been defined to check the global optimality of the Levenshtein bound. For the sake of completeness and self-containing of this article, we define them once again.

Let the numbers $\alpha_0 < \alpha_1 < \cdots < \alpha_{k-1+\varepsilon} = s$ ($-1 \leq \alpha_0$, $\varepsilon \in \{0,1\}$) be all different zeros of the Levenshtein polynomial $f_{2k-1+\varepsilon}^{(s)}(t)$ (see (2.3)). In [**16**, Theorems 4.1 and 4.3], positive weights ρ_i, $i = 0,1,\ldots,k+\varepsilon$, have been defined with the property that, for any real polynomial $f(t)$ of degree at most $2k-1+\varepsilon$ the equality

$$f_0 = \rho_{k+\varepsilon} f(1) + \sum_{i=0}^{k-1+\varepsilon} \rho_i f(\alpha_i) \tag{2.4}$$

holds [**16**, Theorems 4.1 and 4.3]. Obviously, we have $L_{2k-1+\varepsilon}(n,s) = 1/\rho_{k+\varepsilon}$.

Now the test functions are defined by

$$Q_j(n,s) = \rho_{k+\varepsilon} + \sum_{i=0}^{k-1+\varepsilon} \rho_i P_j^{(n)}(\alpha_i) \tag{2.5}$$

for $t_{k-1+\varepsilon}^{1,1-\varepsilon} \leq s < t_k^{1,\varepsilon}$. Until now, they played role only as necessary and sufficient conditions for global optimality of the Levenshtein bounds as given by the next theorem [**6**].

THEOREM 2.5. *The bound* $L_m(n,s)$ *can be improved by a polynomial from* $A_{n,s}$ *of degree at least* $m+1$ *if and only if* $Q_j(n,s) < 0$ *for some* $j \geq m+1$*. Moreover, if* $Q_j(n,s) < 0$ *for some* $j \geq m+1$*, then* $L_m(n,s)$ *can be improved by a polynomial from* $A_{n,s}$ *of degree* j.

3. Methods for finding indexes

We describe three approaches which appear to be useful in investigations of indexes of concrete codes. Our first proposition lies on the necessary conditions for attaining the LPB for spherical codes. The next theorem can be extracted from [**14, 15, 16**].

THEOREM 3.1. *Let* C *be an* (n,M,s) *code and* $f(t) \in A_{n,s}$ *be such that* $M = f(1)/f_0$*. Then* $f_k > 0$ *implies that* C *has index* k.

PROOF. Apply (2.2) and Lemma 2.2(ii). □

The second method uses the test-functions described in subsection 2.4. As we already pointed out they provide necessary and sufficient conditions of optimality of the Levenshtein bounds for spherical codes. It turns out that they deserve once again their name, i.e. they show indexes in some way. By Theorem 2.4, the codes which meet the bound $L_m(n,s)$ are m-designs. Thus, for such codes we are only interested in indexes larger than m.

THEOREM 3.2. *Let* C *be an* $(n, L_m(n,s), s)$ *code. Then* $Q_k(n,s) = 0$ *for some* $k > m$ *implies that* C *has index* k.

PROOF. We construct a polynomial of degree k for application in Theorem 3.1. We can uniquely write

$$P_k^{(n)}(t) = f_m^{(s)}(t)q(t) + h(t),$$

where $\deg(h) \leq m-1$. Since the Levenshtein's polynomial $f_m^{(s)}(t)$ has strictly positive Gegenbauer expansion [**16**, Th. 5.1], we can find large enough real a such that the polynomial

$$f(t) = P_k^{(n)}(t) - h(t) + af_m^{(s)}(t)$$

belongs to $A_{n,s}$. Obviously, $f_k = 1 > 0$ and thus it remains to prove that $f(1)/f_0 = L_m(n,s)$.

Denote

$$\begin{aligned} h(t) &= \sum_{i=0}^{\deg h} h_i P_i^{(n)}(t), \\ f_m^{(s)}(t) &= \sum_{i=0}^{m} f_i' P_i^{(n)}(t), \\ f_m^{(s)}(t)q(t) &= \sum_{i=0}^{k} f_i'' P_i^{(n)}(t). \end{aligned}$$

Using the condition $Q_k(n,s) = 0$ in (2.4) and (2.5) for h_0, we obtain

$$h_0 = -\frac{f_m^{(s)}(1)q(1)}{L_m(n,s)} = -f_0' q(1).$$

Since $h_0 + f_0'' = 0$ and $f_0 = -h_0 + af_0' = (q(1)+a)f_0'$, we calculate

$$\begin{aligned} \frac{f(1)}{f_0} &= \frac{1 - h(1) + af_m^{(s)}(1)}{f_0} \\ &= \frac{f_m^{(s)}(1)(q(1)+a)}{f_0'(q(1)+a)} \\ &= L_m(n,s), \end{aligned}$$

which completes the proof. □

Finally we give a third method for checking if a spherical code has a certain index. It is based on the following statement.

THEOREM 3.3. *For a spherical code $C \subset \mathbf{S}^{n-1}$, the equality*

$$\sum_{y \in C} P_k^{(n)}(\langle x, y \rangle) = 0$$

holds for every point $x \in C$ if and only if C has index k.

PROOF. Suppose that C has index k. Using (2.1) and Lemma 2.2(ii), we consecutively have

$$\begin{aligned}\sum_{y\in C} P_k^{(n)}(\langle x,y\rangle) &= \sum_{y\in C}\frac{1}{r_k}\sum_{j=1}^{r_k} v_{kj}(x)\overline{v_{kj}(y)}\\ &= \frac{1}{r_k}\sum_{j=1}^{r_k} v_{kj}(x)\overline{\sum_{y\in C} v_{kj}(y)}\\ &= 0\end{aligned}$$

Conversely, let $\sum_{y\in C} P_k^{(n)}(\langle x,y\rangle) = 0$ for every point $x \in C$. Summation over all $x \in C$ gives

$$\begin{aligned}0 &= \sum_{x,y\in C} P_k^{(n)}(\langle x,y\rangle)\\ &= \sum_{x,y\in C}\frac{1}{r_k}\sum_{j=1}^{r_k} v_{kj}(x)\overline{v_{kj}(y)}\\ &= \frac{1}{r_k}\sum_{j=1}^{r_k}\sum_{x,y\in C} v_{kj}(x)\overline{v_{kj}(y)}\\ &= \frac{1}{r_k}\sum_{j=1}^{r_k}\left|\sum_{x\in C} v_{kj}(x)\right|^2.\end{aligned}$$

Thus $\sum_{x\in C} v_{kj}(x) = 0$ for every $j = 1, 2, \ldots, r_k$. Since $\{v_{kj}(x)\}_{j=1}^{r_k}$ is a basis of V_k, this implies $\sum_{x\in C} v(x) = 0$ for any function $v(x) \in V_k$. By Lemma 2.2(ii) C has index k. □

Although the conditions in Theorem 3.3 look more complicated and difficult to check, we shall see that it is very powerful for finding indexes of some "good" codes. Actually, in the next two sections we illustrate the action of all three methods we have described above.

4. Indexes of codes on Levenshtein bounds

We firstly deal with codes on Levenshtein bounds, i.e. with $(n, L_m(n,s), s)$ codes. Such codes seem to exist rarely (see [**15**, Table 6.1] for all known such codes). We have checked for nontrivial indexes all known $(n, L_m(n,s), s)$ codes for $n \geq 2$ and $m \geq 2$. For $n \geq 3$ and $s > 0$ (the later means that $|C| > 2n$), all but the codes in the next two theorems have not nontrivial indexes smaller than 20.

We illustrate the application of our first method (Theorem 3.1) on the icosahedron.

THEOREM 4.1. *The icosahedron (any $(3, 12, 1/\sqrt{5})$ spherical code) has nontrivial indexes 8 and 14.*

PROOF. We describe in detail the construction of a fourteenth degree polynomial which can be used in Theorem 3.1. We look for a polynomial $f(t)$, $\deg(f) = 14$, such that

$$f(t) = (t+1)(t - 1\sqrt{5})(t + 1/\sqrt{5})^2 a(t),$$

where $a(t) = \sum_{i=0}^{10} a_i t^i$, $a_{10} = 1$, shall satisfy further restrictions to ensure $f(t) \in A_{3,1/\sqrt{5}}$ and $f(1)/f_0 = 12$. We require $f_{13} = f_{12} = f_{11} = 0$ and this gives

$$a_9 = -\frac{\sqrt{5}}{5} - 1, a_8 = \frac{\sqrt{5}}{5} - \frac{266}{135}, a_7 = \frac{266\sqrt{5}}{675} + \frac{266}{135}.$$

We now solve the equation $f(1)/f_0 = 12$ to obtain

$$a_2 = -\frac{3724\sqrt{5}+1659}{4455} - \frac{\sqrt{5}+5}{5}a_3 - \frac{25+3\sqrt{5}}{15}a_4 - \frac{\sqrt{5}+5}{3}a_5 - \frac{70+11\sqrt{5}}{33}a_6.$$

Using the inequalities $f_{10} \geq 0$ and $f_6 \geq 0$ we have

$$a_6 \geq \frac{-266\sqrt{5}+844}{675}, \; a_6 \leq \frac{-266\sqrt{5}+844}{675},$$

respectively. Therefore, we must take $f_{10} = f_6 = 0$ and $a_6 = (-266\sqrt{5}+844)/675$. The inequalities $f_9 \geq 0$, $f_8 \geq 0$, $f_7 \geq 0$, $f_5 \geq 0$ lead to

$$\begin{aligned}
a_5 &\geq -\frac{844(\sqrt{5}+5)}{3375}, \\
a_4 &\geq -\frac{20332\sqrt{5}+135410}{3375} - \frac{\sqrt{5}+5}{5}a_5, \\
a_3 &\geq -\frac{137572(\sqrt{5}+5)}{286875} - \frac{17\sqrt{5}+163}{85}a_5 - \frac{\sqrt{5}+5}{5}a_4, \\
a_1 &\geq \frac{21341\sqrt{5}}{43875} + \frac{13\sqrt{5}-14}{5}a_3 + \frac{374\sqrt{5}+375}{975}a_5 + \frac{244\sqrt{5}+440}{975}a_4 + \frac{4027}{5265},
\end{aligned}$$

respectively. We choose to take $f_9 = f_8 = f_7 = f_5 = 0$ and the above inequalities become equalities. Therefore, we have found all coefficients of $a(t)$ but a_0.

We must determine a_0 in such a way that $f_4 \geq 0$, $f_3 \geq 0$, $f_2 \geq 0$, $f_1 \geq 0$, $f_0 > 0$ and $a(t) \geq 0$ for $-1 \leq t \leq 1/\sqrt{5}$. The first four conditions are linear, and for the last one we must calculate the minimum of $a(t)$. It turns out that we can take $a_0 = 1/2$. As a final result we obtain the polynomial

$$\begin{aligned}
f(t) &= \frac{5(t+1)}{138}(t-\frac{1}{\sqrt{5}})(t+\frac{1}{\sqrt{5}})^2[3881250t^{10} - (776250\sqrt{5}+3881250)t^9 \\
&\quad +(776250\sqrt{5}-7647500)t^8 + (1529500\sqrt{5}+7647500)t^7 \\
&\quad +(4853000-1529500\sqrt{5})t^6 - (970600\sqrt{5}+4853000)t^5 \\
&\quad +(970600\sqrt{5}-946900)t^4 + (189380\sqrt{5}+946900)t^3 \\
&\quad -(189380\sqrt{5}+13130)t^2 + (13130+2626\sqrt{5})t + 1940625] \\
&= \frac{1280000}{22287}P_{14}(t) + (\frac{49170692}{3059} + \frac{10504\sqrt{5}}{483})P_4(t) \\
&\quad +(\frac{1943251+390751\sqrt{5}}{69})P_3(t) + (\frac{758866141}{24633} + \frac{593251\sqrt{5}}{63})P_2(t) \\
&\quad +\frac{1943251+390751\sqrt{5}}{69}P_1(t) + \frac{1943251+390751\sqrt{5}}{207},
\end{aligned}$$

which works! All this calculation can be easily performed on *Maple*.

To prove that the icosahedron has (nontrivial) index 8, we construct in a similar (but simpler) way the polynomial

$$\begin{aligned} f(t) &= (t+1)(t-\frac{1}{\sqrt{5}})(t+\frac{1}{\sqrt{5}})^2[t^4-\frac{1+\sqrt{5}}{\sqrt{5}}t^3+\frac{3\sqrt{5}-7}{15}t^2+\frac{35+7\sqrt{5}}{75}t+\frac{3}{10}] \\ &= \frac{128}{6435}P_8(t)+(\frac{2468}{34125}+\frac{8\sqrt{5}}{375})P_4(t)+\frac{59+23\sqrt{5}}{375}P_3(t) \\ &\quad +(\frac{4909}{28875}+\frac{91\sqrt{5}}{1125})P_2(t)+\frac{23\sqrt{5}+59}{375}P_1(t)+\frac{23\sqrt{5}+59}{1125}. \end{aligned}$$

□

This result, as well as the results below can be obtained by Theorem 3.2 as well. Actually, computer calculations of test-functions for the "suspicious" cases were our starting point. The remaining cases are described together.

THEOREM 4.2. *a) Let C be a $(5,16,1/5)=(5,L_3(5,1/5),1/5)$ code. Then C has nontrivial index 7.*
b) Let C be a $(6,27,1/4)=(6,L_4(6,1/4),1/4)$ code. Then C has nontrivial index 7.
c) Let C be an $(8,240,1/2)=(8,L_7(8,1/2),1/2)$ code. Then C has nontrivial index 10.
d) Let C be a $(24,196560,1/2)=(24,L_{11}(24,1/2),1/2)$ code. Then C has nontrivial index 14.

PROOF. a) Use the polynomial

$$\begin{aligned} f(t) &= (t-\frac{1}{5})(t+\frac{3}{5})^2(t^4-t^3-\frac{13}{25}t^2+\frac{89}{125}t+\frac{3}{10}) \\ &= \frac{64}{715}P_7(t)+\frac{15702}{56875}P_3(t)+\frac{5286}{15625}P_2(t)+\frac{289648}{1203125}P_1(t)+\frac{984}{15625}. \end{aligned}$$

We have $f(1)/f_0=16$ and Theorem 3.1 can be applied.

b) This code is a tight spherical 4-design which is unique up to isometry [**5**]. We use the polynomial

$$\begin{aligned} f(t) &= (t+1)^2(t-\frac{1}{4})(t+\frac{1}{2})^2(t^2-\frac{11}{4}t+\frac{15}{4}) \\ &= \frac{15}{128}P_7(t)+\frac{231}{128}P_4(t)+\frac{2205}{512}P_3(t)+\frac{569}{128}P_2(t)+\frac{1191}{512}P_1(t)+\frac{1}{2}. \end{aligned}$$

Since $f(1)/f_0=27$, we can apply Theorem 3.1.

c) This code is a tight spherical 7-design which is an antipodal maximal code and is unique up to isometry [**4**] ≅ [**9**, Chapter 14]. We use the polynomial

$$\begin{aligned} f(t) &= (t-\frac{1}{2})t^2(t+\frac{1}{2})^2(t+1)(t^4-\frac{3}{2}t^3+\frac{1}{8}t^2+\frac{9}{16}t+\frac{1}{8}) \\ &= \frac{91}{2048}P_{10}(t)+\frac{129}{2048}P_6(t)+\frac{45}{256}P_5(t)+\frac{357}{2048}P_4(t) \\ &\quad +\frac{39}{256}P_3(t)+\frac{137}{2048}P_2(t)+\frac{3}{128}P_1(t)+\frac{3}{1024} \end{aligned}$$

Since $f(1)/f_0=240$, we can apply Theorem 3.3.

d) This code is a tight spherical 11-design which is formed by the minimum norm vectors in the Leech lattice. It is unique up to isometry [**4**] ≅ [**9**, Chap14].

We use the polynomial

$$f(t) = (t+1)(t-\frac{1}{2})t^2(t+\frac{1}{4})^2(t-\frac{1}{4})^2(t+\frac{1}{2})^2(t^4-\frac{3}{2}t^3+\frac{11}{48}t^2+\frac{13}{32}t+\frac{1}{2}).$$

The interested reader can verify that $f(t) \in A_{24,1/2}$ and $f(1)/f_0 = 196560$. The index 14 for the Leech lattice was already mentioned in [**8**]. □

5. Indexes of other interesting codes

We have attacked by methods from Section 3 many spherical codes which are somewhat good or well known. In our opinion, the 600-cell (the regular polytope in four dimensions with Schläfli symbols $\{3,3,5\}$) holds the record (together with its dual – the 120-cell). It is an 120-point spherical 11-design on $\mathbf{S}^3$ which is also an antipodal maximal $(4, 120, \cos(\pi/5))$ code. This code and design is unique up to isometry [**7**].

THEOREM 5.1. *The 600-cell has nontrivial indexes 14, 16, 18, 22, 26, 28, 34, 38, 46 and 58.*

PROOF. This is done by using Theorem 3.3 and a simple code in *Maple* for calculation of $\sum_{y \in C} P_k^{(n)}(\langle x, y\rangle)$. It is well known that the 600-cell is distance invariant. More precisely, if x is an arbitrary point of C, then there are 1, 12, 20, 12, 30, 12, 20,12 points of C with inner products -1, $-\frac{\sqrt{5}+1}{4}$, $-\frac{1}{2}$, $-\frac{\sqrt{5}-1}{4}$, 0, $\frac{\sqrt{5}-1}{4}$, $\frac{1}{2}$, $\frac{\sqrt{5}+1}{4}$, respectively. □

REMARK 5.2. After versions of this paper were circulating, we were made aware by Nikolay Andreev that the nontrivial indexes of the 600-cell were also found in [**2, 3**].

6. Upper bounds on the nontrivial indexes

So far we were interested in finding non-trivial indexes of interesting spherical codes. A natural question arises if we have found all such indexes for these codes.

It is known that $\lim_{k\to\infty} P_k^{(n)}(t) = 0$ for all $n \geq 3$ and $t \in (-1, 1)$. This combined with Theorem 3.3 shows that an arbitrary spherical code of dimension at least 3 can not have infinitely many non-trivial indexes, because ($x \in C$)

$$\lim_{k\to\infty}\left(|C \cap \{x, -x\}| + \sum_{y \in C\setminus\{x,-x\}} P_k^{(n)}(\langle x, y\rangle)\right) = |C \cap \{x, -x\}| \geq 1.$$

Using estimations on the convergence of $P_k^{(n)}(t)$ for particular values of t and n, one can obtain upper bounds for the non-trivial indexes of concrete codes (since the inner products and the distance distribution are known). We illustrate this by showing that all nontrivial indexes of the 600-cell are these which were already shown in Theorem 5.1. The nontrivial indexes of its dual polytope, the 120-cell, can be found in the same way.

Code	Name, Reference	$\tau(C)$	Antipodal	Nontrivial indexes
$(2, 2r, \cos\frac{\pi}{r})$	Regular $2r$-gon	$2r-1$	Yes	$j > 2r$, $r \not| j$, j odd
$(2, 2r+1, \cos\frac{2\pi}{2r+1})$	Regular $(2r+1)$-gon	$2r$		$j > 2r$, $(2r+1) \not| j$
$(3, 4, -1/3)$	Regular tetrahedron	2		5
$(3, 6, 0)$	Octahedron	3	Yes	No
$(3, 8, 1/3)$	Cube	3	Yes	No
$(3, 12, 1/\sqrt{5})$	Icosahedron	5	Yes	8,14
$(3, 20, \sqrt{5}/3)$	Dodecahedron	5	Yes	8,14
$(3, 30, \cos(\pi/5))$	Icosidodecahedron	5	Yes	8,14
$(3, 60, \frac{9\sqrt{5}+80}{109})$	Truncated icosahedron [**12**]	5	Yes	8,14
$(3, 24, \approx 0.72307)$	Snub cube	3		5,7,11
$(3, 24, \approx 0.75038)$	Improved snub cube [**13, 17**]	7		11
$(4, 120, \cos(\pi/5))$	600-cell [**10**]	11	Yes	14,16,18,22,26, 28,34,38,46,58
$(4, 600, \frac{\sqrt{5}+1}{8})$	120-cell [**10**]	11	Yes	14,16,18,22,26, 28,34,38,46,58
$(5, 16, 1/5)$		3		7
$(6, 27, 1/4)$		4		7
$(8, 240, 1/2)$		7	Yes	10
$(24, 196560, 1/2)$	Leech lattice [**9**]	11	Yes	14

TABLE 1. Non-trivial indexes of certain spherical codes.

LEMMA 6.1. *The polynomials $P_{2i}^{(4)}(t)$ take the following values:*

$$\begin{aligned} P_{2i}^{(4)}(0) &= \frac{a}{2i+1}, \text{ if } a \equiv 1-2i (mod\ 4),\ a \in \{\pm 1\}, \\ P_{2i}^{(4)}(\frac{1}{2}) &= \frac{b}{2i+1}, \text{ if } b \equiv 1-i (mod\ 3),\ b \in \{0, \pm 1\}, \\ P_{2i}^{(4)}(\frac{\sqrt{5}+1}{4}) + P_{2i}^{(4)}(\frac{\sqrt{5}-1}{4}) &= \frac{c}{2i+1}, \text{ if } c \equiv 2-i (mod\ 5),\ c \in \{0, \pm 1, \pm 2\}. \end{aligned}$$

PROOF. By induction. □

THEOREM 6.2. *If k is a nontrivial index of the 600-cell then $k \le 58$, i.e. Theorem 5.1 gives all nontrivial indexes of the 600-cell.*

PROOF. The distance distribution of the 600-cell was given in the proof of Theorem 5.1, and since $P_{2i}^{(4)}(t)$ are even functions, we have to estimate

$$2(1 + 15P_{2i}^{(4)}(0) + 20P_{2i}^{(4)}(\frac{1}{2}) + 12P_{2i}^{(4)}(\frac{\sqrt{5}+1}{4}) + 12P_{2i}^{(4)}(\frac{\sqrt{5}-1}{4})).$$

It is easy to see that the number $2i$ is an index of the 600-cell if and only if $2i + 1 + 15a + 20b + 12c = 0$, with a, b, c as defined in Lemma 6.1. Thus we have $2i = -(1 + 15a + 20b + 12c) \le 58$. □

Similar calculations imply that all indexes from the previous sections which were summarized in Table 1, are actually all nontrivial indexes of the codes we have considered. In the cases when the described technique does not work, i.e.

when the dimension is two and the codes are the regular k-gons, we use the facts that $P_l^{(2)}(\cos\phi) = \cos l\phi$ and $\sum_{i=0}^{k-1}\cos(2\pi il/k) \neq 0$ iff $l|k$.

THEOREM 6.3. *Table 1 shows all nontrivial indexes of the codes in its first column. All other codes from* [**15**, Table 6.1] *do not have nontrivial indexes.*

Acknowledgment. The authors would like to thank Nikolay Andreev for his comments on a preliminary version of this paper.

References

[1] M.Abramowitz, I.A.Stegun, *Handbook of Mathematical Functions*, Dover, New York, 1965.

[2] N.N.Andreev, *Minimal 11-design on the 3-dimensional sphere*, Math. Zam., to appear (in Russian); English translation in Math. Notes, to appear.

[3] N.N.Andreev, V.A.Yudin, *Problems of approximation theory in discrete geometry*, Math. Res. (Advances in Multivariate Approximations, W.Haussmann, K.Jetter and M.Reimer, Eds.) **107** (1999), 19–32.

[4] E.Bannai, N.J.A.Sloane, *Uniqueness of certain spherical codes*, Canad. J. Math. **33** (1981), 437–449.

[5] P.G.Boyvalenkov, *Computing distance distribution of spherical designs*, Lin. Alg. Appl. **226/228**, 1995, 277–286.

[6] P.G.Boyvalenkov, D.P.Danev, S.P.Bumova, *Upper bounds on the minimum distance of spherical codes*, IEEE Trans. Inform. Theory **42** (1996), 1576–1581.

[7] P.G.Boyvalenkov, D.P.Danev, *Uniqueness of the 120-point spherical 11-design in four dimensions*, to appear in Archiv der Mathematik.

[8] P.G.Boyvalenkov, S.Nikova, *Improvements of the lower bounds on the size of some spherical designs*, Math. Balk. **12**, 1998, 151–160.

[9] J.H.Conway, N.J.A.Sloane, *Sphere Packings, Lattices and Groups*, Springer – Verlag, New York, 1988.

[10] H.S.M.Coxeter, *Regular polytopes*, Dover, New York, 1973.

[11] P.Delsarte, J.-M.Goethals, J.J.Seidel, *Spherical codes and designs*, Geom. Dedicata **6**, 1977, 363–388.

[12] J.-M.Goethals, J.J.Seidel, *The football*, Nieuw Archief Wisk. **29**, 1981, 50–58.

[13] R.H.Hardin, N.J.A.Sloane, *McLaren's improved cube and other new spherical designs in three dimensions*, Discr. Comp. Geom. **15**, 1996, 429–441.

[14] V.I.Levenshtein, *On bounds for packings in n-dimensional Euclidean space*, Soviet Math. Dokladi **20**, 1979, 417–421.

[15] V.I.Levenshtein, *Universal bounds for codes and designs*, Chapter 6 in *Handbook of Coding Theory*, V.Pless and W.C.Huffman, Eds., Elsevier Science B.V., 1998.

[16] V.I.Levenshtein, *Designs as maximum codes in polynomial metric spaces*, Acta Appl. Math. **25**, 1992, 1–82.

[17] A.D.McLaren, *Optimal numerical integration on a sphere*, Math. Comp. **17**, 1963, 361–383.

INSTITUTE OF MATHEMATICS AND INFORMATICS, BULGARIAN ACADEMY OF SCIENCES, 8 G. BONCHEV STR., SOFIA 1113, BULGARIA
E-mail address: peter@moi.math.bas.bg

DEPT. OF ELECTRICAL ENGINEERING, LINKÖPING UNIVERSITY, SE-581 83 LINKÖPING, SWEDEN
E-mail address: danyo@isy.liu.se

DELFT UNIVERSITY OF TECHNOLOGY, TWI/ITS, MEKELWEG 4, 2600 GA DELFT, PO BOX 5031, THE NETHERLANDS
E-mail address: peterkazakov@yahoo.com

DIMACS Series in Discrete Mathematics
and Theoretical Computer Science
Volume **56**, 2001

Codes over $\mathbb{Z}_{p^n}$ and Association Schemes

Paul Camion

ABSTRACT. Association schemes are the appropriate context to obtain all possible p-ary generalizations, p a prime, of binary codes from codes over $\mathbb{Z}_4$ given by the Gray map. The results come within the scope of the author's Chapter 18 *Codes and Association Schemes* in the Handbook of Coding Theory edited by Vera Pless, W. Cary Huffman and Richard A, Brualdi, Elsevier Science Publisher B.V., Amsterdam, 1998. We obtain a construction which generalizes the Gray map when $\mathbb{Z}_4$ is replaced with $\mathbb{Z}_{p^k}$ or any other Abelian group of size p^k. Merging of Abelian groups are to be considered and it turns out that all mergings of $(\mathbb{Z}_8,+)$ and $(\mathbb{F}_2^3,\oplus)$, which are Abelian schemes, with three classes are isomorphic. In pursuit of generalizing the Gray map, we are also led to investigate automorphisms of an Abelian scheme and from there to exhibit permutation groups on the set of units of the ring $\mathbb{Z}_{p^k}$.

1. Introduction

The results in [**13**] shed a new light on properties discovered in [**29**] and [**30**] of some unrestricted codes. Besides, apart from Carlet's paper on codes over Z_{2^k} [**6**], there exist contributions by A. A. Nechaev et al [**25, 26, 27, 28**], T. Honold and I. Landjev [**14, 15, 16, 17, 18, 19, 20**], and recently by M. Greferath and S. E. Schmidt [**12**]. Let us furthermore mention the result by Sălăgean-Mandache [**31**] stating that there are no new Gray isometries from Z_{p^k} to Z_p^k.
We here endeavour to broaden the breakthrough.

Le X be an Abelian group. A partition $\{X_0 = \{0\}, X_1, \ldots, X_n\}$ of X defines an association scheme (X, R) invariant under translation if the relations $R_i = \{(x,y)|x - y \in X_i\}$, $i = 0, 1, \ldots, n$ are those of an association scheme, as defined by R.C.Bose and T.Shimamoto [**2**]. Then (X, R) is called an Abelian scheme which essentially is a kind of Schur ring [**1**, pages 104-113]. It appears that the Gray map σ is an isomorphism from a self-dual Abelian scheme defined on the group $\mathcal{F} = (\mathbb{Z}_4, +)$ to the Hamming scheme $\mathbb{F}_2^2 = H(2,2)$ consisting of all binary vectors of length 2. Let us denote by ρ the extension from $\mathcal{F}$ to $\mathcal{F}^m$ of σ. Then for any subgroup Y of $\mathcal{F}^m$, $\rho(Y)$ is an unrestricted but regular code in the Hamming scheme $H(2m, 2)$. Another consequence is that the MacWilliams transform of the weight enumerator of Y is the weight enumerator of its dual Y° and of $\rho(Y^\circ)$ as

1991 *Mathematics Subject Classification.* Primary 94B05, 94B25, 05E30; Secondary 20B25, 16S50.

well. A similar situation can be obtained when $\mathbb{Z}_4$ is replaced with $\mathbb{Z}_{p^k}$ or any other Abelian group of size p^k. Then an identification of the sets $\mathbb{Z}_{p^k}$ and $\mathbb{F}_p^k$ is pointed out that yields a merging which is an Abelian scheme as well for the Abelian group $(\mathbb{Z}_{p^k}, +)$ as for $(\mathbb{F}_p^k, \oplus)$. That Abelian scheme has k classes. That identification permits the automorphism group of $(\mathbb{Z}_{p^k}, +)$ as well as a subgroup of that of $(\mathbb{F}_p^k, \oplus)$ to act as a permutation group on the same set $\mathbb{F}_p^{k*}$ and it is shown that the permutation group generated in this way is properly contained in the symmetric group on $\mathbb{F}_p^{k*}$ and contains transpositions. For $p = 2$ and $k = 3$ it is shown that no Abelian scheme on $\mathbb{Z}_8$ can be isomorphic to $H(3, 2)$. However an Abelian scheme on the Abelian group $(\mathbb{Z}_4, +) \times (\mathbb{Z}_2, +)$ is pointed out which is isomorphic to $H(3, 2)$.

2. Abelian schemes

2.1. Definitions and notations. For the definition of an association scheme, we refer to [**4**, Section 1]. An association scheme with n classes on a finite set X is there denoted by (X, R) where $R = \{R_0, R_1, \ldots, R_n\}$ is a family of $n+1$ relations on X verifying the conditions $A_1, \ldots, A_5$ below.

A_1 The diagonal relation $\{(x, x) | x \in X\}$ is the relation R_0 of R.

A_2 The family R forms a partition of $X \times X$ i.e.,

$$X \times X = R_0 \cup \ldots \cup R_n, \ R_i \cap R_j = \emptyset \quad \text{if } i \neq j.$$

A_3 The reciprocal relation R_i^T of R_i, $R_i^T = \{(x, y) | (y, x) \in R_i\}$ also belongs to $R, i = 0, \ldots, n$. We thus have $R_i^T = R_{i'}$, for some $i' \in \{0, 1, \ldots, n\}$.

A_4 For any triple of integers $i, j, k \in \{0, 1, \ldots, n\}$ the number of $z \in X$ such that $(x, z) \in R_i$ and $(z, y) \in R_j$ does not depend on the choice of (x, y) in R_k. This constant number is denoted by p_{ij}^k. (Notice that k here is a superscript used for avoiding three subscripts. We will not need any other exception to the usual notation.)

A_5 We have that $p_{ij}^k = p_{ji}^k$ for all i, j, k.

An **Abelian scheme** is an association scheme (X, R) on an Abelian group X such that there exists a partition $\{X_0 = \{0\}, X_1, \ldots, X_n\}$ of X which defines R by

$$(x, y) \in R_i \leftrightarrow x - y \in X_i \tag{2.1}$$

In such a scheme, $X_1, \ldots, X_n$ are called the **Abelian classes** of (X, R).

In what follows we use the notations in [**4**] and in particular in [**4**, Section 4] for what concerns association schemes on an Abelian group. An Abelian scheme is essentially a Schur ring. An account on S-rings is given in [**1**, pages 104-113]. This is brought to the fore in the statement of [**4**, Theorem 4.3] which states that a partition $\{X_0 = \{0\}, X_1, \ldots, X_n\}$ of X defines an association scheme by (2.1) if and only if the subalgebra generated by $\{Z^{X_0}, Z^{X_1}, \ldots, Z^{X_n}\}$ in the group algebra $\mathbb{C}X$ has dimension $n+1$. More on Abelian schemes can be found in [**5**].

Definition 2.1. When a finite set X is equipped with the structure of an association scheme (X, R), then a subset Y of X is called a **code**.

Definition 2.2. When X is an Abelian group, a subgroup Y of X is called an **additive code**.

2.2. Regular codes and Abelian schemes. We here recall [**4**, Definition 4.69].

DEFINITION 2.3. A code Y in an Abelian scheme (X, R), with Abelian classes $X_1, \ldots X_n$, is called **regular** when for $y \in Y$ the integer $|\{x \in Y | x - y \in X_i\}|$ does not depend on a particular choice of y.

The two following remarks are essential for the sequel.

REMARK 2.4. In particular if Y is an additive code of X, then Y is a regular code. Indeed $\forall y \in Y$, we have that $\lambda_i = |\{x \in Y | x - y \in X_i\}| = |(Y - y) \cap X_i| = |Y \cap X_i|$ is independent of the choice of $y \in Y$.

DEFINITION 2.5. Given two association schemes (X, T) and (X, R) with t and n classes respectively, $t \geq n$, we say that (X, T) **is covered by** (X, R), and we write $(X, T) \preceq (X, R)$ when $R_i = \bigcup_{\ell \in I_i} T_\ell$, where $\{I_0, I_1, \ldots, I_n\}$ is a partition of $\{0, \ldots, t\}$. In other words the partition T is a refinement of R.

REMARK 2.6. Let (X, T) and (X, R) be two Abelian schemes such that (X, R) **covers** (X, T) (Definition 2.5). Then by definition, every Abelian class of (X, T) is contained in an Abelian class of (X, R). As a result, if Y is a regular code of (X, T), it also is a regular code of (X, R):

$$\lambda_i = |\{x \in Y | x - y \in X_i\}| = \sum_{T_j \subset X_i} |\{x \in Y | x - y \in T_j\}| = \sum_{T_j \subset X_i} \mu_j.$$

2.3. Isomorphism of Abelian schemes. A bijection σ from X to E is an isomorphism of (X, R) to (E, T) if the image of every relation of R is a relation of T. More formally we have the following definition which is Definition 4.54 in [**4**].

DEFINITION 2.7. Let (X, R) and (E, T) be two association schemes with classes $R_1, R_2, \ldots, R_n$ and $T_1, T_2, \ldots, T_n$ respectively. Then (X, R) and (E, T) are **isomorphic** if the following conditions are satisfied. There exists a bijection σ from X to E inducing a bijection τ from the set of subsets of $X \times X$ to those of $E \times E$, such that $\tau(R_i) = T_{\pi i}$, $i = 0, 1, \ldots, n$, for some permutation π on $\{0, 1, \ldots, n\}$ which leaves 0 fixed.

Now let σ be an isomorphism from the Abelian scheme (X, R) to an Abelian scheme (E, T). Clearly a group isomorphism from X to E which maps the Abelian classes $X_1, \ldots, X_n$ of (X, R) to the Abelian classes $E_1, \ldots, E_n$ of (E, T), respectively, always is an Abelian scheme isomorphism, by (2.1). However we want this isomorphism σ not to be a group isomorphism in general. Let $\sigma(0) = y$. Then $x \mapsto \sigma(x) \ominus y$, where $\ominus$ is the minus sign for the group on the set E, also is an isomorphism from (X, R) to (E, T) since (E, T) is **invariant under translation**, by (2.1). We can then assume that $\sigma(0)$ is the zero element of the group E. Now if X_i is the class of (X, R) defining R_i by $((x, y) \in R_i \leftrightarrow x - y \in X_i)$ then $X_i = \{x | (x, 0) \in R_i\}$. But $\sigma(X_i)$ is the set of points $\sigma(x) \in E$ which verifies $(\sigma(x), 0) \in \sigma(R_i)$. Under our assumption, $\sigma(R_i) = T_i$ where T_i is a relation of T, after possibly reindexing those relations. Thus $\sigma(X_i)$ is the set $\{y | (y, 0) \in T_i\}$; $\sigma(X_i)$ is the Abelian class E_i of (E, T) corresponding to T_i.
We then have the following theorem.

THEOREM 2.8. *let (X, R) and (E, T) be two Abelian schemes with Abelian classes $X_1, \ldots, X_s$ and $E_1, \ldots, E_s$, respectively. If there exists a bijection σ from*

X to E which is an isomorphism from (X, R) to (E, T), then such an isomorphism exists that maps 0 to 0 and each Abelian class of (X, R) to an Abelian class of (E, T).

Let us now point out a simple though important remark about isomorphic association schemes and regular codes.

REMARK 2.9. In the sequel it is assumed that **an isomorphism σ of an Abelian scheme to another maps 0 to 0**. If Y is an additive code then it is a regular code of (X, R). Now $\sigma(Y)$ is an unrestricted code in (E, T). However $\sigma(Y)$ is regular. Indeed for whatever $u \in \sigma(Y)$, $u = \sigma(y)$, we have that the set $\{(x, y)|(x, y) \in R_i\}$ maps to the set $\{(z = \sigma(x), u)|(z, u) \in T_i\}$. Thus those sets have the same size. This entails that the integer

$$|\{z \in \sigma(Y)|z - u \in E_i\}| = |\{x \in Y|x - y \in X_i\}|,$$

does not depend on a particular choice of u.

2.4. Duality.

2.4.1. *The dual scheme of an Abelian scheme.* It is known that an Abelian group X is isomorphic with its character group X' which is the group of homomorphisms of X into the multiplicative group of the complex numbers. By the Theorem of Tamaschke Delsartes [**4**, Theorem 4.5], to an Abelian scheme (X, R) correponds a dual scheme (X', R') defined with respect to a symmetric matrix $S \in \mathbb{C}(X, X')$ of group characters of X, given by $S(x, y) =< x, y >$, for $(x, y) \in X \times X'$, also called the **character table** of X. Corollary 4.51 in [**4**], which is essentially contained in the material of [**7**] provides a usefull tool to construct the dual scheme (X', R') on the data of the Abelian scheme (X, R). We use it in Section 7 where it is stated as Corollary 7.2.

DEFINITION 2.10. An Abelian scheme (X, R) is **self-dual** if the dual scheme (X', R') with respect to the symmetric matrix $S \in \mathbb{C}(X, X')$ of group characters of X is identified with (X, R) by some bijection from X to X'.

REMARK 2.11. Definition 4.66 in [**4**] requires the bijection in Definition 2.10 to be a group isomorphism. In view of Definition 2.7 that should apparently not be required. An example of an Abelian scheme which is not isomorphic to its dual for Definition 2.7 is however given in [**4**, Example 4.116].

REMARK 2.12. Given two Abelian schemes (X, R) and (X, T) on the same Abelian group X, does there exist an isomorphism from (X, R) to (X, T), which would not be a group automorphism. The answer is yes.

Here is an example.

EXAMPLE 2.13. The group is $(\mathbb{F}_2^3, \oplus)$, shortly denoted by $X_\oplus$, where $\oplus$ is the addition modulo 2 on the binary 3-vectors. The Abelian classes of an Abelian scheme are
$X_1 = \{001, 011, 101, 111\}, X_2 = \{010, 110\}, X_3 = \{100\}$. The isomorphism σ, which is not linear is as follows.

$$001 \mapsto 110, 011 \mapsto 100, 101 \mapsto 111, 111 \mapsto 101,$$

$$010 \mapsto 011, 110 \mapsto 010, 100 \mapsto 001.$$

It comes as a byproduct of Section 9 where $\nu_\oplus \circ \sigma \circ \mu_+ \circ \sigma^{-1}$ is the isomorphism from $(X_\oplus, R_\sigma)$ to an Abelian scheme on $X_\oplus$ obtained by taking for μ_+ the multiplication by 3 modulo 8 and for $\nu_\oplus$ the linear map τ of relation (7.3) in Section 7.2.

2.4.2. *The dual of a subgroup.* We consider an Abelian group X together with the group X' of its characters. We now define the dual of a subgroup Y of X.

DEFINITION 2.14. We observe that the subset Y° of X' defined by

$$Y^\circ = \{x' \in X' | < x, x' >= 1, \forall x \in Y\} \tag{2.2}$$

is itself a group. It is called **the dual** of Y in X'.

REMARK 2.15. After Definition 2.2, Y is an additive code of X and Y° is an additive code of X'.

2.5. Isomorphism and Kronecker product of Abelian schemes.

2.5.1. *Kronecker product.* Let $X = X^{(1)} \times X^{(2)}$ be the product of the Abelian groups $X^{(1)}$ and $X^{(2)}$ on which the Abelian schemes $(X^{(1)}, R^{(1)})$ and $(X^{(2)}, R^{(2)})$ are given with n_1 classes and n_2 classes respectively. The Kronecker product of Abelian schemes is dealt with in [**4**, Definition 4.84 and Section 4.10]. We only need to know that the Kronecker product $(X = X^{(1)} \times X^{(2)}, R = R^{(1)} \otimes R^{(2)})$ is the Abelian scheme on X given by the Abelian classes $X_i^{(1)} \times X_j^{(2)}$, $(i,j) \neq (0,0)$; $i = 0, 1, \dots, n_1$; $j = 0, 1, \dots, n_2$. If there is an isomorphism σ_i from $(X^{(i)}, R^{(i)})$ to the Abelian scheme $(E^{(i)}, T^{(i)})$, $i = 1, 2$, then an isomorphism σ from (X, R) to (E, T), where $E = E^{(1)} \times E^{(2)}$, $T = T^{(1)} \otimes T^{(2)}$ is given by the mapping $X \to E$ defined by $(x, y) \mapsto (\sigma_1(x), \sigma_2(y))$. Indeed after suitably reindexing the classes, $\sigma(X_i^{(1)} \times X_j^{(2)}) = E_i^{(1)} \times E_j^{(2)}$.

2.5.2. *Kronecker product and regular codes.* The general construction of unrestricted regular codes relies on the following theorem.

THEOREM 2.16. *Let σ_i be an isomorphism from the Abelian scheme $(X^{(i)}, R^{(i)})$ to the Abelian scheme $(E^{(i)}, T^{(i)})$, $i = 1, 2$. The Kronecker product $(X^{(1)} \times X^{(2)}, R^{(1)} \otimes R^{(2)})$ (resp. $(E^{(1)} \times E^{(2)}, T^{(1)} \otimes T^{(2)})$) is denoted by (X, R) (resp. (E, T)). Let Y be an additive code of (X, R) and let (E, S) be an Abelian scheme on the group E that covers (E, T). Then the isomorphism σ from (X, R) to (E, T) defined by $\sigma(x, y) = (\sigma_1(x), \sigma_2(y))$, $\forall (x, y) \in X^{(1)} \times X^{(2)} = X$ maps the additive code Y to a regular code of (E, S).*

Proof: This is a consequence of Remark 2.6. □

2.5.3. *The Gray map as an isomorphism of association schemes.* Let us first bring to the fore one of the very attractive properties of the Gray map from $\mathbb{Z}_4$ to $\mathbb{F}_2^2 = H(2, 2)$ that we want to find in a more general map from $\mathbb{Z}_{p^k}$ to the Hamming scheme $H(k, p) = \mathbb{F}_p^k$, defined on the Abelian group $(\mathbb{F}_p^k, \oplus)$, where $\mathbb{F}_p$ denotes the prime field with p elements and where $\oplus$ is the addition modulo p of p-ary vectors.

A mapping $\sigma = (\sigma_1, \dots, \sigma_m)$ from $\mathbb{Z}_4^m$ to $\mathbb{F}_2^{2m}$ is defined as follows. Choose for every $i = 1, \dots, m$ a bijection σ_i from $\mathbb{Z}_4$ to $\mathbb{F}_2^2$ by either $0 \mapsto 00, 1 \mapsto 10, 3 \mapsto 01, 2 \mapsto 11$ or $0 \mapsto 00, 1 \mapsto 01, 3 \mapsto 10, 2 \mapsto 11$.

A basic fact is the following. Denote the Abelian group $(\mathbb{Z}_4, +)$ by $\mathcal{F}$. Then $\mathcal{F}_1 = \{1, 3\}$ and $\mathcal{F}_2 = \{2\}$ are the **classes**, [**4**, Definition 4.34] of a **self-dual Abelian scheme** on $\mathcal{F}$, according to Definition 2.10. We also denote $\{0\}$ by $\mathcal{F}_0$.

Let us denote that scheme by $(\mathcal{F}, K)$. Now any mapping σ_i as defined above is an isomorphism from $(\mathcal{F}, K)$ to the binary Hamming scheme $H(2,2)$. To verify this easy but basic fact we only have to observe that

$$x - y \in X_i \subset \mathbb{Z}_4 \leftrightarrow \sigma(x) \oplus \sigma(y) \in \sigma(X_i) \subset \mathbb{F}_2^2,$$

where $\sigma(X_i)$ appears to be the Abelian class of the Hamming scheme $H(2,2)$ consisting of all weight i vectors.

We denote by $\mathbb{C}(\mathcal{F}, +)$ the group algebra over $\mathbb{C}$ of the Abelian group $(\mathbb{Z}_4, +)$ and by $\mathbb{C}(\mathcal{F}, \oplus)$ that of the Abelian group $(\mathbb{F}_2^2, \oplus)$. The set $\mathbb{Z}_4$ is identified with $\mathbb{F}_2^2$ by one of the 2 possible mappings described above, and next denoted by $\mathcal{F}$.

With the notations in [**4**, Section 4] we can write:

$$Z^{\mathcal{F}_1} Z^{\mathcal{F}_1} = 2Z^{\mathcal{F}_0} + 2Z^{\mathcal{F}_2}, \; Z^{\mathcal{F}_1} Z^{\mathcal{F}_2} = 2Z^{\mathcal{F}_1}, \; Z^{\mathcal{F}_2} Z^{\mathcal{F}_2} = 2Z^{\mathcal{F}_0}.$$

Those relations hold both in $\mathbb{C}(\mathcal{F}, +)$ and in $\mathbb{C}(\mathcal{F}, \oplus)$. This is because the subalgebras generated by $Z^{\mathcal{F}_1}$ and $Z^{\mathcal{F}_2}$ on the one hand in $\mathbb{C}(\mathcal{F}, +)$ and in the other hand in $\mathbb{C}(\mathcal{F}, \oplus)$ are isomorphic despite these group algebras are not isomorphic. Indeed $Z^1 Z^3 = Z^0$ in $\mathbb{C}(\mathcal{F}, +)$ while $Z^{01} Z^{10} = Z^{11}$ in $\mathbb{C}(\mathcal{F}, \oplus)$. The Bose-Mesner-Delsarte, **shortly** ***BMD*** **algebra**, [**4**, Definition 2.14] of each concerned Abelian scheme is a representation of such a subalgebra and both BMD algebra are isomorphic since their association schemes are isomorphic.

The converse is not true in general: if the BMD algebras of two association schemes are isomorphic, then those association schemes are not isomorphic, in general. This is pointed out in Section 2 of [**22**] where an example from [**3**] is quoted. Here we have a more specific example of two Abelian schemes whose BMD algebras are isomorphic and which are not isomorphic.

EXAMPLE 2.17. Let $\mathcal{F}$ be the additive group $(\mathbb{Z}_8, +)$ and let the Abelian classes of $(\mathcal{F}, K)$ be $\mathcal{F}_1 = \{1,3,5,7\}$, $\mathcal{F}_2 = \{2,6\}$, $\mathcal{F}_3 = \{4\}$. The mapping $\pi: 1 \mapsto 1, 3 \mapsto 5, 5 \mapsto 3, 7 \mapsto 7, 2 \mapsto 2, 6 \mapsto 6, 4 \mapsto 4$ is an automorphism of the BMD algebra of $(\mathcal{F}, K)$ since that algebra is the representation of the subalgebra (Schur ring) generated by $Z^{\mathcal{F}_1}, Z^{\mathcal{F}_2}$ and $Z^{\mathcal{F}_3}$ in the group algebra $\mathbb{C}\mathcal{F}$. However it is not an automorphism of $(\mathcal{F}, K)$ since

$$(3,1) \in K_2, (5,3) \in K_2, \text{ and } (\pi(3), \pi(1)) \in K_3, \; (\pi(5), \pi(3)) \in K_2$$

The Gray map provides an application of theorem 2.16 relying on Remark 2.6. For the Hamming scheme $H(2m, 2)$ covers the Kronecker power $H(2,2)^m_\otimes$. Indeed for $\alpha \in \mathbb{Z}_4$, let us define $w(\alpha)$ as $0, 1$ or 2 according as α belongs to $\mathcal{F}_0, \mathcal{F}_1$ or $\mathcal{F}_2$ respectively. Let $x = (x_1, \ldots, x_m)$ be a vector in $\mathbb{Z}_4^m$. Then $\sigma(x)$ belongs to the class $\sigma_1(X_{w(x_1)}) \times \ldots \times \sigma_m(X_{w(x_m)})$ of $H(2,2)^m_\otimes$ and that Abelian class is contained in the Hamming class of $H(2m, 2)$ consisting of all vectors of weight $\sum_{i=0}^n w(x_i)$.

After Theorem 2.16 then σ maps any additive code Y to a regular code of $H(2m, 2)$.

2.6. Kronecker product and more general unrestricted regular codes. We here introduce an application of Theorem 2.16 which does not bring back to known results.

Let $(X^{(1)}, R^{(1)})$ be the Kronecker power $\mathcal{F}^4_\otimes$ and let $(X^{(2)}, R^{(2)})$ be the binary Hamming scheme $H(4,2)$. We consider the Kronecker product $(X^{(1)} \times X^{(2)}, R_1 \otimes R_2)$ and its additive code Y generated by 4 vectors which are the 8-tuples between

the two straight lines in the array below, each one beeing obtained by concatenating a vector of $\mathbb{Z}_4^4$ and a vector of $\mathbb{F}_2^4$. Now let σ_1 be an isomorphism from $(\mathcal{F}, K)_{\otimes}^4$ to $H(2,2)_{\otimes}^4$ and let σ_2 be the identity mapping from $H(4,2)$ to itself. Thus the mapping σ of Theorem 2.16 is an isomorphism from the Kronecker product $(\mathcal{F}, K)_{\otimes}^4 \otimes H(4,2)$ to $H(2,2)_{\otimes}^4 \otimes H(4,2)$. By the same argument as above, it is seen that this last Abelian scheme is covered by $H(m,2)$, $m = 12$. Hence $\sigma(Y)$ is an unrestricted regular code of $H(m,2)$. The 16 vectors of $\sigma(Y)$ are shown on the right side of the array below. **It is not linear since the sum of the first two vectors of $\sigma(Y)$ does not belong to $\sigma(Y)$.** Here is a more detailed description of the construction.
The generator matrix is in two parts. The first part is that of a code of dimension 2 over $\mathbb{Z}_4$. Le us denote its two rows by a and b. The second part is that of a generator matrix over the binary field. It has 4 independent rows which are concatenated with $a, b, 2a$ and $2b$, respectively. All additions are to be performed over $\mathbb{Z}_4$ for the first generator matrix and then the corresponding binary 8-tuple are obtained by the Gray map. The binary 4-tuples obtained from the second generator matrix are then concatenated with the binary 8-tuple to yield the code.

0	0	0	0	0	0	0	0	0	0	0	0	0	0	0	0	0	0	0	0
1	0	1	1	1	0	1	0	0	1	0	0	0	1	0	1	1	0	1	0
0	1	1	3	0	1	0	1	0	0	0	1	0	1	1	0	0	1	0	1
2	0	2	2	0	1	0	0	1	1	0	0	1	1	1	1	0	1	0	0
0	2	2	2	1	0	0	0	0	0	1	1	1	1	1	1	1	0	0	0
1	1	2	0	1	1	1	1	0	1	0	1	1	1	0	0	1	1	1	1
3	0	3	3	1	1	1	0	1	0	0	0	1	0	1	0	1	1	1	0
1	2	3	3	0	0	1	0	0	1	1	1	1	0	1	0	0	0	1	0
2	1	3	1	0	0	0	1	1	1	0	1	1	0	0	1	0	0	0	1
0	3	3	1	1	1	0	1	0	0	1	0	1	0	0	1	1	1	0	1
2	2	0	0	1	1	0	0	1	1	1	1	0	0	0	0	1	1	0	0
3	1	0	2	1	0	1	1	1	0	0	1	0	0	1	1	1	0	1	1
1	3	0	2	0	1	1	1	0	1	1	0	0	0	1	1	0	1	1	1
3	2	1	1	0	1	1	0	1	0	1	1	0	1	0	1	0	1	1	0
2	3	1	3	1	0	0	1	1	1	1	0	0	1	1	0	1	0	0	1
3	3	2	0	0	0	1	1	1	0	1	0	1	1	0	0	0	0	1	1

The code is a regular unrestricted code with weight enumerator

$$\mathcal{W}_Y(U,V) =$$

$$U^{12} + 2U^7V^5 + 6U^5V^7 + 6U^6V^6 + U^4V^8.$$

Computing the dual Y° of Y is possible help to [**4**, Remark 4.17] together with Definition 2.14. In its turn $\sigma(Y^\circ)$ is a regular unrestricted code in $H(m,2)$.

3. Towards more general constructions

We here sketch what leads to one of the processes that can be pursued. Denote the Abelian group $(\mathbb{Z}_{p^k}, +)$ by $\mathcal{F}$. We need an Abelian scheme $(\mathcal{F}, K)$ with classes $K_1, \dots, K_s$ isomorphic with an Abelian scheme defined on a code C of rank k, additive code of $(\mathbb{F}_p^\ell, \oplus)$, $\ell \geq k$. The isomorphism $\sigma : \mathcal{F} \mapsto C$ will generalize the Gray map. We must have that $\sigma(K_j) = C_{i_j}$, $j = 1, \dots, s$ where C_{i_j} is the set of codewords of C with weight i_j. The code C, being linear of rank k has a generator

matrix G whose columns (possibly repeated) are in $\mathbb{F}_p^k$. Now each codeword of C is obtained by G from a unique linear form, element of the dual of $\mathbb{F}_p^k$. This defines an isomorphism γ of the Abelian group $(C, \oplus)$ to $\mathbb{F}_p^k$ and thus $\gamma(C_{i_j})$, $j = 1, \ldots, s$ is an Abelian class of an Abelian scheme $((\mathbb{F}_p^k, \oplus), T)$. Finally that Abelian scheme is then itself isomorphic with $(\mathcal{F}, K)$. Fortunately, we will prove here a useful theorem, derived from a general form of the theorem of Goethals-Van Tilborg, as stated in [**4**, Section 5.1.5], which relies on an important theorem of Delsarte: [**4**, Theorem 5.13]. That theorem will assure that if there exists an Abelian scheme on $(\mathbb{F}_p^k, \oplus)$, then there is a generator matrix G, with possibly repeated columns, defining an code C in $\mathbb{F}_p^\ell$ with the required property. This means that an Abelian scheme is defined on C and its class $(C_{i_j})_{1\le j\le s}$ precisely consists of all codewords of weight w_j, $j = 1, \ldots, s$. To reach our goal, we will need moreover that the Abelian scheme obtained be self-dual.
We still have to construct such an Abelian scheme $(\mathcal{F}, K)$ on $\mathcal{F}$ that will be isomorphic with an Abelian scheme on that other Abelian group $(\mathbb{F}_p^k, \oplus)$. This is possible, help to the following result.

The Hecke scheme $\mathcal{H}(X)$ on an Abelian group X is given by the Abelian classes $X_g = \{g\}$, $g \in X \setminus \{0\}$. It is described in [**4**, Section 4.1.2].

LEMMA 3.1. *Let $(\mathcal{F}, K)$ and $(\mathcal{E}, L)$ be any two Abelian schemes with s classes defined on the Abelian groups $\mathcal{F}$ and $\mathcal{E}$ respectively. If there exists a mapping τ from $\mathcal{F}$ to $\mathcal{E}$ which is an isomorphism of those schemes, then $\mathcal{F}$ and $\mathcal{E}$ are identified by τ and $(\mathcal{F}, K)$ (resp. $(\mathcal{E}, L)$) covers both Hecke schemes $\mathcal{H}(\mathcal{F})$ and $\mathcal{H}(\mathcal{E})$ (Section 4.1. of* [**4**]*), it thus covers the* **merging** *(Section 4.13 of* [**4**]*) of those schemes.*

An algorithm for constructing the merging of two schemes defined on the same set X, and in particular of two Hecke schemes is given in [**4**, Section 4.13.1]. We will then be able to construct an appropriate merging of the Hecke scheme $\mathcal{H}(\mathbb{Z}_{p^k}, +)$ with the Hecke scheme $\mathcal{H}(\mathbb{F}_p^k, \oplus)$ for every p and k. Moreover for $p = 2$ and $k = 3$, using Maple, we construct the merging corresponding to every bijection from $\mathbb{Z}_8$ to $\mathbb{F}_2^3$ mapping 0 to 0. Those mergings directly yield the isomorphic Abelian schemes sought for. We then can describe all possible generalizations of the Gray map when replacing 4 by 8.
We will find out that there does not exist an Abelian scheme on the Abelian group $(\mathbb{Z}_8, +)$ isomorphic with the Hamming scheme $H(3, 2)$, for $k = 3$ and $p = 2$. We however observe that the Abelian group $\mathcal{L} = (\mathbb{Z}_4, +) \times (\mathbb{Z}_2, +)$ contains such an Abelian scheme. The class corresponding to that of the weight one vectors in $H(3, 2)$ is $\mathcal{L}_1 = \{(1, 0), (3, 0), (0, 1)\}$, that corresponding to the weight 2 vectors is $\mathcal{L}_2 = \{(1, 1), (3, 1), (2, 0)\}$ and that to the weight 3 vector is $\mathcal{L}_3 = \{(2, 1)\}$. To verify that $\mathcal{L}_1, \mathcal{L}_2, \mathcal{L}_3$ are Abelian classes on the group $\mathcal{L}$ we use the following theorem, which is theorem 4.30 in [**4**].

THEOREM 3.2. *Let $X_0 = \{0\}, X_1, \ldots, X_n$ be a partition of the Abelian group X. Then the relations R_i defined by*

$$(x, y) \in R_i \text{ if } x - y \in X_i, \ i = 0, 1, \ldots, n$$

define an Abelian scheme if and only if the subalgebra $\mathcal{A}$ of $\mathbb{C}X$ generated by $\{Z^{X_0}, Z^{X_1}, \ldots, Z^{X_n}\}$ has dimension $n + 1$.

We don't write here the whole verification. For example we have that

$$Z^{\mathcal{L}_1} Z^{\mathcal{L}_2} = 2Z^{\mathcal{L}_1} + 3Z^{\mathcal{L}_3}.$$

The bijection from $\mathcal{L}$ to $\mathbb{F}_2^3$ is the Gray map for the first component and the identity map for the second. The fact that the association scheme is isomorphic to the Hamming scheme $H(3,2)$ is then readily verified.

4. Abelian schemes and additive codes over $\mathbb{F}_q$

In order to construct the generator matrix G of the code C of $(\mathbb{F}_p^\ell, \oplus)$ with the required properties, we need some machinery.

4.1. First and second eigenmatrices.

NOTATION 4.1. Let us denote by (X, R) an association scheme with n classes $R_1, \ldots, R_n$. Its BMD algebra is the commutative (n+1)-dimensional matrix subalgebra $\mathcal{A}$ of the matrix algebra $\mathbb{C}(X,X)$ generated over the complex numbers by the adjacency matrices $D_0, D_1, \ldots, D_n$ of the relations $R_0, R_1, \ldots, R_n$, respectively. The $n+1$ primitive idempotents of $\mathcal{A}$ are denoted $J_0, J_1, \ldots, J_n$. For an Abelian scheme on the group X we use the notation

$$D_z(x,y) = \begin{cases} 1 & \text{if } x-y=z, \\ 0 & \text{otherwise.} \end{cases}$$

Those matrices generate the Hecke representation $\mathcal{B}$ of the group algebra $\mathbb{C}X$. The adjacency matrices can be written as

$$D_i = \sum_{z \in X_i} D_z. \tag{4.1}$$

PROPERTY 4.2. The primitive idempotents of $\mathcal{A}$ consist of a basis of $\mathcal{A}$ and we have

$$D_k = \sum_{i=0}^{n} P_k(i) J_i, \; k = 0, \ldots, n,$$

and

$$J_k = |X|^{-1} \sum_{i=0}^{n} Q_k(i) D_i \tag{4.2}$$

where $P_k(i)$ (resp. $Q_k(i)$) denotes the entry in row i and column k of P (resp. of Q).

DEFINITION 4.3. The matrix P (resp. the matrix Q) is the **first eigenmatrix** of the scheme (resp. **second eigenmatrix**).

Those matrices can be computed with the matrix S of group characters of X. Thus $\forall (x,y) \in X \times X'$, we have $S(x,y) = < x, y >$ and we have the important relation (Theorem 2.4 of [**8**] and Theorem 2.2 of [**10**]) numbered (53) in [**4**].

$$\forall y \in X_i', \; P_k(i) = \sum_{x \in X_k} \bar{S}(x,y) = \sum_{x \in X_k} < -x, y > . \tag{4.3}$$

and

$$\forall x \in X_i, Q_k(i) = \sum_{y \in X_k'} < x, y > . \tag{4.4}$$

Relation (44) of [**4**] is

$$\forall z \in X', \; D_u S_z = < -u, z > S_z. \tag{4.5}$$

4.2. Idempotents in Abelian schemes and multisets. When (X, R) is an Abelian scheme, its BMD algebra $\mathcal{A}$ is a subalgebra of the Hecke representation $\mathcal{B}$ of the group algebra $\mathbb{C}X$. To avoid heavy notations we use the same symbol $\mathcal{A}$ for the BMD algebra of an Abelian scheme, which is a subalgebra of the matrix algebra $\mathbb{C}(X, X)$ and the subalgebra of the group algebra $\mathbb{C}X$ that $\mathcal{A}$ represents. We remind the following property.

PROPERTY 4.4. Let $\mathcal{A}$ be a BMD subalgebra of $\mathbb{C}X$ and denote by $\mathbf{u}_0, \mathbf{u}_1, \ldots,$ $\mathbf{u}_n$ the primitive idempotents of $\mathcal{A}$. They can be written as

$$\mathbf{u}_k = |X|^{-1} \sum_{i=0}^{n} Q_k(i) Z^{X_i}. \tag{4.6}$$

Moreover for the Fourier coefficients of $\mathbf{u}_k$, we have that

$$\hat{u}_{k,z} = \begin{cases} 1 & \text{if } z \in -X'_k \\ 0 & \text{otherwise.} \end{cases} \tag{4.7}$$

Proof: Since D_x represents Z^x in the matrix representation of $\mathbb{C}X$, then by (4.1) D_i represents Z^{X_i}. Thus relations (4.2) are the matrix representation of relations (4.6). The values of $\hat{u}_{k,z}$ follow from $PQ = |X|I$ by relations (4.2). □

We now need a formal definition of generator matrices with repeated columns.

DEFINITION 4.5. A **preprojective multiset** E_π of $\mathbb{F}_q^k \setminus \{0\}$ is a mapping π from $\mathbb{F}_q^k \setminus \{0\} \to \mathbb{N}$ such that

$$\pi(x) = i \leftrightarrow (\forall \alpha \in \mathbb{F}_q^*, \pi(\alpha x) = i)$$

DEFINITION 4.6. The **projective multiset** E_{π^*} associated with the preprojective multiset E_π of $\mathbb{F}_q^k \setminus \{0\}$ is a mapping π^* from $\mathbb{F}_q^k \setminus \{0\} \to \mathbb{N}$ defined as follows. For each i such that $\pi^{-1}(i) \neq \emptyset$ a nonzero vector x is chosen in a single one dimensional space contained in $\pi^{-1}(i) \bigcup \{0\}$. Then $\pi^*(x) = i$ and $\pi^*(y) = 0$ for every $y \neq x$ in the one dimensional space containing x.

DEFINITION 4.7. The **preprojective code** $C(E_\pi)$ of a preprojective multiset E_π is a linear code of **length** $\sum_{x \in \mathbb{F}_q^k \setminus \{0\}} \pi(x)$. The columns of its generator matrix G_{E_π} are the elements x of $\mathbb{F}_q^k \setminus \{0\}$ such that $\pi(x) > 0, x$ being repeated $\pi(x)$ times.

DEFINITION 4.8. The **projective code** $C(E_{\pi^*})$ of a projective multiset E_{π^*} is a linear code of length $\sum_{x \in \mathbb{F}_q^k \setminus \{0\}} \pi(x)/(q-1)$ associated with a projective multiset E_{π^*}. The columns of its generator matrix $G_{E_{\pi^*}}$ are the elements x of $\mathbb{F}_q^k \setminus \{0\}$ such that $\pi^*(x) > 0, x$ being repeated $\pi^*(x)$ times.

NOTATION 4.9. For a preprojective multiset E_π (projective multiset E_{π^*}), the group algebra element $\sum_{x \in \mathbb{F}_q^k \setminus \{0\}} \pi(x) Z^x$ is denoted by Z^{E_π} (the group algebra element $\sum_{x \in \mathbb{F}_q^k \setminus \{0\}} \pi^*(x) Z^x$ is denoted by $Z^{E_{\pi^*}}$).

NOTATION 4.10. We denote by $(z.x)_{x \in G_{E_\pi}}$ the codeword obtained by applying the rowvector z to the generator matrix G_{E_π}.

The following lemma is essentially Lemma 4.76 of [**4**].

LEMMA 4.11. *Let E_π be a preprojective multiset. Denoting Z^{E_π} by* **a***, we have that the weight w_z of the codeword $(z.x)_{x \in G_{E_\pi}}$, is related to the Fourier coefficient $\hat{a}_z$ of* **a** *by*

$$qw_z = (\sum_{x \in \mathbb{F}_q^k \setminus \{0\}} \pi(x) - \hat{a}_z)(q-1).$$

DEFINITION 4.12. A preprojective multiset E_π (projective multiset E_{π^*}) has **rank** k if

$$rank\{x | x \in \mathbb{F}_q^k, \pi(x) > 0\} = k \; (rank\{x | x \in \mathbb{F}_q^k, \pi^*(x) > 0\} = k).$$

DEFINITION 4.13. A preprojective multiset E_π of $\mathbb{F}_q^k$ is **compatible with the classes** $E_1, \ldots, E_s$ of a partition of $\mathbb{F}_q^k \setminus \{0\}$ if

$$\forall x, y \in E_i, \pi(x) = \pi(y), \; i = 1, \ldots, s.$$

DEFINITION 4.14. The **restriction** (C, R^C) of an Abelian scheme (X, R) to an additive code C with weights $i_0 = 0, i_1, \ldots, i_s$ is defined by the partition $\{C_0, C_{i_1}, \ldots, C_{i_s}\}$ where $C_{i_\nu} = C \cap X_{i_\nu}$ as follows:

$$(x, y) \in R_\nu^C \; if \; x - y \in C_{i_\nu}, \; \nu = 0, 1, \ldots, s$$

In the sequel we only consider the case where q is a prime p.

LEMMA 4.15. *To any Abelian scheme (E, T) of $\mathbb{F}_p^k$ with Abelian classes $E_1, \ldots, E_s$, such that $\mathbb{F}_p^* E_i = E_i$, $i = 1, \ldots, s$, corresponds a preprojective multiset E_π of rank k compatible with the partition $E_1, \ldots, E_s$ of $\mathbb{F}_p^k \setminus \{0\}$ such that the preprojective code C_{E_π} has s nonzero weights.*

Proof: First observe that $\mathbb{F}_p^* X_k' = X_k'$, $\forall k$. Indeed by (4.6) and (4.7) we have that $\hat{u}_{k,z'} = \hat{u}_{k,z}$ for $z' = \alpha z$, $\alpha \in \mathbb{F}_p^*$, $\hat{u}_{k,z'}$ being a complex onjugate of $\hat{u}_{k,z}$. Then by relations (4.4) all $Q_k(i)$ in relation (4.6) are rational integers and there exist integers α_k such that $\alpha_k \mathbf{u}_k$ has integer coefficients and moreover

$$\mathbf{a} = \sum_{k=0}^{s} \alpha_k \mathbf{u}_k$$

has s distinct Fourier coefficients. Indeed those coefficients are $\alpha_0, \alpha_1, \ldots, \alpha_s$. There clearly corresponds to **a** a multiset E_π compatible with with the partition $E_1, \ldots, E_s$ such that $Z^{E_\pi} = \mathbf{a}$, since $\mathbf{u}_0 = |X|^{-1} \sum_{i=0}^{n} Z^{X_i}$ and we are free to make $\alpha_0 > 0$ large enough for all coefficients in **a** to be positive. Let us see that there exists such a multiset E_π with rank k such that $\pi(x) = 0$ for some $x \in \mathbb{F}_p^k \setminus \{0\}$. For a partition of $\mathbb{F}_p^k$ into two sets S_1 and S_2, since $p^{k-1} + (p^{k-1} - 1) < p^k$, then we have $\text{rank}(S_1) = k$ or $\text{rank}(S_2) = k$. Now assume that **a** does not define a multiset of rank k and let β is the largest coefficient in **a**. Then $\beta \mathbf{u}_0 - \mathbf{a}$ necessarily defines a multiset of rank k. □

NOTATION 4.16. For any subset F of a set E we denote by ψ_F the column-vector of $\mathbb{C}^E$ with components

$$\psi_F(x) = \begin{cases} 1 & \text{if } x \in F \\ 0 & \text{otherwise} \end{cases} \tag{4.8}$$

THEOREM 4.17. *The following statements are equivalent.*

(i) *There exists a preprojective multiset* E_π *of* $\mathbb{F}_p^k$, *of rank* k, *such that the code* $C(E_\pi)$, *of length* n, *has* s *weights, and the restriction of* $H(n,q)$ *to* $C(E_\pi)$ *is a subscheme.*

(ii) *There exists an Abelian scheme* (E,T) *on* $(\mathbb{F}_p^k,+)$ *with Abelian classes* $E_1,\ldots,E_s$.

Proof: $(i) \Longrightarrow (ii)$ Consider the generator matrix G_{F_π} of the linear code $\mathcal{C}(F_\pi)$. To each coset of the dual $C^\perp(F_\pi)$ of $C(F_\pi)$ corresponds a syndrom of $C(F_\pi)$ which is a linear combination of the columns of G_{F_π}. By [**4**, Theorem 5.10] there are exactly $s+1$ distinct coset weight distributions including the weight distribution of $C^\perp(F_\pi)$. By [**4**, Theorem 5.13] the **distribution partition**, [**4**, Definition 5.7], which may be considered a partition $\{E_0=\{0\},E_1,\ldots,E_s\}$ of the set of syndroms of $C(F_\pi)$, defines an Abelian scheme on $(\mathbb{F}_p^k,+)$.
$(ii) \Longrightarrow (i)$ The given Abelian scheme is denoted by $(\mathbb{F}_p^k,T)$. Let us denote by $E'_1,\ldots,E'_s$ the Abelian classes of the dual $((\mathbb{F}_p^k)',T')$ of $(\mathbb{F}_p^k,T)$. We consider the BMD algebra $\mathcal{A}$ of $(\mathbb{F}_p^k,T)$ and its Fourier image $\hat{\mathcal{A}}$ which is generated by $\psi_0,\psi_{E'_1},\psi_{E'_2},\ldots,\psi_{E'_s}$. We thus know that the Fourier coefficient $\sum_{x\in E_i} < x,z >$ is a constant $r_{i,\nu}$ for $z\in E'_\nu,\nu,i=1,2,\ldots,s$. Those constants moreover are, by (4.3), the entries $P_i(\nu)$ of the first eigenmatrix of $(\mathbb{F}_p^k,T)$ since $-E_i=E_i$. By Lemma 4.15 there exists a preprojective multiset E_π of rank k compatible with the partition $E_1,\ldots,E_s$ of $\mathbb{F}_p^k\setminus\{0\}$. Let us denote by α_i the common integer $\pi(x)$, $x\in E_i$. The weights of the corresponding code $Y=C(E_\pi)$ are, by Lemma 4.11

$$w_\nu = (\sum_{x\in\mathbb{F}_q^k\setminus\{0\}} \pi(x) - \sum_{i=1}^{s}\alpha_i r_{i,\nu})(q-1)/q, \nu=1,2,\ldots,s.$$

The restriction of $H(n,q)$ to $C(E_\pi)$ is a subscheme isomorphic with $((\mathbb{F}_p^k)',T')$. □

REMARK 4.18. If (i) holds, then the restriction (C,R^C) of $H(n,q)$ to $C(E_\pi)$ is isomorphic to the dual (E',T') of (E,T) with Abelian classes $E'_1,\ldots,E'_s$. We recall that the invoked [**4**, Theorem 5.13] is essentially Theorem 6.11 of [**7**].

COROLLARY 4.19. Under the assumption (i) of the theorem, the multiset E_π is compatible with the partition $E_1,\ldots,E_s$ of $\mathbb{F}_p^k$.

Proof: For if x is a column of G_{F_π} then $x\in E_i$ for some $i\in[1,s]$. Thus x is a syndrom whose corresponding coset in $\mathbb{F}_p^n$ contains a vector of weight one. But by construction of E_i, the number of weight one vectors, say α_i, in the coset corresponding to the syndrom y does not depend of the choice of y in E_i. Indeed two syndroms in E_i correspond to cosets of $\mathcal{C}(F_\pi)$ with the same weight distribution, by [**4**, Corollary 5.12]. Then $\pi(y)=\alpha_i$, $\forall y\in E_i$. □

5. Weight enumerators

Another attractive property of the Gray map is here pointed out. The notion of **weight** which is familiar in Hamming schemes is generalized in Abelian schemes as follows by [**4**, Definition 4.69] here reproduced.

DEFINITION 5.1. For an Abelian scheme (X,R) an element $x\in X$ has **weight i** if it belongs to X_i. Then if Y is regular and contains 0, we have

$$a_i = |\{x\in Y | x-0\in X_i\}| = |Y\cap X_i|,$$

and the inner distribution **a** of Y is here the **weight distribution** $(|Y \cap X_0|, |Y \cap X_1|, \ldots, |Y \cap X_n|)$ of Y.

REMARK 5.2. The classes of (X, R) are not ordered in a general Abelian scheme. We draw the reader's attention on the fact that the word **weight** was however retained for the classes, as in the case of Hamming schemes.

5.1. Complete weight enumerators and regular unrestricted codes. The Delsarte extension of an association scheme is described in [**4**, Section 4.11]. We recall a definition in [**4**, Section 5, Definition 5.49].

DEFINITION 5.3. Let there be given an Abelian scheme $(\mathcal{F}, K)$ with Abelian classes $\mathcal{F}_1, \mathcal{F}_2, \ldots, \mathcal{F}_s$. The **complete weight enumerator** of a code Y of (X, R), the extension of length m of $(\mathcal{F}, K)$, is the polynomial

$$\mathcal{CW}_Y(T) = \sum_{k_0 + k_1 + \ldots + k_s = m} a_{\mathbf{k}} T_0^{k_0} T_1^{k_1} \ldots T_s^{k_s}$$

where $a_{\mathbf{k}}$, with $\mathbf{k} = (k_0, k_1, \ldots, k_s)$, is the number of $u \in Y$ with k_r components in the class $\mathcal{F}_r$ of $(\mathcal{F}, K), r = 0, 1, \ldots, s$.

If we denote by (X, R) the extension of degree m of $(\mathcal{F}, K)$, then a class $X_{\mathbf{k}}$ of (X, R) is the set of vectors of $\mathcal{F}^m$ with k_r components in $\mathcal{F}_r$, $r = 0, 1, \ldots, s$, $\mathbf{k} = (k_0, k_1, \ldots, k_s)$. Now let us assume that $\mathcal{F}$ is an additive code C with s nonzero weights in a Hamming scheme $H(\ell, p)$ and that the class $\mathcal{F}_r$ is the set of all codewords of C with equal weight w_r, $r = 0, 1, \ldots, s$. Then clearly for an unrestricted code Y of X, being regular implies being regular for the Hamming distance.

5.2. A MacWilliams transform for unrestricted codes. Several weight enumerators may be defined for a code Y in an extension (X, R) of length m of an Abelian scheme $(\mathcal{F}, K)$ [**4**, Section 5.3]. If Y is an additive code in such an Abelian scheme (X, R), then there exists a dual code Y° and the Macwilliams transform of a weight enumerator of Y yields the weight enumerator of Y°. In the case where $\mathcal{F}$ is $(\mathbb{Z}_4, +)$, then the MacWilliams transform of such an additive code Y gives the weight enumerator of Y° which also is the weight enumerator of $\sigma(Y^\circ)$, which is a regular code in its turn. This holds because (X, R) is self-dual (Definition 2.10). We will verify this below in a more general setting.

This is another reason to consider isomorphisms of Abelian schemes in order to generalize the Gray map to $\mathbb{Z}_{p^n}$.

5.3. The general situation.

5.3.1. *The dual of the Delsarte extension.* Let us recall [**4**, Theorem 4.135].

THEOREM 5.4. *If each of the association schemes* $(X^{(j)}, R^{(j)}), j = 1, \ldots, m$ *has a dual, then the Kronecker product* $(X, T) = (X^{(1)} \times \ldots \times X^{(m)}, R^{(1)} \otimes \ldots \otimes R^{(m)})$ *has a dual which is the Kronecker product of the duals of the* $(X^{(j)}, R^{(j)})$.

The theorem entails that if the association scheme $(X^{(j)}, R^{(j)})$ is identical to $(\mathcal{F}, K)$, $j = 1, \ldots, m$ and if $(\mathcal{F}, K)$ is self-dual, then the extension of length m of $(\mathcal{F}, K)$ is self-dual. Indeed if $(X^{(j)}, R^{(j)}) = (\mathcal{F}, K)$, $j = 1, \ldots, m$, then the dual (X', T') of (X, T) is the Kronecker power $(\mathcal{F}, K)^m_\otimes$. If $(\mathcal{F}, K)$ has a dual $(\mathcal{F}', K')$ then the identification of $\mathcal{F}$ and $\mathcal{F}'$ extends to an identification of X and X'. Thus $(X', T') = (X, T)$. The fact that the Delsarte extension of length m of $(\mathcal{F}, K)$ is self-dual is then a consequence of [**4**, Property 4.106].

5.3.2. *The analytic MacWilliams transform for Abelian schemes.* Given the complete weight enumerator of a code Y (Definition 5.3), Theorem 5.51 of [**4**] yields the complete weight enumerator of the dual code Y°. Notice that the Abelian scheme $(\mathcal{F}, K)$ here considered is any Abelian scheme with s classes.

THEOREM 5.5. *Let there be given the complete weight enumerator*

$$\mathcal{CW}_Y(T) = \sum_{k_0+k_1+\ldots+k_s=m} a_{\mathbf{k}} T_0^{k_0} T_1^{k_1} \ldots T_s^{k_s}$$

of an additive code Y in the extension of length m of $(\mathcal{F}, K)$. Then the complete weight enumerator $\mathcal{CW}_{Y^\circ}(T')$ of the dual Y° of Y is

$$|Y|^{-1} \sum_{k_0+k_1+\ldots+k_s=m} a_{\mathbf{k}} \prod_{r=0}^{s} \left(\sum_{j=0}^{s} Q_j(r) T_j' \right)^{k_r}$$

where Q is the second eigenmatrix of the association scheme $(\mathcal{F}, K)$.

In the case of a Gray map, we substitute B_0 for T_0, B_1 for T_1 and B_1^2 for T_2 in

$$\mathcal{CW}_Y(T) = \sum_{k_0+k_1+k_2=m} a_{\mathbf{k}} T_0^{k_0} T_1^{k_1} T_2^{k_2}$$

which is the weight enumerator of the linear code Y of $\mathbb{Z}_4^m = \mathcal{F}^m$. We thus obtain the Hamming weight enumerator of $\sigma(Y)$. Now the MacWilliams transform in Theorem 5.5 gives the weight enumerator of Y°. The substitution above in that complete weight enumerator yields $\sigma(Y^\circ)$.

We endeavour to carry that construction from the pair $\{\mathbb{Z}_4, \mathbb{F}_2^2\}$ to pairs $\{\mathbb{Z}_{p^k}, \mathbb{F}_p^\ell\}$ where ℓ is a suitable integer, $\ell \geq k$.

6. Unrestricted regular p-ary codes corresponding to linear codes over $\mathbb{Z}_{p^k}$

6.1. Linear codes which are subschemes. The Gray map. Let $\mathcal{F}$ be the Abelian group $(\mathbb{Z}_{p^k}, +)$. The first condition for the seeked correspondance is that we dispose of an bijective mapping σ from $\mathcal{F}$ to a p-ary additive code C, subgroup of $\mathbb{F}_p^\ell$, $\ell \geq k$. The mapping σ is not required to be a linear isomporphism. It should however be an isomorphism from an association scheme to another.

We have an Abelian scheme $(\mathcal{F}, K)$ with s classes $\mathcal{F}_1, \ldots, \mathcal{F}_s$ and an isomorphism from it to the restriction (C, R^C) to C of the hamming scheme $H(\ell, p)$ denoted $((\mathbb{F}_p^\ell, +), R)$. It is an additive code with s weights in $\mathbb{F}_p^\ell$. The class $\sigma(\mathcal{F}_j)$ of (C, R^C) will be the set of vectors with common weight i_j, $j = 1, \ldots, s$.
Clearly if such an isomorphism exists, to a linear code Y of $\mathcal{F}^m$ will correspond an unrestricted code $\sigma(Y)$, subset of C^m, itself a subset of $\mathbb{F}_p^{\ell m}$. The Macwilliams transform of the complete weight enumerator

$$\mathcal{CW}_Y(T) = \sum_{k_0+k_1+\ldots+k_s=m} a_{\mathbf{k}} T_0^{k_0} T_1^{k_1} \ldots T_s^{k_s} \tag{6.1}$$

of Y yields, by Theorem 5.5, the complete weight enumerator

$$\mathcal{CW}_{Y^\circ}(T') = \sum_{k_0+k_1+\ldots+k_s=m} a_{\mathbf{k}} T_0'^{k_0} T_1'^{k_1} \ldots T_s'^{k_s} \tag{6.2}$$

of Y°.

Now, **if $(\mathcal{F}, K)$ is self-dual**, substituting $B_j^{i_j}$ for T_j, $j = 1, \ldots, s$ in (6.1) (resp. $B_j^{'i_j}$ for T'_j, $j = 1, \ldots, s$ in (6.2)) will give the weight enumerator of $\sigma(Y)$ (resp. of $\sigma(Y^\circ)$). For $p = 2$, every code is either preprojective or projective (Definitions 4.7 and 4.8). For general p we require C to be a preprojective code. A projective code C^* can afterwards be easily obtained from C. Now C has a generator matrix G_{E_π} corresponding to a preprojective multiset E_π of rank k of $\mathbb{F}_p^k$, since C has size p^k and, as an additive code is a linear code with dimension k over $\mathbb{F}_p$. There thus exists a linear isomorphism γ from C to $\mathbb{F}_p^k$ which defines, on the Abelian group $(\mathbb{F}_p^k, \oplus)$ an Abelian scheme denoted $(\mathbb{F}_p^k, T)$ isomorphic with $(\mathcal{F}, K)$. We are now able to invoque Theorem 4.17. We will use Lemma 3.1.

Actually if $(\mathcal{F}, K)$ were not self-dual then (6.2) would be a weight enumerator for the dual code as well. However the classes of $\mathcal{F}'$ being denoted by $\mathcal{F}'_0, \mathcal{F}'_1, \ldots, \mathcal{F}'_s$, then $a_{\mathbf{k}}$ would be the number of $u \in Y^\circ$, for $\mathbf{k} = (k_0, k_1, \ldots, k_s)$ with k_r components in the class $\mathcal{F}'_r$, $r = 0, 1, \ldots, s$.

6.2. New isomorphisms for new constructions. In order to generalize the Gray map to $\mathbb{Z}_{p^k}$ we need, as we have just seen, a projective linear code C^* of $\mathbb{F}_p^\ell$, $\ell > k$ of dimension k. We consider the preprojective code C which yields C^*. Clearly if (C^*, R^{C^*}) is a restricted Abelian scheme, so is (C, R^C). We will then only be concerned with (C, R^C) with the view of using Theorem 4.17.

By a theorem of Delsarte stated as Theorem 5.13 in [**4**], the dual of (C, R^C) is the coset scheme $(C', \bar{R}')$. The set of syndromes of C° inherits the structure of $(C', \bar{R}')$. The Abelian classes of that scheme, each of which being a class of syndromes, are denoted by $E_1, \ldots, E_s$. To make this clear let us say that $E_1, \ldots, E_s$ are the Abelian classes of $(\mathbb{F}_p^k, T)$, and that the Abelian classes of its dual scheme $((\mathbb{F}_p^k)', T')$ are $E'_1, \ldots, E'_s$. Taking into account that $(\mathbb{F}_p^k, T)$ is isomorphic with its dual $((\mathbb{F}_p^k)', T')$, then if there exists an isomorphism from an Abelian scheme $(\mathcal{F}, K)$ to (C, R^C), $(\mathcal{F}, K)$ also is isomorphic with $((\mathbb{F}_p^k, \oplus), T)$ that is simply denoted by (E, T). By Lemma 3.1, if such an isomorphism exists, then $(\mathcal{F}, K)$ covers the merging of the Hecke scheme $\mathcal{H}(\mathcal{F})$ and the Hecke scheme $\mathcal{H}(E)$. Such a merging may be the Abelian scheme sought-after. Here is an example.

EXAMPLE 6.1. In order to construct a merging, we need to identify $\mathbb{Z}_{p^k}$ with $\mathbb{F}_p^k$. Let us consider the case where $p = 2$ and $k = 3$. The identification is as follows.

$$0 \mapsto 000, 1 \mapsto 100, 3 \mapsto 110, 5 \mapsto 101, 7 \mapsto 111, 2 \mapsto 010, 4 \mapsto 001, 6 \mapsto 011.$$

The merging of the two Hecke schemes is computed with the algorithm described in [**4**, Section 4.13.1]. It is given by its Abelian classes

$$\mathcal{F}_1 = \{1, 3, 5, 7\}, \mathcal{F}_2 = \{2, 6\}, \mathcal{F}_3 = \{4\}.$$

We have in mind that such a merging is an Abelian scheme for both $(\mathbb{Z}_{p^k}, +)$ and $(\mathbb{F}_p^k, \oplus)$. For clarity we denote it by $(\mathcal{F}, K)$ (resp. (E, T)) when $(\mathbb{Z}_{p^k}, +)$ (resp. $(\mathbb{F}_p^k, \oplus)$) is concerned.

The dual of $(\mathcal{F}, K)$ is $(\mathcal{F}', K')$ where

$$\mathcal{F}'_1 = \{1, 3, 5, 7\}, \mathcal{F}'_2 = \{2, 6\}, \mathcal{F}'_3 = \{4\},$$

by Theorem 4.65 of [**4**] applied to $(\mathcal{F}, K)$ with the set $\{1, 3, 5, 7\}$ as automorphism group of $(\mathbb{Z}_{p^k}, +)$. Using Corollary 4.51 of [**4**], we construct the dual

(E', T') of (E, T) with the character table $< x, y >= (-1)^{x.\tau(y)}$, $x, y \in \mathbb{F}_2^3$ where $\tau(a_1, a_2, a_3) = (a_3, a_2, a_1)$. We obtain

$$E_1' = \{100, 110, 101, 111\}, E_2' = \{010, 011\}, E_3' = \{001\}.$$

In other words, (E, T) is seen to be self-dual in the sense of Definition 4.56 of [**4**]. A generator matrix for the code C can be taken as

$$\begin{bmatrix} 1 & 1 & 1 & 1 & 0 \\ 0 & 1 & 0 & 1 & 0 \\ 0 & 0 & 1 & 1 & 1 \end{bmatrix}, \tag{6.3}$$

its columns being the vectors of $E_1' \cup E_3'$. A desired mapping σ corresponding to the Gray map is thus

$$0 \mapsto 00000, 1 \mapsto 00111, 3 \mapsto 01101, 5 \mapsto 11001, 7 \mapsto 10011, 2 \mapsto 01010,$$

$$6 \mapsto 10100, 4 \mapsto 11110.$$

Let us notice that the columns of the generator matrix obtained from (6.3) by deleting the last column makes up an Abelian class of $(\mathcal{F}, K)$ for $\mathcal{F} = (\mathbb{F}_2^3, \oplus)$. That generator matrix thus corresponds to a projective multiset E_π compatible (Definitions 4.6 and 4.13) with the classes $\mathcal{F}_1, \mathcal{F}_2, \mathcal{F}_3$ of $(\mathcal{F}, K)$. Then each set of linear forms $\mathcal{F}_1', \mathcal{F}_2', \mathcal{F}_3'$ of $(\mathcal{F}, K)$ corresponds to codewords of $C(E_\pi)$ with equal weights. Those weights are 2 for $\mathcal{F}_1'$ and $\mathcal{F}_2'$ and 4 for $\mathcal{F}_3'$. The restriction of $H(4, 2)$ to $C(E_\pi)$ is no longer an association scheme. However the mapping obtained from $(\mathcal{F}, K)$ to $C(E_\pi)$ is such that the image of each class $\mathcal{F}_i$, $i = 1, 2, 3$ is a set of codewords with equal weight. Then clearly Remark 2.9 extends to such a situation and if Y is an additive code of $(\mathcal{F}, K)_\otimes^m$, then denoting by ρ the extension from to $\mathcal{F}$ to to $\mathcal{F}^m$ of σ we have that $\rho(Y)$ is a regular code of $C(E_\pi)^m$ and from there an unrestricted regular code of $H(4m, 2)$, by Remark 2.6. This mapping σ has been introduced by Claude Carlet in [**6**]. The following example is an application of Theorem 7.1 in next section. The class $X_{\oplus 1}$ of $(X_\oplus, R_\sigma)$ consists of the columns of the generator matrix (6.4), for $p = 3$ and $k = 2$.

EXAMPLE 6.2. For $p = 3$ and $k = 2$, a generator matrix for the projective code C is

$$\begin{bmatrix} 1 & 1 & 1 \\ 0 & 1 & 2 \end{bmatrix} \tag{6.4}$$

A desired mapping σ corresponding to the Gray map is thus

$$0 \mapsto 000, 1 \mapsto 012, 2 \mapsto 021, 3 \mapsto 111, 4 \mapsto 120, 5 \mapsto 102,$$

$$6 \mapsto 222, 7 \mapsto 201, 8 \mapsto 210$$

7. A general identification of $\mathbb{Z}_{p^k}$ with $\mathbb{F}_p^k$

7.1. A priviledged mapping. The mapping that we consider here, yet quite natural will show very attractive regarding the structure of association schemes. We consider the set X consisting of the integers $0, 1, \ldots, p^k - 1$, where p is a prime, either equipped with the structure of the Abelian group $(\mathbb{Z}_{p^k}, +)$ or of the Abelian group $(\mathbb{F}_p^k, \oplus)$. We recall that addition (resp. subtraction) in the Abelian group $\mathbb{F}_p^k$, is denoted by $\oplus$ (resp. by $\ominus$). In the first situation the Abelian group is designated by X_+ and in the latter by $X_\oplus$. The common underlying set X may be obtained by the mapping σ from $(\mathbb{Z}_{p^k}, +)$ to $\mathbb{F}_p^k$, defined as

$$c = c_0 + c_1 p + \ldots + c_{k-1}p^{k-1} \to \sigma(c) = (c_0, c_1, \ldots, c_{k-1}). \tag{7.1}$$

In the sequel c is identified with $\sigma(c)$ and the common underlying set of X_+ and $X_\oplus$ may be denoted by X when it is clear by context wether c is considered an integer or an k-tuple with components in $\mathbb{F}_p$. In the latter situation c will be dealt with as a column-vector, i.e. $[c_0, c_1, \ldots, c_{k-1}]^T$.

7.2. The central class scheme on the Abelian group $(\mathbb{Z}_{p^k}, +)$. This is a particular instance of Example 3 in Section 2.4 of [**7**] which is quoted as Example 5.60 in [**4**]. Let X_0 denote the set $\{0\}$. The set X_i consits of all elements in the set $p^{i-1}\mathbb{Z}_{p^k} \setminus p^i\mathbb{Z}_{p^k}$, $i = 0, \ldots, k-1$. In other words X_i is the set of all vectors c such that $c_0 = c_1 = \ldots = c_{i-2} = 0$, and $0 < c_{i-1} < p$. The group of units of $\mathbb{Z}_{p^k}$ acting on the ring $\mathbb{Z}_{p^k}$ is the group of automorphisms of $(\mathbb{Z}_{p^k}, +)$. Since every X_i, $i = 0, 1, \ldots, k$, is an orbit of that group, then by a property stated as theorem 4.65 of [**4**], $X_1, X_2, \ldots, X_k$ are the classes of a self-dual Abelian scheme that will be denoted by (X_+, R). Notice that the dual scheme is defined with respect to the orthogonal matrix $[< x, y >= \theta^{xy}]$ where θ is a primitive p^k-th root of unity in $\mathbb{C}$ and xy is the product of x and y in the ring $(\mathbb{Z}_{p^k}, +)$. Now the class R_i of that scheme consists of all couples $(x, y) \in X \times X$ such that $x - y \in X_i$. In other words

$$(x, y) \in R_i \text{ if and only if } x_0 = y_0, .., x_{i-2} = y_{i-2}, x_{i-1} \neq y_{i-1}. \tag{7.2}$$

More precisely the $(i-1)$-th coefficient of $x - y$ or of $x \ominus y$ is $x_{i-1} - y_{i-1}$ or $p + x_{i-1} - y_{i-1}$ according as $x_{i-1} > y_{i-1}$ or $x_{i-1} < y_{i-1}$. Indeed if $x_{i-1} < y_{i-1}$ then in $\mathbb{Z}_{p^k}$ we can write

$$\begin{aligned} x - y &= pp^{i-1} + (p-1)p^i + \ldots (p-1)p^{k-1} - y + x \\ &= (p - y_{i-1} + x_{i-1})p^{i-1} + \beta p^i \in X_i. \end{aligned}$$

7.3. An Abelian scheme on the Abelian group $(\mathbb{F}_p^k, \oplus)$. Let us consider the group G of all lower triangular invertible matrices with size $k \times k$ over $\mathbb{F}_p$. Clearly the orbits of $X_\oplus$ under the action of that group are the sets X_i, $i = 0, 1, \ldots, k$. Thus by Property 4.45 of [**4**], $X_1, \ldots, X_k$ are the classes of an Abelian scheme on $X_\oplus$. The relations of that Abelian scheme are the same as those of (X_+, R). Indeed we also have that $x \ominus y \in X_i$ if and only if $x_0 = y_0, .., x_{i-2} = y_{i-2}, x_{i-1} \neq y_{i-1}$. We denote that Abelian scheme on $X_\oplus$ by $(X_\oplus, R_\sigma)$.

THEOREM 7.1. *The Abelian schemes (X_+, R) and $(X_\oplus, R_\sigma)$ are isomorphic.*

Proof: This is a consequence of Definition 2.7. □

The matrix S of group characters of $X_\oplus$ also called the character table of $X_\oplus$ is defined as follows. Let η be a primitive p-th root of unity.

$$\forall (x, y) \in X_\oplus \times X'_\oplus,\ S(x, y) =< x, y >= \eta^{x.\tau(y)},$$

where $x.z$ denotes the scalar product of x and z and

$$\tau(x_1, \ldots, x_k) = (x_k, \ldots, x_1). \tag{7.3}$$

We will show that $(X_\oplus, R_\sigma)$ is self-dual with respect to that orthogonal matrix S. With this aim in view we will have to diagonalize the BMD algebra of $(X_\oplus, R_\sigma)$ with S.

The fact that $(X_\oplus, R_\sigma)$ is self-dual makes it possible for every p and k to generalize the Gray map to every group $(\mathbb{Z}_{p^k}, +)$, by constructing an isomorphism from (X_+, R) to a restricted scheme (C, R^C) where C is a k-weight projective code. We here refer the reader to Sections 4, 5 and 6.

7.4. Duality. Let us first recall a classical result stated as [**4**, Corollary 4.51]. We here again use Notation 4.16.

COROLLARY 7.2. Let $\{X_0 = \{0\}, X_1, \dots, X_k\}$ be a partition of the Abelian group X. Those classes are the Abelian classes of a scheme (X, R) if and only if there exists a partition $\{X'_0 = \{0\}, X'_1, \dots, X'_k\}$ of the character group X' such that the Fourier transform of Z^{X_i} can be written as $\sum_{\ell=0}^{k} \lambda_{i\ell}\psi_{X'_\ell}$. The classes $X'_1, X'_2, \dots, X'_k$ are then those of the dual scheme with respect to the matrix S of group characters defining that Fourier transform.

We have our eye on showing that the Abelian scheme $(X_\oplus, R_\sigma)$ is self-dual. We know that any Abelian group X is isomorphic to its character group X'. We now denote $X_\oplus$ by X and we identify X' with X. In view of Corollary 7.2 we define a partition of the set X' of characters of X as $\{X'_0, X'_1, \dots, X'_k\}$, $X'_i = X_i$, $i = 0, \dots, k$. We will show that the Fourier transform of Z^{X_i}, for that given character table, can be written as $\sum_{k=0}^{k} \lambda_{ik}\psi_{X'_k}$. We next use the following Lemma.

NOTATION 7.3. Let E be an k-elements set and π a permutation on E. We denote by P_π the permutation matrix such that, with Notation 4.16, we have, for $a, b \in E$, $P_\pi\psi_a = \psi_b$ whenver $\pi(a) = b$. We denote by A^T the transpose of the matrix A. We here recall the easy Lemma 4.89 of [**4**].

LEMMA 7.4. *Let π be a permutation on X and let A be a matrix in $\mathbb{C}(X, X)$. If P_π denotes the permutation matrix in $\mathbb{C}(X, X)$ such that $P_\pi\psi_a = \psi_{\pi a}$, then we have*

$$A(\pi(a), \pi(b)) = (P_\pi^T A P_\pi)(a, b).$$

We consider below the permutation γ on the set $\{0, \dots, k-1\}$ defined by $\gamma(i) = k-1-i$, $i = 0, \dots, k-1$. Informally, the permutation matrix P_γ has all 1 entries in the second diagonal and 0 everywhere else.

LEMMA 7.5. *Let μ be a lower triangular $k \times k$ matrix and P_γ the $k \times k$ permutation matrix defined by*

$$P_\gamma(i, j) = \delta_{i,k-j-1}, \ i = 0, \dots, k-1. \tag{7.4}$$

Then $P_\gamma \mu^T P_\gamma$ is the **lower triangular matrix** *defined by*

$$P_\gamma \mu^T P_\gamma(i, j) = \mu^T(k-1-i, k-1-j).$$

Proof: By Lemma 7.4, we have

$$\mu^T(\gamma(i), \gamma(j)) = P_\gamma \mu^T P_\gamma(i, j).$$

Since μ^T is upper triangular, then $P_\gamma \mu^T P_\gamma$ is lower triangular. □

COROLLARY 7.6. The mapping $\mu \to \mu' = P_\gamma \mu^T P_\gamma$ is an antiautomorphism of G. As μ runs over G, then $\mu' = P_\gamma \mu^T P_\gamma$ acts transitively on the set X_i, $i = 0, \dots, k$.

Proof: Let M be any lower triangular matrix and let μ be the matrix defined by $\mu(j,i) = M(k-1-i, k-1-j)$. Thus $\mu = (P_\gamma M P_\gamma)^T$, by the lemma, and then $\mu' = P_\gamma \mu^T P_\gamma = M$. □

COROLLARY 7.7. Let μ be a lower triangular matrix. Then $< \mu x, z > = < x, \mu' z >$, where $\mu' = P_\gamma \mu^T P_\gamma$.

Proof: In matrix notations we have

$$< \mu x, z >= \eta^{\mu x . \tau(z)} = \eta^{\mu x . P_\gamma z} =$$

$$\eta^{x.\mu^T P_\gamma z} = \eta^{x.P_\gamma \mu' z} =< x.\mu' z > \text{,where } \mu' = P_\gamma \mu^T P_\gamma.$$

□

LEMMA 7.8.

$$\forall y \in X_i', \sum_{x \in X_\ell} < x, y >= \lambda_{\ell i},$$

where $\lambda_{\ell i}$ is given by (7.5).

Proof:

We have $\forall \mu \in G, \mu X_\ell = X_\ell$, then

$$\forall \mu \in G, \forall z \in X_i', \sum_{x \in X_\ell} < x, z >= \sum_{\mu^{-1} x \in X_\ell} < x, z >$$

$$= \sum_{x \in X_\ell} < \mu x, z >= \sum_{x \in X_\ell} < x, \mu' z >,$$

by Corollary 7.7. But by Corollary 7.6, to every $z \in X_i'$ corresponds a $\mu \in G$ such that $\mu' z = p^{i-1}$. Since $\sum_{0 \le i \le p} \eta^i = 0$, $\sum_{1 \le i \le p} \eta^i = -1$, and since $(p-1)p^{k-\ell}$ is the size of X_ℓ, we have that

$$\lambda_{\ell i} = \begin{cases} 0 & \text{if } \ell < k-i+1 \\ -p^{k-\ell} & \text{if } \ell = k-i+1 \\ (p-1)p^{k-\ell} & \text{if } \ell > k-i+1, \end{cases} \tag{7.5}$$

$\ell, i = 0, \ldots, k$. □

REMARK 7.9. For $\ell = 1$, then λ_{1i} is 0 if $i < k$ else $-p^{k-1}$.

We thus have obtained the following result.

THEOREM 7.10. *The Abelian scheme $(X_\oplus, R_\sigma)$ is self-dual with respect to the orthogonal matrix S.*

THEOREM 7.11. *The dual of (X_+, R) with respect to the orthogonal matrix $[< x, y >= \theta^{xy}]$ of Section 7.2 is isomorphic to the dual of $(X_\oplus, R_\sigma)$ with respect to the orthogonal matrix S.*

Proof: Let σ be the Abelian scheme isomorphism from (X_+, R) to $(X_\oplus, R_\sigma)$ in the statement of Theorem 7.1. Let us denote by ψ_1 the automorphism of Section 7.2 from (X_+, R) to (X_+', R') and by ψ_2 the automorphism of theorem 7.10 from $(X_\oplus, R_\sigma)$ to its dual. Then $\psi_2 \circ \sigma \circ \psi_1^{-1}$ is an isomorphism from (X_+', R') to the dual of $(X_\oplus, R_\sigma)$. □

7.5. Merging of Association Schemes. We denote by $\mathcal{H}(X_+)$ (resp. $\mathcal{H}(X_\oplus)$) the Hecke schemes, defined in [**4**, Section 4.1.2] on the Abelian group $X_+ = (\mathbb{Z}_{p^k}, +)$ (resp. $X_\oplus = \mathbb{Z}_{p^k}$).

THEOREM 7.12. *Denoting by X the common underlying set of X_+ and $X_\oplus$, a partition of X is defined as*

$$\forall \alpha \in]0,p[, \forall i = 0,..,k-1, X_{\alpha,i} = \{x | x_{i-1}p^{i-1} + x_i p^i + \ldots + x_{k-1}p^{k-1}, x_{i-1} = \alpha\},$$

where $x_i, \ldots x_{k-1} \in [0,p[$. Then the classes $X_{\alpha,i}, 0 < \alpha < p$, $i = 1, \ldots, k$ are the Abelian classes of the merging $(X_+, K) = (X_\oplus, K)$ of $\mathcal{H}(X_+)$ and $\mathcal{H}(X_\oplus)$.

Proof: Let Y be the Abelian class of the merging of $\mathcal{H}(X_+)$ and $\mathcal{H}(X_\oplus)$that contains the element $y = \alpha p^{i-1}$. Let us denote by R_Y the set $\{(a,b) | a - b \in Y\}$. Now consider the set E_y of all couples (a,b) such that $a - b = y$. We have in particular $(a,b) \in E_y$ for $a = p^i$, $b = -\alpha p^{i-1} + p^i = (p-\alpha)p^{i-1}$, $0 < p - \alpha < p$. Now $a \ominus b = \alpha p^{i-1} + p^i \in Y$. By induction we assume that $\alpha p^{i-1} + y_i p^i + \ldots + y_{k-1}p^{k-1} \in Y$, where $y_i, \ldots, y_{k-1} \in [0,p[$. We then see by a similar argument that $\alpha p^{i-1} + (y_i + 1)p^i + \ldots + y_{k-1}p^{k-1} \in Y$. Hence $X_{\alpha,i} \subset Y$.
We now ascertain that $Y \subset X_{\alpha,i}$. In other words we have to show that $X_{\alpha,i}$ is a class of the merging sought for. By the algorithm in Secion 4.13.1 of [**4**], we have to show that $a - b \in X_{\alpha,i}$ if and only if $a \ominus b \in X_{\alpha,i}$.

First if $a - b = \alpha p^{i-1} + \beta p^i$, $0 < \alpha < p$, we write $a = \alpha' p^{i-1} + \beta' p^i$, $b = \alpha'' p^{i-1} + \beta'' p^i$, $\alpha', \alpha'', \beta', \beta'' \geq 0$. If $\alpha' > \alpha''$, then $\alpha = \alpha' - \alpha''$ else $\alpha = p + (\alpha' - \alpha'')$. In both cases the $(i-1)$-th component of $a \ominus b$ is α, since for $\alpha'' \in \mathbb{F}_p$, $-\alpha'' = p - \alpha$. Now if $a \ominus b \in X_{\alpha,i}$ with $a = \alpha' p^{i-1} \oplus \beta' p^i, b = \alpha'' p^{i-1} \oplus \beta'' p^i$, then we also have that $\alpha = \alpha' - \alpha''$ if $\alpha' > \alpha''$, else $\alpha = \alpha' + (p - \alpha'')$ and $a - b$ still is in $X_{\alpha,i}$. □

REMARK 7.13. The merging of two Abelian schemes is an Abelian scheme, by Corollary 4.110 of [**4**]. We here have another way of showing that the classes $X_{\alpha,i}$ are Abelian classes. Indeed those classes are the orbits of $(\mathbb{Z}_{p^k}, +)$ under the action of the subgroup of units $X_{1,0}$. Thus (X_+, K) clearly is a self-dual Abelian scheme as was (X_+, R). Moreover the same argument as that used for proving theorem 7.10 holds here while replacing lowertriangular matrices with lowertriangular matrices with an all-one diagonal. We then have the following.

THEOREM 7.14. *The Abelian scheme $(X_\oplus, K)$ is self-dual with respect to the orthogonal matrix S.*

8. Specific examples of unrestricted codes over $\mathbb{F}_p$ from additive codes over $\mathbb{Z}_{p^k}$

8.1. Examples for $p \neq 2$. We only deal here with a simple example of a projective code which generalizes Example 6.2. More sophisticated examples can be worked out based on the techniques mentioned in Section 2.6, Section 3 and Section 4. For an easier understanding, this section has been written in order to be possibly read independently. We consider a class X_ℓ of $(X_\oplus, R_\sigma)$ defined in Section 7.3.

Back to Definition 4.8, we can see that to every X_ℓ corresponds a projective code $C(X_\ell^*)$. We will confine ourselve to investigating $C(X_1^*)$. Notice that X_1, has rank k and $|X_1| = (p-1)p^{k-1}$.

THEOREM 8.1. *The projective code $C(X_1^*)$ is an Abelian group on which an Abelian scheme is defined as follows. Its Abelian classes are $C_1 \dots, C_k$ where C_i is the set of codewords of the form*

$$\tau(x).X_1^* = ((\tau(x).z)_{z\in X_1^*} | x \in X_i'). \tag{8.1}$$

That Abelian scheme is isomorphic to (X_+, R).

Proof: Indeed $C(X_1^*)$ inherits the structure of Abelian scheme of the dual of $(X_\oplus, R)$. More precisely we have that

$$\forall (x,z) \in X' \times X', x \ominus z \in X_i' \leftrightarrow \tau(x).X_1^* \ominus \tau(z).X_1^* \in C_i,$$

by (4.3) and Lemma 4.11. □

NOTATION 8.2. We denote by γ the mapping $(\mathbb{Z}_{p^k}, +) \mapsto \mathcal{C}(X_1^*)$ defined by $x \to \gamma(x) = \tau(x).X_1^*$.

PROPERTY 8.3. The Hamming weights of $C(X_1^*)$ are given by

$$\forall x \in X_i',\ w_H(\gamma(x)) = p^{k-2}(p-1),\ i = 1, \dots, k-1;\ \text{else}\ w_H(\gamma(x)) = p^{k-1}.$$

Proof: We use Lemma 4.11 to determine the weights of $\mathcal{C}(X_1)$ with $|X_1| = (p-1)p^{k-1}$. The element Z^{X_ℓ} is denoted by $\mathbf{a}$ and for $z \in X_i'$, then the Fourier coefficient $\hat{a}_z$ of $\mathbf{a}$ is $\lambda_{\ell i}$ given in Lemma 7.8. Here $\pi(x) = 1, \forall x \in X_1$, else 0. □

THEOREM 8.4. *Let us define a weight on the elements of $\mathbb{Z}_{p^k}$ as follows. We make $w(x) = p^{k-1}$ for $x \in X_k, x \neq 0$ else $w(x) = (p-1)p^{k-2}$. Now the distance $d(x,y)$ is defined as $d(x,y) = w(x-y)$. Then γ is an isometry from $\mathbb{Z}_{p^k}$ onto $C(X_1^*)$ equipped with the Hamming distance d_H:*

$$\forall x, y \in, \mathbb{Z}_{p^k},\ d(x,y) = d_H(\gamma(x), \gamma(y)).$$

That isometry can be extended to $\mathbb{Z}_{p^k}^m$ by

$$\forall \mathbf{x} = (x_1, \dots, x_m) \in \mathbb{Z}_{p^k}^m, \gamma(x) = (\gamma(x_1), \dots, \gamma(x_m)),$$

while

$$d_H(\gamma(\mathbf{x}), \gamma(\mathbf{y})) = \sum_{i=1}^{m} d_H(\gamma(x_i), \gamma(y_i)) = \sum_{i=1}^{m} d(x_i, y_i) = d(\mathbf{x}, \mathbf{y}).$$

Proof: Since $w(x)$ is a constant on each class $X_i,\ i = 0, 1, \dots, k$, then the distance $d(x,y)$ only depends on which class $x - y$ belongs. But by theorem 8.1, we have that

$$x - y \in X_j \leftrightarrow x \ominus y \in X_j' \leftrightarrow \gamma(x) \ominus \gamma(y) \in C_j.$$

□

COROLLARY 8.5. Let C be an additive code of $\mathbb{Z}_{p^k}^m$, then $\gamma(C)$ is a regular unresricted code of $\mathbb{F}_p^{p^{k-1}m}$. More precisely it has the same distance distribution as C.

Proof: Indeed an additive code is regular. □

REMARK 8.6. The first eigenvalues of $(X_\oplus, R)$ are the same as those of (X_+, R), since those association schemes are isomorphic. They can be computed with the characters of the Abelian group $(\mathbb{Z}_{p^k}, +)$. Notice that the first eigenvalues $\lambda_{\ell i} =$

$P_\ell(i)$ of that scheme are, by (4.3), which is given by Theorem 2.4 of [**8**] and Theorem 2.2 of [**10**],

$$P_\ell(i) = \sum_{x \in X_\ell} < -x, z >= \sum_{x \in X_\ell} < x, z >= \lambda_{\ell i},\ \forall z \in X_i'. \tag{8.2}$$

We can thus compute the same eigenvalues by using the character table of the group $(\mathbb{Z}_{p^n}, +)$. Denoting by θ a p^k-th root of unity in $\mathbb{C}$, we thus have for every $z \in p^{i-1}\mathbb{Z}_{p^k} \setminus p^i\mathbb{Z}_{p^k}$,

$$\lambda_{\ell i} = \sum_{x \in p^{\ell-1}\mathbb{Z}_{p^k} \setminus p^\ell \mathbb{Z}_{p^k}} \theta^{xz \bmod p^k},\ i = 0, \dots, k-1,$$

where $\lambda_{\ell i}$ is given by (4.8). Following the authors of [**9**], we observe that those numbers are some of the Ramanujan's trigonometric sums.

8.2. Examples for $p = 2$.

8.2.1. *Gray maps from $\mathbb{Z}_{2^k}$ with k weights.* A generalization of the generator matrix (6.3) is as follows. Let G_k be the generator matrix for the code of rank k. Then the general form of G_k, with k rows is

$$G_k = \begin{bmatrix} & 0 \dots 0 \\ X_1 & 0 \dots 0 \\ & G_{k-2} \end{bmatrix},$$

where X_1 is the set of 2^{k-1} columns with the form $(c_0, \dots, c_{k-1})^T$ with $c_0 = 1$. Clearly such a generator matrix is that of the code $C(E_\pi)$ of a preprojective multiset of $\mathbb{F}_p^k$ compatible with the classes $X_{\oplus 1}, \dots, X_{\oplus k}$ of $(X_\oplus, \oplus)$. Thus X_1 is the set of 2^{k-1} columns with the form $(c_0, \dots, c_{k-1})^T$ for $c_0 = 1$. The codeword corresponding to the linear form $[1, 0, \dots, 0]^T$ which is the only one in the set X_k' has weight 2^{k-1}. Those corresponding to linear forms in X_{k-1}' have weight 2^{k-2}. Then in general, with the notation $w_k(i)$ for the weight of a codeword for the generator matrix G_k and corresponding to a linear from in X_i', we have

$$w_k(k) = 2^{k-1}, w_k(k-1) = 2^{k-2},$$

$$w_k(k-2) = 2^{k-2} + w_{k-2}(k-2), \dots, w_k(1) = 2^{k-2} + w_{k-2}(1)$$

The weights are distinct, by induction. For the length n_k of the code with generator matrix G_k, we have the recurrence $n_k = 2^{k-1} + n_{k-2}$. Then $n_{2t+1} = 2^{2t} + 2^{2t-2} + \dots + 1$ and $n_{2t} = 2^{2t-1} + 2^{2t-3} + \dots + 2$.

8.2.2. *An extended cyclic code over $\mathbb{Z}_8$.* We find in [**35**, Lemma 5.2] an easy tool to handle the Hensel Lemma for factoring polynomials over a Galois ring. We see that, in $\mathbb{Z}_8[x]$:

$$x^7 - 1 = (x-1)(x^3 + 6x^2 + 5x + 7)(x^3 + 3x^2 + 2x + 7).$$

Notice that this is a natural way of extending over $\mathbb{Z}_8$ the factorization over $\mathbb{Z}_4$ of $X^7 - 1$ which introduces the remarkable construction of Kerdock and Preparata codes in [**13**]. We then consider the generator matrix G

$$\begin{bmatrix} 7 & 5 & 6 & 1 & 0 & 0 & 0 & 5 \\ 0 & 7 & 5 & 6 & 1 & 0 & 0 & 5 \\ 0 & 0 & 7 & 5 & 6 & 1 & 0 & 5 \\ 0 & 0 & 0 & 7 & 5 & 6 & 1 & 5 \end{bmatrix}$$

and we compute with Maple the weight enumerator given in Defintion 5.3 were the class $\mathcal{F}_i$ is here the class X_{+i} of the Abelian scheme (X_+, R) introduced in Section 7.2. Here $p = 2$ and $k = 3$. The result is

$$T_0^8 + 14T_0^4T_3^4 + 112T_0^3T_2^4T_3 + 224T_0^3T_1^4T_2 + 672T_0^2T_1^4T_2T_3 + 112T_0T_2^4T_3^3 +$$

(8.3)

$$672T_0T_1^4T_2T_3^2 + 896T_0T_1^4T_2^3 + 896T_1^4T_2^3T_3 + 224T_1^4T_2T_3^3 + 256T_1^8 + 16T_2^8 + T_3^8.$$

The term $672T_0^2T_1^4T_2T_3$, for example shows that there are 672 codewords with 4 components in $\{1, 3, 5, 7\}$, 1 component in $\{2, 6\}$ and 1 component equal to 4. Now we compute the second eigenmatrix of (X_+, R) by using the character table $[< x, y >= \theta^{xy}]$ introduced in Section 7.2.

$$\begin{bmatrix} 1 & 4 & 2 & 1 \\ 1 & 0 & 0 & -1 \\ 1 & 0 & -2 & 1 \\ 1 & -4 & 2 & 1 \end{bmatrix}$$

After the substitution described in Theorem 5.5, where $|Y| = 4096$ we obtain the weight enumerator of the dual code Y° over $\mathbb{Z}_8$ which is the same as (8.3) as expected. Now the binary expansion of that code for the map

$$0 \mapsto 0000, 1 \mapsto 0011, 3 \mapsto 0110, 5 \mapsto 1100, 7 \mapsto 1001, 2 \mapsto 0101, 6 \mapsto 1010,$$
$$4 \mapsto 1111$$

is a $(32, 2^{12}, 10)$ unrestricted code. The binary weight enumerator is thus

$$B_0^{32} + 224B_0^{22}B_1^{10} + 112B_0^{20}B_1^{12} + 1568B_0^{18}B_1^{14} + 286B_0^{16}B_1^{16}$$
$$+1568B_0^{14}B_1^{18} + 112B_0^{12}B_1^{20} + 224B_0^{10}B_1^{22} + B_1^{32}.$$

As for the map

$$0 \mapsto 00000, 1 \mapsto 00111, 3 \mapsto 01101, 5 \mapsto 11001, 7 \mapsto 10011,$$
$$2 \mapsto 01010, 6 \mapsto 10100, 4 \mapsto 11110$$

we have a $(40, 2^{12}, 12)$ unrestricted code whose weight enumerator is

$$B_0^{40} + 112B_0^{28}B_1^{12} + 224B_0^{26}B_1^{14} + 30B_0^{24}B_1^{16} + 1568B_0^{22}B_1^{18} + 112B_0^{20}B_1^{20}$$
$$+1568B_0^{18}B_1^{22} + 256B_0^{16}B_1^{24} + 224B_0^{14}B_1^{26} + B_0^8B_1^{32}.$$

Here $B_0^2B_1^3$ substitute for T_1, $B_0^3B_1^2$ for T_2, $B_0B_1^4$ for T_4 and B_0 for T_0.

We can also map $\mathbb{Z}_8^8$ to $H(3,2)^8$, using the Abelian scheme on $\mathcal{L} = (\mathbb{Z}_4, +) \times (\mathbb{Z}_2, +)$ described in the end of Section 3. Indeed a homomorphism from the group $(\mathbb{Z}_8, \oplus)$ to the group $\mathcal{L}$ maps an additive code to an additive code. Then the map from $\mathcal{L}$ to $H(3,2)$ defined at the end of Section 3 yields a regular unrestricted code in $H(3,2)^8$. The first map has a kernel of 16 codewords. Thus the weight enumerator of the regular code obtained in $H(24,2)$ is

$$B_0^{24} + 14B_0^{16}B_1^8 + 112B_0^{14}B_1^{10} + 112B_0^{14}B_1^{10} + 17B_0^8B_1^{16}.$$

9. A group acting on the set of bijections

We here consider passing from a bijection from $\mathbb{Z}_{p^k}$ to $\mathbb{F}_p^k$ to another yielding a merging which would be an isomorphic Abelian scheme. Let σ be the mapping from $X_+ = \mathbb{Z}_{p^k}$ to $X_\oplus = \mathbb{F}_p^k$ defined by (7.1). Let us denote by μ_+ an automorphism of the Abelian group $(X_+, +)$ and by $\mu_\oplus$ an automorphism of $(X_\oplus, \oplus)$. Then the sets $\mu_+(X_1) = X_1, \ldots, \mu_+(X_k) = X_k$ (resp. $\mu_\oplus(\sigma(X_1)), \ldots, \mu_\oplus(\sigma(X_k))$ are the Abelian classes of an Abelian scheme on $(X_+, +)$ (resp. $(X_\oplus, \oplus)$) isomorphic to (X_+, R) (resp. to $(X_\oplus, R_\sigma)$), as observed in Section 2.3. As a consequence the mapping $\mu_\oplus \circ \sigma \circ \mu_+$ is an isomorphism from (X_+, R) to an Abelian scheme on $(X_\oplus, \oplus)$. Thus the group $\mathbb{Z}_{p^k}^* \times GL(k,p)$ acts on the set of bijections yielding isomorphic mergings with k classes. We denote by $\Sigma(k,p)$ the orbit of σ under the action of that group.

EXAMPLE 9.1. Example 2.13 is obtained for $p = 2$ and $k = 3$ with for μ_+ the multiplication by 3 modulo 8 and for $\mu_\oplus$ the mapping τ defined in relation (7.3). The size of $\Sigma(k,p)$ can be determined if the subgroup consisting of all couples $\mu_\oplus, \mu_+$ such that $\mu_\oplus \circ \sigma \circ \mu_+ = \sigma$ is known. For $p = 2$ and $k = 3$ then besides the identity, $\mu_\oplus \circ \sigma \circ \mu_+ = \sigma$ only for $\mu_\oplus = \begin{bmatrix} 1 & 0 & 0 \\ 0 & 1 & 0 \\ 1 & 0 & 1 \end{bmatrix}$ and where μ_+ is the multiplication by 5 modulo 8. This is easily established because $\mu_\oplus$ should be a lower triangular matrix in order that $\mu_\oplus^{-1} \circ \sigma = \sigma \circ \mu_+$ and multiplication by 5 modulo 8 is the only μ_+ that operates like that matrix.

Since the size of $GL(3,2)$ is $7.6.4 = 168$ then the size of $\Sigma(3,2)$ is $\frac{4.168}{2} = 336$.

9.1. All possible Gray maps for $p = 2$ and $k = 3$. With maple we have computed all possible mergings of the Hecke scheme $\mathcal{H}(X_+)$ with the Hecke scheme $\mathcal{H}(X_\oplus)$. This is done by applying the merging algorithm of [**4**, Section 4.13.1] to every bijection leaving 0 fixed from $\mathbb{Z}_{2^3}$ to $\mathbb{F}_2^3$. There are 1008 trivial merging, i.e. yielding the Abelian scheme with only one class $\{1,2,3,4,5,6,7\}$. Then 672 bijections yield Abelian schemes with two classes: $\{1,3,5,7\}$ and $\{2,6,4\}$. Finally 336 bijections yield Abelian schemes with three classes: $\{1,3,5,7\}, \{2,6\}$ and $\{4\}$. Starting from the mappping σ, which yields the Abelian classes $\{100,110,101,111\}$, $\{010,011\}$ and $\{001\}$ on $(\mathbb{F}_2^3, \oplus)$, we let $\mathbb{Z}_{p^k}^* \times GL(k,p)$ act on σ and we obtain $|\Sigma(3,2)| = 336$ distinct bijections leaving 0 fixed from $\mathbb{Z}_{2^3}$ to $\mathbb{F}_2^3$ for which the mergings yield three Abelian classes. This means that every possible bijection for which the merging has three classes is obtained by letting $\mathbb{Z}_{p^k}^* \times GL(k,p)$ act on σ. Notice that since the group $\mathbb{Z}_{2^3}^*$ preserves the classes $\{1,3,5,7\}, \{2,6\}$ and $\{4\}$ then all possible partitions into three Abelian classes are obtained by letting act $GL(3,2)$ on $\mathbb{F}_2^3$. Thus the sum of binary vectors in the class of size 4 always is 0. The computer program shows that for all mergings with the form $\{1,3,5,7\}, \{2,6,4\}$, then the sum of binary vectors in the class of size 4 also is 0. Thus there essentially exists one code C as that obtained by the generator matrix (6.3) and one as that introduced by C. Carlet in [**6**].

10. A class of permutation groups

10.1. The group $MG(k,p)$.

NOTATION 10.1. Let us denote by $\nu_\oplus$ (resp. μ_+) the generic element of the subgroup of lower triangular matrices in $GL(k,p)$ (resp. the generic element of the automorphisms of $(\mathbb{Z}_{p^k},+)$). The automorphism group of $(X_\oplus,\oplus)$ generated by those $\nu_\oplus$ is denoted by $Aut^*(X_\oplus,\oplus)$.

THEOREM 10.2. *The permutation $\mu_+ \circ \sigma^{-1} \circ \nu_\oplus \circ \sigma$ is an automorphism of the Abelian scheme (X_+,R).*

Proof: We have that $\nu_\oplus(X_{\oplus i}) = X_{\oplus i}$, $i = 0,1,\ldots,k$. Such a mapping is an automorphism of the Abelian scheme $(X_\oplus,R_\sigma)$, since

$$\forall (x,y) \in R_{\sigma,i} \Rightarrow x \ominus y \in X_{\oplus i} \Rightarrow \nu_\oplus(x) \ominus \nu_\oplus(y) \in X_{\oplus i} \Rightarrow (\nu_\oplus(x),\nu_\oplus(y)) \in R_{\sigma,i}.$$

But if $(\nu_\oplus(x),\nu_\oplus(y))$ belongs to $R_{\sigma,i}$, then, by definition of $R_{\sigma,i}$, $(\sigma^{-1}\circ\nu_\oplus(x),\sigma^{-1}\circ\nu_\oplus(y))$ belongs to R_i and it turns out that the permutation defined on X_+ by $\sigma^{-1}\circ\nu_\oplus\circ\sigma$ is an automorphism of (X_+,R). This also holds for the permutation $\mu_+\circ\sigma^{-1}\circ\nu_\oplus\circ\sigma$ where μ_+ is an automorphism of $(X_+,+) = (\mathbb{Z}_{p^k},+)$. □

Every X_{+i}, $i = 0,\ldots,k$ is a **fixed block** [**36**, §6] under the action of the group generated by

$$\{\mu_+\circ\sigma^{-1}\circ\nu_\oplus\circ\sigma\}_{\nu_\oplus\in Aut^*(X_\oplus,\oplus),\mu_+\in Aut(X_+,+)}.$$

That group will be denoted by $MG(k,p)$. Here MG is for merging group. That group contains a subgroup isomorphic to the group of units of $\mathbb{Z}_{p^k}$ and another isomorphic to a linear group on $\mathbb{F}_p^k$. This is a curious feature.

Example 2.17 suggest the following.

THEOREM 10.3. *The group $MG(k,p)$ acts on $\mathbb{Z}_{p^k}^*$ as a permutation group on X_{+1} which is properly contained in the symmetric group if $k\ge 2$ for $p>2$ and $k\ge 3$ for $p=2$.*

We show that the permutation $(p^{k-2}+1,p^{k-1}+1)$, which preserves X_{+1} is not an automorphism of (X_+,R). We have that $(p^{k-2}+1,1)$ is in R_{k-1} as well as $(p^{k-1}+1,p^{k-2}+1)$, since $p^{k-2}(p-1)\in X_{+(k-1)}$. But $(p^{k-1}+1,1)\in R_k$ and $(p^{k-2}+1,p^{k-1}+1)\in R_{k-1}$, which was to be shown.

REMARK 10.4. To show that $MG(k,p)$ is not the alternating group either, we exhibit in next theorem 10.6 transpositions which are automorphisms of the Abelian scheme (X_+,R).

For the group $MG(k,p)$ not to be trivial, its action should not be reduced to that of $Aut^*(X_\oplus,\oplus)$ on $(\mathbb{Z}_{p^k},+)$, i.e. the action of

$$\{\sigma^{-1}\circ\nu_\oplus\circ\sigma\}_{\nu_\oplus\in Aut^*(X_\oplus,\oplus)}.$$

We consider the following example which shows that the action of $MG(k,p)$ on $\mathbb{Z}_{32}^*$ is not reduced to linear maps on $\mathbb{F}_2^5$.

EXAMPLE 10.5. Let $p=2$ and $k=5$. we consider the matrix $\nu_\oplus$ of the operator that maps $\{1,3,5,9,17\}$ to $\{3,9,15,27,19\}$.

$$\nu_\oplus = \begin{bmatrix} 1&1&1&1&1\\ 1&0&1&1&1\\ 0&0&1&0&0\\ 0&1&1&1&0\\ 0&0&0&1&1 \end{bmatrix} \begin{bmatrix} 1&1&1&1&1\\ 0&1&0&0&0\\ 0&0&1&0&0\\ 0&0&0&1&0\\ 0&0&0&0&1 \end{bmatrix} = \begin{bmatrix} 1&0&0&0&0\\ 1&1&0&0&0\\ 0&0&1&0&0\\ 0&1&1&1&0\\ 0&0&0&1&1 \end{bmatrix}$$

Then $\sigma^{-1} \circ \nu_{\oplus} \circ \sigma$ acts on $\mathbb{Z}_{32}^*$ as follows:

$$(1,3,9,27)(5,15,29,7)(11,17,19,25)(13,23,21,31).$$

We now consider on $\mathbb{Z}_{32}^*$ the permutation $\mu_+ : x \mapsto 3x \bmod 32$:

$$(1,3,9,27,17,19,25,11)(5,15,13,7,21,31,29,23)$$

which also maps the basis $\{1,3,5,9,17\}$ to $\{3,9,15,27,19\}$ but maps 7 on 21 instead of 5, for example.

10.2. The group $ZF(k,p)$.

THEOREM 10.6. *The transposition $(x, x+\alpha p^{k-1}), x \in \mathbb{Z}_{p^k} \setminus \{0\},\ \alpha < p$ is an automorphism of the Abelian scheme (X,R).*

Proof: For $\alpha < p$ and $0 < i < k :\ X_i + \alpha p^{k-1} = X_i$. Indeed if $x \in X_i$ then $x_{i-1} \neq 0$ and for $y = x + \alpha p^{k-1}$, then $y_\ell = x_\ell,\ \ell = 0, \ldots, i-1,\ y_{i-1} \neq 0$ and thus $y \in X_i$, by relation (7.2). We thus have that if $(x,y) \in R_i,\ 0 < i < k$, then $(x + \alpha p^{k-1}, y)$ (resp. $(y, x + \alpha p^{k-1})$ belongs to R_i, since $x - y + \alpha p^{k-1}$ is in X_i (resp. $y - x - \alpha p^{k-1}$) is in X_i. This is for $0 < i < k$. Now a couple belongs to R_k if and only if it is written $(x + \beta p^{k-1}, x),\ \beta \neq 0$. If x is replaced with $y = x + \alpha p^{k-1}$, then $(y + \beta p^{k-1}, y)$ still is in R_k. □

NOTATION 10.7. The group obtained from $MG(k,p)$ by extending that automorphism group of (X,R) with the transpositions introduced in theorem 10.6 is denoted by $ZF(k,p)$. Here ZF is for $\mathbb{Z}$ and $\mathbb{F}_p$.

10.3. A decreasing chain of permutation groups. In next statement we only deal with the case $p = 2$ for simplicity. The general situation needs considering projective points.

THEOREM 10.8. *The automorphism group of relation R_1 acts on $\mathbb{Z}_{2^k}$ as the symmetric group.*

Proof: Let $x, y \in \mathbb{Z}_{2^k}^* = X_{+1}$ and $y = x + z$. Then $z = 2z', z' \in \mathbb{Z}_{2^k}$. By the same argument as that in the proof of theorem 10.6, the transposition $(x, x+z)$ is seen to preserve R_1. Since (x,y) is an automorphism of R_1 for any two x and y of X_{+1}, then the whole symmetric group acts on X_{+1} as an automorphism of group of R_1. □

NOTATION 10.9. We denote by $ZF_t(k,p)$ the group of permutations of X_+ which are automorphisms of each relation in the set $\{R_1, \ldots, R_t\},\ t \in [1,k]$.

Thus $ZF(k,p) = ZF_k(k,p)$.

THEOREM 10.10. *The generators of $ZF_t(k,p)$ are given by:*

1 *The linear maps on $X_\oplus$ defined by all invertible matrices whose first t rows are respectively*

$$(a_0, 0, \ldots, 0), (a_0, a_1, 0, \ldots, 0), \ldots, (a_0, \ldots, a_{t-1}, 0, \ldots, 0).$$

2 *The units of $\mathbb{Z}_{p^k}$ acting on X_+.*
3 *The transpositions with the form $(x, x + p^i z),\ x \in X_\ell,\ \ell \leq t,\ i \geq t-1$.*

References

[1] Bannai, E., Ito, T., (1984) Algebraic Combinatorics, The Benjamin / Cummings Publishing Company, Inc.

[2] Bose, R.C. and Shimamoto, T. (1952) *Classification and analysis of partially balanced incomplete block designs with two associate classes*, J. Amer. statist. Assoc. **47**,151-184.

[3] Brouwer, A.E, Cohen A.M. and Neumaier A. (1989) Distance-Regular Graphs, Springer-Verlag Berlin Eidelberg.

[4] Camion, P., (1998) *Codes and Association Schemes*, Vol II, chap. 18, in Handbook of Coding theory, edited by Vera Pless, W. Cary Huffman and Richard A, Brualdi, Elsevier Science Publisher B.V., Amsterdam, 1998.

[5] Camion, P., Courteau, B, Montpetit, A (1999) *Metric Coset Schemes Revisited*, Ann. Inst. Fourier, Grenoble **49**, 3, 829-859.

[6] Carlet, C. (1998) $\mathbb{Z}_{2^k}$*-Linear Codes* IEEE Transaction on Information theory, Vol **44**, No 4, 1543-1547.

[7] Delsarte, P. (1973) *An Algebraic Approach to Association Schemes in Coding*, Philips Res. Repts Suppl. **10**

[8] Delsarte, P. (1973) *Four fundamental parameters of a code and their combinatorial significance*, Inform. Control **23**, 407-438.

[9] Ericson T.,Simonis J., Tarmanen H., Zinoviev V. (1997) *F-Partition sof Cyclic groups* AAEC-C **8**,387-393.

[10] Goethals J-M.(1979) *Association Schemes*, in Algebraic Coding theory and Applications, edited by G.Longo, CISM courses and Lectures N0.**258**, Springer-Verlag Wien, New-York.

[11] Godsil, C. D. (1993) Algebraic Combinatorics, Chapman and Hall,New York, London.

[12] M. Greferath and S. E. Schmidt. (1999) *Gray isometries over finite chain rings* IEEE Transaction on Information theory, Vol **45**, No 7, 2522-2524.

[13] A.Roger Hammons, Jr, P.Vijay Kumar, A.R. Calderbank and N.J.A. Sloane, Patrick Solé (1994) *The* $\mathbb{Z}_4$*-Linearity of Kerdock, Preparata, Goethals and Related Codes*, IEEE Trans. Inf. theory **40**.301-319.

[14] T. Honold and I. Landjev (1998) *Linear codes over finite chain rings* In Optimal Codes and Related Topics, 116-126, Sozopol, Bulgaria.

[15] T. Honold and A. A. Nechaev (1998) *Weighted modules and representtion of codes* Proceedings of the ACCT 6, Pskov, Russia, 123-129.

[16] T. Honold and I. Landjev (1998) *Linearly Representable Codes over Chain Rings* Proceedings of the ACCT 6, Pskov, Russia, 135-141.

[17] T. Honold and I. Landjev (1999) *All Reed-Muller codes are linearly representable over the ring of dual numbers over* $\mathbb{Z}_2$ IEEE Transaction on Information theory, Vol **45**, No 2, 700-701.

[18] T. Honold and I. Landjev (1999) *Linearly representable codes over chain rings* Abhandlungen aus dem mathematischen Seminar der Universität Hamburg,Vol **69**, 187-203.

[19] T. Honold and A. A. Nechaev (1999) *Weighted modules and representtion of codes* Problems of Information Transmission, **35**, 3, 205-223.

[20] T. Honold and A. A. Nechaev (2000) *Linear Codes over Finite Chain Rings* The Electronic Journal of Combinatorics **7** #R11.

[21] Helleseth T. and Kumar P. V.(1998) *Sequences with Low Correlation*, in Handbook of Coding theory, edited by Vera Pless, W. Cary Huffman and Richard A, Brualdi. , Elsevier Science Publisher B.V., Amsterdam.

[22] Jaeger, F., Matsumoto M., Nomura, K., (1998) *Bose-Mesner Algebras Related with Type II matrices and Spin Models* J. Algebraic Combin. **8**, No 1, 39-72.

[23] Lang, S. (1984) Algebra, Addison Wesley, Menlo Park, California.

[24] Marcus, M. and Minc, H.(1992), A survey of matrix theory and matrix inequalities, Dover Publications, New-York.

[25] A. A. Nechaev (1996) *Linear codes over modules and over spaces. MacWilliams identities* In Proceeding of the IEEE Int. Symp. Inf. Theory and Appl., 35-38, Victoria B. C. Canada.

[26] A. A. Nechaev and A. S. Kuzmin (1996) *Linearly pressentable codes* In Proceeding of the IEEE Int. Symp. Inf. Theory and Appl., 31-34, Victoria B. C. Canada.

[27] A. A. Nechaev and A. S. Kuzmin *Formal duality of linear presentable codes over a Galois field*

[28] A. A. Nechaev, A. S. Kuzmin and V. T. Markov (1995) *Linear codes over finite rings and modules* Preprint N 1995-6-1, Center of New Information Technologies, Moscow State University.

[29] Semakov N.V., Zinoviev V.A. and Zaitsev C.V. (1971) *Uniformly packed codes*, Probl. Peredach. Inform., **7**, N0 **1**, 38-50

[30] Semakov N.V., Zinoviev V.A. and Zaitsev C.V. (1972) *On duality between Preparata and Kerdock codes*, in " Fifth All-union Conf. on Coding theory" Moscow-Gorky, part 2, 55-58.

[31] A. Sălăgean-Mandache (1999) *On the Isometries between* $\mathbb{Z}_{p^k}$ *and* $\mathbb{Z}_p^k$ IEEE Transaction on Information theory, Vol **45**, No 6, 2146-2148.

[32] MacWilliams, F.J. and Sloane N.J.A. (1977) The theory of Error-Correcting Codes, North-Holland.

[33] Tamaschke O. (1963) *Zur theoremrie der Permutationsgruppen mit regulärer Untergruppen*, Math. Z. **80** , 328-352

[34] Tamaschke O. (1970) Schur-Rings, B.I-Hochschulskripten **735** a^*, Bibliographisches Institut, Mannheim.

[35] Wan Zhe-Xian (1997) Quaternary Codes, World Scientific.

[36] Wielandt,H. (1964) Finite Permutation Groups, Academic Paperback.

Centre National de la Recherche Scientifique, Equipe Combinatoire, Paris 6, 175 rue du Chevaleret, F-75013 Paris

E-mail address: `Paul.Camion@ecp6.jussieu.fr,Camion-Plaisir@wanadoo.fr`

DIMACS Series in Discrete Mathematics
and Theoretical Computer Science
Volume **56**, 2001

Bent, resilient Functions and the Numerical Normal Form

Claude Carlet and Philippe Guillot

Abstract. We carry on the study of the representation of Boolean functions as polynomials with numerical coefficients (the N.N.F.). We give a simple characterization of resilient functions by means of the degree of this polynomial representation. The N.N.F.leads also to a characterization of bent functions as integer solutions of a particular quadratic equation. We relate this representation to an affine invariant of Boolean functions. This invariant is more precise than the algebraic degree and allows another characterization of bent functions.

1. Preliminaries

Let n be any positive integer. We are interested, from coding theoretic and cryptographic viewpoint, in Boolean (*i.e.* $\{0,1\}$-valued) functions on the set $\mathbb{F}_2^n$ of all binary words of length n. This set is a vector space of dimension n over the two elements field $\mathbb{F}_2$. For any Boolean function f on $\mathbb{F}_2^n$, let $\chi_f = (-1)^f = 1 - 2f$ denote the $\{-1,1\}$-valued function corresponding to f. Since $\{0,1\}$ can be viewed either as a subset of $\mathbb{Z}$ or as the field $\mathbb{F}_2$, we need to distinguish between the addition in $\mathbb{Z}$ denoted by $+$ and the addition in $\mathbb{F}_2$ denoted by $\oplus$. The vector addition in $\mathbb{F}_2^n$ will be denoted $+$.

1.1. Numerical Normal Form. The recent progresses in the study of bent functions ([**4, 5, 6, 11, 9**]) have shown that they are better viewed as integer-valued functions than as $\mathbb{F}_2$-valued functions. This viewpoint has led to several new characterizations of bent functions, which may yield to their classification under the action of the general affine group. This change of viewpoint has led the authors to study in [**7**] a representation of Boolean functions (and more generally of real-valued functions) similar to the Algebraic Normal Form, but with numerical coefficients (the N.N.F.). This representation was already used at that time in the literature (in the field of circuit complexity for instance, see [**1, 3, 14, 15**]). The authors ignored this fact. Anyway, they made a systematic study of the properties of the N.N.F. and of the relations between the N.N.F. and all the other representation used in coding and in crytpography (such systematic work had not been done until

1991 *Mathematics Subject Classification.* 06E30, 11T71, 94A60.

Key words and phrases. Boolean Functions, Resilient Functions, Nonlinearity, Stream Cipher.

then). Moreover, they obtained a characterization of bent functions by means of the N.N.F.

We show in the present paper that the N.N.F. allows a simple characterization of resilient functions as well. We also give a characterization, with a single formula, of the fact that an integer vector $(\lambda_u)_{u\in\mathbb{F}_2^n}$ corresponds to the N.N.F. of a bent function. We finally derive an affine invariant generalizing the algebraic degree, which leads to a second characterization of bent functions.

DEFINITION 1.1. Let f be a real-valued function on $\mathbb{F}_2^n$. We call Numerical Normal Form (N.N.F.) of f, the following expression of f as a polynomial with real coefficients

$$f(x_1,\dots,x_n) = \sum_{u\in\mathbb{F}_2^n} \lambda_u \left(\prod_{i=1}^{n} x_i^{\,u_i}\right) = \sum_{u\in\mathbb{F}_2^n} \lambda_u\, x^u. \tag{1.1}$$

We call numerical degree of f the (global) degree of its N.N.F.

In the above formula, the vector u has binary coefficients and thus, the N.N.F. is a polynomial expression in which each variable has degree at most 1.

Any real-valued function admits a unique Numerical Normal Form. Moreover, a function is integer-valued if and only if its N.N.F. has integer coefficients (see [**7**]).

The N.N.F. can be viewed in the general framework of Möbius transform (see [**17**], [**20**]). $\mathbb{F}_2^n$ is a lattice, with partial order relation $\preceq$ defined by:

$$\forall u=(u_1,\dots,u_n), v=(v_1,\dots,v_n)\in\mathbb{F}_2^n \quad u\preceq v \iff \forall i\in\{1,\dots,n\}\ u_i\le v_i$$

The N.N.F. coefficients of a function f are deduced from the values of the so called *Möbius transform* of f:

$$\forall u\in\mathbb{F}_2^n \quad \lambda_u = (-1)^{\mathrm{w}(u)} \sum_{x\in\mathbb{F}_2^n \,|\, x\preceq u} (-1)^{\mathrm{w}(x)} f(x).$$

EXAMPLE 1.2 (Quadratic functions). If $f(x)=\sum_{u\in\mathbb{F}_2^n}\lambda_u x^u$ is quadratic, then λ_u is either null or equal to ±1 if $u=0$ and to $\pm 2^i$, where $\frac{\mathrm{w}(u)-2}{2}\le i\le \mathrm{w}(u)-1$ otherwise (see [**7**]).

EXAMPLE 1.3 (Flat indicator). Let E be a flat of $\mathbb{F}_2^n$ and $f(x)=\sum_{x\in\mathbb{F}_2^n}\lambda_u x^u$ be the N.N.F. of the indicator of E (*i.e.* the Boolean function whose support is E). Since one has:

$$\lambda_u = (-1)^{\mathrm{w}(u)} \sum_{x\in E \,|\, x\preceq u} (-1)^{\mathrm{w}(x)}$$

λ_u is either equal to $\pm 2^k$ (where k is the dimension of the flat $\{x\in E\,|\,x\preceq u\}$) if $\mathrm{w}(x)$ is constant on $\{x\in E\,|\,x\preceq u\}$, or null otherwise. Notice that we have $\dim E+\mathrm{w}(u)-n\le k\le \dim E$, since $\dim\{x\in E\,|\,x\preceq u\} = \dim E + \dim\{x\in \mathbb{F}_2^n\,|\,x\preceq u\} - \dim(E+\{x\in\mathbb{F}_2^n\,|\,x\preceq u\})$.

1.2. Fourier transform. Let f be any real-valued function on $\mathbb{F}_2^n$. The *Fourier transform* of f is by definition the function $\widehat{f}$ defined by

$$\forall u\in\mathbb{F}_2^n \quad \widehat{f}(u) = \sum_{x\in\mathbb{F}_2^n} f(x)(-1)^{u\cdot x},$$

where $u\cdot x$ denotes the usual dot product $u\cdot x = u_1x_1\oplus\cdots\oplus u_nx_n$. The Fourier transform is a very important tool for studying Boolean functions. It corresponds to the expression of f in the orthogonal basis of the so called Walsh functions $\chi_u : x \mapsto (-1)^{u \cdot x}$. The Parseval's equality states that for any real-valued function f:

$$\sum_{u \in \mathbb{F}_2^n} \widehat{f}^2(u) = 2^n \sum_{x \in \mathbb{F}_2^n} f^2(x).$$

Let us recall now the link between the N.N.F. and the Fourier transform (see [**7**]). Let $f(x) = \sum_{u \in \mathbb{F}_2^n} \lambda_u x^u$ be any real-valued function f. For each vector $a \in \mathbb{F}_2^n$, the coefficients λ_u of the N.N.F. of f and the values of the Fourier transform of f are related through following relations (see [**7**]):

$$\widehat{f}(a) = (-1)^{\mathrm{w}(a)} \cdot \sum_{u \in \mathbb{F}_2^n \mid a \preceq u} 2^{n-\mathrm{w}(u)} \lambda_u; \tag{1.2}$$

$$\widehat{\chi_f}(a) = 2^n \delta_0(a) + (-1)^{\mathrm{w}(a)+1} \sum_{u \in \mathbb{F}_2^n \mid a \preceq u} 2^{n-\mathrm{w}(u)+1} \lambda_u, \tag{1.3}$$

where δ_0 denotes the function equal to 1 on the zero vector and 0 elsewhere. Notice that relation (1.2) for $a = 0$ says that the mean of f (*i.e.* the weight of f divided by the size of $\mathbb{F}_2^n$) equals the value of the N.N.F. of f at the vector $(\frac{1}{2}, \ldots, \frac{1}{2})$, which is the mean of all binary vectors. Notice also that, according to (1.2), the numerical degree of f equals the largest possible weight of a such that $\widehat{f}(a) \neq 0$.

The converse of relation (1.2) is (see [**7**]):

$$\lambda_u = 2^{-n}(-2)^{\mathrm{w}(u)} \sum_{x \in \mathbb{F}_2^n \mid u \preceq x} \widehat{f}(x). \tag{1.4}$$

1.3. Resilient functions. A Boolean function f on $\mathbb{F}_2^n$ is called k-th order correlation immune (resp. k-resilient) with $1 \leq k \leq n$ if for every non-zero word (resp. every word) b of Hamming weight smaller than or equal to k the value of $\widehat{\chi_f}(b)$ is null (see [**21**]). The cryptographic Boolean functions used as combining functions of several Linear Feedback Shift Registers in stream ciphers must satisfy this criterion with high order to resist the correlation attack (see [**19**]).

1.4. Bent functions. A Boolean function on $\mathbb{F}_2^n$ (n even) is called *bent* if the Fourier transform of χ_f has constant magnitude equal to $2^{\frac{n}{2}}$. Bent functions are fascinating extremal objects related to various problems of combinatorial theory, coding and cryptography (see [**18**], [**8**], [**10**], [**11**], [**2**], [**5**], [**6**]). Many problems about these functions such as counting and classifying them under the action of the affine group still remain open.

If a Boolean function f is bent, then the Boolean function $\widetilde{f}$ defined on $\mathbb{F}_2^n$ by:

$$\widehat{\chi_f}(a) = 2^{\frac{n}{2}}(\chi_{\widetilde{f}})(a)$$

is also bent. It is called the *dual* of f. The dual of $\widetilde{f}$ is f itself. The mapping $f \mapsto \widetilde{f}$ is an isometry (the Hamming distance between two bent functions is equal to the Hamming distance between their duals).

1.5. Affine invariants. As the number of Boolean functions is incredibly huge, it is interesting to classify them into classes of functions which have similar properties. This is usually done through the action of affine groups.

Let us recall that the affine group of $\mathbb{F}_2^n$ is the group of all affine isomorphisms on $\mathbb{F}_2^n$ together with the law of composition. The general affine group of $\mathbb{F}_2^n$ is the direct product of the affine group of $\mathbb{F}_2^n$ and of the group of affine forms on $\mathbb{F}_2^n$ together with the addition.

These two groups act on the set $\mathcal{B}$ of all Boolean functions as follows: for any affine isomorphism A, any affine form ℓ and any $f \in \mathcal{B}$:

$$\begin{aligned} A \cdot f &: x \mapsto (f \circ A)(x) \\ (A, \ell) \cdot f &: x \mapsto (f \circ A)(x) \oplus \ell(x) \end{aligned}$$

Two Boolean functions f and g are said to be *affinely equivalent* if they belong to the same orbit for the action of the affine group. Moreover, a mapping φ from the set $\mathcal{B}$ to any set E is called an *affine invariant* (respectively a *general affine invariant*) if it is constant on the orbits of $\mathcal{B}$ for the action of the affine group (respectively the general affine group) of $\mathbb{F}_2^n$.

EXAMPLE 1.4. The weight (*i.e.* the number of nonzero values) is an affine invariant, but it is not a general affine invariant.

EXAMPLE 1.5. The multiset of the magnitudes of the Fourier transform is a general affine invariant.

2. A characterization of resilient functions by means of their N.N.F.

PROPOSITION 1. *A Boolean function f on $\mathbb{F}_2^n$ is k-resilient if and only if the function $f(x) \oplus x_1 \oplus \cdots \oplus x_n$ has numerical degree at most $n-k-1$.*

PROOF. Denote by $g(x)$ the function $f(x) \oplus x_1 \oplus \cdots \oplus x_n = f(x) \oplus (w(x) \bmod 2)$. For each vector $a \in \mathbb{F}_2^n$, denote by $\bar{a}$ the componentwise complement of a equal to $a \oplus (1, \ldots, 1)$. We have $\widehat{\chi_f}(a) = \widehat{\chi_g}(\bar{a})$. Thus f is k-resilient if and only if, for each word b of weight greater than or equal to $n-k$, the number $\widehat{\chi_g}(b)$ is null. Consider the N.N.F. of g:

$$g(x_1, \ldots, x_n) = \sum_{u \in \mathbb{F}_2^n} \lambda_u \, x^u.$$

According to relations (1.3) and (1.4) applied to g, we have for nonzero a:

$$\widehat{\chi_g}(a) = (-1)^{\mathrm{w}(a)+1} \sum_{u \in \mathbb{F}_2^n \,|\, a \preceq u} 2^{n-\mathrm{w}(u)+1} \lambda_u,$$

and for nonzero u:

$$\lambda_u = 2^{-n}(-2)^{\mathrm{w}(u)-1} \sum_{x \in \mathbb{F}_2^n \,|\, u \preceq x} \widehat{\chi_g}(x).$$

We deduce that $\widehat{\chi_g}(b)$ is null for every word b of weight greater than or equal to $n-k$ if and only if g has numerical degree at most $n-k-1$. □

3. A characterization of bent functions by means of their N.N.F.

In [**5**] is given a characterization of bent functions by means of their Möbius transform relative to the dual order relation $\succeq$. In [**7**] is obtained a similar characterization by means of the N.N.F. We recall briefly its proof for self completeness. Relation (1.3) implies:

$$\chi_{\tilde{f}}(x) = 2^{\frac{n}{2}} \delta_0(x) + (-1)^{\mathrm{w}(x)+1} \sum_{u \in \mathbb{F}_2^n \,|\, x \preceq u} 2^{\frac{n}{2}-\mathrm{w}(u)+1} \lambda_u.$$

From $\widetilde{f} = \frac{1-\chi_{\widetilde{f}}}{2}$, we deduce then:

$$\widetilde{f}(x) = \frac{1}{2} - 2^{\frac{n}{2}-1}\delta_0(x) + (-1)^{\mathrm{w}(x)} \sum_{u\in\mathbb{F}_2^n \mid x\preceq u} 2^{\frac{n}{2}-\mathrm{w}(u)}\lambda_u.$$

We have $x \preceq u \Leftrightarrow \overline{u} \preceq \overline{x}$. Thus, changing u into $\overline{u}$, we have $x \preceq \overline{u} \Leftrightarrow u \preceq \overline{x} \Leftrightarrow \prod_{i=1}^n (1-x_i)^{u_i} = 1$, and we obtain the N.N.F. of $\widetilde{f}$ by expanding the following relation:

(3.1)

$$\widetilde{f}(x) = \frac{1}{2}x^{(0,\dots,0)} - 2^{\frac{n}{2}-1}\prod_{i=1}^n(1-x_i) + (-1)^{\mathrm{w}(x)}\sum_{u\in\mathbb{F}_2^n} 2^{\mathrm{w}(u)-\frac{n}{2}}\lambda_{\overline{u}}\prod_{i=1}^n(1-x_i)^{u_i}.$$

We deduce that, for every u such that $\mathrm{w}(u) > \frac{n}{2}$, $u \neq (1,\dots,1)$, the coefficient of x^u in the N.N.F. of $\widetilde{f}$ is divisible by $2^{\mathrm{w}(u)-\frac{n}{2}}$.

THEOREM 3.1. *Let f be a Boolean function on $\mathbb{F}_2^n$. Consider its N.N.F.*

$$f(x) = \sum_{u\in\mathbb{F}_2^n} \lambda_u\, x^u.$$

Then f is bent if and only if it satisfies:

- *for every u such that $\frac{n}{2} < \mathrm{w}(u) < n$, the coefficient λ_u is divisible by $2^{\mathrm{w}(u)-\frac{n}{2}}$;*
- *$\lambda_{(1,\dots,1)} \equiv 2^{\frac{n}{2}-1} \bmod 2^{\frac{n}{2}}$.*

PROOF. According to Lemma 1 of [**4**], f is bent if and only if, for every $a \in \mathbb{F}_2^n$, $\widehat{f}(a) \equiv 2^{\frac{n}{2}-1} \bmod 2^{\frac{n}{2}}$. Thus, according to relation (1.2), conditions 1. and 2. are sufficient for a Boolean function f to be bent.

Conversely, assume that f is bent. The observation above on the N.N.F. of the dual of a bent function, applied to $\widetilde{f}$ (whose dual is f) shows that condition 1 is necessary. Condition 2 is also necessary since $\widehat{f}(1,\dots,1) = (-1)^n\lambda_{(1,\dots,1)}$. □

4. Characterization of bent functions by means of a quadratic Diophantine equation

A necessary and sufficient condition for a real-valued function to be Boolean, is that all its values are equal to their squares. Since this condition has to be satisfied for the 2^n vectors of $\mathbb{F}_2^n$, it has high computational complexity. But in the case of integer valued functions, there exists a simpler necessary and sufficient condition: an integer-valued function f is Boolean if and only if:

$$\sum_{x\in\mathbb{F}_2^n} f^2(x) = \sum_{x\in\mathbb{F}_2^n} f(x) \tag{4.1}$$

i.e. if and only if $\widehat{f^2}(0) = \widehat{f}(0)$. We deduce, by applying relation (1.2) with $a = 0$ to the functions $f^2(x_1,\dots,x_n) = \sum_{v,v'\in\mathbb{F}_2^n} \lambda_v\,\lambda_{v'}\,x^{v\vee v'}$ (where $v \vee v'$ denotes the componentwise logical or of v and v') and $f(x_1,\dots,x_n) = \sum_{u\in\mathbb{F}_2^n} \lambda_u\,x^u$ that the $\mathbb{Z}$-polynomial $\sum_{u\in\mathbb{F}_2^n} \lambda_u\,x^u$ is the N.N.F. of a Boolean function if and only if the λ_u's satisfy

$$\sum_{u\in\mathbb{F}_2^n} 2^{n-\mathrm{w}(u)} \sum_{v,v'\in\mathbb{F}_2^n \mid u=v\vee v'} \lambda_v\,\lambda_{v'} = \sum_{u\in\mathbb{F}_2^n} 2^{n-\mathrm{w}(u)}\lambda_u. \tag{4.2}$$

This statement means that the components of the vector $(\lambda_u)_{u\in\mathbb{F}_2^n}$ of $\mathbb{Z}^{2^n}$ are the coefficients of the N.N.F. of a Boolean function if and only if they satisfy the quadratic Diophantine equation given by relation (4.2). Thus, the number of solutions of this equation is equal to 2^{2^n}.

A similar characterization exists for bent functions. Consider the basis $\left(\phi_u\right)_{u\in\mathbb{F}_2^n}$ of the real vector space of the real valued functions on $\mathbb{F}_2^n$ defined by

$$\begin{cases} \phi_u = x^u & \text{if } \mathrm{w}(u) \leq \frac{n}{2}; \\ \phi_u = 2^{\mathrm{w}(u)-\frac{n}{2}} x^u & \text{if } \frac{n}{2} < \mathrm{w}(u) < n; \\ \phi_{(1,\dots,1)} = 2^{\frac{n}{2}-1} x^{(1,\dots,1)}. \end{cases}$$

Theorem 3.1 means that bent functions are characterized by integer coefficients in the decomposition relatively to the above basis and by an odd coefficient of $\phi_{(1,\dots,1)}$. Note that according to relation (1.3) with $a = (1,\dots,1)$, this coefficient is then necessarily equal to ± 1.

More precisely, let $\left(z_u\right)_{u\in\mathbb{F}_2^n}$ be an integer vector. Then the function f defined by $f(x) = \sum_{u\in\mathbb{F}_2^n} z_u\phi_u(x)$ is integer valued. It is bent if and only if it is Boolean and $z_{(1,\dots,1)} = \pm 1$. By applying the characterization given by relation (4.1), one yields that f is Boolean if and only if the z_u's satisfy

$$\sum_{x\in\mathbb{F}_2^n}\sum_{u\in\mathbb{F}_2^n} z_u\phi_u(x)\sum_{v\in\mathbb{F}_2^n} z_v\phi_u(x) = \sum_{x\in\mathbb{F}_2^n}\sum_{u\in\mathbb{F}_2^n} z_u\phi_u(x)$$

$$\sum_{u,v\in\mathbb{F}_2^n} z_u z_v\left(\sum_{x\in\mathbb{F}_2^n}\phi_u(x)\phi_v(x)\right) = \sum_{u\in\mathbb{F}_2^n}\left(\sum_{x\in\mathbb{F}_2^n}\phi_u(x)\right).$$

In matrix notation indexed by $\mathbb{F}_2^n$, let Q denote the matrix whose general u,v-term is $\sum_{x\in\mathbb{F}_2^n}\phi_u(x)\phi_v(x)$ and ℓ the vector whose general u-term is $\sum_{x\in\mathbb{F}_2^n}\phi_u(x)$. Then one has proved the following result:

PROPOSITION 2. *Let $z = \left(z_u\right)_{u\in\mathbb{F}_2^n}$ be an integer vector. The function $f(x) = \sum_{u\in\mathbb{F}_2^n} z_u\phi_u(x)$ is bent if and only if*

1. $z_{(1,\dots,1)} = \pm 1$;
2. *the vector z is solution of the quadratic equation $x^tQx = \ell x$.*

Moreover, every bent function is obtained this way.

5. An affine invariant related to the N.N.F.

This section is devoted to the study of a general affine invariant which generalizes the notion of degree of a Boolean function in a similar way as the generalized distance of codes does with the minimum distance. This invariant is similar to the invariant introduced by HOU in [**10**], but thanks to the N.N.F., its definition appears naturally. Moreover, it allows to characterize bent functions in a rather simple way.

DEFINITION 5.1. Let f be a Boolean function on $\mathbb{F}_2^n$. We call *generalized degree* of f the sequence $\left(d_i\right)_{i\geq 1}$ defined as follows:

- d_1 is the algebraic degree of f;
- for every $i > 1$, d_i is the smallest integer $d > d_{i-1}$ such that, for every word u of weight greater than d, the coefficient λ_u of x^u in the N.N.F. of f is multiple of 2^i.

EXAMPLE 5.2. The generalized degree of any nonzero affine function is the sequence of all positive integers.

REMARK 5.3. By definition, the sequence $(d_i)_{i\geq 1}$ satisfies $d_{i+1} > d_i$, for all integer $i \geq 1$. Thanks to this condition, we shall prove that the generalized degree is an affine invariant. If we do not impose this condition, we have the sequence $\delta(f) = (\delta_i)_{i\geq 1}$ whose term δ_i is the smallest integer d such that for any word u of weight greater than d, the coefficient λ_u of x^u in the N.N.F. of f is divisible by 2^i. This sequence is not an affine invariant. A counter example is given by the following functions: $f(x_0, x_1, x_2) = x_0x_1$ and $g(x_0, x_1, x_2) = x_0(x_1 \oplus x_2) = x_0x_1 \oplus x_0x_2 = x_0x_1 + x_0x_2 - 2x_0x_1x_2$. The functions f and g are affinely equivalent, but one has $\delta(f) = (2, 2, \dots)$ and $\delta(g) = (2, 3, 3, \dots)$. With Definition 5.1, the generalized degree of f and g are both equal to the sequence $(2, 3, 4, \dots)$.

THEOREM 5.4. *The generalized degree is an affine invariant. Moreover, it is a general affine invariant if we restrict ourselves to the functions of algebraic degree greater than 1.*

PROOF. There are two steps in the proof. The first step is in proving that for any affine isomorphism A, one has $d_i(f \circ A) = d_i(f)$ for all integer i, and the second step is to prove that for any affine form ℓ, one has similarly $d_i(f \oplus \ell) = d_i(f)$.

For the first step, it is clearly sufficient to prove $d_i(f \circ A) \leq d_i(f)$ since applying this inequality to the function $f \circ A^{-1}$ yields the converse inequality. To prove $d_i(f \circ A) \leq d_i(f)$ for every i, it suffices to prove $\delta_i(f \circ A) \leq d_i(f)$ for every i (see definition of δ_i in remark 5.3). Indeed, if $d_i(f \circ A) \neq \delta_i(f \circ A)$ then $i > 1$ and $d_i(f \circ A) = d_{i-1}(f \circ A) + 1$. By induction $d_{i-1}(f \circ A) \leq d_{i-1}(f) < d_i(f)$. This implies $d_i(f \circ A) \leq d_i(f)$.

Let $f = \sum_{u\in\mathbb{F}_2^n} \lambda_u x^u$ be the N.N.F. of f. As the function $x \mapsto x^u$ is the indicator of the flat $[u, (1, \dots, 1)] = \{x \in \mathbb{F}_2^n \mid x \succeq u\}$, one has

$$f \circ A(x) = \sum_{u\in\mathbb{F}_2^n} \lambda_u \phi_{[u,(1,\dots,1)]}\big(A(x)\big) = \sum_{u\in\mathbb{F}_2^n} \lambda_u \phi_{A^{-1}\left([u,(1,\dots,1)]\right)}(x).$$

For all vector u, let $\sum_{v\in F_2^n} \mu_{u,v} x^v$ be the N.N.F. of the function $\phi_{A^{-1}\left([u,(1,\dots,1)]\right)}$. We have $f \circ A(x) = \sum_{u,v\in\mathbb{F}_2^n} \lambda_u \mu_{u,v} x^v$, *i.e.* the coefficient of x^v in the N.N.F. of $f \circ A$ is equal to $\sum_{u\in\mathbb{F}_2^n} \lambda_u \mu_{u,v}$.

To prove $\delta_i(f \circ A) \leq d_i(f)$ for every i, it suffices to prove that for each $i > 1$ and each v of weight greater than $d_i(f)$, the value of $\sum_{u\in\mathbb{F}_2^n} \lambda_u \mu_{u,v}$ is a multiple of 2^i. To do that, we show that all the terms of this sum of integers are multiple of 2^i.

- if $\mathrm{w}(u) > d_i(f)$, then by definition, λ_u is a multiple of 2^i;
- suppose now that $\mathrm{w}(u) \leq d_i(f)$ and let $k = d_i(f) - \mathrm{w}(u)$. We have $\mathrm{w}(u) > d_i(f) - k - 1$. From condition $d_{i+1}(f) > d_i(f)$, we have $\mathrm{w}(u) > d_{i-k-1}(f)$, thus λ_u is multiple of 2^{i-k-1}. But, $A^{-1}\big([u, (1, \dots, 1)]\big)$ being a $(n - \mathrm{w}(u))$-dimensional flat, from the results of example 1.3, $\mu_{u,v}$ is multiple of $2^{\mathrm{w}(v)-\mathrm{w}(u)}$. Finally, $\sum_{u\in\mathbb{F}_2^n} \lambda_u \mu_{u,v}$ is multiple of $2^{i-k+\mathrm{w}(v)-1-\mathrm{w}(u)} = 2^{i-1+\mathrm{w}(v)-d_i(f)}$. As we have $\mathrm{w}(v) > d_i(f)$, this quantity is multiple of 2^i.

Similarly, for the second step, it is sufficient to prove that for any affine form ℓ, we have $\delta_i(f \oplus \ell) \leq d_i(f)$. Thus, one has to establish that for each $i > 1$ and each vector u of weight $> d_i(f)$, the N.N.F. x^u coefficient of $f \oplus \ell$ is multiple of 2^i. We can assume without loss of generality that the affine form ℓ is equal to 1 or x_1.

If $\ell = 1$, then $f \oplus 1 = 1 - f$ and for each nonzero vector u, the N.N.F. x^u coefficient of $1 - f$ is equal to $-\lambda_u$ which is a multiple of 2^i by definition.

If $\ell = x_1$, then $f \oplus x_1 = f + x_1 - 2x_1 f$. Let u' denote the vector deduced from u by replacing the first term u_1 by zero. The N.N.F. x^u coefficients of $f \oplus x_1$ are

$$\begin{cases} \lambda_u & \text{if } u_1 = 0, \\ 1 - \lambda_u - 2\lambda_0 & \text{if } u = (1, 0, \dots, 0), \\ -\lambda_u - 2\lambda_{u'} & \text{if } u_1 = 1 \text{ and } u' \neq (0, \dots, 0). \end{cases}$$

In each case where $\mathrm{w}(u) \geq d_i(f)$, these values are multiples of 2^i (for instance if $u_1 = 1$, we have $\mathrm{w}(u') = \mathrm{w}(u) - 1$ and $d_i(f) > d_{i-1}(f)$). □

REMARK 5.5. The N.N.F. of f can be written $P(x) = \sum_{i=1}^{n} 2^{i-1} P_i(x)$, where the P_i's have odd coefficients. Then $d_i(f)$ is equal to $\max(d_{i-1}(f) + 1, d^\circ P_i)$, where we take $d^\circ P_i = 0$ if P_i is null. Thus, the sequence $d_1, \dots, d_i$ can be derived from the polynomial $P(x) \bmod 2^i$, which has an intermediate complexity between those of the A.N.F. $P_1(x)$ and of the N.N.F.

REMARK 5.6. One could think that the generalized degree shares the property $d(f \oplus g) \leq \max(d(f), d(g))$ with the classical degree $d = d_1$, but this is false. Take for instance, $f(x) = x_1 x_2$ and $g(x) = x_3 x_4$. They both have generalized degrees $2, 3, \dots$. But $f(x) \oplus g(x) = x_1 x_2 + x_3 x_4 - 2x_1 x_2 x_3 x_4$ has generalized degree $2, 4, \dots$.

We show now with the example of quadratic functions that the generalized degree is strictly better that the usual algebraic degree. As shown in [**16**], the set of quadratic functions has $\lfloor \frac{n}{2} \rfloor + 1$ orbits under the action of the general affine group. The degree does not distinguish these orbits but the generalized degree does.

PROPOSITION 3. *The knowledge of the orbit containing a given quadratic function can be directly derived from its generalized degree.*

Indeed, according to Theorem 5.4, the generalized degree of a quadratic function f of rank $2r$ is equal to the generalized degree of the function $\bigoplus_{i=1}^{r} x_{2i-1}\, x_{2i}$. By using an argument of inclusion-exclusion, one has (see [**7**]):

$$\bigoplus_{i=1}^{r} x_{2i-1}\, x_{2i} = \sum_{k=1}^{r} (-2)^{k-1} \sum_{1 \leq i_1 < \cdots < i_k \leq r} x_{2_{i_1}-1} x_{2_{i_1}} \cdots x_{2_{i_k}-1} x_{2_{i_k}}.$$

Thus, the generalized degree of f is equal to:

$$d_i(f) = \begin{cases} 2i & \text{if } i \leq r; \\ r + i - 1 & \text{if } i > r. \end{cases}$$

OPEN PROBLEM 5.7. Classify under the action of the general affine group all the functions whose generalized degree is a given sequence of consecutive integers.

Finally, we show now that the generalized degree is enough to characterize bent functions:

THEOREM 5.8. *A Boolean function $f(x) = \sum_{u \in \mathbb{F}_2^n} \lambda_u x^u$ on $\mathbb{F}_2^n$ (n even) is bent if and only if the two following conditions are satisfied:*

1. $d_{\frac{n}{2}-1}(f) \leq n-2$ *and*
2. $d_{\frac{n}{2}}(f) = n$.

PROOF. Assume first that f is bent. To show that condition 1 is fulfilled, we shall prove that it satisfies the following condition 1': for each i such that $1 \leq i < \frac{n}{2}$, we have $d_i(f) \leq \frac{n}{2}+i-1$. This condition is in fact equivalent to condition 1, since we know by definition that for each i, we have $d_{i-1}(f) < d_i(f)$. Notice that condition 1' for $i = 1$ corresponds to the bound on the degrees of bent functions given by Rothaus in [**18**]. Condition 1' is clearly equivalent to: for every $1 \leq i < \frac{n}{2}$ and every word u such that $\mathrm{w}(u) > \frac{n}{2}+i-1$, the number λ_u is divisible by 2^i, which is true, according to Theorem 3.1. Since $\lambda_{(1,\dots,1)} \equiv 2^{\frac{n}{2}-1} \bmod 2^{\frac{n}{2}}$, we have $d_{\frac{n}{2}}(f) \geq n$ and thus $d_{\frac{n}{2}}(f) = n$ since $d_{\frac{n}{2}-1}(f) < n-1$.

Conversely, condition 1 implies condition 1', as above, which implies that for every u such that $\frac{n}{2} < \mathrm{w}(u) < n$, the coefficient λ_u is divisible by $2^{\mathrm{w}(u)-\frac{n}{2}}$ (take $i = \mathrm{w}(u) - \frac{n}{2}$) and that $\lambda_{(1,\dots,1)}$ is divisible by $2^{\frac{n}{2}-1}$. Condition 2 implies then that $\lambda_{(1,\dots,1)} \equiv 2^{\frac{n}{2}-1} \bmod 2^{\frac{n}{2}}$ and f is bent according to Theorem 3.1. □

6. Conclusion

We have shown in this paper that the representation of Boolean functions through real polynomials (the N.N.F.) leads to an affine invariant generalizing the algebraic degree. This affine invariant is a powerful tool for classifying functions. For instance, it allows to characterize bent functions in a very simple way. There exists simple formulæ applied to the N.N.F. coefficients of any real valued function to characterize Boolean and bent functions. This makes this representation much more efficient than the Fourier representation to construct a general Boolean function with prescribed combinatorial properties such as bentness or high nonlinearity.

References

[1] R. Beigel *The polynomial method in circuit complexity*, Proc. 8th Structure, p. 82-95 (1993)

[2] C. Carlet *Two new classes of bent functions*, EUROCRYPT' 93, Advances in Cryptology, Lecture Notes in Computer Science 765, p. 77-101 (1994)

[3] J. Bruck *Harmonic analysis of polynomial threshold functions*, SIAM J. Disc. Math. Vol. 3, No 2, p. 168-177 (1990)

[4] C. Carlet. *Generalized Partial Spreads*, IEEE Transactions on Information Theory vol 41 (1995) 1482-1487

[5] C. Carlet and P. Guillot. *A characterization of binary bent functions*, Journal of Combinatorial Theory, Series A, Vol. 76, No. 2 (1996) 328-335

[6] C. Carlet and P. Guillot. *An alternate characterization of the bentness of binary functions, with uniqueness*, Designs, Codes and Cryptography, 14, 133-140 (1998).

[7] C. Carlet and P. Guillot. *A new representation of Boolean functions*, Proceedings of AAECC'13, LNCS 1719, pp. 94–103 (1999).

[8] J. F. Dillon. *Elementary Hadamard Difference sets*, Ph. D. Thesis, Univ. of Maryland (1974).

[9] P. Guillot. *The completed class of $\mathcal{GPS}$ covers all bent functions.* Preprint. Paper presented at ISIT'98, Cambridge, USA (1998).

[10] X.D. Hou. *On the coefficients of binary bent functions*, Proceedings of the American Mathematical Society, S 0002-9939(99)05146-1. Article electronically published on August 17, 1999.

[11] X.D. HOU and and P. LANGEVIN. *Results on bent functions*, Journal of Combinatorial Theory, Series A, 80, 232-246 (1997).

[12] J.P.S. Kung. *Source Book in Matroïd Theory*, Birkhaüser (1986).

[13] W. Meier and O. Staffelbach. *Nonlinearity Criteria for Cryptographic Functions*, Advances in Cryptology, EUROCRYPT' 89, Lecture Notes in Computer Science 434, 549-562, Springer Verlag (1990).

[14] D. A. Mix Barrington, R. Beigel and S. Rudich. *Representing Boolean functions as polynomials modulo composite numbers*, Comput. Complexity 4, p. 367-382 (1994)
[15] N. Nisan and M. Szegedy *On the degree of Boolean functions as real polynomials*, Comput. Complexity 4, No 4, p. 301-313 (1994)
[16] F. J. Mac Williams and N. J. Sloane. *The theory of error-correcting codes*, Amsterdam, North Holland 1977.
[17] G-C. Rota. *On the foundations of Combinatorial Theory*; Springer Verlag (1964) ; reprint in [**12**].
[18] O. S. Rothaus. *On bent functions*, J. Comb. Theory, 20A (1976) 300- 305
[19] T. Siegenthaler, *Correlation-immunity of nonlinear combining functions for cryptographic applications, IEEE Transactions on Information theory, V. IT-30, No 5*, 1984 pp. 776–780
[20] J. H. Van Lint. *Coding Theory*, Springer Verlag 201 (1971).
[21] Xiao Guo-Zhen and J. L. Massey. A Spectral *Characterization of Correlation-Immune Combining Functions.* IEEE Trans. Inf. Theory, Vol IT 34, n° 3, pp. 569-571 (1988).

GREYC, Université de Caen, France and INRIA Projet CODES, BP 105, 78153 Le Chesnay Cedex, France

E-mail address: `Claude.Carlet@inria.fr`

URL: `http://www.info.unicaen.fr/~claude/english.html`

Thomson-CSF Communication, 66 rue du Fossé blanc, 92231 Gennevilliers Cedex, France

E-mail address: `Philippe.Guillot@tcc.thomson-csf.com`

DIMACS Series in Discrete Mathematics
and Theoretical Computer Science
Volume **56**, 2001

On Identifying Codes

Gérard D. Cohen, Iiro Honkala, Antoine Lobstein, and Gilles Zémor

ABSTRACT. Fault diagnosis of multiprocessor systems motivates the following graph-theoretic definition. A subset C of points in a connected undirected graph $G = (V, E)$ is called an r-identifying code, if the sets $B_r(v) \cap C$ consisting of all elements of C within distance r from the vertex v are nonempty and different for all $v \in V$. We show that no nontrivial perfect r-identifying codes exist for $r > 1$, and study how small the density of an r-identifying code can be in certain infinite graphs. We also show that the problem of existence of a 1-identifying code of size at most k in a graph is NP-complete.

1. Introduction

Throughout the paper we assume that $G = (V, E)$ is a connected undirected graph, which may be finite or infinite. We say that the distance $d(u, v)$ between two vertices is the number of edges on any shortest path between u and v.

A subset $C \subseteq V$ is called an *r-identifying code* if the sets $B_r(v) \cap C$, $v \in V$, are nonempty and different. Here $B_r(v)$ denotes the ball of radius r centred at v, i.e., the set of all vertices whose distance to v is at most r. When r is given, it is natural to say that a vertex u covers a vertex v (and vice versa) if $d(u, v) \leq r$.

The study of such codes was initiated in the paper by Karpovsky, Chakrabarty and Levitin [**16**]; see also [**17**]. Such codes can be used for quality control. Assume that each vertex of G represents a processor and each edge represents a connection between two processors. Assume further that we know that in such a network all the processors are working properly — possibly with one exception. We wish to locate the faulty processor, if there is one. If we take an r-identifying code, and each processor (codeword) $v \in C$ checks its r-neighbourhood $B_r(v)$, then simply based on the information which of these processors reported back about problems, we can tell which processor is malfunctioning.

After [**16**], the problem has been studied for certain infinite graphs in [**6**], [**5**], [**7**], [**8**] — see Section 3 for a short overview — and for the Hamming spaces in [**3**], [**2**], [**10**], [**14**] and [**15**].

There is another closely related problem, namely that of locating-dominating sets in graphs. A set $S \subseteq V$ is called a *locating-dominating set* if the sets $B_1(v) \cap S$, $v \in V \setminus S$, are all nonempty and different. For this problem, we refer to the book

2000 *Mathematics Subject Classification.* Primary 05C70, 68R10, 94B65, 94C12.

The second author would like to thank the Academy of Finland for financial support.

of Haynes, Hedetniemi and Slater [**13**], and to the papers [**4**], [**9**], [**11**], [**18**], [**19**], [**20**], [**21**], [**22**].

The paper is organized as follows: in Section 2 we show that no nontrivial perfect r-identifying codes exist for $r > 1$. In Section 3 we briefly survey what is known about the density of 1-identifying codes in certain infinite graphs and obtain an improvement on [**6**, Theorem 2.9] about the infinite square lattice (this improvement was stated but not proved in [**6**]). In Section 4 we give lower bounds on the density for $r > 1$. Finally in Section 5 we prove the NP-completeness of the problem of the existence of a 1-identifying code of size at most k in a graph.

2. On perfect identifying codes

Assume that V is finite and $|B_r(v)|$ is independent of v, and denote $B_r = |B_r(v)|$. If C is r-identifying, then $B_r(v) \cap C$ has cardinality one for at most $|C|$ vertices, and therefore

$$B_r \cdot |C| \geq |C| + 2(|V| - |C|),$$

leading to the bound

$$|C| \geq \frac{2|V|}{B_r + 1} \tag{1}$$

from [**16**].

When inequality (1) holds as an equality, it is natural to call C a *perfect r-identifying code*. In other words, C is perfect, if and only if each of the $|C|$ codewords is covered only by itself and the remaining $|V| - |C|$ vertices are covered by two codewords each. More generally, we take this as the definition:

DEFINITION 2.1. *In an arbitrary connected graph G, an r-identifying code C is called perfect, if $B_r(v) \cap C$ has cardinality one, if v is a codeword, and cardinality two, if v is not a codeword.*

There are lots of examples of graphs where there are perfect 1-identifying codes, e.g., a path of even length (i.e., with an odd number of vertices), an even cycle of length at least six or an infinite path (infinite in one or two directions), and the code in Figure 1.

If $G = (V, E)$ is any graph, then $C \subseteq V$ is a perfect 1-identifying code, if and only if i) there are no edges between vertices in C, ii) every $v \in V \setminus C$ is adjacent to exactly two vertices in C, and iii) no two vertices v are adjacent to the same two vertices in C.

We call a code C *trivial*, if $C = V$. In this section we prove that in any connected graph G, no nontrivial perfect r-identifying code exists, unless $r = 1$. Of course, for every graph, $C = V$ is trivially a perfect 0-identifying code.

LEMMA 2.1. *Let G be an arbitrary connected graph. If C is a perfect r-identifying code in G, then for all distinct codewords $c, c' \in C$, we have $d(c, c') \geq 2r$.*

PROOF. If $d(c, c')$ is at most $2r - 1$, there are two consecutive vertices, say x and x', on any given shortest path from c to c' that are both within distance r from both c and c'. Then at least one of x and x' is covered by a third codeword, and C cannot be perfect. □

THEOREM 2.1. *If G is an arbitrary connected graph and C is a nontrivial perfect r-identifying code in G, then $r = 1$.*

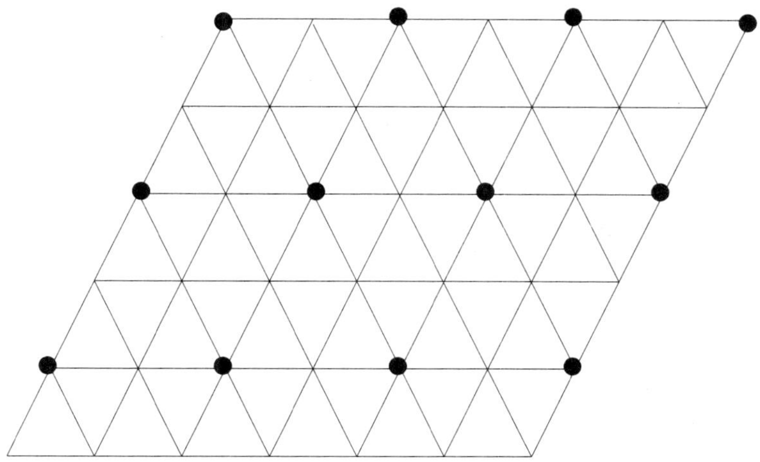

FIGURE 1. A perfect 1-identifying code for the triangular lattice.

PROOF. Assume $r \geq 2$. Consider a codeword $c \in C$. By Lemma 2.1, all other codewords must have distance at least $2r$ to c. Let v be a neighbour of c. The distance from v to any codeword other than c is at least $2r - 1$, and hence at least $r + 1$ since $r \geq 2$. But then c and v would both only be covered by c, and we could not separate between them. □

3. On the density of the best 1-identifying codes on certain grids

In this and the next sections we consider certain specific infinite graphs: the triangular lattice, the square lattice, the square lattice with diagonals, and the hexagonal grid.

Figures 1, 2, 3 and 4 show the best-known 1-identifying codes in the triangular lattice [**16**], square lattice [**5**], square lattice with diagonals [**7**] and the hexagonal grid [**8**].

We can think of all these grids in the real plane, and denote by Q_n the set of vertices $(x, y) \in V$ with $|x| \leq n$ and $|y| \leq n$. Then the density of C is defined as

$$D = D(C) = \limsup_{n \to \infty} \frac{|C \cap Q_n|}{|Q_n|}.$$

For triangular lattice, the code in Figure 1 has optimal density [**16**]. In this section we shall prove that for the square lattice the density of any 1-identifying code must be at least $15/43$, a fact mentioned without proof in [**6**]. The proof relies heavily on the arguments used in [**6**]. For the square lattice with diagonals, the best-known lower bound on the density is $2/9$ [**7**], and for the hexagonal grid it is $16/39$ [**8**].

We now prove the new lower bound for the square lattice.

Let G be the graph T_{nm} with vertex set $V = \mathbb{Z}_n \times \mathbb{Z}_m$ and edge set $E = \{\{u, v\} : u - v = (\pm 1, 0) \text{ or } u - v = (0, \pm 1)\}$. Assume that C is a 1-identifying code in T_{nm} with $n, m \geq 30$.

We use the following notations. Let $L_{\geq 3}$ denote the set of points covered by at least three codewords. Let C' denote the set of codewords c such that if $c \in B_1(x) \cap C$ then $|B_1(x) \cap C| \leq 2$.

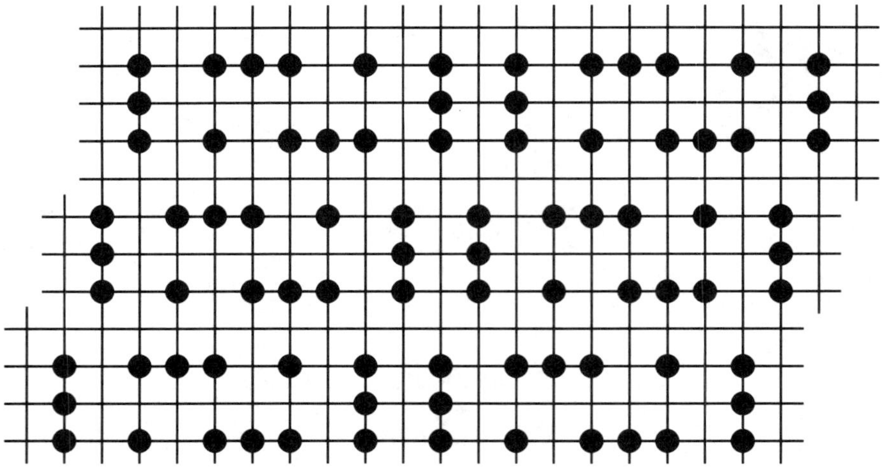

FIGURE 2. A 1-identifying code for the square lattice with the best-known density, i.e., 0.35.

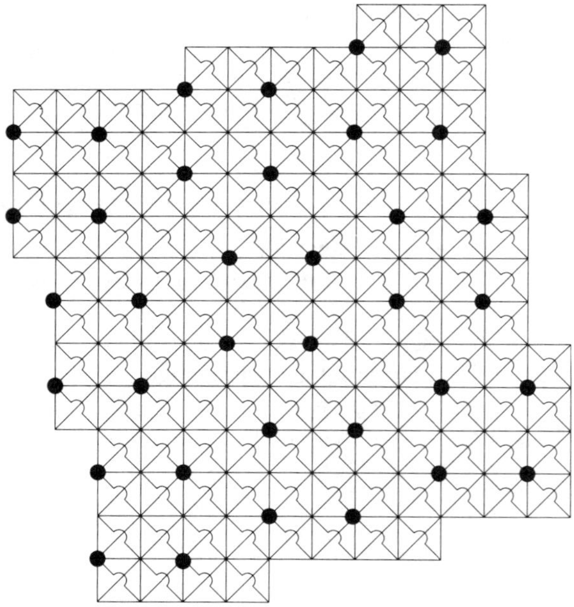

FIGURE 3. A 1-identifying code for the square lattice with diagonals, with density 4/17.

Define an auxiliary bipartite graph Γ whose vertex set is $C' \cup L_{\geq 3}$ and there is an edge from $c' \in C'$ to $x \in L_{\geq 3}$ if and only if $x \in C$ and the Euclidean distance between x and c' equals 2 or $\sqrt{5}$.

In [**6**] it is proved that

- an element of C' has degree at least two in Γ;
- an element of $L_{\geq 3}$ has degree at most three in Γ;

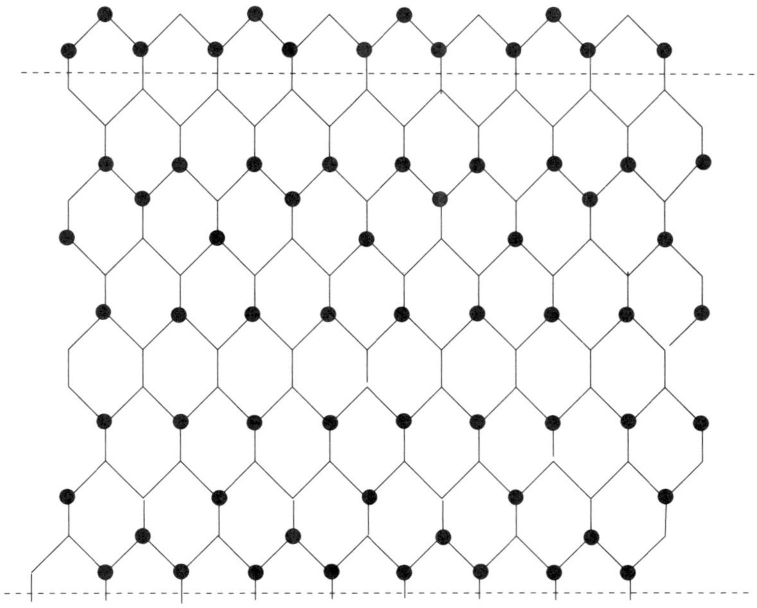

FIGURE 4. A 1-identifying code for the hexagonal grid with the best-known density, i.e., 3/7.

- if an element of $L_{\geq 3}$ has degree three in Γ, then this element is like the point $f3$ in Figure 5 (a) in which case the degrees of $e5$ and $g5$ are both at least four, or like $f3$ in Figure 5 (b) in which case the degrees of $e5$ and $g1$ are both at least four.

In Figure 5 crosses denote non-codewords, white squares codewords and black squares elements of C'.

Denote by A the set of points in $L_{\geq 3}$ with degree exactly three in Γ and B the set $L_{\geq 3} \setminus A$, and by X the set of points in C' with degree at most three in Γ and Y the set $C' \setminus X$. The new item that we prove is that d, the number of edges between Y and B in Γ is at least a, the number of edges between X and A in Γ.

LEMMA 3.1. $d \geq a$.

PROOF. Based on the previous discussion, each edge between A and X has as its end point in A an element which is of either of the following two types:

1) a point like $f3$ in Figure 5 (a), which belongs to $L_{\geq 3}$, has degree three, and for which $f1$ has degree less than four, $e5$ and $g5$ have degree at least four;

2) a point like $f3$ in Figure 5 (b), which belongs to $L_{\geq 3}$, has degree three, and for which $g5$ has degree less than four, $e5$ and $g1$ have degree at least four.

Consider first a point $f3$ of type 1). If $h3$ is a codeword, then $g3$ is in $L_{\geq 3} \cap C$ and at least one of its neighbours in Γ (i.e., $g5$) has degree at least four. Moreover, the degree of $g3$ is two: otherwise $h1$ is in C' and by applying the proof of [**6**, Lemma 2.4] to $g3$ we see that $f1$ has degree at least four contrary to our assumption. We call $g3$ a *friend* of $f3$.

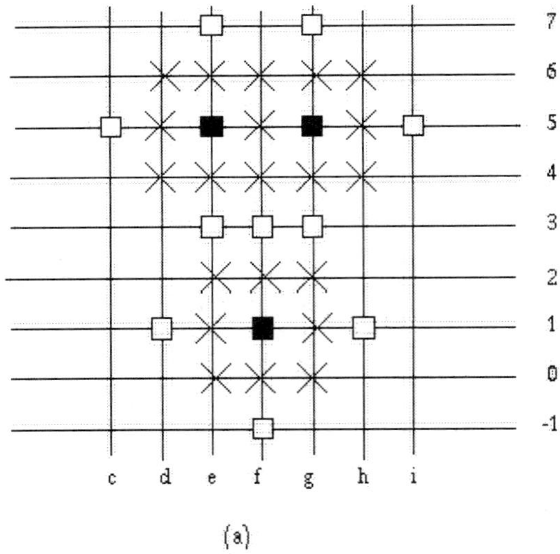

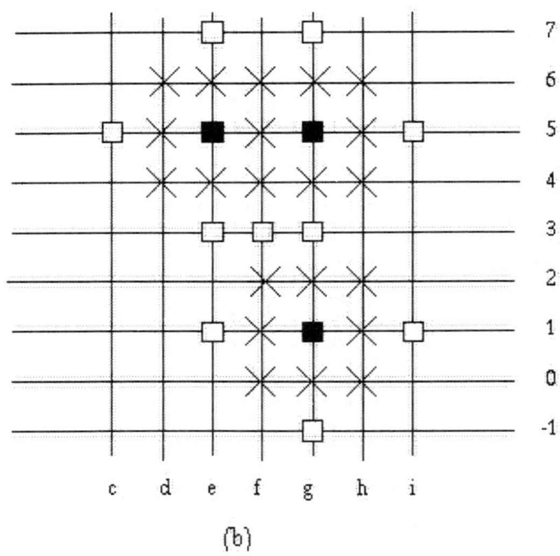

FIGURE 5. Possible locations for three elements of C' adjacent in Γ to $f3 \in L_{\geq 3}$.

If $h3$ is not a codeword, then $i5$, $i4$ and $i3$ all are ($i3$ because we have to be able to separate between $g2$ and $h3$). Now $i4$ is in $L_{\geq 3} \cap C$, has degree at most two ($k3$ and $k5$ cannot both be in C', otherwise $h4$ and $j4$ could not be separated) and has a neighbour in Γ of degree at least four (i.e., $g5$). We call $i4$ a *friend* of $f3$.

Whatever $h3$, we have shown that $f3$ has a friend. We find a second friend by going to the left instead of right in Figure 5 (a).

In the same way as in the proof of [**6**, Lemma 2.6] one verifies that a point can be a friend to at most two points of type 1). So we have shown that the number of friends is at least the number of points of type 1). Since each friend belongs to B and has a neighbour in Γ of degree at least four, the number of edges between B and Y is at least the number of points of type 1).

Consider now a codeword $f3$ of type 2). Now $e2$ is a codeword (otherwise we could not separate between $f2$ and $f4$) and we call it the *mate* of $f3$. It has a neighbour in Γ, $g1$, which has degree at least four. It belongs to $L_{\geq 3} \cap C$ and has degree at most two (indeed, if $c1$ and $c3$ are in C', then neither $c4$ nor $d3$ are codewords and nobody covers $d4$). Notice that $e2$ cannot be the mate of $d1$: otherwise $c3$ would be in C' and again nobody would cover $d4$. On the other hand, $d3$ can also have $e2$ as its mate, but then the neighbours of $d3$ are $c1$, $c5$ and $e5$ and from the proof of [**6**, Lemma 2.4] we know that $c1$ has degree at least four and therefore in this case *both the neighbours of* $e2$ *in* Γ *have degree at least four.* Since all the mates belong to B, we have shown that the number of edges between B and Y is at least the number of points of type 2).

It can furthermore be verified that a mate cannot be a friend. For instance, $e2$ cannot be a friend of $e1$: if $e1$ is of type 1), then $g1$ would have degree less than four contrary to our assumption.

Therefore the two constructed edge sets are disjoint and we have shown that $d \geq a$. □

In the proof of [**6**, Lemma 2.7] it was shown that

$$4|C'| \leq 3|A| + \frac{8}{3}|B| + \frac{1}{3}a - \frac{1}{3}d.$$

Now using Lemma 3.1 we see that $4|C'| \leq 3|L_{\geq 3}|$. Using the same argument as in [**6**] we then get $43|C| \geq 15|V|$.

THEOREM 3.1. *The minimum density of a* 1*-identifying code of the infinite square lattice satisfies* $D \geq 15/43$.

4. The case $r > 1$

In this section we consider the case when $r > 1$.

Now the volume of a ball of radius r equals

- $2r^2 + 2r + 1$ in the square lattice;
- $3r^2 + 3r + 1$ in the triangular lattice;
- $(2r+1)^2$ in the square lattice with diagonals;
- $\frac{3}{2}r^2 + \frac{3}{2}r + 1$ in the hexagonal grid.

Let us assume that C is an r-identifying code in any of these four graphs, and denote by D_r its density. So in the above four types of grid, (1) gives a lower bound which is in $1/r^2$. We now establish a lower bound in $1/r$, using the following general idea.

We pick a path $\mathcal{A} \subseteq V$ with $n+1$ elements: $\mathcal{A} = \{v_0, v_1, \dots, v_n\}$ (for instance, in the case of the square lattice, the square lattice with diagonals and the triangular lattice, $\mathcal{A}$ will be a line segment $\{(0,0), (1,0), \dots, (n,0)\}$) and we want this path to have the following properties:

(i) the union of the balls of radius r centred at the elements of $\mathcal{A}$ is a tile. We denote $B_r(\mathcal{A}) = \cup_{i=0}^{n} B_r(v_i)$, and we require that $|B_r(\mathcal{A})| = n(\alpha r + \beta) + \gamma r^2 + \delta r + \varepsilon$, where $\alpha, \beta, \gamma, \delta$ and ε are constants;

(ii) any ball of radius r centred at an element of $B_r(\mathcal{A})$ intersects $\mathcal{A}$ in (a) either a subpath $\{v_i, v_{i+1}, \dots, v_{i+t}\}$, (b) or a subset of type $\{v_j, v_{j+2}, \dots, v_{j+2s}\}$, $i, t, i+t, j, s, j+2s \in [0, n]$.

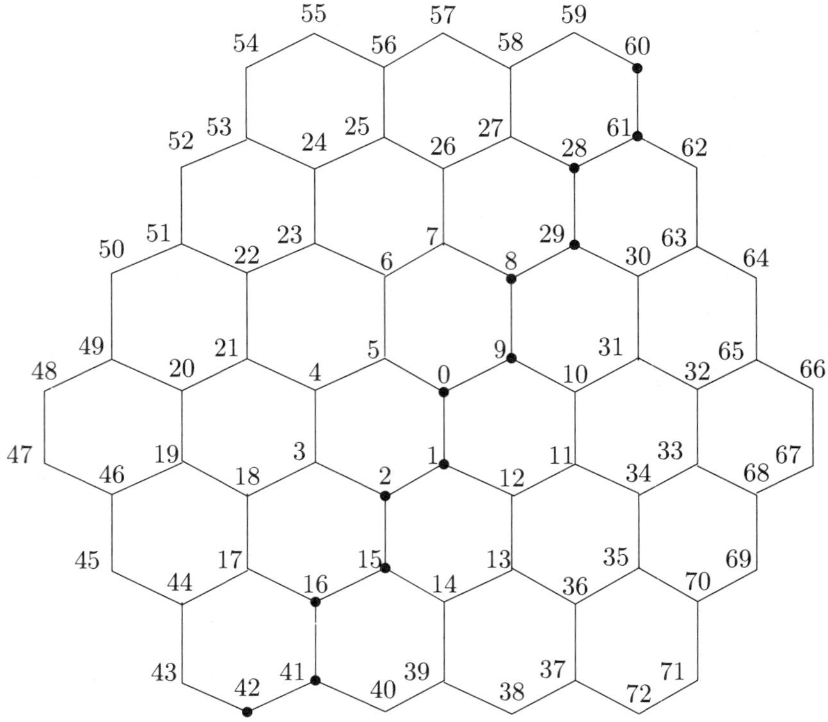

FIGURE 6. The numbering of the points in the hexagonal grid.

Remark. Since ultimately we'll let n tend to infinity for fixed r, we can replace condition (i) by the alternative

(i') $B_r(\mathcal{A})$ is contained in a tile $\mathcal{T}$ of size $n(\alpha r + \beta) + f(r)$, where f depends only on r.

Using (ii), we will derive an estimate on the minimum number of codewords in $B_r(\mathcal{A})$ which are necessary to identify the elements of $\mathcal{A}$, as a fraction of n:

$$|C \cap B_r(\mathcal{A})| \geq n/x.$$

If we can tile the entire grid using $B_r(\mathcal{A})$ or $\mathcal{T}$ as a tile, we see that we obtain:

$$D_r \geq \frac{n}{x|B_r(\mathcal{A})|} \geq \frac{n}{x|\mathcal{T}|}. \tag{2}$$

By letting n tend to infinity in (2), we obtain

$$D_r \geq \frac{1}{x(\alpha r + \beta)}. \tag{3}$$

We need now to be more specific. Let $X = \{\{v_0, v_1\}, \{v_1, v_2\}, \ldots, \{v_{n-1}, v_n\}\}$, and $Y = Y' \cup Y''$, with $Y' = \{\{v_0, v_2\}, \{v_2, v_4\}, \ldots\}$ and $Y'' = \{\{v_1, v_3\}, \{v_3, v_5\}, \ldots\}$. The sets X, Y, Y' and Y'' have cardinalities n, $n-1$, $\lfloor n/2 \rfloor$ and $\lfloor (n-1)/2 \rfloor$, respectively. For each of these pairs, there has to be a *separating* codeword $c \in C \cap B_r(\mathcal{A})$, which covers one point of the pair but not the other.

Consider first the square and triangular lattices, and the square lattice with diagonals. Choose $\mathcal{A}$ to be a line segment: $\mathcal{A} = \{(0,0), (1,0), \ldots, (n,0)\}$. Here, we

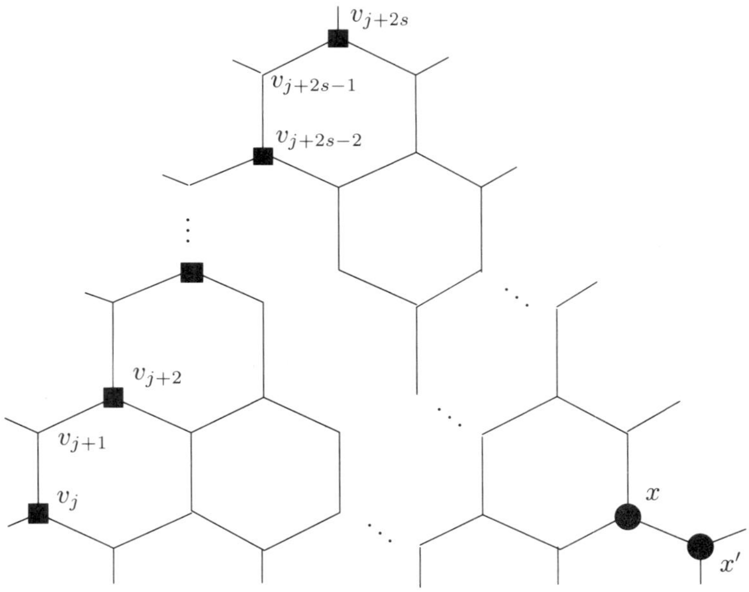

FIGURE 7. Black squares represent the intersection of $\mathcal{A}$ with a ball of radius $2s$ centred at x or radius $2s+1$ centred at x'.

are always in case (a): any ball intersects $\mathcal{A}$ in $\{v_i, v_{i+1}, \ldots, v_{i+t}\}$. Therefore a codeword $c \in C \cap B_r(\mathcal{A})$ can separate *at most* two pairs in X, namely $\{v_{i-1}, v_i\}$ and $\{v_{i+t}, v_{i+t+1}\}$. Hence

$$|X|/2 = n/2 \leq |C \cap B_r(\mathcal{A})| \leq |C \cap \mathcal{T}|.$$

Of course, c can also separate at most two pairs in Y' and two pairs in Y''. However, since these sets are smaller than X, it is unfavourable to take them into account here, in contrast to the case of the hexagonal grid that we now study.

Take $\mathcal{A} = \{\ldots, 60, 61, 28, 29, 8, 9, 0, 1, 2, 15, 16, 41, 42, \ldots\}$ (see Figure 6 for the numbering of the points in the hexagonal grid). A ball centred at $c \in C \cap B_r(\mathcal{A})$ can intersect $\mathcal{A}$ in way (a) or (b): $B_r(c) \cap \mathcal{A} = \{v_i, v_{i+1}, \ldots, v_{i+t}\}$ or $B_r(c) \cap \mathcal{A} = \{v_j, v_{j+2}, \ldots, v_{j+2s}\}$. Studying the shape of a ball in the hexagonal grid shows that in the latter case, the only possibility is given by Figure 7, where we can see that $s = \lfloor r/2 \rfloor (\geq 1)$.

If $B_r(c) \cap \mathcal{A} = \{v_i, v_{i+1}, \ldots, v_{i+t}\}$, then c can separate at most

– two pairs in X, $\{v_{i-1}, v_i\}$ and $\{v_{i+t}, v_{i+t+1}\}$;

– four pairs in Y, $\{v_{i-2}, v_i\}$, $\{v_{i-1}, v_{i+1}\}$, $\{v_{i+t-1}, v_{i+t+1}\}$ and $\{v_{i+t}, v_{i+t+2}\}$.

If $B_r(c) \cap \mathcal{A}$ is of type $\{v_j, v_{j+2}, \ldots, v_{j+2s}\}$, then c can separate at most

– two pairs in Y, $\{v_{j-2}, v_j\}$ and $\{v_{j+2s}, v_{j+2s+2}\}$;

– $2s+2$ pairs in X, $\{v_{j-1}, v_j\}$, $\{v_j, v_{j+1}\}$, $\ldots$, $\{v_{j+2s-1}, v_{j+2s}\}$ and $\{v_{j+2s}, v_{j+2s+1}\}$.

If we consider only X, we can say that

$$|C \cap B_r(\mathcal{A})| \geq |X|/(2s+2) = n/(2s+2), \tag{4}$$

the most unfavourable case being if we are always in case (b).

If we consider only Y, we obtain

$$|C \cap B_r(\mathcal{A})| \geq |Y|/4 = (n-1)/4, \tag{5}$$

the most unfavourable case being now (a). Since we consider large n, (5) is as good as (4) when $s = 1$, and better when $s > 1$.

We can improve on (5) if we take both X and Y into account. In case (a), c can separate at most six pairs in $X \cup Y$, in case (b), $2s + 4$.

If $s = 1$ ($r = 2$ or 3), we obtain

$$|C \cap B_r(\mathcal{A})| \geq |X \cup Y|/6 = (2n-1)/6.$$

However, for these small values of r, it turns out that the bound in $1/r^2$, (1), remains better. And if $s > 1$ ($r > 3$), we obtain no better lower bound than $n/4$, i.e., (5).

Summarizing, we obtain a lower bound on $|C \cap B_r(\mathcal{A})|$ which is in $n/2$ for the first three lattices; and for the hexagonal grid: $n/3$ for $r = 2$ or 3, $n/4$ otherwise.

We end in detail the case of the square lattice, the other three grids not being much more difficult to handle.

In the square lattice, $B_r(\mathcal{A})$ is a hexagon whose vertices are $(n+r, 0)$, $(n, -r)$, $(0, -r)$, $(-r, 0)$, $(0, r)$ and (n, r); it is a tile which contains $(2r+1)(n+1) + 2r^2 = (2r+1)n + 2r^2 + 2r + 1$ lattice points (using (i') we can replace it with a slightly larger rectangle). Therefore, we have the following result.

THEOREM 4.1. *In the infinite square lattice, the density of any r-identifying code is at least* $\frac{1}{2(2r+1)}$.

For the remaining three grids we have:

THEOREM 4.2. *In the infinite triangular lattice, the density of any r-identifying code is at least* $\frac{1}{2(2r+1)}$.

THEOREM 4.3. *In the infinite square lattice with diagonals, the density of any r-identifying code is at least* $\frac{1}{2(2r+1)}$.

THEOREM 4.4. *In the infinite hexagonal grid, the density of any r-identifying code is at least* $\frac{1}{4(r+1)}$ *for* $r \geq 4$.

5. Complexity

We show that the following problem is NP-complete (see [**1**] or [**12**] for a complete discussion of NP-completeness):

NAME: 1-Identifying Code (1-IC).
INSTANCE: A connected graph $G = (V, E)$ and an integer $k \leq |V|$.
QUESTION: Is there a 1-identifying code $C \subseteq V$ of size at most k?

Of course, there is a 1-identifying code $C \subseteq V$ of size *at most* k if and only if there is one of size k: indeed, adding elements to a 1-identifying code always gives a 1-identifying code.

We were inspired by the proof in [**9**] for the related problem of locating-dominating sets. Obviously, the problem 1-IC is in NP: given a subset C of V, it is polynomial in time (with respect to $|V|$, the size of the problem), to check that C is a 1-identifying code of size at most k, for instance by studying the pairs of vertices at distance at most two from one another.

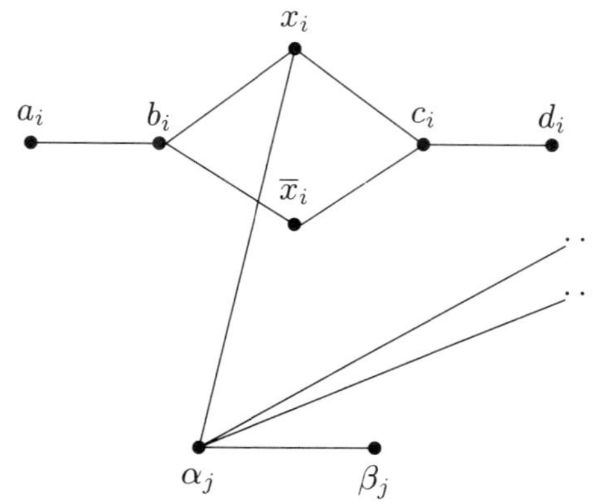

FIGURE 8. Construction of the graph G: here $\mathcal{C}_j = \{u_{j,1}, u_{j,2}, u_{j,3}\}$ and $u_{j,1} = x_i$.

Next, we polynomially reduce the well-known NP-complete 3-satisfiability problem (3-SAT) to 1-IC.

NAME: 3-satisfiability (3-SAT).

INSTANCE: A collection $\mathcal{E}$ of clauses over a set X of variables, where each clause contains exactly three distinct literals.

QUESTION: Can $\mathcal{E}$ be satisfied, i.e., is there a truth assignment for the variables of X such that each clause contains at least one true literal?

We consider any instance of 3-SAT, $\mathcal{E} = \{\mathcal{C}_1, \mathcal{C}_2, \dots, \mathcal{C}_m\}$ over the set of variables $X = \{x_1, x_2, \dots, x_n\}$, where each clause contains exactly three literals taken in $U = \{x_1, x_2, \dots, x_n, \bar{x}_1, \bar{x}_2, \dots, \bar{x}_n\}$. From $\mathcal{E}$, we construct a graph $G = (V, E)$ and an integer k such that $\mathcal{E}$ can be satisfied if and only if V contains a 1-identifying code of size at most k.

For each variable x_i of X, we construct the graph $G_{x_i} = (V_{x_i}, E_{x_i})$ which contains six vertices and six edges (see Figure 8):

$$V_{x_i} = \{a_i, b_i, x_i, \bar{x}_i, c_i, d_i\},$$

$$E_{x_i} = \{\{a_i, b_i\}, \{b_i, x_i\}, \{b_i, \bar{x}_i\}, \{x_i, c_i\}, \{\bar{x}_i, c_i\}, \{c_i, d_i\}\}.$$

For each clause $\mathcal{C}_j = \{u_{j,1}, u_{j,2}, u_{j,3}\}$ ($u_{j,1}$, $u_{j,2}$, $u_{j,3}$ in U), we construct the graph $G_{\mathcal{C}_j} = (V_{\mathcal{C}_j}, E_{\mathcal{C}_j})$ which consists of two vertices, α_j and β_j, and one edge, $\{\alpha_j, \beta_j\}$, to which we add the set of three edges $E'_{\mathcal{C}_j} = \{\{\alpha_j, u_{j,1}\}, \{\alpha_j, u_{j,2}\}, \{\alpha_j, u_{j,3}\}\}$.

Finally, G is the graph with vertex set

$$V = \Big(\bigcup_{i=1}^{n} V_{x_i}\Big) \cup \Big(\bigcup_{j=1}^{m} V_{\mathcal{C}_j}\Big)$$

and edge set

$$E = \Big(\bigcup_{i=1}^{n} E_{x_i}\Big) \cup \Big(\bigcup_{j=1}^{m} E_{\mathcal{C}_j}\Big) \cup \Big(\bigcup_{j=1}^{m} E'_{\mathcal{C}_j}\Big),$$

and we set $k = 3n + m$. This construction is polynomial in $n + m$, the size of 3-SAT, since $|V| = 6n + 2m$.

If $\mathcal{E}$ can be satisfied, we can construct a 1-identifying code $C \subseteq V$, of size equal to k, for instance as follows: for all i between 1 and n, b_i, c_i and whichever of x_i

and $\bar{x}_i$ is true, belong to C; for all j between 1 and m, α_j belongs to C, which has size $3n + m = k$. A brief study shows that all vertices are covered and that, with the assumption that each clause contains at least one true literal, no two vertices have the same set of codewords that cover them.

Now assume that there is a subset C of V, of size at most k, which is a 1-identifying code. Obviously, $|C \cap V_{\mathcal{C}_j}| = 1$ or 2. In all cases, α_j must be covered by a vertex other than β_j and itself.

Next, we consider the intersections $C \cap V_{x_i}$. We show that these intersections each have at least three elements, and that if $|C \cap V_{x_i}| = 3$, then *exactly one* of x_i and $\bar{x}_i$ is in C. Indeed, assume that neither x_i nor $\bar{x}_i$ are codewords. Then the same codewords will cover a_i and b_i. So at least one of x_i and $\bar{x}_i$ is in C, but anyway two more codewords are necessary, to cover a_i and d_i.

As a consequence, C contains at least $m + 3n = k$ elements, hence C contains *exactly* k elements. This implies that, for all i, $|C \cap V_{x_i}| = 3$; furthermore, we know that in this case, $|C \cap \{x_i, \bar{x}_i\}| = 1$. Thus, setting $x_i =$ True if $C \cap \{x_i, \bar{x}_i\} = \{x_i\}$ and $x_i =$ False if $C \cap \{x_i, \bar{x}_i\} = \{\bar{x}_i\}$ is a valid truth assignment for the variables of X. Finally, since α_j must be covered by a vertex other than β_j and itself, at least one of $u_{j,1}$, $u_{j,2}$, $u_{j,3}$ belongs to C, which shows that there is at least one true literal in $\mathcal{C}_j$. So we have proved

THEOREM 5.1. *The problem 1-IC is NP-complete.*

References

[1] J. P. Barthélemy, G. Cohen, A. Lobstein: *Algorithmic Complexity and Communication Problems*, London: University College of London, 1996.

[2] U. Blass, I. Honkala, S. Litsyn: On binary codes for identification, *J. Combin. Des.*, vol. 8, pp. 151–156, 2000.

[3] U. Blass, I. Honkala, S. Litsyn: Bounds on identifying codes, *Discrete Math.*, to appear.

[4] D. I. Carson: On generalized location domination. In: Graph Theory, Combinatorics, and Applications, Proceedings of the Seventh Quadrennial International Conference on the Theory and Applications of Graphs, 1992 (eds. Y. Alavi and A. Schwenk), Wiley, pp. 161–179, 1995.

[5] G. Cohen, S. Gravier, I. Honkala, A. Lobstein, M. Mollard, C. Payan, G. Zémor: Improved identifying codes for the grid, *Electron. J. Combin.*, Comments to 6(1), R19, 1999.

[6] G. Cohen, I. Honkala, A. Lobstein, G. Zémor: New bounds for codes identifying vertices in graphs, *Electron. J. Combin.*, vol. 6(1), R19, 1999.

[7] G. Cohen, I. Honkala, A. Lobstein, G. Zémor: On codes identifying vertices in the two-dimensional square lattice with diagonals, *IEEE Trans. Comp.*, to appear.

[8] G. Cohen, I. Honkala, A. Lobstein, G. Zémor: Bounds for codes identifying vertices in the hexagonal grid, *SIAM J. Discrete Math.*, to appear.

[9] C. J. Colbourn, P. J. Slater, L. K. Stewart: Locating dominating sets in series parallel networks, *Congr. Numer.*, vol. 56, pp. 135–162, 1987.

[10] G. Exoo: Computational results on identifying t-codes, preprint.

[11] A. Finbow, B. L. Hartnell: On locating dominating sets and well-covered graphs, *Congr. Numer.*, vol. 65, pp. 191–200, 1988.

[12] M. R. Garey, D. S. Johnson: *Computers and Intractability, a Guide to the Theory of NP-Completeness*, New York: Freeman, 1979.

[13] T. W. Haynes, S. T. Hedetniemi, P. J. Slater: *Fundamentals of Domination in Graphs,* New York: Marcel Dekker, 1998.

[14] I. Honkala: On the identifying radius of codes. In: Proceedings of the 7th Nordic Combinatorial Conference (eds. T. Harju and I. Honkala), Turku, 1999.

[15] I. Honkala, T. Laihonen, S. Ranto: On codes identifying sets of vertices in Hamming spaces, *Designs, Codes and Cryptography*, submitted.

[16] M. G. Karpovsky, K. Chakrabarty, L. B. Levitin: On a new class of codes for identifying vertices in graphs, *IEEE Trans. Inform. Theory*, vol. 44, pp. 599–611, 1998.

[17] M. G. Karpovsky, K. Chakrabarty, L. B. Levitin, D. R. Avresky: On the covering of vertices for fault diagnosis in hypercubes, *Inform. Process. Lett.*, vol. 69, pp. 99–103, 1999.

[18] D. F. Rall, P. J. Slater: On location-domination numbers for certain classes of graphs, *Congr. Numer.*, vol. 45, pp. 97–106, 1984.

[19] P. J. Slater: Leaves of trees, *Congr. Numer.*, vol. 14, pp. 549–559, 1975.

[20] P. J. Slater: Domination and location in acyclic graphs, *Networks*, vol. 17, pp. 55–64, 1987.

[21] P. J. Slater: Dominating and reference sets in a graph, *J. Math. Phys. Sci.*, vol. 22, pp. 445–455, 1988.

[22] P. J Slater: Locating dominating sets and locating-dominating sets, In: Graph Theory, Combinatorics, and Applications, Proceedings of the Seventh Quadrennial International Conference on the Theory and Applications of Graphs, 1992 (eds. Y. Alavi and A. Schwenk), Wiley, pp. 1073–1079, 1995.

G. Cohen, A. Lobstein, G. Zémor, École nationale supérieure des télécommunications, Centre national de la recherche scientifique, 46 rue Barrault, 75634 Paris Cédex 13, France.

E-mail address: cohen@infres.enst.fr

E-mail address: lobstein@infres.enst.fr

E-mail address: zemor@infres.enst.fr

I. Honkala, University of Turku, 20014 Turku, Finland.

E-mail address: honkala@utu.fi

DIMACS Series in Discrete Mathematics
and Theoretical Computer Science
Volume **56**, 2001

Ternary Hamming and Binary Perfect Covering Codes

Italo J. Dejter and Kevin T. Phelps

ABSTRACT. We establish that the perfect dominating set of the 13-cube found by Östergård and Weakley (as a counterexample to Weichsel's uniformity conjecture) exists very naturally in the ternary Hamming code of length 13 and has a very special geometric-combinatorial structure. This approach provides very short proofs of its characteristic properties.

1. Introduction

Given $1 < n \in \mathcal{Z}$, the n-cube Q_n is the graph with vertex set $V = \{0,1\}^n$ and edge set $E = \cup_{i=0}^{n-1} f^i$, where $f^i = \{(v,w) \in V \times V : (v-w)_j = 1$ if and only if $j = i, (j = 0, \dots, n-1)\}$, for $i = 0, \dots, n-1$. The edges of f^i are said to have (or to lie along) the *direction* i, for $i = 0, \dots, n-1$. Given a graph G, we say that $S \subseteq V(G)$ is a *perfect dominating set* (or PDS) of G if every $v \in V(G) \setminus S$ is neighbor of exactly one vertex of S. Weichsel [**6**] showed that each connected component of the subgraph induced by a PDS in Q_n is a subcube and conjectured that S is uniform, i.e. all such components are isomorphic. Östergård and Weakley [**3, 4**] showed that the conjecture is false, by exhibiting a PDS S in Q_{13} whose induced subgraph has as connected components 26 4-cubes and 288 isolated vertices (which correspond to the words with no zero coordinates of the ternary Hamming code) and whose automorphism group is the general linear group $GL(3,3)$.

Perfect dominating sets satisfy an analog of the sphere-packing bound and can be thought of as generalizations of perfect codes. Let $S \subseteq V$ be a PDS in Q_n. Recall that each component of S is an i-cube, where $0 \leq i \leq n$, [**6**]. For such a fixed i, let X_i be the number of i-cubes which are components of S. A sphere of radius 1 about any vertex in such an i-cube will cover (or "dominate") exactly $n-i$ vertices not in this i-cube. For S to be a PDS, the spheres must cover every vertex of Q_n and their intersections with $Q_n \setminus S$ must be disjoint. For these intersections to be disjoint the Hamming distance between any two vertices in different components of S must be at least 3. Equivalently, the distance between any two components of S is at least 3. Equivalently, no vertex of $Q_n \setminus S$ can be dominated twice, that is by two different vertices of S.

DEFINITION 1.1. Perfect Domination Bound.

1991 *Mathematics Subject Classification.* Primary 94B05; Secondary 05C99.

If no vertex of $Q_n \setminus S$ can be dominated twice, then

$$\sum_{i=0}^{i=n} 2^i(n-i+1)X_i \leq 2^n.$$

PROPOSITION 1.1. *A set S of vertices in the n-cube is a PDS if and only if the distance between any two components is at least 3 and the number of i-cube components satisfies the Perfect Domination Bound with equality.*

The proof is elementary and left to the reader.

Remark: Recall that Weischel [**6**] showed that a necessary condition for a perfect dominating set in the n-cube to have an i-cube as a component was that $n-i \equiv 1$ *or* 3 (mod 6). In other words, in the Perfect Domination Bound above X_i must be zero whenever this numerical condition is not met. This puts a significant limitation on the set of numerically feasible solutions for a PDS. This may help for example in the tasks suggested in our Conclusions below. ∎

The i-cube components of a PDS can be represented by codewords over a ternary alphabet. The vertices of an i-cube component all agree on $n-i$ coordinates. Take any vertex in such a component and put another symbol, *, in the i coordinates which form the i-cube. We then have one codeword in $\{*, 0, 1\}$ for each component. The distance required between ternary codewords is not the standard Hamming distance however.

DEFINITION 1.2. *(0,1,*)-Distance*

$$D_{(0,1,*)}(x,y) = |\{j; \{x_j, y_j\} = \{0,1\}\}|$$

The name (0,1,*)-Distance arises for example in [**5**]. The (0,1,*)-distance differs from the Hamming distance $HD(x,y)$ in that we ignore disagreements which involve the symbol *. Thus,

$$D_{(0,1,*)}(x,y) \leq HD(x,y).$$

Now we think of PDS's in terms of ternary codes with minimum Hamming distance at least 3. Instead of the alphabet $\{*, 0, 1\}$ we will use the more natural ternary alphabet $\{0, 1, 2\}$ with 0,1,2 taking the place of *,0,1, respectively. We show that the Östergård-Weakley PDS S lives very naturally inside the ternary Hamming Code of length 13 and has a very special geometric-combinatorial structure.

2. Ternary Hamming Code

As we remarked in the Introduction, the problem of finding perfect dominating sets can be reduced to finding appropriate ternary codes. For length $n = 13$, one feasible solution to the Perfect Domination Bound would involve 26 codewords of weight 9 and 288 codewords of weight 13. It is natural to look for such a ternary code inside the ternary Hamming code of length 13. Information about ternary Hamming codes can be found in MacWilliams and Sloane, [**2**], chapters 5, 6.

The *support* of a codeword is the set of nonzero coordinates of the codeword. If two codewords in a ternary code have the same support, then their Hamming distance is the (0,1,*)-distance. There are 288 codewords of weight 13 in the ternary Hamming code and obviously the minimum distance is at least 3. (Östergård had previously observed this.) Denote this set of codewords by S_{13}.

The dual of the Hamming code has 27 words: the zero word and 26 words of weight 9. The sum and difference of any two codewords is again a codeword and thus has weight 9 (or 0).

Consider the words of weight 9 in the dual of the Hamming code (which is a subcode of the Hamming code because the rows of a parity check matrix are orthogonal also to themselves). Take any such word without loss of generality (0000111111111). A word that is not a multiple must have the form (0111000111222) because the two words can be rows of a parity check matrix which has *different* projective points as columns. So, any two such words of weight 9 have (0,1,*)-distance 3. A word of weight 13 must be orthogonal to (0000111111111). If there are not three 2's among the final nine coordinates, these must all be 1. Then the initial part (a, b, c, d) is orthogonal to (0111) and (1012) and hence does not have weight 4, a contradiction. So, a word of weight 13 has (0,1,*)-distance at least 3 to a word of weight 9.

THEOREM 2.1. *The ternary code S consisting of the 288 words of weight 13 in the ternary Hamming code and the 26 words of weight 9 in the dual code, forms a perfect dominating set.*

The automorphism group of the ternary code S is the same as that of the ternary Hamming code since any automorphism must be an automorphism of the dual code (S_9), and the words of weight 13 (S_{13}).

3. Conclusion

The optimal perfect dominating set in the 4-cube exists as a subcode of the ternary Hamming code of length 4. With the result of this paper, it is tempting to conjecture that the PDS ternary codes of length $n = (3^r - 1)/(3 - 1)$ can always be found inside the ternary Hamming code. Before venturing into such a conjecture it would be worthwhile to first establish this result for $r = 4$ and $r = 5$, ($n = 40$ and $n = 121$).

Of course, the Remark after Proposition 1.1 must be taken into account for these purposes.

Acknowledgements. P. R. J. Östergård has communicated initially that there are 288 codewords of weight 13 in the ternary Hamming code. We are also indebted to our anonymous referee for his helpful corrections and suggested improvements.

References

[1] Blake I. F. and Mullin R. C., The Mathematical Theory of Coding, Academic Press, New York, 1975.

[2] McWilliams F. J and Sloane N. J. A., The Theory of Error-Correcting Codes, North-Holland, Amsterdam, 1977.

[3] Östergård P. R. J., Weakley W. D., *Constructing Covering Codes with Given Automorphism*, Des. Codes and Cryptogr., **16**(1999), 65-73.

[4] Östergård P. R. J., Weakley W. D., *On Perfect Dominating Sets in n-Cubes*, XXIX South Eastern Conf. on Combin., Graph Theory and Comp., Florida Atlantic University, abstract (1998).

[5] van Lint J. H., (0,1,*)-*Distance Problems in Combinatorics*, in Surveys in Combinatorics, (Glasgow 1985), 113-135, London Math. Soc. Lect. Notes Ser. 103, Cambridge Univ. Press, Cambridge UK, 1985.

[6] Weichsel P. M., *Dominating Sets in n-Cubes*, Journal of Graph Theory, **18**(1994), 479-488.

DIMACS Series in Discrete Mathematics
and Theoretical Computer Science
Volume **56**, 2001

A Riemann hypothesis analogue for self-dual codes

Iwan Duursma

ABSTRACT. We formulate a Riemann hypothesis analogue for self-dual codes. If the hypothesis is true for a sufficiently large class of self-dual codes, this would imply an improvement of asymptotic upper bounds for the minimum distance of self-dual codes [**4**]. Whether the hypothesis holds for a given code depends on its weight enumerator. Verification for codes with a known weight enumerator suggests that the hypothesis holds in many cases, in particular for good self-dual codes, but not for self-dual codes in general. Weight enumerators of good self-dual codes satisfy strong linear constraints. In this paper, we show for a special case how the Riemann hypothesis analogue follows from such linear constraints. The general case is left as an open problem.

1. Introduction

Much of the arithmetic information about a number field or an algebraic curve over a finite field is contained in its zeta function. Properties of the zeta function of an algebraic curve over a finite field can be found in [**8**], [**9**]. For a linear code, the main arithmetic invariant is its weight distribution. General references for linear codes are [**10**], [**7**]. In [**3**], we defined for a linear code its zeta function. It contains precisely the same information as the weight distribution of a linear code, but presents it in a form that resembles the classical zeta functions. In the analogy that appears between codes and curves, there is a correspondence between the MacWilliams identities for codes and the functional equation for curves. And the Hasse-Weil bound for curves can be interpreted for codes as an upper bound for the minimum distance. It is a natural open question whether the analogy extends to the Riemann hypothesis for curves, which was proved by André Weil in 1948 and which says that all zeros of the zeta function have the same absolute value. Such a "Riemann hypothesis analogue for linear codes" implies tighter asymptotic upper bounds for self-dual codes [**4**]. The RH analogue can be verified directly for known weight enumerators. And we find many examples for which it holds (e.g. all known extremal codes, including the binary and ternary Golay codes). For a hypothetical binary code of type $[72, 36, 16]$ (whose existence is undecided), the weight enumerator is known and the zeros of its zeta polynomial (not of its weight enumerator) are depicted in the complex plane in Figure 1. All zeros have the same absolute value. The aim of this paper is to replace the ad-hoc verification of

1991 *Mathematics Subject Classification.* Primary 94B65; Secondary 11R58.

the RH analogue for a particular code by an argument that proves it for a larger class of codes. By using linear programming we can prove that the RH holds for all sufficiently good binary self-dual codes of length 12. The class of all self-dual codes of length 12 is too large to expect a RH analogue. But when we apply additional linear constraints, such as minimum distance at least four, then we can show that the zeros of the zeta function of the code all have the same absolute value. We believe that a similar result can be obtained for any length, but if true this would require additional arguments and the method we present here would have to be modified. As an illustration we show that the Nordstrom-Robinson code is optimal, a fact that in itself is well-known, but that is proved here by showing that it is optimal with respect to the analogue of the Hasse-Weil bound for curves. Many ideas presented here require further study. There are obvious observations that require an explanation and there are some deeper problems. A few of the latter problems are listed as open problem.

2. Weight enumerators

Let C be a linear code of length n and minimum distance d over the finite field of q elements. Let A_i be the number of words of weight i in C. The weight enumerator of the code C is defined as

$$A_C(x,y) = x^n + \sum_{i=d}^{n} A_i x^{n-i} y^i \tag{2.1}$$

The weight enumerator of the dual code $C^\perp$ is obtained with the MacWilliams transform

$$A_{C^\perp}(x,y) = \frac{1}{|C|} A_C(x+(q-1)y, x-y) \tag{2.2}$$

A code with $A_{C^\perp} = A_C$ is called formally self-dual. To avoid dealing with degenerate cases we will assume throughout that both C and its dual $C^\perp$ have effective length n, or equivalently, that both have minimum distance at least two. The Singleton upper bound on the size of a code has the form

$$|C| \leq q^{n+1-d}$$

The bound is attained for Maximum Distance Separable codes (MDS codes). Combination of the Singleton bound for a code and its dual gives

$$d + d^\perp \leq n+2 \tag{2.3}$$

A weight enumerator for which $d+d^\perp = n+2$ will be called MDS weight enumerator. An MDS weight enumerator satisfies $(d-1)+(d^\perp-1) = n$ independent linear conditions and is uniquely determined by its length n and its minimum distance d. We denote it by $M_{n,d}$. Explicit expressions for $M_{n,d}$ were obtained in [**1**], [**5**], [**6**] and can be found in [**7**]. From [**3**], [**4**] we have the following generating function.

PROPOSITION 2.1. *The q-ary MDS weight enumerator $M_{n,d}$, for $d = 2, \ldots, n$, is given by the generating function*

$$\frac{1}{(1-T)(1-qT)} (y(1-T)+xT)^n = \frac{1}{q-1} \sum_{a=0}^{\infty} (M_{n,n-a}(x,y) - x^n) T^a$$

In general, an MDS weight enumerator $M_{n,d}$ may have negative coefficients and does not correspond to an actual code. It will often be the case that we are not interested what code if any realizes a given weight enumerator. In such cases, we omit the index C in A_C and by a slight abuse talk about weight enumerators of length n and minimum distance d. Without a code that realizes them, MDS weight enumerators are still interesting for the following reasons. First, as a family they are closed under the MacWilliams transform and under puncturing and shortening. Furthermore, any given weight enumerator of length n can be expressed with respect to a basis of MDS weight enumerators of length n. That is, we may write an arbitrary weight enumerator $A(x, y)$ of length n and minimum distance d as

$$A(x,y) = a_0 M_{n,d} + a_1 M_{n,d+1} + \cdots + a_r M_{n,d+r} \tag{2.4}$$

for some $r \geq 0$. By taking the MacWilliams transform of (2.4) term by term it is not hard to see that $r = n + 2 - d - d^{\perp}$. For a self-dual code, $r = 2g$ is even and the integer g is called the *genus* of the code.

3. Zeta functions

We will be interested in the polynomial $P(T) = a_0 + a_1 T + \ldots + a_r T^r$, where the coefficients are given by (2.4). It is clear that the polynomial is completely determined by the weight enumerator $A(x, y)$ and vice versa. We call it the *zeta polynomial* of the code. This definition, based on (2.4) and MDS weight enumerators, has the advantage of being entirely coding theoretic. The next definition is equivalent but does not mention MDS weight enumerators explicitly. It shows a closer analogy with the zeta function of a curve [**3**].

DEFINITION 3.1. For a q-ary linear code of length n and minimum distance d with weight enumerator $A(x, y)$, the *zeta polynomial*

$$P(T) = a_0 + a_1 T + \ldots + a_r T^r$$

is defined as the unique polynomial of degree at most $n-d$ such that the generating function

$$\frac{P(T)}{(1-T)(1-qT)}\,(y(1-T)+xT)^n = \frac{1}{q-1}\sum_{a=0}^{\infty}(A_{n,n-a}(x,y) - x^n)T^a \tag{3.1}$$

has $A_{n,d} = A(x, y)$. The quotient $Z(T) = P(T)/((1-T)(1-qT))$ is called the *zeta function* of the linear code.

We give a different transformation from weight enumerator to zeta function that in many situations is preferable. It uses the following normalized version of the weight enumerator.

DEFINITION 3.2. Associate to the weight enumerator $A(x, y)$ the polynomial

$$a(t) = \frac{1}{q-1}\sum_{i=d}^{n}\frac{A_i}{\binom{n}{i}}\,t^{i-d}.$$

THEOREM 3.3 ([**4**]). *For a given weight enumerator $A(x, y)$, the zeta polynomial $P(T)$ and the normalized weight enumerator $a(t)$, when considered as series, are related by*

$$\frac{P(T)}{(1-T)(1-qT)}(1-T)^{d+1} \equiv a\Big(\frac{T}{1-T}\Big) \pmod{T^{n-d+1}}$$

The theorem shows that the leading coefficients in the normalized version $m_{n,d}(t)$ of the MDS weight enumerator $M_{n,d}(x,y)$ do not depend on the length n but only on the minimum distance d. This justifies the following definition

DEFINITION 3.4. For $d = 1, 2, \ldots$, define the series

$$m_d(t) = \lim_{n \to \infty} m_{n,d}(t)$$

For later use we formulate

LEMMA 3.5. *The series $m_d(t)$, for $d = 1, 2, \ldots$, correspond to rational functions*

$$m_d(t) = \frac{1}{(1-(q-1)t)(1+t)^{d-1}}$$

They can be computed recursively via

$$m_1(t) = \frac{1}{1-(q-1)t} \qquad m_{d+1}(t) = m_d(t)/(1+t) \tag{3.2}$$

PROOF. Use Theorem 3.3 with $P(T) = 1$, $a(t) = m_d(t)$, and apply the substitution

$$T = t/(1+t) \qquad t = T/(1-T)$$

□

4. A Riemann hypothesis analogue for self-dual codes

We first give a relation between the minimum distance d of a code and the zeros of the zeta polynomial $P(T)$ of the code. The relation serves as a motivation to study properties of the zeros. Information about the zeros will translate into upper bounds for the minimum distance. The approach seems justified for two reasons. First, similar techniques are well-known in the theory of algebraic curves, where they lead to sharp asymptotic upper bounds for the number of rational points on an algebraic curve over a finite field [**2**]. Secondly, the set of zeros of the zeta polynomial of a code shows a rich structure. Much of this structure has still to be analyzed and to be captured by general theorems.

To derive the required relation we use Theorem 3.3, though it can also be derived directly from Definition 3.1. In the theorem, substitution of $P(T) = a_0(1+aT+\cdots)$ gives

$$a_0(1+aT)(1+T)(1+qT)(1-(d+1)T)$$
$$\equiv \frac{1}{q-1}\left(\frac{A_d}{\binom{n}{d}} + \frac{A_{d+1}}{\binom{n}{d+1}}T\right) \pmod{T^2}$$

Comparison of the linear coefficients on both sides shows that $A_{d+1} \geq 0$ if and only if

$$d+1 \leq q+1+a \tag{4.1}$$

Therefore, estimates for the reciprocal zeros of $P(T)$, and in particular for their sum $-a$, yield upper bounds on the minimum distance of a linear code. We remark that for an algebraic curve with zeta polynomial $P(T)$ the analogous property (4.1) says $N = q+1+a$, where N is the number of rational points of the curve. For self-dual codes we have the following additional information about the zeros. For a self-dual code of genus g, the zeta polynomial is of degree $2g$ and satisfies ([**3**])

$$P(1/qT)q^gT^{2g} = P(T)$$

Therefore, the reciprocal zeros different from $\pm\sqrt{q}$ appear in pairs $(\alpha, q/\alpha)$. For zeta functions of curves it is known that all reciprocal zeros have the same absolute value $\sqrt{q}$, a property known as the Riemann hypothesis for curves. The Riemann hypothesis is crucial in obtaining upper bounds for the number of rational points on an algebraic curve. It yields immediately that

$$N \leq q + 1 + 2g\sqrt{q}$$

We will need to verify the Riemann hypothesis for self-dual codes if we want to derive in a similar way upper bounds for the minimum distance of a self-dual code.

DEFINITION 4.1. A self-dual code satisfies the *analogue of the Riemann hypothesis* (has the RH property) if all the reciprocal zeros of its zeta polynomial have the same absolute value $\sqrt{q}$.

As an illustration, Figure 1 gives the zeros of the zeta polynomial of the binary extremal code of length 72. It is not known if such a code exists but its weight enumerator is uniquely determined by the parameters $[72, 36, 16]$ of the code and the zeta polynomial can therefore be computed. The zeros all have the same absolute value $1/\sqrt{2}$ and the code has the RH property.

OPEN PROBLEM 4.2. Prove or disprove that all extremal weight enumerators have the RH property.

An example of a doubly even self-dual code (not extremal) that does not satisfy the Riemann hypothesis is the set of words $(a|b|c)$ where a, b, c are arbitrary words of the $[8, 4, 4]$ extended Hamming code.

5. Codes of length 12

We restrict ourselves to a special case to show that information about the zeros of a zeta function can be obtained through linear programming. Let $M_{n,d}$ be the weight enumerator of a hypothetical MDS code of length n and minimum distance d (as in Section 2). We will prove

PROPOSITION 5.1 (Riemann hypothesis for [12,6,4] codes). *Let $A(x, y)$ be a self-dual weight enumerator of length 12 and minimum distance 4, so that $A(x, y)$ can be written*

$$A(x, y) = a_0M_{12,4} + a_1M_{12,5} + \cdots + a_6M_{12,10}$$

Then the polynomial $p(T) = a_0 + a_1T + \cdots + a_6T^6$ has complex zeros

$$\alpha_1, \bar{\alpha}_1 = \frac{1}{\sqrt{2}}e^{\pm i\theta_1}, \qquad \frac{4}{15}\pi < \theta_1 < \frac{8}{15}\pi$$

$$\alpha_2, \bar{\alpha}_2 = \frac{1}{\sqrt{2}}e^{\pm i\theta_2}, \qquad \frac{9}{15}\pi < \theta_2 < \frac{11}{15}\pi$$

$$\alpha_3, \bar{\alpha}_3 = \frac{1}{\sqrt{2}}e^{\pm i\theta_3}, \qquad \frac{13}{15}\pi < \theta_3 < \frac{14}{15}\pi$$

PROOF. Indeed the weight enumerator $A(x,y)$ is a linear combination of the MDS weight enumerators $M_{12,4}, M_{12,5}, \ldots, M_{12,10}$. The leading weight enumerator $M_{12,4}$ belongs to a $[12,9,4]$ code and the trailing weight enumerator $M_{12,10}$ to its dual code of type $[12,3,10]$. Application of the MacWilliams transform term by term gives

$$A^{\perp}(x,y) = 8a_0 M_{12,10} + 4a_1 M_{12,9} + \cdots + \frac{1}{8} a_6 M_{12,4}.$$

Therefore, the weight enumerator $A(x,y)$ is self-dual if and only if

$$a_6 = 8a_0, \quad a_5 = 4a_1, \quad a_4 = 2a_2. \tag{5.1}$$

The coefficients are the same for the normalized weight enumerator

$$a(t) = a_0 m_4(t) + a_1 m_5(t) t + \cdots + a_6 m_{10}(t) t^6.$$

The normalized weight enumerators $m_4, m_5, \ldots, m_{10}$ are given by Lemma 3.5 and in matrix form the coefficients of $a(t)$ are given by

$$(a_0, a_1, \ldots, a_6) \begin{bmatrix} 1 & -2 & 4 & -6 & 9 & -12 & 16 & -20 & 25 \\ 0 & 1 & -3 & 7 & -13 & 22 & -34 & 50 & -70 \\ 0 & 0 & 1 & -4 & 11 & -24 & 46 & -80 & 130 \\ 0 & 0 & 0 & 1 & -5 & 16 & -40 & 86 & -166 \\ 0 & 0 & 0 & 0 & 1 & -6 & 22 & -62 & 148 \\ 0 & 0 & 0 & 0 & 0 & 1 & -7 & 29 & -91 \\ 0 & 0 & 0 & 0 & 0 & 0 & 1 & -8 & 37 \end{bmatrix} \tag{5.2}$$

Note that the computation of the coefficients in the matrix is particularly straightforward with the recurrence (3.2). The coefficients of $a(t)$ are

$$\frac{A_4}{\binom{12}{4}}, \frac{A_5}{\binom{12}{5}}, \ldots, \frac{A_{12}}{\binom{12}{12}}$$

The factor $1/(q-1)$, that needs to be included in general, disappears for binary codes. Thus we find the transformation between the zeta polynomial and the weight enumerator, originally defined in Definition 3.1, in a different form as a linear transformation between vectors. The coefficients of $a(t)$ are all non-negative. Using (5.1), we find the following nine constraints in four variables

$$(a_0, a_1, a_2, a_3) \begin{bmatrix} 1 & -2 & 4 & -6 & 9 & -12 & 24 & -84 & 321 \\ 0 & 1 & -3 & 7 & -13 & 26 & -62 & 166 & -434 \\ 0 & 0 & 1 & -4 & 13 & -36 & 90 & -204 & 426 \\ 0 & 0 & 0 & 1 & -5 & 16 & -40 & 86 & -166 \end{bmatrix} \geq 0$$

These constraints will determine to some extent the zeros of the zeta polynomial $P(T)$. We divide $p(T)$ by $(\sqrt{2}T)^3$ so that the quotient takes real values on the circle $|T| = 1/\sqrt{2}$. Let $t = \sqrt{2}\,T$.

$$\begin{aligned} p(T)/(\sqrt{2}T)^3 &= \frac{a_0}{t^3} + \frac{a_1}{\sqrt{2}t^2} + \frac{a_2}{2t} + \frac{a_3}{2\sqrt{2}} + \frac{a_2 t}{2} + \frac{a_1 t^2}{\sqrt{2}} + \frac{a_0 t^3}{1} \\ &= \frac{a_3}{2\sqrt{2}} + \frac{a_2}{2}(t^{-1} + t) + \frac{a_1}{\sqrt{2}}(t^{-2} + t^2) + a_0(t^{-3} + t^3). \end{aligned}$$

After the substitution $t = e^{i\theta}$ the right hand side becomes

$$f(\theta) = \frac{a_3}{2\sqrt{2}} + \frac{a_2}{2}(2\cos(\theta)) + \frac{a_1}{\sqrt{2}}(2\cos(2\theta)) + a_0(2\cos(3\theta)).$$

For the real valued function $f(\theta)$ we have a real zero between any two θ, θ' such that $f(\theta)$ and $f(\theta')$ are of opposite sign. To determine the sign of $f(\theta)$ we use a subset of the constraints.

$$(a_0, a_1, a_2, a_3)\begin{bmatrix} -12 & 24 & -84 & 321 \\ 26 & -62 & 166 & -434 \\ -36 & 90 & -204 & 426 \\ 16 & -40 & 86 & -166 \end{bmatrix} = (\alpha_9, \alpha_{10}, \alpha_{11}, \alpha_{12}) \geq 0$$

Then

$$\begin{aligned} f(\theta) &= (\alpha_9, \alpha_{10}, \alpha_{11}, \alpha_{12})\frac{1}{462}\begin{bmatrix} 154 & 693 & 1683 & 2805 \\ 154 & 462 & 869 & 1320 \\ 70 & 140 & 176 & 221 \\ 14 & 28 & 44 & 64 \end{bmatrix}\begin{bmatrix} 2\cos(3\theta) \\ \sqrt{2}\cos(2\theta) \\ \cos(\theta) \\ \frac{1}{2\sqrt{2}} \end{bmatrix} \\ &= \alpha_9 f_9(\theta) + \alpha_{10} f_{10}(\theta) + \alpha_{11} f_{11}(\theta) + \alpha_{12} f_{12}(\theta) \end{aligned}$$

Thus $f(\theta)$ lies in the cone generated by the functions $f_9, f_{10}, f_{11}, f_{12}$. Figure 2 shows that any such function has three real zeros in the interval $(0, \pi)$. And any formally self-dual code of type $[12, 6, 4]$ satisfies the analogue of the Riemann hypothesis. Moreover

$$f_9(\theta), f_{10}(\theta), f_{11}(\theta), f_{12}(\theta) > 0, \qquad \text{for } \theta = \frac{4}{15}\pi, \frac{11}{15}\pi, \frac{13}{15}\pi$$

$$f_9(\theta), f_{10}(\theta), f_{11}(\theta), f_{12}(\theta) < 0, \qquad \text{for } \theta = \frac{8}{15}\pi, \frac{9}{15}\pi, \frac{14}{15}\pi$$

which confirms the positions of the zeros as given by Proposition 5.1. □

6. Codes of genus three

A self-dual code is of genus $g = 3$ if $n+2-2d = 2g = 6$, in other words if it is of type $[n, n/2, n/2-2]$. For $n = 12$, we find the $[12, 6, 4]$ codes of the previous section. In this section, we use the result of the previous section to prove the following

PROPOSITION 6.1 (Upper bound for genus three codes). *Binary formally self-dual weight enumerators of length n and minimum distance $d = n/2 - 2$ do not exist for $n > 16$.*

REMARK 6.2. The case $n = 16$ is realized by the distance distribution

$$A(x, y) = x^{16} + 112\, x^{10}y^6 + 30\, x^8y^8 + 112\, x^6y^{10} + y^{16}$$

of the non-linear Nordstrom-Robinson code and the proposition confirms that this is optimal. The corresponding zeta polynomial is

$$P(T) = (1 + 4\,T + \frac{55}{6}\,T^2 + 15\,T^3 + \frac{55}{3}\,T^4 + 16\,T^5 + 8\,T^6)/(143/2)$$

To relate weight enumerators such as the previous to codes of length 12 we puncture and shorten them till the required length is reached. To guarantee that the result is a self-dual weight enumerator we need to puncture and shorten an equal number of times. To make the process independent of the choice of the coordinates at which we puncture or shorten, we puncture or shorten at all coordinates and average the result. A word of weight w appears in $n-w$ of the n shortened codes, in each case with weight w. The average of the n shortened codes is therefore obtained by acting on $A(x,y)$ with

$$\mathcal{S} = (\frac{\partial}{\partial_x})/n$$

A word of weight w contributes to all of the punctured codes, w times as a word of weight $w-1$ and $n-w$ times as a word of weight w. The average of the punctured codes is therefore obtained by acting on $A(x,y)$ with

$$\mathcal{P} = (\frac{\partial}{\partial_x} + \frac{\partial}{\partial_y})/n$$

The operation

$$\mathcal{PS} = (\frac{\partial^2}{\partial_x^2} + \frac{\partial^2}{\partial_x \partial_y})/n(n-1)$$

sends a self-dual weight enumerator of given genus to a self-dual weight enumerator of same genus and of length two less. The reason that we can apply the results of the previous section to longer codes is the following

PROPOSITION 6.3 ([**4**]). *The zeta function of a code is invariant under the operations $\mathcal{P}$ (puncturing and averaging) and $\mathcal{S}$ (shortening and averaging).*

Thus any binary formally self-dual weight enumerator of length $n > 12$ and genus $g = 3$ can be punctured and shortened to a self-dual weight enumerator of type [12,6,4]. This leaves the zeta polynomial of the code unchanged. The intervals in Proposition 5.1 determine the following range for the sum $-a$ of the reciprocal roots of the zeta polynomial,

$$\begin{aligned} a &= -2\sqrt{2}(\cos(\theta_1) + \cos(\theta_2) + \cos(\theta_3)) \\ &\leq -2\sqrt{2}(\cos(\frac{8}{15}\pi) + \cos(\frac{11}{15}\pi) + \cos(\frac{14}{15}\pi)) < 4.96 \end{aligned}$$

We can apply (4.1),

$$d+1 \leq q+1+a < 7.96$$

And therefore $d \leq 6$ and $n \leq 16$.

Note that the Riemann hypothesis alone would give

$$d+1 \leq q+1+a \leq 3+6\sqrt{2} < 11.49$$

This last estimate may be considered as an analogue to the Hasse-Weil bound. To improve on this and to obtain the sharp bound $d \leq 6$, we need Proposition 5.1.

OPEN PROBLEM 6.4. Generalize Proposition 5.1 and give bounds for the zeros of the zeta polynomial in terms of the parameters n and d for sufficiently good formally self-dual codes.

It is interesting to compare self-dual codes of genus three with curves of genus three. The Hasse-Weil bound on the number of rational points on such a curve gives $N < 11.49$. This can be improved with the method of explicit formula, which yields that N is at most 7. The maximum is attained by a unique curve (up to isomorphism) that has zeta polynomial

$$P(T) = 1 + 4\,T + 9\,T^2 + 15\,T^3 + 18\,T^4 + 16\,T^5 + 8\,T^6$$

OPEN PROBLEM 6.5. Prove or disprove that a binary self-dual code of genus g has minimum distance at most $N - 1$, where N is the maximum number of rational points on a curve of genus g over the binary field.

References

[1] E.F. Assmus Jr., H.F. Mattson Jr. and R.J. Turyn. Cyclic Codes. AFCRL-65-332, Air Force Cambridge Research Labs, Bedford, Mass., 1965.
[2] V. G. Drinfeld and S. G. Vladuts, Number of points of an algebraic curve, Func. Anal. 17 (1983), pp. 53-54.
[3] I. M. Duursma, Weight distributions of geometric Goppa codes, Trans. Amer. Math. Soc. 351 (1999), pp. 3609-3639.
[4] I. M. Duursma, From weight enumerators to zeta functions, Discrete Applied Mathematics, to appear.
[5] G.D. Forney. *Concatenated Codes.* The M.I.T. Press, Cambridge, Mass., 1966.
[6] T. Kasami, S. Lin and W.W. Peterson. Some Results on Weight Distributions of BCH Codes. *IEEE Trans. Inf. Theor.*, 12:274, 1966.
[7] F.J. MacWilliams and N.J.A. Sloane. *The theory of error-correcting codes.* North Holland, Amsterdam, 1978.
[8] C.J. Moreno. *Algebraic curves over finite fields.* Tracts in Mathematics. Cambridge Univ. Press, Cambridge, England, 1991.
[9] H. Stichtenoth. *Algebriac function fields and codes.* Universitext. Springer-Verlag, New York, 1993.
[10] J.H. van Lint. *Introduction to coding theory*, volume 86 of *GTM.* Springer, Berlin, 1982.

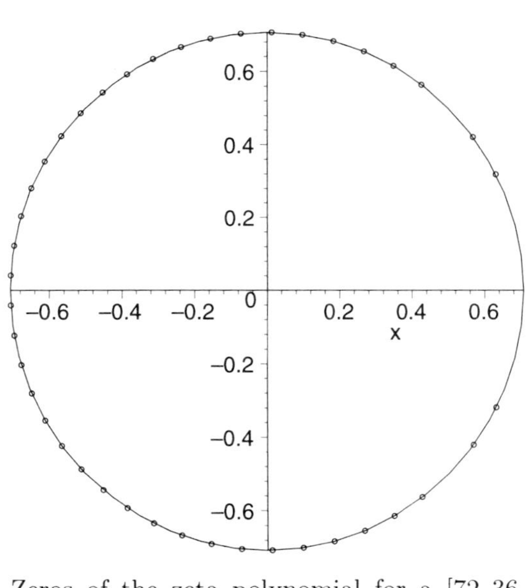

FIGURE 1. Zeros of the zeta polynomial for a $[72, 36, 16]$ code. (Section 4)

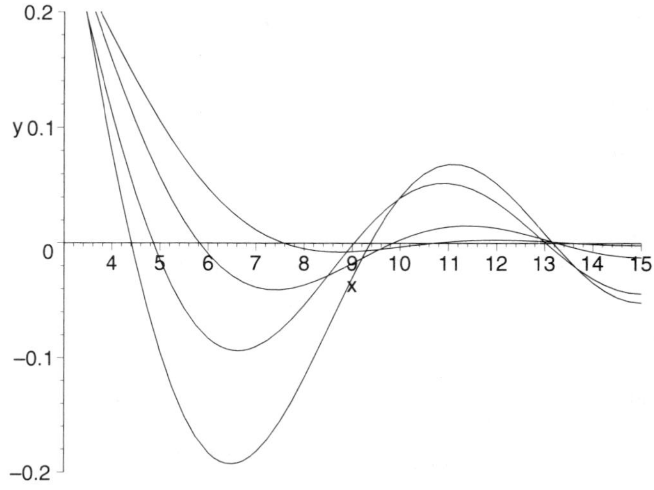

FIGURE 2. Vertically scaled functions $f_9, f_{10}, f_{11}, f_{12}$. Horizontal axis marked at multiples of $\pi/15$. (Section 5)

DEPARTEMENT DE MATHÉMATIQUES, UNIVERSITÉ DE LIMOGES, 87060 LIMOGES, FRANCE

Current address: Department of Mathematics, University of Illinois at Urbana-Champaign, Urbana, Illinois 61801

E-mail address: duursma@math.uiuc.edu

DIMACS Series in Discrete Mathematics
and Theoretical Computer Science
Volume **56**, 2001

On Perfect Codes in the Johnson Scheme

Tuvi Etzion

ABSTRACT. In his pioneer work from 1973 Delsarte has conjectured that no nontrivial perfect codes exist in the Johnson scheme. Up to date this conjecture remains wide open although many efforts to prove it were done during the past 25 years. We will survey the parameters for which the nonexistence of perfect codes is known. Some new results which we will present will enable to exclude other sets of parameters.

1. Introduction

In his well known work [4], Delsarte has presented the connections between association schemes and coding theory. He mainly considered two schemes, the Hamming scheme and the Johnson scheme. In chapter 4 of his work, Delsarte has considered perfect codes in these schemes. He has mentioned that only "very few" perfect codes exist in the Hamming scheme, see [7,8,11,12]. Delsarte has conjectured that in the Johnson scheme there exist no nontrivial perfect codes. For more than 25 years efforts were done to make progress towards proving this conjecture. Some important results were proved and for many sets of parameters we know that perfect codes do not exist in the Johnson scheme. But, the conjecture is far from being settled. In this paper we will survey some of the known knowledge, and provide some new results concerning the existence of perfect codes in the Johnson scheme.

In the Johnson scheme, we are given two integers, n and w, such that $0 \leq w \leq n$. Given a code C, all its codewords have length n and constant weight w. Two codewords u and v are in distance (J – *distance*) d apart if there are exactly d positions in which u has 1 values and v has 0 values. Obviously, there are exactly d other positions in which u has 0 values and v has 1 values. With the Johnson scheme we associate the Johnson graph $J(n, w)$. The vertex set, V_w^n, of the Johnson graph consists of all w-subsets of a fixed n-set. Two such w-subsets are adjacent if and only if their intersection has size $w - 1$. A code C of such w-subsets is called e-perfect code in $J(n, w)$ (or in the Johnson scheme) if the e-spheres of all the codewords of C form a partition of V_w^n. In other words, C is an e-perfect code if for

---

1991 *Mathematics Subject Classification.* Primary 51E10, 94B25.

This research was supported in part by grant no. 88/99–1 of the Israeli Science Foundation

each element $v \in V_w^n$, there exists a unique element $c \in C$, such that the J-distance between v and c is less or equal e. There are some trivial perfect codes in $J(n, w)$.

1. V_w^n is 0-perfect.
2. Any $\{v\}$, $v \in V_w^n$, is w-perfect.
3. if $n = 2w$, w odd, any pair of disjoint w subsets is e-perfect with $e = \frac{1}{2}(w-1)$.

It is conjectured that this set of codes are the only perfect codes in $J(n, w)$. Bannai[1] proved the nonexistence of e-perfect codes in $J(2w-1, w)$ and $J(2w+1, w)$ for $e \geq 2$. Hammond[6] extended the result and showed that $J(n, w)$ cannot contain a nontrivial perfect code for $n \in \{2w-2, 2w-1, 2w+1, 2w+2\}$. Generalizations of Lloyd's theorem[1,2,4] and other interesting ideas[9], didn't lead to significant results. A significant improvement was made by Roos[10] who showed the following result.

THEOREM 1.1. *If an e-perfect code in* $J(n, w)$, $n \geq 2w$, *exists then* $n \leq (w-1)(2e+1)/e$.

Etzion[5] has used Steiner systems and a new concept of configuration distribution, which is the analog for weight distribution in the Hamming scheme, to show many sets of parameter for which e-perfect codes cannot exist in the Johnson scheme. In this paper we survey some of Etzion's results and improve on some of them to show more sets of parameters in which e-perfect codes cannot exist in the Johnson scheme.

The rest of this paper is organized as follows. In Section 2 we survey the results from[5] and present the new results which are direct continuation of the results in[5]. In Section 3 we examine the theorems of Section 2 to show that the range in which e-perfect codes can exist is considerably reduced.

2. On the structure of perfect codes in $J(n, w)$

Perfect codes in the Johnson scheme have strong connection to constant weight codes and Steiner systems. For this purpose we need to use the Hamming metric. Two binary words u and v of the same length n have Hamming distance ($H-$ *distance*) d if they differ in exactly d positions. Note that two words $u, v \in V_w^n$ have J-distance d if and only if their H-distance is $2d$. A code C has minimum H-distance d if for any two codewords $u, v \in C$, the H-distance between u and v is at least d. All the lemmas and theorems which follow are given without proof. Lemmas 2.1 and 2.2 are taken from[5].

LEMMA 2.1. *If* C *is an e-perfect code in the Johnson scheme then its minimum H-distance is* $4e+2$.

An (n, d, w) code is a code of length n, constant weight w to all the codewords, and minimum H-distance d. $A(n, d, w)$ denote the maximum size of an (n, d, w) code. An extensive survey on the lower bounds on $A(n, d, w)$ can be found in[3].

LEMMA 2.2. *If* C *is an e-perfect code in* $J(n, w)$ *then* $A(n, 4e+2, w) = |C|$.

A Steiner system $S(t, k, n)$ is a collection of k-subsets (called *blocks*) taken from an n-set such that each t-subset of the n-set is contained in exactly one block. The following theorem is well known, e.g.[8, p. 60]

THEOREM 2.3. *A necessary condition that a Steiner system $S(t,k,n)$ exists is that the numbers* $\frac{\binom{n-i}{t-i}}{\binom{k-i}{t-i}}$, *be integers for* $0 \le i \le t$.

Henceforth, let $N = \{1, 2, ..., n\}$ be the n-set. From a Steiner system $S(t,k,n)$ we construct constant weight code on n coordinates as follows. From each block B we construct a codeword with 1's in the positions of B and 0's in the positions of $N \setminus B$. This construction leads to the following well known theorem[3]

THEOREM 2.4. $A(n, 2(k-t+1), k) = \frac{n(n-1) \cdots (n-t+1)}{k(k-1) \cdots (k-t+1)}$ *if and only if a Steiner system* $S(t,k,n)$ *exists.*

From Theorem 2.4 and Lemmas 2.1 and 2.2 we immediately infer the following result.

LEMMA 2.5. *If C is an e-perfect code in $J(n,w)$ and it is also a Steiner system then it is a Steiner system $S(w-2e, w, n)$.*

The next lemma is a simple observation of a considerable use.

LEMMA 2.6. *The complement of an e-perfect code in $J(n,w)$ is an e-perfect code in $J(n, n-w)$.*

Finally, we need a few more definitions which we will use in the proofs of the nonexistence theorems in the sequel. For a given partition of N into two subsets A and B such that $|A| = k$ and $|B| = n-k$, let *configuration* (i,j) consists of all vectors with weight i in the positions of A and weight j in the positions of B. For an e-perfect code C in $J(n,w)$ we say that $u \in C$ $J-cover$ $v \in V_w^n$ if the J-distance between u and v is less or equal e. For a given two subsets u and v we say that u $C-cover$ v if v is a subset of u (this is our usual understanding of the word cover).

In the sequel we will use a mixed language of sets notation and vectors notation. It should be understood from the context which one we are using, and the translation between the two different notations. Theorems 2.7 through 2.11 which follow were proved in[5]

THEOREM 2.7. *If an e-perfect code in $J(n,w)$ exists then a Steiner system $S(e+1, 2e+1, w)$ and a Steiner system $S(e+1, 2e+1, n-w)$ exist.*

THEOREM 2.8. *If an e-perfect code in $J(n,w)$, which is noi a Steiner system $S(w-2e, w, n)$ exists, then for some k, $0 \le k \le e-1$, a Steiner system $S(2, 2e-k+1, n-w+2e-2k)$ exists.*

THEOREM 2.9. *If an e-perfect code in $J(n,w)$, which is not a Steiner system $S(n-w-2e, n-w, n)$ exists, then for some k, $0 \le k \le e-1$, a Steinersystem $S(2, 2e-k+1, w+2e-2k)$ exists.*

THEOREM 2.10. *Except for the Steiner systems $S(1,w,n)$ and $S(w,w,n)$, there are no more Steiner systems which are also perfect codes in the Johnson scheme.*

THEOREM 2.11. *There are no e-perfect codes in* $J(2w + e + 1, w)$.

We are now in a position to state a couple of new results. First, we present and alternative and shorter proof to Theorem 1.1.

THEOREM 2.12. *If an e-perfect code in* $J(n, w)$ *exists then* $n \leq (w - 1)(2e + 1)/e$.

PROOF. Assume C is an e-perfect code in $J(n, w)$. We partition N into two subsets A and B, such that $|A| = n - w + 1$, $|B| = w - 1$, and a vector of the $(e + 1, w - e - 1)$ configuration is a codeword. Clearly, the J-distance between a vector from configuration $(e + 1, w - e - 1)$ and a vector from configuration $(e + 1 - i, w - e - 1 + i)$, $i > 0$, is less than $2e + 1$ and hence C doesn't have any codeword from configuration $(e + 1 - i, w - e - 1 + i)$. Therefore, all the vectors from configuration $(1, w - 1)$ are J-covered by codewords from configuration $(e + 1, w - e - 1)$. To J-cover each vector from configuration $(1, w - 1)$ exactly once we must have exactly $\frac{n-w+1}{e+1}$ codewords from configuration $(e + 1, w - e - 1)$. Since the minimum J-distance of C is $2e + 1$, two codewords from configuration $(e + 1, w - e - 1)$ cannot intersect in the zeroes of part B. Hence, $w - 1 \geq \frac{n-w+1}{e+1}e$, which is equivalent to $n \leq (w - 1)(2e + 1)/e$. □

THEOREM 2.13. *If an e-perfect code in* $J(n, w)$ *exists and* $n < (w-1)(2e+1)/e$ *then an* $S(2, e + 2, n - w + 2)$ *exists.*

PROOF. Assume C is an e-perfect code in $J(n, w)$. Again, as in the proof of Theorem 2.12, we partition N into two subsets A and B, such that $|A| = n - w + 1$, $|B| = w - 1$, and there are $\frac{n-w+1}{e+1}$ codewords from configuration $(e + 1, w - e - 1)$. Since $n < (w - 1)(2e + 1)/e$, i.e., $\frac{n-w+1}{e+1}e < w - 1$, we have at least one coordinate in B which have ones in all the codewords from configuration $(e + 1, w - e - 1)$. We remove this coordinate from B to obtain B_1 and join it to A to obtain A_1. $|A_1| = n - w + 2$, $|B_2| = w - 2$, and C doesn't have any codeword from configuration $(e+2-i, w-e-2+i)$, $i > 0$. Therefore, all the vectors from configuration $(2, w-2)$ are J-covered by codewords from configuration $(e + 2, w - e - 2)$. Since each vector from configuration $(2, w - 2)$ must be J-covered by exactly one codeword from configuration $(e + 2, w - e - 2)$, it follows that part A_1 of the codewords from configuration $(e + 2, w - e - 2)$ forms a Steiner system $S(2, e + 2, n - w + 2)$. □

COROLLARY 2.14. *If an e-perfect code in* $J(n, w)$ *exists and* $w \leq n - w$, *then an* $S(2, e + 2, w + 2)$ *exists.*

3. Applications

The theorems mentioned in Section 2 make it possible to reduce the range in which perfect codes in the Johnson scheme can exist. If an e-perfect code in $J(n, w)$ exists then by Theorem 2.7, Steiner systems $S(e + 1, 2e + 1, w)$ and $S(e + 1, 2e + 1, n - w)$ exist. By the divisibility conditions of Theorem 2.3, we have that $e + 1$ should divide $w - e$ and $n - w - e$ and hence we have $w \equiv e \ (mod\ e + 1)$. This divisibility condition was also obtained by Martin[9]. This condition itself limits the range in which e-perfect codes can exist. Combining this condition with the nonexistence of e-perfect codes in $J(2w+e+1,w)$ obtained in Theorem 2.11, we have the following theorem of[5].

THEOREM 3.1. *There are no perfect codes in* $J(2w+p,w)$, *p prime.*

Henceforth, without loss of generality, we assume that $w \leq n-w$. We will also consider e-perfect codes which are not Steiner systems since by Theorem 2.10 all the e-perfect codes which are Steiner systems are already known. Examining carefully of the divisibility conditions of Theorem 2.3 for $S(e+1,2e+1,w)$, $S(e+1,2e+1,n-w)$, $S(2,e+2,w+2)$, and possibly $S(2,e+2,n-w+2)$ (which exist by Theorem 2.13 if $n<(w-1)(2e+1)/e$), leads to many other similar results about non existence of e-perfect codes in several graphs $J(n,w)$. Some of these results are mentioned in[5]. On the otherhand, one might try to show when e-perfect code cannot exist in $J(n,w)$ for a given e.

If $e=1$ then the existence of $S(2,3,w)$, $S(2,3,w+2)$, $S(2,3,n-w)$, and $S(2,3,n-w+2)$ (implied by Theorem 2.8), imply that $w \equiv 1\ (mod\ 6)$ and $n-w \equiv 1\ (mod\ 6)$.

If $e=2$ then the existence of $S(3,5,w)$ and $S(2,4,w+2)$ imply that $w \equiv 2,\ 26$ or $50\ (mod\ 60)$ and therefore $n<(w-1)(2e+1)/e$ which implies the existence of $S(2,4,n-w+2)$. Since also $S(3,5,n-w)$ exists, we have that $n-w \equiv 2,\ 26$ or $50\ (mod\ 60)$.

If $e=3$ then the existence of $S(4,7,w)$ and $S(2,5,w+2)$ imply that $w \equiv 3,\ 23,\ 43$ or $63\ (mod\ 140)$ and therefore $(w-1)(2e+1)/e \equiv 18\ (mod\ 20)$. Since $S(4,7,n-w)$ exists and hence $S(2,5,n-w-2)$ exists, we have $n-w \equiv 3$ or $7\ (mod\ 20)$ and therefore $n<(w-1)(2e+1)/e$ which implies the existence of $S(2,5,n-w+2)$. Hence, we also have that $n-w \equiv 3,\ 23,\ 43$ or $63\ (mod\ 140)$.

As another example how these results limit the range in which e-perfect codes in the Johnson scheme may exist, we consider the case where the necessary conditions given by Theorem 2.3 for $S(2,e+2,v)$ are $v \equiv 1$ or $e+2\ (mod\ [e+1][e+2])$. Since both $S(2,e+2,w-e+1)$ and $S(2,e+2,w+2)$ exist, we have that $w \equiv e\ (mod\ [e+1][e+2])$. Since $S(2,e+2,n-w-e+1)$ exists, we have that $n-w \equiv e$ or $2e+1\ (mod [e+1][e+2])$. Now, we want to examine whether the bound $n \leq (w-1)(2e+1)/e$ can hold with equality. When e is even, $w \equiv e\ (mod\ [e+1][e+2])$ is even and hence e does not divide $(w-1)(2e+1)$. It follows that $n<(w-1)(2e+1)/e$. When e is odd, since $w \equiv e\ (mod\ [e+1][e+2])$, we have $(w-1)(2e+1)/e \equiv \frac{e^2+3}{2}+4e\ (mod\ [e+1][e+2])$. Clearly, $n=w+(n-w) \equiv 2e$ or $3e+1\ (mod\ [e+1][e+2])$ and hence $n<(w-1)(2e+1)/e$. Therefore, by Theorem 2.13 an $S(2,e+2,n-w+2)$ also exists, which together with the existence of $S(2,e+2,n-w-e+1)$ implies that $n-w \equiv e\ (mod\ [e+1][e+2])$. Note, that the condition that the necessary condition for $S(2,e+2,v)$ is $v \equiv 1$ or $e+2\ (mod\ [e+1][e+2])$ is met in many cases, e.g., when $e+2$ is and odd prime.

Finally, another application can be found by a combination with the results of Martin[9]. He has proved that if a 1-perfect code exists in $J(n,w)$ then there exist two integers r and s such that $w=rs+1$, $n=2rs+r-s+1$. He has found that for $2 \leq s \leq 10$, 1-perfect codes cannot exist. For $s=1$, we have $n=3w-3$, which is possible only for $w = 3$ (it corresponds to the 1-perfect code $S(1,3,6)$), since otherwise $w \equiv 1 \ (mod\ 6)$, $n - w \equiv 1 \ (mod\ 6)$, and hence $n \equiv 2 \ (mod\ 6)$.

References

[1] E. Bannai, *Codes in bi-partite distance-regular graphs*, Journal London Math. Soc. **2** (1977), 197–202.

[2] E. Biggs, *Perfect codes in graphs*, Journal of Combinatorial Theory, Series B **15** (1973), 289–296.

[3] A. E. Brouwer, J. B. Shearer, N. J. A. Sloane, W. D. Smith, *A new table of constant weight codes*, IEEE Trans. on Information Theory **IT-36** (1990), 1334–1380.

[4] P. Delsarte, *An algebraic approach to association schemes and coding theory*, Philips Journal Res **10** (1973), 1–97.

[5] T. Etzion, *On the nonexistence of perfect codes in the Johnson scheme*, SIAM Journal on Discrete Mathematics **9** (1996), 201–209.

[6] P. Hammond, *On the non-existence of perfect codes and nearly perfect codes*, Discrete Mathematics **39** (1982), 105–109.

[7] J.H. van Lint, *Nonexistence theorems for perfect error-correcting-codes*, Computers in Algebra and Number Theory, SIAM-AMS Proceedings **IV** (1971).

[8] F.J. MacWilliams and N.J.A. Sloane, *The Theory of Error-Correcting Codes*, North-Holland, Amsterdam, 1977.

[9] W. J. Martin, *Completely Regular Subsets*, Ph.D. Thesis, University of Waterloo, 1992.

[10] C. Roos, *A note on the existence of perfect constant weight codes*, Discrete Mathematics **47** (1983), 121–123.

[11] A. Tietäväinen, *On the nonexistence of perfect codes over finite fields*, SIAM J. Appl. Math. **24** (1973), 88–96.

[12] V.A. Zinov'ev and V.K. Leont'ev, *The nonexistence of perfect codes over Galois fields*, Probl. Control and Info. Theory **2** (1973), 123–132.

COMPUTER SCIENCE DEPARTMENT, TECHNION, HAIFA 32000, ISRAEL
E-mail address: etzion@cs.technion.ac.il

DIMACS Series in Discrete Mathematics
and Theoretical Computer Science
Volume **56**, 2001

On Perfect Constant Weight Codes

Tuvi Etzion and Jack van Lint

ABSTRACT. We generalize the construction of perfect ternary constant weight codes given by Van Lint and Tolhuizen to alphabets of size $q = 2^k + 1$ and we show that the construction does not work for $q = 4$.

I. Introduction

In [1] Van Lint and Tolhuizen considered as space R the subset of all words of fixed weight w in $\mathbb{F}_3^n$ for given w and n. Hamming distance of two words in R was defined in the usual way. A subset C of R is called a *perfect single error correcting code* if every word in R has distance 0 or 1 to *exactly* one word in C. In [1] it is shown that if such a code exists, then $w+1$ must be a power of 2. If $n = w$ one simply has a binary perfect code (over the alphabet $\{1, 2\}$). A construction is given for the case $n = w + 1$ and larger n turn out to be impossible.

Now, let $t = 2^k$ and consider an alphabet Q of $q = t + 1$ symbols. We wish to generalize the construction of [1] and [2] by replacing $\mathbb{F}_3$ by Q. Again, C will be a perfect single error correcting code in the space R of all words of length n and weight w over Q. The following results are straightforward generalizations of results in [1] with similar proofs.

The fact that C is perfect is characterized by equality in the following inequality which holds for all codes C of length n, constant weight w and minimum distance 3 :

$$|C| \leq \frac{\binom{n}{w} t^w}{(t-1)w+1}. \tag{1.1}$$

Let the subcode C_i of C be defined by

$$C_i := \{\mathbf{c} \in C \mid c_i = 0\}, \quad (i = 1, \ldots, n). \tag{1.2}$$

Lemma 1 *If C is a perfect single error correcting constant weight code over Q with parameters n and w, then the shortened codes, defined by deleting the i-th coordinate in the codes C_i of (1.2), are also perfect (with parameters $n-1$ and w).*

1991 *Mathematics Subject Classification.* 94B25.

Research of the first author was supported in part by grant no. 88/99–1 of the Israeli Science Foundation

Corollary 1 *If the code of Lemma 1 exists, then* $(t-1)w+1$ *must be a power of* t.

II. A construction

For the alphabet Q we take $\mathbb{F}_t \cup \{\infty\}$. At the end of the construction we interchange the symbols 0 and ∞ which will have the effect that all codewords have weight w. For $k > 1$ the construction only works if we take $w = t + 1 = q$ and $n = w + 1$.

We consider the cyclic t-ary Hamming code $\mathcal{H}$ of length w defined by

$$(c_0, c_1, \dots, c_{w-1}) \in \mathcal{H} \Leftrightarrow c_0 + c_1\beta + \cdots + c_{w-1}\beta^{w-1} = 0,$$

where β is a *primitive* w-th root of unity in $\mathbb{F}_t^2$. It is well known that $\mathcal{H}$ has minimum distance 3 and that the extended code $\overline{\mathcal{H}}$ has minimum distance 4.

For $0 \le i < w$ we define $\overline{\mathcal{H}_i}$ to be the coset of $\overline{\mathcal{H}}$ containing $(0, 0, \dots, 0, 1, 0, \dots, 0)$ where the 1 is in position i.

The code $\mathcal{T}$ is obtained from $\overline{\mathcal{H}}$ by replacing the symbol in the final position by the symbol ∞. Similarly, the code $\mathcal{T}_i$ is obtained from $\overline{\mathcal{H}_i}$ by replacing the coordinate in position $i + 1 \pmod{w}$ (in each word) by the symbol ∞. Clearly $\mathcal{T}$ and each $\mathcal{T}_i$ have minimum distance 3.

We define $C := \mathcal{T} \cup \mathcal{T}_0 \cup \cdots \cup \mathcal{T}_{w-1}$. To show that C has minimum distance 3 we have to take pairs of codewords $(\mathbf{a}', \mathbf{b}')$ where $\mathbf{a}'$ is in $\mathcal{T}$ or some $\mathcal{T}_i$ and $\mathbf{b}'$ is in some $\mathcal{T}_j$ and show that $d(\mathbf{a}', \mathbf{b}') \ge 3$. Here $\mathbf{a}'$ is derived from a word $\mathbf{a} \in \overline{\mathcal{H}}$ and similarly $\mathbf{b}'$ from a word $\mathbf{b} := (b_0, \dots, b_{w-1}, b_\infty) \in \overline{\mathcal{H}}$. Because $\overline{\mathcal{H}}$ is linear we may take $\mathbf{a} = \mathbf{0}$.

For the subcodes containing $\mathbf{a}'$ and $\mathbf{b}'$ we must distinguish three cases.
Case 1 : $\mathcal{T}$ and $\mathcal{T}_i$. We have

$$\mathbf{a}' = (0, 0, \dots, 0, \infty) \text{ and } \mathbf{b}' = (b_0, \dots, b_{i-1}, b_i + 1, \infty, b_{i+2}, \dots, b_{w-1}, b_\infty).$$

If $\mathbf{b} = \mathbf{0}$, then $d(\mathbf{a}', \mathbf{b}') = 3$ (from positions i, $i+1$ and ∞). If $\mathbf{b} \ne \mathbf{0}$, then again positions $i+1$ and ∞ differ and since $w(\mathbf{b}') \ge 4$ there is at least one more position where $\mathbf{a}'$ and $\mathbf{b}'$ differ.

Case 2 : $\mathcal{T}_i$ and $\mathcal{T}_{i+1}$. We compare

$$(0, \dots, 0, 1, \infty, 0, 0, \dots, 0) \text{ and } (b_0, \dots, b_{i-1}, b_i, b_{i+1} + 1, \infty, b_{i+3}, \dots, b_\infty).$$

If $\mathbf{b} = \mathbf{0}$ then clearly $d(\mathbf{a}', \mathbf{b}') = 3$. If $\mathbf{b} \ne \mathbf{0}$, then again $w(\mathbf{b}') \ge 4$ guarantees $d(\mathbf{a}', \mathbf{b}') \ge 4$.

Case 3 : $\mathcal{T}_i$ and $\mathcal{T}_j$ where $j > i+1$. It is now clear that we only have to worry about the possibility that $\overline{\mathcal{H}}$ contains a word of weight 4 that has as nonzero coordinates ones in positions i and j, a coordinate ξ in position $i+1$ and a coordinate η in position $j+1$. Since this word has a 0 in position ∞, the sum of its coordinates is 0, that is $\xi = \eta$. However,

$$1 + \xi\beta + \beta^{j-i} + \xi\beta^{j-i+1} = (1 + \xi\beta)(1 + \beta^{j-i}) \ne 0$$

and therefore such a word does not exist in $\overline{\mathcal{H}}$.

We have proved the following theorem.

Theorem 1 *Let $t = 2^k$, $w = q = t+1$, and $n = w+1$. There exists a perfect single error correcting constant weight code of length n and weight w over an alphabet of size q.*

For length greater than $q+1$, the extended Hamming code does not have minimum distance 4. So, the idea of the construction does not work (except of course if $k = 1$, the case of [1]). However, a perfect code with parameters corresponding to Lemma 1 could still exist. We have not been able to show nonexistence. In the next section, we treat the smallest unknown case, namely alphabet-size 4, weight 4 and length 5.

III. Quaternary codes

For any q, the code $\{(1,0),(0,1)\}$ is a perfect single error correcting constant weight code with weight $w = 1$. We consider alphabet size $q = 4$. For $n = 4$ we have the perfect ternary code over the alphabet $\{1,2,3\}$ interpreted as a quaternary code over $\{0,1,2,3\}$ where the words have constant weight $w = n = 4$. These are the trivial examples. The first interesting case is $n = 5$.

In this case we again have the results of Lemma 1 and its corollary. Here $t = 3$. For the sake of convenience we use as alphabet $\mathbb{F}_3$ with a special symbol adjoined which we call 3. (As before, the roles of 0 and the extra symbol have been interchanged.) To construct a perfect code of length 5 we now have to find five codes C_i $(1 \le i \le 5)$ where in C_i the i-th coordinate is 3 and deleting this coordinate yields a perfect ternary code C_i' of length 4. Such a perfect ternary code is essentially unique, namely a translate of a Hamming code. There are eight different linear codes that are all equivalent to the Hamming code and each has eight cosets (besides itself). So, there are 72 possible choices for the code C_i'.

Nonexistence

The eight different Hamming codes $A_0, B_0, \ldots, H_0$ have as generator matrices

$$\begin{pmatrix} 0 & 1 & 1 & 1 \\ 1 & 0 & 1 & 2 \end{pmatrix}, \quad \begin{pmatrix} 0 & 1 & 1 & 1 \\ 1 & 0 & 2 & 1 \end{pmatrix}, \quad \begin{pmatrix} 0 & 1 & 1 & 2 \\ 1 & 0 & 1 & 1 \end{pmatrix}, \quad \begin{pmatrix} 0 & 1 & 1 & 2 \\ 1 & 0 & 2 & 2 \end{pmatrix},$$

$$\begin{pmatrix} 0 & 1 & 2 & 1 \\ 1 & 0 & 1 & 1 \end{pmatrix}, \quad \begin{pmatrix} 0 & 1 & 2 & 1 \\ 1 & 0 & 2 & 2 \end{pmatrix}, \quad \begin{pmatrix} 0 & 1 & 2 & 2 \\ 1 & 0 & 2 & 1 \end{pmatrix}, \quad \begin{pmatrix} 0 & 1 & 2 & 2 \\ 1 & 0 & 1 & 2 \end{pmatrix}.$$

The coset of the code X_0, containing $(00xy)$, where xy is the ternary representation of the integer j $(0 \le j \le 8)$ is called X_j $(X = A, B, \ldots, H)$.

If we wish to construct a quaternary constant weight code with $n = 5$, $w = 4$, we may without loss of generality take C_5' to be A_0. Since C_5 contains (00003), none of the codes C_i', $(1 \le i \le 4)$ can be X_0 or an X_j with $1 \le j \le 2$ because then C_i would contain a word of weight 2 with a nonzero coordinate in position 5. (This contradicts $d = 3$.)

We shall now list restrictions on the candidates for the codes C_i' $(1 \le i \le 4)$. To explain the notation we treat the first case in detail.

Since $(0111) \in A_0$ we have $(01113) \in C_5$ and hence C_1' may not contain a word $(111*)$ because $d(01113),(3111*)) = 2$. This implies that $C_1' \notin \{A_j : 6 \le j \le 8\} \cup \{C_j : 6 \le j \le 8\}$ and also not in one of the codes B_j, D_j, E_j, H_j with $3 \le j \le 5$, Similarly, none of the C_i' $(2 \le i \le 4)$ is among E_j, F_j, G_j, H_j $(6 \le j \le 8)$.

This restriction is listed as follows

$$(0111) \in A_0 \Rightarrow \begin{cases} C_1' \notin (A, C; 6, 7, 8) \cup (B, D, E, H; 3, 4, 5) \\ C_2' \notin (E, F, G, H; 6, 7, 8) \\ C_3' \notin (E, F, G, H; 6, 7, 8) \\ C_4' \notin (E, F, G, H; 6, 7, 8) \end{cases}.$$

Similarly, we find

$$(0222) \in A_0 \Rightarrow \begin{cases} C_1' \notin (A, C; 3, 4, 5) \cup (B, D, E, H; 6, 7, 8) \\ C_2' \notin (E, F, G, H; 3, 4, 5) \\ C_3' \notin (E, F, G, H; 3, 4, 5) \\ C_4' \notin (E, F, G, H; 3, 4, 5) \end{cases}.$$

So, the only candidates for C_1' are the codes F_j and G_k. Similarly, the only candidates for C_2' turn out to be A_l and C_m. There are four possibilities to be considered. We look at the combination of F and A.

For $i = 1, 2, 3$ consider the words in F_0 with second coordinate i and similarly for A_0. These yield words of type $(3 * ipq) \in C_1$ derived from F_0 and words of type $(*3irs)$ in C_2 derived from A_0. The differences $(p, q) - (r, s)$ contain all nine possible pairs. Hence no choice of j and l is compatible.

Other cases are treated in the same way. This analysis was easily carried out by hand. So, in this case there is no perfect code.

References

1. J. H. van Lint and L. M. G. M. Tolhuizen, *On Perfect Ternary Constant Weight Codes*, Designs, Codes and Cryptography **18** (1999), 231–234.
2. M. Svanström, *A Class of Perfect Ternary Constant–Weight Codes*, Designs, Codes and Cryptography **18** (1999), 223–229.

DIMACS Series in Discrete Mathematics
and Theoretical Computer Science
Volume **56**, 2001

On Additive $GF(4)$ Codes

Philippe Gaborit, W. Cary Huffman, Jon-Lark Kim, and Vera Pless

ABSTRACT. In this paper we discuss the status of the classification of the extremal additive codes over $GF(4)$ which are self-dual under the trace inner product. These codes are of interest because of their connection to quantum codes. There are exactly 5 inequivalent codes of length 8, exactly 8 of length 9, at least 56 of length 10, and exactly 1 of length 11. We discuss the methods of classification, the concepts of lengthening and shortening, and connections to binary codes.

1. Introduction

An *additive code $\mathcal{C}$ over $GF(4)$ of length n* is an additive subgroup of $GF(4)^n$. As $\mathcal{C}$ is a free $GF(2)$-module, it has size 2^k for some $0 \leq k \leq 2n$. We call $\mathcal{C}$ an $(n, 2^k)$ code. It has a basis, as a $GF(2)$-module, consisting of k basis vectors; a *generator matrix* of $\mathcal{C}$ will be a $k \times n$ matrix with entries in $GF(4)$ whose rows are a basis of $\mathcal{C}$. Interest in additive codes over $GF(4)$ has arisen because of their correspondence to quantum codes as described in [**3**]. There is a natural inner product arising from the trace map, on the additive codes induced by this correspondence. If we let $GF(4) = \{0, 1, \omega, \overline{\omega}\}$ where $\overline{\omega} = \omega^2 = 1 + \omega$, the *trace* map $\mathrm{Tr} : GF(4) \to GF(2)$ is given by

$$\mathrm{Tr}(x) = x + x^2.$$

In particular $\mathrm{Tr}(0) = \mathrm{Tr}(1) = 0$ and $\mathrm{Tr}(\omega) = \mathrm{Tr}(\overline{\omega}) = 1$. The *conjugate* of $x \in GF(4)$, denoted $\overline{x}$, is the image of x under the Frobenius automorphism; in other words, $\overline{0} = 0$ $\overline{1} = 1$, and $\overline{\overline{\omega}} = \omega$. We now define the *trace inner product* of two vectors $\mathbf{x} = x_1 x_2 \cdots x_n$ and $\mathbf{y} = y_1 y_2 \cdots y_n$ in $GF(4)^n$ to be

$$\mathbf{x} \star \mathbf{y} = \sum_{i=1}^{n} \mathrm{Tr}(x_i \overline{y_i}).$$

EXAMPLE 1.1. Let $\mathcal{G}_6$ be the $[6, 3, 4]$ *hexacode* whose generator matrix as a linear $GF(4)$-code is

$$\begin{bmatrix} 1\,0\,0\,1\,\omega\,\omega \\ 0\,1\,0\,\omega\,1\,\omega \\ 0\,0\,1\,\omega\,\omega\,1 \end{bmatrix}.$$

2000 *Mathematics Subject Classification.* Primary 94B60; Secondary 94B05, 94B25.

This is also an additive code; thinking of $\mathcal{G}_6$ as an additive code, it has generator matrix

$$\begin{bmatrix} 1 & 0 & 0 & 1 & \omega & \omega \\ \omega & 0 & 0 & \omega & \overline{\omega} & \overline{\omega} \\ 0 & 1 & 0 & \omega & 1 & \omega \\ 0 & \omega & 0 & \overline{\omega} & \omega & \overline{\omega} \\ 0 & 0 & 1 & \omega & \omega & 1 \\ 0 & 0 & \omega & \overline{\omega} & \overline{\omega} & \omega \end{bmatrix}.$$

If $\mathcal{C}$ is an additive code, its *dual*, denoted $\mathcal{C}^\perp$, is the additive code $\{\mathbf{x} \in GF(4)^n \mid \mathbf{x}\star\mathbf{c} = 0 \text{ for all } \mathbf{c} \in \mathcal{C}\}$. If $\mathcal{C}$ is an $(n, 2^k)$ code, then $\mathcal{C}^\perp$ is an $(n, 2^{2n-k})$ code. As usual, $\mathcal{C}$ is *self-orthogonal* if $\mathcal{C} \subseteq \mathcal{C}^\perp$ and *self-dual* if $\mathcal{C} = \mathcal{C}^\perp$. In particular, if $\mathcal{C}$ is self-dual, $\mathcal{C}$ is an $(n, 2^n)$ code. The code $\mathcal{G}_6$ in Example 1.1 is self-dual as an additive code. (Any $GF(4)$-linear code that is self-orthogonal under the Hermitian inner product is a self-orthogonal additive code under the trace inner product.)

As usual, the *weight* $\text{wt}(\mathbf{c})$ of $\mathbf{c} \in \mathcal{C}$ is the number of nonzero components of $\mathbf{c}$. The minimum weight d of $\mathcal{C}$ is the smallest weight of any nonzero codeword in $\mathcal{C}$. If $\mathcal{C}$ is an $(n, 2^k)$ additive code of minimum weight d, $\mathcal{C}$ is called an $(n, 2^k, d)$ code. $\mathcal{C}$ is *Type II* if $\mathcal{C}$ is self-dual and all codewords have even weight; as we see later Type II codes of length n exist only if n is even. If $\mathcal{C}$ is self-dual but some codeword has odd weight (in which case the code cannot be $GF(4)$-linear), the code is *Type I* (see [**11**, section 4.2]). There is a bound on the minimum weight of an additive self-dual code [**11**, Theorem 33]. If d_I, d_{II} is the minimum distance of an additive self-dual Type I, Type II code, respectively, of length $n > 1$, then

$$d_I \le \begin{cases} 2\left\lfloor \frac{n}{6} \right\rfloor + 1 & \text{if } n \equiv 0 \pmod 6 \\ 2\left\lfloor \frac{n}{6} \right\rfloor + 3 & \text{if } n \equiv 5 \pmod 6 \\ 2\left\lfloor \frac{n}{6} \right\rfloor + 2 & \text{otherwise} \end{cases} \tag{1}$$

$$d_{II} \le 2\left\lfloor \frac{n}{6} \right\rfloor + 2. \tag{2}$$

A code that meets the appropriate bound is called *extremal.* Type II codes meeting the bound d_{II} have a unique weight enumerator. This property is not true for Type I codes, to have a unique weight enumerator they would have to meet the $\lfloor n/2 \rfloor + 1$ bound (cf [**9**]), which is never reached except for the hexacode.

We say that two additive codes $\mathcal{C}_1$ and $\mathcal{C}_2$ are *equivalent* provided there is a map sending the codewords of $\mathcal{C}_1$ onto the codewords of $\mathcal{C}_2$ where the map consists of a permutation of coordinates followed by a scaling of coordinates by elements of $GF(4)$ followed by conjugation of some of the coordinates. In terms of generator matrices, if G_1 is a generator matrix of $\mathcal{C}_1$, then there is a generator matrix of $\mathcal{C}_2$ obtained from G_1 by permuting columns, scaling columns, and then conjugating some of the resulting columns. Notice that permuting coordinates, scaling coordinates, and conjugating some coordinates of a self-orthogonal (or self-dual) code does not change self-orthogonality (or self-duality). The *automorphism group* of $\mathcal{C}$, denoted $\text{Aut}(\mathcal{C})$, consists of all maps which permute coordinates, scale coordinates, and conjugate coordinates that send codewords of $\mathcal{C}$ to codewords of $\mathcal{C}$.

The purpose of this paper is to examine the classification of the extremal Type I and II additive codes.

2. Results

We summarize what is currently known about extremal (or optimal) Type I and II codes in Table 1 partly from [**8**]. In the table, codes have length 'n'; 'd_I' is

n	d_I	num$_I$	d_{II}	num$_{II}$	n	d_I	num$_I$	d_{II}	num$_{II}$
2	1^O	1	2^E	1	10	4^E	≥ 51	4^E	≥ 5
3	2^E	1	–	–	11	5^E	1	–	–
4	2^E	1	2^E	2	12	5^E	≥ 7	6^E	1
5	3^E	1	–	–	13	5°	≥ 5	–	–
6	3^E	1	4^E	1	14	6^E	?	6^E	≥ 490
7	3°	3	–	–	15	6^E	≥ 2	–	–
8	4^E	2	4^E	3	16	6^E	≥ 3	6^E	≥ 4
9	4^E	8	–	–					

TABLE 1. Type I and II Codes of Length $2 \leq n \leq 16$

the largest minimum weight for which a Type I code is known to exist while 'd_{II}' is the largest minimum weight for which a Type II code is known to exist. The superscript 'E' indicates that the code is extremal; the superscript 'O' indicates the code is not extremal but no code of the given type can exist with a larger minimum weight (and so the code is *optimal*). The number of inequivalent Type I and II codes of the given minimum weight is listed under 'num$_I$' and 'num$_{II}$', respectively; when the number in the column is exact (without $\geq$), the classification of those codes is complete.

The numbers for $2 \leq n \leq 7$ come from the work of Höhn [**9**]. Höhn also found that there are exactly 3 inequivalent extremal Type II codes of length 8; we found that there are exactly 2 inequivalent extremal Type I codes of length 8. We also found the entries for $n = 9$, 10, and 11. The unique extremal Type II code of length 12 is called the *dodecacode* [**3**]; as we will see, using the fact that the extremal self-dual code of length 11 is known to be unique, it is easy to show that the extremal Type II code of length 12 is also unique. It is known [**11**] that there is no extremal length 13 codes. We found that there are at least 5 optimal self-dual codes of length 13 and at least 3 extremal Type I codes of length 16; there are exactly 4 extremal Hermitian self-dual $GF(4)$-linear codes of length 16 [**5**] showing that there are at least 4 extremal Type II codes of length 16. The Type II entry for length $n = 14$ comes from [**1**].

3. Techniques

In this section, we describe the techniques used to produce some of the results in the previous section.

3.1. The Balance Principle. We notice that a result found in [**10**] applies to additive self-dual codes with minor modifications. This is sometimes called the Balance Principle.

THEOREM 3.1. *Let $\mathcal{C}$ be a self-dual additive $(n, 2^n)$ code. Let $n = n_1 + n_2$. We can choose a generator matrix G for $\mathcal{C}$ of the following form:*

$$G = \begin{bmatrix} A & O \\ O & B \\ D & E \end{bmatrix}. \tag{3}$$

Here A is a $k_1 \times n_1$ matrix where $[A \ \ O]$ generates the largest subcode of $\mathcal{C}$ which is zero on the right-most n_2 coordinates. Similarly B is a $k_2 \times n_2$ matrix where $[O \ \ B]$

generates the largest subcode of $\mathcal{C}$ which is zero on the left-most n_1 coordinates. (O is the appropriate size zero matrix.) Let $\mathcal{A}$ be the additive code of length n_1 generated by A, $\mathcal{A}_D$ the additive code of length n_1 generated by the rows of A and D, $\mathcal{B}$ the additive code of length n_2 generated by B, and $\mathcal{B}_E$ the additive code of length n_2 generated by the rows of B and E. The following hold.

(i) $n - k_2 = 2n_1 - k_1$ *and* $n - k_1 = 2n_2 - k_2$.

(ii) $\mathcal{A}^\perp = \mathcal{A}_D$ *and* $\mathcal{B}^\perp = \mathcal{B}_E$.

We apply the Balance Principle in the following manner. Suppose we wish to construct a self-dual additive $(n, 2^n, d)$ code $\mathcal{C}$. By rearranging coordinates, we may assume that there is a minimum weight codeword whose nonzero components are the left-most $n_1 = d$. Then by scaling the coordinates appropriately, we may assume that the first row of A is all 1's. There is at most one more row of A and that row can have only ω's and $\overline{\omega}$'s since otherwise there would exist a nonzero word in $\mathcal{C}$ of weight smaller than d, a contradiction. By conjugation of columns, we may assume that row has all ω's; as the two rows must be orthogonal under the trace inner product, d must be even by recalling that $\mathrm{Tr}(\omega) = 1$. Thus we can always choose A to be either

$$(4)\quad (a)\quad A = \begin{bmatrix} 1 & 1 & 1 & \cdots & 1 \end{bmatrix} \quad \text{or} \quad (b)\quad A = \begin{bmatrix} 1 & 1 & 1 & \cdots & 1 \\ \omega & \omega & \omega & \cdots & \omega \end{bmatrix}$$

with (b) occurring only when d is even. Thus we have $k_1 = 1$ or $k_1 = 2$. This determines the value for k_2. As a special case, if $n_1 = n_2 = n/2$, then $k_1 = k_2$ by *(i)* in Theorem 3.1. This applies to the following example.

EXAMPLE 3.2. If $\mathcal{C}$ is a self-dual additive $(8, 2^8, 4)$ code, then the Balance Principle shows that a generator matrix for $\mathcal{C}$ can be assumed to be of the form

$$\left[\begin{array}{cccc|cccc} 1 & 1 & 1 & 1 & & & & \\ \hline & & & & 1 & 1 & 1 & 1 \\ \hline 1 & 0 & 0 & 0 & & & & \\ 0 & 1 & 0 & 0 & & & & \\ 0 & 0 & 1 & 0 & & & E & \\ \omega & \omega & 0 & 0 & & & & \\ \omega & 0 & \omega & 0 & & & & \\ \omega & 0 & 0 & \omega & & & & \end{array}\right] \quad \text{or} \quad \left[\begin{array}{cccc|cccc} 1 & 1 & 1 & 1 & & & & \\ \omega & \omega & \omega & \omega & & & & \\ \hline & & & & 1 & 1 & 1 & 1 \\ & & & & \omega & \omega & \omega & \omega \\ \hline 1 & 1 & 0 & 0 & & & & \\ \omega & \omega & 0 & 0 & & & E & \\ 1 & 0 & 1 & 0 & & & & \\ \omega & 0 & \omega & 0 & & & & \end{array}\right].$$

Using these two forms, self-duality, the minimum weight, and equivalence of codes, MAGMA([**2**]) was used to fill in the possible entries for E and thereby produce exactly 5 inequivalent self-dual additive $(8, 2^8, 4)$ codes, 3 being of Type II.

EXAMPLE 3.3. Suppose we want to produce a self-dual additive $(9, 2^9, 4)$ code $\mathcal{C}$. We let a generator matrix for $\mathcal{C}$ have form (3). If A has form (4)(a), up to equivalence there are 3 choices for the 2×5 matrix B; if A has form (4)(b), up to equivalence there is only 1 choice for the 3×5 matrix B. Given A, D can be explicitly written down and E can be determined by MAGMA. These lead to exactly 8 inequivalent codes.

EXAMPLE 3.4. Suppose we want to produce a self-dual additive $(11, 2^{11}, 5)$ code $\mathcal{C}$. A generator matrix for $\mathcal{C}$ has form (3) where this time A has form (4)(a) only and D can explicitly be written down. Up to equivalence there are 2 choices for the 2×6 matrix B. Using MAGMA E can be determined. This leads to exactly 1 code up to equivalence.

3.2. Equivalence of Codes. We determine the equivalence or inequivalence of two additive codes over $GF(4)$ by operations on binary codes. Let $\mathcal{C}$ be an additive code over $GF(4)$. The transformation from $\mathcal{C}$ into a binary code is done by applying the map β: $0 \to 000, 1 \to 011, \omega \to 101, \overline{\omega} \to 110$ to the coordinates of $\mathcal{C}$. The map β sends an $(n, 2^k)$ additive code $\mathcal{C}$to $\beta(\mathcal{C})$, a $[3n, k]$ binary code.

The six different operations on a column of an additive code which do not change the weight (multiplication by a constant with or without the conjugation) are then expressed on F_2^3 as the six possible permutations of 3 elements.

Let Ω be the $(n, 2^{2n})$ additive code containing all the elements of $GF(4)^n$. The automorphism group of $\beta(\Omega)$ is isomorphic to the permutation group $S_3^n.S_n$. The straightforward lemma of [**3**] follows:

LEMMA 3.5. *Let $\mathcal{C}$ be a $(n, 2^k)$ additive code over $GF(4)$. Then $Aut(\mathcal{C})$ is isomorphic to $Aut(\beta(\mathcal{C})) \cap Aut(\beta(\Omega))$.*

The following lemma will be mainly used in the proof of Lemma 3.7.

LEMMA 3.6. *Let $\mathcal{B}_1$, $\mathcal{B}_2$ be binary codes with a permutation P such that $\mathcal{B}_1 P = \mathcal{B}_2$. Any permutation Q satisfies $\mathcal{B}_1 Q = \mathcal{B}_2$ if and only if $Q \in Aut(\mathcal{B}_1)P$, a right coset of $Aut(\mathcal{B}_1)$ in the symmetric group on the length of $\mathcal{B}_1$.*

Proof: Suppose that there exists a permutation Q such that $\mathcal{B}_1 Q = \mathcal{B}_2$. Then we have $\mathcal{B}_1 P = \mathcal{B}_2 = \mathcal{B}_1 Q$, implying $\mathcal{B}_1 Q P^{-1} = \mathcal{B}_1$. Hence $QP^{-1} \in \mathrm{Aut}(\mathcal{B}_1)$, i.e. $Q \in \mathrm{Aut}(\mathcal{B}_1)P$. Conversely suppose $Q \in \mathrm{Aut}(\mathcal{B}_1)P$. Then $Q = AP$ for some $A \in \mathrm{Aut}(\mathcal{B}_1)$. Hence $\mathcal{B}_1 Q = (\mathcal{B}_1 A)P = \mathcal{B}_1 P = \mathcal{B}_2$, as required. □

Suppose $\mathcal{C}_1$ and $\mathcal{C}_2$ are additive codes over $GF(4)$ of length n associated to binary codes $\beta(\mathcal{C}_1) = \mathcal{B}_1$ and $\beta(\mathcal{C}_2) = \mathcal{B}_2$ of length $3n$. If M is a map with $\mathcal{C}_1 M = \mathcal{C}_2$, where M permutes columns, scales columns, and conjugates columns, then let $\widetilde{M}$ be the associated permutation on the $3n$ columns of $\mathcal{B}_1$. Then immediately we see that $\mathcal{C}_1$ is equivalent to $\mathcal{C}_2$ as additive codes over $GF(4)$ if and only if there is a permutation $\widetilde{M}$ on $3n$ elements and $\widetilde{M} \in \mathrm{Aut}(\beta(\Omega))$ with $\mathcal{B}_1 \widetilde{M} = \mathcal{B}_2$.

If two additive codes are equivalent, clearly their binary images by the map β are also equivalent. In particular if the binary images of two additive codes are not equivalent, neither are the additive codes. The following lemma relates the equivalence of additive codes over $GF(4)$ to the equivalence of binary codes by the map β.

LEMMA 3.7. *Let $\mathcal{C}_1$ and $\mathcal{C}_2$ are additive codes over $GF(4)$ of length n associated to binary codes $\beta(\mathcal{C}_1) = \mathcal{B}_1$ and $\beta(\mathcal{C}_2) = \mathcal{B}_2$ of length $3n$. Suppose P is a permutation on $3n$ elements such that $\mathcal{B}_1 P = \mathcal{B}_2$. $\mathcal{C}_1$ is equivalent to $\mathcal{C}_2$ as additive codes over $GF(4)$ if and only if $Aut(\mathcal{B}_1)P \cap Aut(\beta(\Omega)) \neq \emptyset$.*

Proof: Suppose $\mathcal{C}_1$ is equivalent to $\mathcal{C}_2$. Then there exists a map M such that $\mathcal{C}_1 M = \mathcal{C}_2$. So we have

$$\mathcal{B}_1 \widetilde{M} = \mathcal{B}_2 \Longleftrightarrow \widetilde{M} \in \mathrm{Aut}(\mathcal{B}_1)P \quad \text{by Lemma 3.6.}$$

Since $\widetilde{M} \in \mathrm{Aut}(\beta(\Omega))$ it follows that

$$\widetilde{M} \in \mathrm{Aut}(\mathcal{B}_1)P \cap \mathrm{Aut}(\beta(\Omega)) \neq \emptyset.$$

Conversely suppose $\mathrm{Aut}(\mathcal{B}_1)P \cap \mathrm{Aut}(\beta(\Omega)) \neq \emptyset$. Let $\widetilde{M} \in \mathrm{Aut}(\mathcal{B}_1)P \cap \mathrm{Aut}(\beta(\Omega))$. Then $\widetilde{M} = AP$ for some $A \in \mathrm{Aut}(\mathcal{B}_1)$. It follows from Lemma 3.6 that $\mathcal{B}_1 \widetilde{M} = \mathcal{B}_2$.

Since $\widetilde{M} \in \mathrm{Aut}(\beta(\Omega))$, we have that $\mathcal{C}_1 M = \mathcal{C}_2$. Hence $\mathcal{C}_1$ and $\mathcal{C}_2$ are equivalent as additive codes. This completes the proof. □

This method was written with the MAGMA [**2**] system. Although it is a way of checking the equivalence of two additive codes, the search for $\mathrm{Aut}(\mathcal{B}_1)P \cap \mathrm{Aut}(\beta(\Omega)) \neq \emptyset$ of Lemma 3.7 slows considerably the process and makes it too long for codes of lengths over 16.

In fact we also developed a concept of *restricted* equivalence which consists in checking intermediate fillings of the matrix E (as described in Example 3.2). The method used is almost the same as the method described in [**7**]. We considered intermediate fillings of the matrix E. We kept only one possible intermediate filling when different intermediate codes corresponding to different fillings were equivalent, as additive codes, through equivalence maps which preserved the first four or five columns.

Eventually this process enabled us to considerably decrease the number of e-quivalences to check.

3.3. Lengthening and Shortening. If $\mathcal{C}$ is a self-dual additive $(n, 2^n, d)$ code, we can obtain a code of length $n-1$ by a process we call *shortening.* We use this term even though it differs from the usual meaning as our notion can be regarded as a generalization. Let G be a generator matrix of $\mathcal{C}$. As $\mathcal{C}$ is self-dual, no column of G is all 0's. Choose any column of G, say the ith one. The entries in column i can be any of 0, 1, ω, or $\overline{\omega}$. By row reducing G (to obtain another generator matrix we call G again), we can make all the entries in column i equal to 0 except for one or two entries; if two they would be two of the three values 1, ω, or $\overline{\omega}$. The *shortened code of* $\mathcal{C}$ *on coordinate* i, denoted $\mathcal{C}'$ is the code with generator matrix G' obtained from G by eliminating one row of G with a nonzero entry in column i and then eliminating column i. [If there is only one nonzero entry in column i of G, then $\mathcal{C}'$ is $\mathcal{C}$ shortened in the usual sense.] Clearly $\mathcal{C}'$ is a self-dual additive $(n-1, 2^{n-1}, d')$ code with $d' \geq d-1$ since $\mathrm{Tr}(x\overline{x}) = \mathrm{Tr}(1) = 0$, where $x = 1, \omega$, or $\overline{\omega}$.

EXAMPLE 3.8. Let

$$G = \begin{bmatrix} 1 & 0 & 1 \\ 0 & 1 & 1 \\ \omega & \omega & \omega \end{bmatrix}.$$

We obtain two inequivalent codes by shortening on the first column. They have the following generator matrices:

$$\begin{bmatrix} 0 & 1 \\ 1 & 1 \end{bmatrix} \quad \text{or} \quad \begin{bmatrix} 1 & 1 \\ \omega & \omega \end{bmatrix}.$$

We would get the same codes shortening on the second column. To shorten on the third column, we would replace G by

$$\begin{bmatrix} 1 & 0 & 1 \\ 1 & 1 & 0 \\ \omega & \omega & \omega \end{bmatrix}.$$

Shortening yields the same codes as before.

EXAMPLE 3.9. If

$$G = \begin{bmatrix} 1 & 0 & 0 \\ 0 & 1 & 0 \\ 0 & 0 & 1 \end{bmatrix},$$

then no matter which column we shorten on, we obtain a shortened code with generator matrix

$$\begin{bmatrix} 1 & 0 \\ 0 & 1 \end{bmatrix},$$

which is the usual shortened code. It also is the same as the first of the shortened codes in Example 3.8.

From Examples 3.8 and 3.9, we see that inequivalent codes can have equivalent shortened codes.

We can reverse shortening by *lengthening* a self-dual additive $(n-1, 2^{n-1}, d')$ code $\mathcal{C}'$ to obtain a self-dual additive $(n, 2^n, d)$ code $\mathcal{C}$. To do this, take a generator matrix G' of $\mathcal{C}'$ and adjoin an arbitrary n^{th} row $\mathbf{x}$ at the bottom. Then add an nth column on the right according to the following scheme: First place a 1 in the nth row. Second place a 0 or ω in each of the first $n-1$ rows so that each of the new length n rows is orthogonal to the length n bottom row. (Note that if we begin with two orthogonal rows of length $n-1$ and want to adjoin a nonzero element to one of the rows, we must adjoin the same nonzero element to the other row if we wish to create two orthogonal rows of length n.) The resulting matrix G, which we denote by

(5) $$G = \left[\begin{array}{c|c} & 0 \\ G' & \text{or} \\ & \omega \\ \hline \mathbf{x} & 1 \end{array}\right],$$

clearly generates a self-dual additive $(n, 2^n, d)$ code with $d \leq d' + 1$. If the new row of length $n-1$ added to the bottom of G' is already in $\mathcal{C}'$, the lengthened code is a direct sum of $\mathcal{C}'$ and the $(1, 2, 1)$ code generated by $[1]$.

EXAMPLE 3.10. Let

$$G' = \begin{bmatrix} 0 & 1 \\ 1 & 1 \end{bmatrix}.$$

If we adjoin the row $[\omega\ \omega]$, then

$$G = \begin{bmatrix} 0 & 1 & \omega \\ 1 & 1 & 0 \\ \omega & \omega & 1 \end{bmatrix}.$$

Note that by multiplying column 3 by $\overline{\omega}$ and then conjugating column 3, we obtain a generator matrix for the $(3, 2^3)$ code of Example 3.8.

EXAMPLE 3.11. Let

$$G' = \begin{bmatrix} 1 & 1 \\ \omega & \omega \end{bmatrix}.$$

If we adjoin the row [1 0], then

$$G = \begin{bmatrix} 1 & 1 & 0 \\ \omega & \omega & \omega \\ 1 & 0 & 1 \end{bmatrix},$$

which is the code $(3, 2^3)$ of Example 3.8.

Examples 3.8 and 3.9 show that equivalent codes can be obtained by lengthening inequivalent codes.

We want to expand the concept of lengthening. Instead of (5) we could use the following generator matrices to lengthen:

$$(6) \quad \left[\begin{array}{c|c} & 0 \\ G' & \text{or} \\ & \overline{\omega} \\ \hline \mathbf{x} & 1 \end{array}\right], \left[\begin{array}{c|c} & 0 \\ G' & \text{or} \\ & 1 \\ \hline \mathbf{x} & \omega \end{array}\right], \left[\begin{array}{c|c} & 0 \\ G' & \text{or} \\ & \overline{\omega} \\ \hline \mathbf{x} & \omega \end{array}\right], \left[\begin{array}{c|c} & 0 \\ G' & \text{or} \\ & 1 \\ \hline \mathbf{x} & \overline{\omega} \end{array}\right], \text{ or } \left[\begin{array}{c|c} & 0 \\ G' & \text{or} \\ & \omega \\ \hline \mathbf{x} & \overline{\omega} \end{array}\right].$$

However by scaling and/or conjugating the last column of the matrices in (6), we obtain the matrix (5) giving the following result.

LEMMA 3.12. *The codes generated by any of the matrices in* (5) *or* (6) *are equivalent.*

As a result of this lemma we expand our definition of lengthening to allow us to lengthen using any of the six matrices in (5) and (6). Up to equivalence lengthening and shortening are inverse operations. Thus we have the following theorem.

THEOREM 3.13. *Every self-dual additive $(n-1, 2^{n-1})$ code is shortened from a self-dual additive $(n, 2^n)$ code, and every self-dual additive $(n, 2^n)$ code is lengthened from a self-dual additive $(n-1, 2^{n-1})$ code.*

EXAMPLE 3.14. Every self-dual additive $(13, 2^{13}, 5)$ code comes from lengthening a self-dual additive $(12, 2^{12}, d)$ code with $d \geq 4$. Lengthening the dodecacode we obtained 4 inequivalent $(13, 2^{13}, 5)$ codes; all possibilities were attempted. One can also shorten a length 14 code to obtain a length 13 code; shortening the length 14 extended quadratic residue code q_{14} leads to a $(13, 2^{13}, 5)$ code as indicated in [**11**]. The latter is not equivalent to the four codes found from lengthening.

EXAMPLE 3.15. By lengthening q_{14}, we found one self-dual additive $(15, 2^{15}, 6)$ code inequivalent to the known $(15, 2^{15}, 6)$ cyclic code c_{15} found in [**11**].

EXAMPLE 3.16. Every self-dual additive $(16, 2^{16}, 6)$ code comes from lengthening a self-dual additive $(15, 2^{15}, d)$ code with $d \geq 5$. Lengthening c_{15} yielded 3 inequivalent $(16, 2^{16}, 6)$ Type I codes; not all possibilities were attempted.

In [**9**], Höhn shows that a Type II code of length n exists only if n is even. If n is even, he defines (analogously to the situation of doubly-even binary codes of length $8r$ and their length $8r-2$ children [**4**]) a child of a Type II length n code. Beginning with a self-dual additive code of length $n-1$, he also shows how to construct a parent, which will be of Type II. The length $n-1$ children are, in our terminology, shortenings of an even length n code. The parent of a length $n-1$ code $\mathcal{C}'$ is obtained by taking the lengthening vector $\mathbf{x}$ to be a vector not in $\mathcal{C}'$ but orthogonal to the subcode $\mathcal{C}_0$ of $\mathcal{C}'$ consisting of all its even weight vectors. The vector $\mathbf{x}$ must have odd weight; otherwise the code $\mathcal{C}_0$ and vector $\mathbf{x}$ generate an additive Type II code of odd length, a contradiction.

EXAMPLE 3.17. Let

$$G = \begin{bmatrix} 1 & 0 & 0 & 1 \\ 0 & 1 & 0 & 1 \\ 0 & 0 & 1 & 1 \\ \omega & \omega & \omega & \omega \end{bmatrix}.$$

Then G generates a Type II code $\mathcal{C}$ of length 4. By shortening we obtain inequivalent children $\mathcal{C}_1$ and $\mathcal{C}_2$ with generator matrices

$$G_1 = \begin{bmatrix} 1 & 0 & 0 \\ 1 & 1 & 0 \\ 1 & 0 & 1 \end{bmatrix} \quad \text{and} \quad G_2 = \begin{bmatrix} 1 & 0 & 1 \\ 0 & 1 & 1 \\ \omega & \omega & \omega \end{bmatrix}.$$

Rearrange the rows of G_1 to obtain

$$G_1' = \begin{bmatrix} 1 & 1 & 0 \\ 1 & 0 & 1 \\ 0 & 0 & 1 \end{bmatrix}$$

so that the first two rows of G_1' generate the even weight subcode $\mathcal{C}_0$ of $\mathcal{C}_1$. Lengthening using the second form in (6) with $\mathbf{x} = [\omega\ \omega\ \omega]$ gives $\mathcal{C}$ with generator matrix

$$\begin{bmatrix} 1 & 1 & 0 & 0 \\ 1 & 0 & 1 & 0 \\ 0 & 0 & 1 & 1 \\ \omega & \omega & \omega & \omega \end{bmatrix}.$$

Lengthening $\mathcal{C}_2$ using form (5) with $\mathbf{x} = [1\ 0\ 0]$ gives $\mathcal{C}$ with generator matrix

$$\begin{bmatrix} 1 & 0 & 1 & 0 \\ 0 & 1 & 1 & 0 \\ \omega & \omega & \omega & \omega \\ 1 & 0 & 0 & 1 \end{bmatrix}.$$

THEOREM 3.18. *Let $\mathcal{C}'$ be a Type I code of odd length $n-1$. Then there is a unique Type II code $\mathcal{C}$ of length n obtained by lengthening $\mathcal{C}'$.*

Proof: There is a unique $(n-1, 2^{n-2})$ subcode $\mathcal{C}_0$ of $\mathcal{C}'$ consisting of even weight vectors. Let G' be a generator matrix of $\mathcal{C}'$ with the first $n-2$ rows a basis of $\mathcal{C}_0$; the bottom row has odd weight. If we lengthen using (5) and we adjoin 0 to each of the first $n-2$ rows and ω to the last row, then we must choose the lengthening vector $\mathbf{x}$ to be in $\mathcal{C}_0^\perp$; to avoid a code of minimum weight 1, $\mathbf{x} \notin \mathcal{C}'$. Furthermore two choices of $\mathbf{x}$ in the same coset of $\mathcal{C}'$ give the same code $\mathcal{C}$. As there is only one coset of $\mathcal{C}'$ in $\mathcal{C}_0^\perp$ not equal to $\mathcal{C}'$, we have a unique lengthening to a Type II code.□

The code $\mathcal{C}$ in Theorem 3.18 is the parent of $\mathcal{C}'$. Thus the theorem states that an odd length Type I code has a unique Type II extension and that extension is the parent.

EXAMPLE 3.19. By Theorem 3.13, (1), and (2), we know that every self-dual additive $(10, 2^{10}, 4)$ code can be obtained from a self-dual additive $(9, 2^9, d)$ code with $d = 3$ or 4. We lengthened some of the 8 $(9, 2^9, 4)$ codes obtained in Example 3.3 and quickly found 51 Type I codes. Five of the 8 parents of the $(9, 2^9, 4)$ codes were inequivalent $(10, 2^{10}, 4)$ Type II codes; the other three parents had minimum weight 2. It seems to be of little interest to try to enumerate all self-dual additive $(10, 2^{10}, 4)$ codes as the number appears to be very high.

Code	n	d	$\lvert\mathrm{Aut}(C)\rvert$	A_0	A_4	A_5	A_6	A_7	A_8	A_9	A_{10}	A_{11}	A_{12}
QC_8a	8	4	128	1	26	64	72	64	29				
QC_8b	8	4	384	1	26	64	72	64	29				
QC_8c	8	4	1152	1	42	0	168	0	45				
QC_8d	8	4	336	1	42	0	168	0	45				
QC_8e	8	4	6.1344	1	42	0	168	0	45				
QC_9a	9	4	12	1	18	72	120	144	117	40			
QC_9b	9	4	16	1	18	72	120	144	117	40			
QC_9c	9	4	12	1	26	48	136	160	93	48			
QC_9d	9	4	144	1	18	72	120	144	117	40			
QC_9e	9	4	16	1	26	48	136	160	93	48			
QC_9f	9	4	64	1	18	72	120	144	117	40			
QC_9g	9	4	192	1	26	48	136	160	93	48			
QC_9h	9	4	18	1	18	72	120	144	117	40			
QC_11	11	5	18	1	0	66	198	330	495	550	330	78	
QC_12	12	6	648	1	0	0	396	0	1485	0	1980	0	234

TABLE 2. Extremal Self-dual codes for lengths $8, 9, 11$, and 12

While the Type II extension of a Type I code of odd length is unique, this example shows that there may be many more Type I extensions. If we wish to extend Type II codes, we may get many extensions, as Example 3.14 illustrates. The following result was stated in [**3**] without proof.

COROLLARY 3.20. *The dodecacode is the unique* $(12, 2^{12}, 6)$ *Type II code.*

Proof: A $(12, 2^{12}, 6)$ Type II code must be a parent of a self-dual additive $(11, 2^{11}, d)$ code $\mathcal{C}'$ with $d \geq 5$. This code, being of odd length, must be Type I; by (1), $d = 5$. By Example 3.4, $\mathcal{C}'$ is unique, and so there is a unique $(12, 2^{12}, 6)$ Type II code by Theorem 3.18. □

4. Enumeration of Extremal Self-Dual Codes

4.1. Extremal Self-Dual Codes of lengths $8, 9, 11$ **and** 12**.** In this section we enumerate the extremal self-dual additive codes for the lengths 8,9,and 11. We also list the dodecacode, the unique extremal Type II code of length 12. Table 2 gives the different codes and their weight distributions, where 'n' and 'd' stand respectively for the length and the minimum distance. The codes are then enumerated in the particular form described in the previous section.

• n=8:

$$QC\_8a = \begin{bmatrix} 1 & 1 & 1 & 1 & 0 & 0 & 0 & 0 \\ 0 & 0 & 0 & 0 & 1 & 1 & 1 & 1 \\ 1 & 0 & 0 & 0 & \omega & \omega & \omega & \omega \\ 0 & 1 & 0 & 0 & 1 & \omega & \overline{\omega} & 0 \\ 0 & 0 & 1 & 0 & \omega & \overline{\omega} & \overline{\omega} & \omega \\ \omega & \omega & 0 & 0 & 1 & 0 & \omega & \omega \\ \omega & 0 & \omega & 0 & 1 & \omega & \overline{\omega} & 1 \\ \omega & 0 & 0 & \omega & \overline{\omega} & \overline{\omega} & 0 & 1 \end{bmatrix}, \quad QC\_8b = \begin{bmatrix} 1 & 1 & 1 & 1 & 0 & 0 & 0 & 0 \\ 0 & 0 & 0 & 0 & 1 & 1 & 1 & 1 \\ 1 & 0 & 0 & 0 & \omega & \omega & \omega & \omega \\ 0 & 1 & 0 & 0 & \omega & \omega & \overline{\omega} & \overline{\omega} \\ 0 & 0 & 1 & 0 & \omega & \overline{\omega} & \overline{\omega} & \omega \\ \omega & \omega & 0 & 0 & \overline{\omega} & \overline{\omega} & 0 & 1 \\ \omega & 0 & \omega & 0 & 1 & \omega & \overline{\omega} & 1 \\ \omega & 0 & 0 & \omega & \omega & 1 & \overline{\omega} & 1 \end{bmatrix},$$

$$QC\_8c = \begin{bmatrix} 1 & 1 & 1 & 1 & 0 & 0 & 0 & 0 \\ 0 & 0 & 0 & 0 & 1 & 1 & 1 & 1 \\ 1 & 0 & 0 & 0 & 1 & \omega & \omega & 0 \\ 0 & 1 & 0 & 0 & 1 & \overline{\omega} & \overline{\omega} & 0 \\ 0 & 0 & 1 & 0 & \overline{\omega} & 1 & 0 & \omega \\ \omega & \omega & 0 & 0 & \omega & \omega & \omega & \omega \\ \omega & 0 & \omega & 0 & 1 & 1 & \omega & \omega \\ \omega & 0 & 0 & \omega & \overline{\omega} & \overline{\omega} & 1 & 1 \end{bmatrix}, \quad QC\_8d = \begin{bmatrix} 1 & 1 & 1 & 1 & 0 & 0 & 0 & 0 \\ 0 & 0 & 0 & 0 & 1 & 1 & 1 & 1 \\ 1 & 0 & 0 & 0 & 1 & \omega & \omega & 0 \\ 0 & 1 & 0 & 0 & 1 & \omega & 0 & \omega \\ 0 & 0 & 1 & 0 & \overline{\omega} & 1 & 0 & \omega \\ \omega & \omega & 0 & 0 & \omega & 1 & \overline{\omega} & 1 \\ \omega & 0 & \omega & 0 & 1 & 0 & 1 & 0 \\ \omega & 0 & 0 & \omega & \overline{\omega} & \omega & \omega & \overline{\omega} \end{bmatrix}, \quad QC\_8e = \begin{bmatrix} 1 & 1 & 1 & 1 & 0 & 0 & 0 & 0 \\ \omega & \omega & \omega & \omega & 0 & 0 & 0 & 0 \\ 0 & 0 & 0 & 0 & 1 & 1 & 1 & 1 \\ 0 & 0 & 0 & 0 & \omega & \omega & \omega & \omega \\ 1 & 1 & 0 & 0 & 1 & 1 & 0 & 0 \\ \omega & \omega & 0 & 0 & \omega & \omega & 0 & 0 \\ 1 & 0 & 1 & 0 & \omega & \overline{\omega} & \omega & \overline{\omega} \\ \omega & 0 & \omega & 0 & \overline{\omega} & 1 & \overline{\omega} & 1 \end{bmatrix}.$$

- n=9:

$$QC\_9a = \begin{bmatrix} 1 & 1 & 1 & 1 & 0 & 0 & 0 & 0 & 0 \\ 0 & 0 & 0 & 0 & 1 & 1 & 1 & 1 & 1 \\ 0 & 0 & 0 & 0 & \omega & \omega & \omega & \omega & 0 \\ 1 & 0 & 0 & 0 & \overline{\omega} & 0 & \overline{\omega} & \omega & \overline{\omega} \\ 0 & 1 & 0 & 0 & 1 & 1 & 0 & \omega & \omega \\ 0 & 0 & 1 & 0 & 0 & \omega & \overline{\omega} & \overline{\omega} & \omega \\ \omega & \omega & 0 & 0 & 1 & 0 & 0 & \overline{\omega} & \omega \\ \omega & 0 & \omega & 0 & 1 & \omega & 1 & \omega & 1 \\ \omega & 0 & 0 & \omega & 0 & \omega & 0 & \omega & 1 \end{bmatrix}, QC\_9b = \begin{bmatrix} 1 & 1 & 1 & 1 & 0 & 0 & 0 & 0 & 0 \\ 0 & 0 & 0 & 0 & 1 & 1 & 1 & 1 & 1 \\ 0 & 0 & 0 & 0 & \omega & \omega & \omega & \omega & 0 \\ 1 & 0 & 0 & 0 & \overline{\omega} & 0 & \overline{\omega} & \omega & \overline{\omega} \\ 0 & 1 & 0 & 0 & 1 & 1 & 0 & \omega & \omega \\ 0 & 0 & 1 & 0 & 0 & \omega & \overline{\omega} & \overline{\omega} & \omega \\ \omega & \omega & 0 & 0 & 1 & 0 & 0 & \overline{\omega} & \omega \\ \omega & 0 & \omega & 0 & \omega & 0 & 0 & 0 & \overline{\omega} \\ \omega & 0 & 0 & \omega & 0 & 1 & 1 & 0 & 0 \end{bmatrix},$$

$$QC\_9c = \begin{bmatrix} 1 & 1 & 1 & 1 & 0 & 0 & 0 & 0 & 0 \\ 0 & 0 & 0 & 0 & 1 & 1 & 1 & 1 & 1 \\ 0 & 0 & 0 & 0 & \omega & \omega & \omega & \omega & 0 \\ 1 & 0 & 0 & 0 & \overline{\omega} & 0 & \overline{\omega} & \omega & \overline{\omega} \\ 0 & 1 & 0 & 0 & 1 & 1 & 0 & \omega & \omega \\ 0 & 0 & 1 & 0 & \omega & 0 & 0 & \omega & 1 \\ \omega & \omega & 0 & 0 & \omega & 0 & 1 & 1 & \overline{\omega} \\ \omega & 0 & \omega & 0 & \omega & \omega & \omega & 0 & \omega \\ \omega & 0 & 0 & \omega & 1 & 1 & \overline{\omega} & \overline{\omega} & 1 \end{bmatrix}, \quad QC\_9d = \begin{bmatrix} 1 & 1 & 1 & 1 & 0 & 0 & 0 & 0 & 0 \\ 0 & 0 & 0 & 0 & 1 & 1 & 1 & 1 & 1 \\ 0 & 0 & 0 & 0 & \omega & \omega & \omega & \omega & 0 \\ 1 & 0 & 0 & 0 & \overline{\omega} & 0 & \overline{\omega} & \omega & \overline{\omega} \\ 0 & 1 & 0 & 0 & 0 & \overline{\omega} & 1 & 0 & \omega \\ 0 & 0 & 1 & 0 & \omega & 1 & 0 & 1 & \overline{\omega} \\ \omega & \omega & 0 & 0 & \omega & 0 & 0 & 0 & \overline{\omega} \\ \omega & 0 & \omega & 0 & 1 & 0 & 1 & \omega & \omega \\ \omega & 0 & 0 & \omega & 0 & \overline{\omega} & 1 & \omega & 0 \end{bmatrix}, \quad QC\_9e = \begin{bmatrix} 1 & 1 & 1 & 1 & 0 & 0 & 0 & 0 & 0 \\ 0 & 0 & 0 & 0 & 1 & 1 & 1 & 1 & 1 \\ 0 & 0 & 0 & 0 & \omega & \omega & \omega & \omega & 0 \\ 1 & 0 & 0 & 0 & \overline{\omega} & 0 & \overline{\omega} & \omega & \overline{\omega} \\ 0 & 1 & 0 & 0 & 0 & \overline{\omega} & 1 & 0 & \omega \\ 0 & 0 & 1 & 0 & \omega & 1 & 0 & 1 & \overline{\omega} \\ \omega & \omega & 0 & 0 & 1 & 0 & 1 & 0 & 1 \\ \omega & 0 & \omega & 0 & 0 & \omega & 0 & \omega & 1 \\ \omega & 0 & 0 & \omega & 1 & \omega & 1 & \omega & 1 \end{bmatrix},$$

$$QC\_9f = \begin{bmatrix} 1 & 1 & 1 & 1 & 0 & 0 & 0 & 0 & 0 \\ 0 & 0 & 0 & 0 & 1 & 1 & 1 & 1 & 1 \\ 0 & 0 & 0 & 0 & \omega & \omega & \omega & \omega & 0 \\ 1 & 0 & 0 & 0 & \overline{\omega} & 0 & \overline{\omega} & \omega & \overline{\omega} \\ 0 & 1 & 0 & 0 & \omega & 1 & 0 & 1 & \overline{\omega} \\ 0 & 0 & 1 & 0 & 1 & 0 & 1 & \omega & \overline{\omega} \\ \omega & \omega & 0 & 0 & 1 & 0 & 0 & \overline{\omega} & \omega \\ \omega & 0 & \omega & 0 & 1 & \omega & 1 & \omega & 1 \\ \omega & 0 & 0 & \omega & \omega & \omega & \overline{\omega} & 1 & \omega \end{bmatrix}, \quad QC\_9g = \begin{bmatrix} 1 & 1 & 1 & 1 & 0 & 0 & 0 & 0 & 0 \\ 0 & 0 & 0 & 0 & 1 & 1 & 1 & 1 & 1 \\ 0 & 0 & 0 & 0 & \omega & \omega & \omega & \omega & 0 \\ 1 & 0 & 0 & 0 & \overline{\omega} & 0 & \overline{\omega} & \omega & \overline{\omega} \\ 0 & 1 & 0 & 0 & \omega & 1 & 0 & 1 & \overline{\omega} \\ 0 & 0 & 1 & 0 & 1 & 0 & 1 & \omega & \overline{\omega} \\ \omega & \omega & 0 & 0 & 1 & 1 & \omega & \omega & 1 \\ \omega & 0 & \omega & 0 & 1 & \omega & 1 & \omega & 1 \\ \omega & 0 & 0 & \omega & \overline{\omega} & 0 & 0 & \overline{\omega} & 0 \end{bmatrix}, \quad QC\_9h = \begin{bmatrix} 1 & 1 & 1 & 1 & 0 & 0 & 0 & 0 & 0 \\ 0 & 0 & 0 & 0 & 1 & 1 & 1 & 1 & 1 \\ 0 & 0 & 0 & 0 & \omega & \omega & \omega & \omega & 0 \\ 1 & 0 & 0 & 0 & \overline{\omega} & 0 & \overline{\omega} & \omega & \overline{\omega} \\ 0 & 1 & 0 & 0 & \omega & 1 & 0 & 1 & \overline{\omega} \\ 0 & 0 & 1 & 0 & \omega & 0 & 1 & \overline{\omega} & 1 \\ \omega & \omega & 0 & 0 & 1 & 0 & 0 & \overline{\omega} & \omega \\ \omega & 0 & \omega & 0 & \omega & 0 & 0 & 0 & \overline{\omega} \\ \omega & 0 & 0 & \omega & 1 & \omega & 1 & \omega & 1 \end{bmatrix}.$$

- n=11,12:

$$QC\_11 = \begin{bmatrix} 0 & 0 & 0 & 0 & 0 & 0 & 1 & 1 & 1 & 1 & 1 \\ \overline{\omega} & 1 & \omega & 0 & 0 & 0 & \omega & \omega & \omega & \overline{\omega} & 1 \\ 1 & 1 & 1 & 1 & 1 & 1 & 0 & 0 & 0 & 0 & 0 \\ \omega & \omega & \omega & \omega & \omega & \omega & 0 & 0 & 0 & 0 & 0 \\ \overline{\omega} & 1 & \omega & 1 & \omega & \overline{\omega} & 0 & 0 & 0 & 0 & 1 \\ 0 & 0 & 0 & \omega & \overline{\omega} & 1 & 0 & 0 & 0 & \omega & \overline{\omega} \\ 1 & \overline{\omega} & \omega & 0 & 0 & 0 & 1 & \overline{\omega} & \omega & 0 & 0 \\ \omega & 1 & \overline{\omega} & 0 & 0 & 0 & \omega & 1 & \overline{\omega} & 0 & 0 \\ 0 & 0 & 0 & 1 & \overline{\omega} & \omega & \omega & \overline{\omega} & 1 & 0 & 0 \\ 0 & 0 & 0 & \omega & 1 & \overline{\omega} & 1 & \omega & \overline{\omega} & 0 & 0 \\ 1 & \omega & \overline{\omega} & 0 & 0 & 0 & 0 & 0 & 0 & \overline{\omega} & \omega \end{bmatrix}, QC\_12 = \begin{bmatrix} 0 & 0 & 0 & 0 & 0 & 0 & 1 & 1 & 1 & 1 & 1 & 1 \\ 0 & 0 & 0 & 0 & 0 & 0 & \omega & \omega & \omega & \omega & \omega & \omega \\ 1 & 1 & 1 & 1 & 1 & 1 & 0 & 0 & 0 & 0 & 0 & 0 \\ \omega & \omega & \omega & \omega & \omega & \omega & 0 & 0 & 0 & 0 & 0 & 0 \\ 0 & 0 & 0 & 1 & \omega & \overline{\omega} & 0 & 0 & 0 & 1 & \omega & \overline{\omega} \\ 0 & 0 & 0 & \omega & \overline{\omega} & 1 & 0 & 0 & 0 & \omega & \overline{\omega} & 1 \\ 1 & \overline{\omega} & \omega & 0 & 0 & 0 & 1 & \overline{\omega} & \omega & 0 & 0 & 0 \\ \omega & 1 & \overline{\omega} & 0 & 0 & 0 & \omega & 1 & \overline{\omega} & 0 & 0 & 0 \\ 0 & 0 & 0 & 1 & \overline{\omega} & \omega & \omega & \overline{\omega} & 1 & 0 & 0 & 0 \\ 0 & 0 & 0 & \omega & 1 & \overline{\omega} & 1 & \omega & \overline{\omega} & 0 & 0 & 0 \\ 1 & \omega & \overline{\omega} & 0 & 0 & 0 & 0 & 0 & 0 & \overline{\omega} & \omega & 1 \\ \overline{\omega} & 1 & \omega & 0 & 0 & 0 & 0 & 0 & 0 & 1 & \overline{\omega} & \omega \end{bmatrix}.$$

4.2. Examples of Extremal Self-Dual Codes for lengths $10, 12$ **and** 13. In this section we present in Table 4 codes constructed by the lengthening technique described in section 3.

For length 10, we constructed some lengthened codes of the 8 extremal self-dual codes of length 9. We found 51 extremal Type I codes and 5 extremal Type II codes (Example 3.19). For length 12, there is only one code of minimum distance 6, and Type I codes are extremal for minimum distance 5. We constructed all possible lengthened codes of the unique extremal self-dual code of length 11 and besides the dodecacode we found 7 extremal Type I codes. The same technique applied to the dodecacode lead to 4 extremal Type I codes of length 13 (Example 3.14).

There presumably exist other extremal codes for these lengths since extremal codes may also be constructed from codes with smaller distance. For instance extremal Type I codes of length 10 could also be constructed from Type I codes of length 9 and minimum distance 3.

In the following we give the generator matrices of the 5 extremal length 10 Type II codes that we found. We also list in Table 3 the lengthened codes obtained for lengths 12 and 13. We define these codes only by giving the vectors 'x', which when added to the dodecacode or QC_11 by the lengthening technique of Example 3.10, give the corresponding lengthened codes.

$$QC\_10a = \begin{bmatrix} 1&1&1&1&0&0&0&0&0&0\\ 1&0&0&0&\omega&1&\omega&\overline{\omega}&\omega&0\\ 0&0&0&0&\omega&\omega&\omega&\omega&0&0\\ 1&1&0&0&\omega&1&\overline{\omega}&0&1&0\\ \omega&1&0&\omega&1&\overline{\omega}&0&0&\overline{\omega}&0\\ 0&1&1&0&1&\overline{\omega}&\overline{\omega}&1&0&0\\ \omega&\omega&1&0&1&\omega&\overline{\omega}&0&0&0\\ 0&\omega&\omega&0&0&\omega&1&1&\overline{\omega}&0\\ \omega&0&0&\omega&0&\omega&0&\omega&1&1\\ 0&0&0&0&0&\omega&0&\omega&\omega&\omega \end{bmatrix}, QC\_10b = \begin{bmatrix} 1&1&1&1&0&0&0&0&0&0\\ 1&0&0&0&\omega&1&\omega&\overline{\omega}&\omega&0\\ 0&0&0&0&\omega&\omega&\omega&\omega&0&0\\ 1&1&0&0&\omega&1&\overline{\omega}&0&1&0\\ 0&1&1&0&1&\overline{\omega}&\overline{\omega}&1&0&0\\ \omega&\omega&1&0&1&\omega&\overline{\omega}&0&0&0\\ \omega&0&\omega&0&\omega&0&0&0&\overline{\omega}&0\\ \omega&0&0&\omega&0&1&1&0&0&0\\ \omega&\omega&0&0&1&0&0&\overline{\omega}&\omega&1\\ 0&0&0&0&0&1&1&0&\overline{\omega}&\omega \end{bmatrix},$$

$$QC\_10c = \begin{bmatrix} 1&1&1&1&0&0&0&0&0&0\\ 1&0&0&0&\omega&1&\omega&\overline{\omega}&\omega&0\\ 0&0&0&0&\omega&\omega&\omega&\omega&0&0\\ 1&0&1&0&1&1&\overline{\omega}&\overline{\omega}&0&0\\ 0&1&0&0&0&\overline{\omega}&1&0&\omega&0\\ \omega&0&1&\omega&\omega&\omega&1&\overline{\omega}&\overline{\omega}&0\\ \omega&\omega&0&0&\omega&0&0&0&\overline{\omega}&0\\ \omega&0&\omega&0&1&0&1&\omega&\omega&0\\ \omega&0&0&\omega&0&\overline{\omega}&1&\omega&0&1\\ 0&0&0&0&0&0&\omega&\omega&\overline{\omega}&\omega \end{bmatrix}, QC\_10d = \begin{bmatrix} 1&1&1&1&0&0&0&0&0&0\\ 1&0&0&0&\omega&1&\omega&\overline{\omega}&\omega&0\\ 0&0&0&0&\omega&\omega&\omega&\omega&0&0\\ 1&1&0&0&1&1&\overline{\omega}&\overline{\omega}&0&0\\ 0&1&1&0&\overline{\omega}&1&1&\overline{\omega}&0&0\\ \omega&\omega&1&0&0&0&1&1&1&0\\ 0&\omega&\omega&0&0&\omega&1&1&\overline{\omega}&0\\ 0&0&\omega&\omega&\overline{\omega}&0&\omega&\overline{\omega}&\overline{\omega}&0\\ \omega&0&0&\omega&\omega&\omega&\overline{\omega}&1&\omega&1\\ 0&0&0&0&0&\omega&0&\omega&\omega&\omega \end{bmatrix}, QC\_10e = \begin{bmatrix} 1&1&1&1&0&0&0&0&0&0\\ 1&0&0&0&\omega&1&\omega&\overline{\omega}&\omega&0\\ 0&0&0&0&\omega&\omega&\omega&\omega&0&0\\ 1&1&0&0&1&1&\overline{\omega}&\overline{\omega}&0&0\\ 0&1&1&0&0&1&1&\omega&\omega&0\\ \omega&\omega&1&0&\overline{\omega}&0&1&0&\overline{\omega}&0\\ 0&\omega&0&\omega&0&\omega&1&1&\overline{\omega}&0\\ \omega&0&\omega&0&\omega&0&0&0&\overline{\omega}&0\\ \omega&0&0&\omega&1&\omega&1&\omega&1&1\\ 0&0&0&0&0&1&1&\omega&0&\omega \end{bmatrix}.$$

5. Connections with Binary Codes

Höhn [9] presents two constructions that begin with a self-dual additive $(n, 2^n)$ code and produce a self-dual binary code of length $4n$. We describe these two (Constructions A and B) and add a third (C). If $\mathcal{C}$ is a self-dual additive $(n, 2^n)$ code, let $\widehat{\mathcal{C}}$ be the binary linear $[4n, n]$ code obtained from $\mathcal{C}$ by replacing each $GF(4)$ component by a 4-tuple in $GF(2)^4$ as follows: $0 \to 0000$, $1 \to 0110$, $\omega \to 1010$, $\overline{\omega} \to 1100$. Let d_4 be the $[4, 1]$ binary linear code $\{0000, 1111\}$. Finally let $(d_4^n)_0$ be the $[4n, n-1]$ binary linear code consisting of all codewords of weights divisible by 8 from the $[4n, n]$ code d_4^n.

Code	x											
QC_12a	0	0	0	0	0	0	0	1	ω	$\overline{\omega}$	1	
QC_12b	0	0	0	0	0	0	0	ω	$\overline{\omega}$	ω	$\overline{\omega}$	
QC_12c	0	0	0	0	0	0	0	ω	$\overline{\omega}$	$\overline{\omega}$	ω	
QC_12d	0	0	0	0	0	0	1	0	$\overline{\omega}$	1	$\overline{\omega}$	
QC_12e	0	0	0	0	0	0	ω	1	0	ω	1	
QC_12f	0	0	0	0	ω	0	0	0	1	$\overline{\omega}$	ω	
QC_12g	0	0	0	0	ω	0	0	0	$\overline{\omega}$	$\overline{\omega}$	1	
QC_13a	0	0	0	0	0	0	0	0	1	1	$\overline{\omega}$	$\overline{\omega}$
QC_13b	0	0	0	0	0	0	0	1	ω	0	ω	1
QC_13c	0	0	0	0	0	0	0	1	ω	0	ω	$\overline{\omega}$
QC_13d	0	0	0	0	0	1	0	0	1	0	1	1

TABLE 3. Lengthened codes of lengths 12 and 13

Code	n	d	$\lvert\mathrm{Aut}(C)\rvert$	A_0	A_4	A_5	A_6	A_7	A_8	A_9	A_{10}	A_{11}	A_{12}	A_{13}
QC_10a	10	4	48	1	30	0	300	0	585	0	108			
QC_10b	10	4	32	1	30	0	300	0	585	0	108			
QC_10c	10	4	1440	1	30	0	300	0	585	0	108			
QC_10d	10	4	256	1	30	0	300	0	585	0	108			
QC_10e	10	4	18	1	30	0	300	0	585	0	108			
QC_12a	12	5	2	1	0	48	188	432	765	1040	972	528	122	
QC_12b	12	5	4	1	0	40	212	424	725	1080	980	504	130	
QC_12c	12	5	2	1	0	48	188	432	765	1040	972	528	122	
QC_12d	12	5	4	1	0	40	212	424	725	1080	980	504	130	
QC_12e	12	5	6	1	0	40	212	424	725	1080	980	504	130	
QC_12f	12	5	4	1	0	40	212	424	725	1080	980	504	130	
QC_12g	12	5	1	1	0	48	188	432	765	1040	972	528	122	
QC_13a	13	5	8	1	0	15	236	356	1197	1530	2012	1956	650	239
QC_13b	13	5	24	1	0	15	236	356	1197	1530	2012	1956	650	239
QC_13c	13	5	6	1	0	15	236	356	1197	1530	2012	1956	650	239
QC_13d	13	5	2	1	0	15	236	356	1197	1530	2012	1956	650	239

TABLE 4. Extremal Self-dual Codes for lengths $n = 10, 12$ and 13

Construction A: $\rho_A(\mathcal{C}) = \widehat{\mathcal{C}} + d_4^n$

Construction B: Assume that n is even. $\rho_B(\mathcal{C}) = \widehat{\mathcal{C}} + (d_4^n)_0 + e_B$ where e_B is the $[4n, 1]$ code generated by

$$\begin{cases} 0001\ 0001\ \cdots 0001\ 0001 & \text{if } n \equiv 0 \pmod 4 \\ 0001\ 0001\ \cdots 0001\ 1110 & \text{if } n \equiv 2 \pmod 4 \end{cases}.$$

Construction C: Assume that n is even. $\rho_C(\mathcal{C}) = \widehat{\mathcal{C}} + (d_4^n)_0 + e_C$ where e_C is the $[4n, 1]$ code generated by

$$\begin{cases} 0001\ 0001\ \cdots 0001\ 1110 & \text{if } n \equiv 0 \pmod 4 \\ 0001\ 0001\ \cdots 0001\ 0001 & \text{if } n \equiv 2 \pmod 4 \end{cases}.$$

The next result is a straightforward extension of Lemma 2 of [**9**].

$\mathcal{C}$	$\lvert\text{Aut}(\mathcal{C})\rvert$	$\lvert\text{Aut}(\rho_B(\mathcal{C}))\rvert$	$\lvert\text{Aut}(\rho_C(\mathcal{C}))\rvert$
$\mathcal{G}_6$	$2^4 \cdot 3^3 \cdot 5$	$2^{10} \cdot 3^3 \cdot 5 \cdot 7 \cdot 11 \cdot 23$	$2^{10} \cdot 3^3 \cdot 5$
$\mathcal{C}_1$	$2^7 \cdot 3^2$	$2^{15} \cdot 3^2 \cdot 5 \cdot 7$	$2^{15} \cdot 3^2$
$\mathcal{C}_2$	$2^4 \cdot 3 \cdot 7$	$2^{12} \cdot 3 \cdot 7$	$2^{12} \cdot 3 \cdot 7$
$\mathcal{C}_3$	$2^7 \cdot 3^2 \cdot 7$	$2^{15} \cdot 3^2 \cdot 5 \cdot 7 \cdot 31$	$2^{15} \cdot 3^2 \cdot 7$

TABLE 5. Some Automorphism Group Orders

THEOREM 5.1. *Let $\mathcal{C}$ be a self-dual additive code. The following hold.*

(i) $\widehat{\mathcal{C}}$ is a self-orthogonal code.
(ii) $\rho_A(\mathcal{C})$, $\rho_B(\mathcal{C})$, and $\rho_C(\mathcal{C})$ are self-dual binary codes.
(iii) If $\mathcal{C}$ is Type II, $\rho_A(\mathcal{C})$ and $\rho_B(\mathcal{C})$ are doubly-even whereas if $\mathcal{C}$ is Type I, $\rho_A(\mathcal{C})$ and $\rho_B(\mathcal{C})$ are singly-even.
(iv) $\rho_C(\mathcal{C})$ is singly-even.

We give a proof of the following, which we have alluded to throughout the paper and was first proved in [**9**] in a more complicated manner.

THEOREM 5.2. *A self-dual additive Type II code of odd length n cannot exist.*

Proof: If $\rho_A(\mathcal{C})$ is a binary self-dual doubly-even code of length $4n$, then n is even as $4n$ must be a multiple of 8. The result follows by Theorem 5.1(iii). □

EXAMPLE 5.3. Let $\mathcal{G}_6$ be the $(6, 2^6, 4)$ hexacode given in Example 1.1. We verified that $\rho_B(\mathcal{G}_6)$ is the $[24, 12, 8]$ binary Golay code and $\rho_C(\mathcal{G}_6)$ is the $[24, 12, 6]$ *odd Golay code.* It is interesting to note that beginning with the unique extremal self-dual additive code of length 6, we obtain the unique extremal self-dual doubly-even and singly-even binary codes of length 24 using constructions B and C. Note also by Table 5 that $\rho_C(\mathcal{G}_6)$ has automorphism group of size 2^6 times that of $\mathcal{G}_6$.

EXAMPLE 5.4. By Example 3.2 there are exactly three $(8, 2^8, 4)$ Type II codes $\mathcal{C}_i (i = 1, 2, 3)$ with generator matrices G_i respectively QC_8c, QC_8d, and QC_8e.

Applying Construction B to each of these codes produces three of the five inequivalent self-dual doubly-even binary $[32, 16, 8]$ codes [**4**]. The codes $\rho_B(\mathcal{C}_i)$ are respectively C83 (or $2g_{16}$), C84 (or $8f_4$), and C82 (or r_{32}) in the notation of [**4**]. Notice that the extended quadratic residue code is not in this list but the second order Reed–Muller code is. Applying Construction C to each produces all three of the inequivalent self-dual singly-even binary $[32, 16, 8]$ codes [**6**]. Analogous to Example 5.3, $\rho_C(\mathcal{C}_i)$ has automorphism group of size 2^8 times that of $\mathcal{C}_i$ as shown in Table 5.

References

[1] C. Bachoc, "Harmonic weight enumerators of nonbinary codes and MacWilliams identities," preprint, 1999.

[2] W. Bosma and J. Cannon, *Handbook of Magma functions*, Sydney, 1995.

[3] A. R. Calderbank, E. M. Rains, P. W. Shor, and N. J. A. Sloane, "Quantum error correction via codes over $GF(4)$," *IEEE Trans. Inform. Theory* **IT-44** (1998), 1369–1387.

[4] J. H. Conway and V. Pless, "On the enumeration of self-dual codes," *J. Comb. Theory* **28A** (1980), 26–53.

[5] J. H. Conway, V. Pless, and N. J. A. Sloane, "Self-dual codes over $GF(3)$ and $GF(4)$ of length not exceeding 16," *IEEE Trans. Inform. Theory* **IT-25** (1979), 312–322.

[6] J. H. Conway and N. J. A. Sloane, "A new upper bound on the minimal distance of self-dual codes," *IEEE Trans. Inform. Theory* **IT-36** (1990), 1319–1333.
[7] J. Fields, P. Gaborit, W. C. Huffman and V. Pless, "On the classification of extremal even formally self-dual codes of length 20 and 22", preprint, 1998.
[8] P. Gaborit, W. C. Huffman, J.-L. Kim, and V. Pless "On the classification of extremal additive codes over $GF(4)$," *Proceedings of the 37th Allerton Conference on Communication, Control, and Computing*, UIUC Sep. 1999, 535-544.
[9] G. Höhn, "Self-dual codes over the Kleinian four group," preprint, 1996. An updated version in http://xxx.lanl.gov/(math.CO/0005266).
[10] V. Pless, *Introduction to the Theory of Error-Correcting Codes*, 3rd ed. New York: J. Wiley & Sons, 1998.
[11] E. M. Rains and N. J. A. Sloane, "Self-dual codes," in *Handbook of Coding Theory*, ed. V. S. Pless and W. C. Huffman. Amsterdam: Elsevier, 1998, pp. 177–294.

LACO, Université de Limoges, 87000 Limoges, France
E-mail address: gaborit@unilim.edu

Department of Mathematical and Computer Sciences, Loyola University, Chicago, IL 60626, USA
E-mail address: wch@math.luc.edu

Department of Mathematics, Statistics, and Computer Science, 322 SEO, University of Illinois–Chicago, 851 S. Morgan, Chicago, IL 60607-7045, USA
E-mail address: jlkim@math.uic.edu

Department of Mathematics, Statistics, and Computer Science, 322 SEO, University of Illinois–Chicago, 851 S. Morgan, Chicago, IL 60607-7045, USA
E-mail address: pless@math.uic.edu

DIMACS Series in Discrete Mathematics
and Theoretical Computer Science
Volume **56**, 2001

Integral zeroes of Krawtchouk polynomials

Laurent Habsieger

ABSTRACT. We present upper bounds for $N_q(n, N)$, the number of integer zeros of Krawtchouk polynomials $K_n(x, q, N)$; for instance, for q a prime and $n \gg \log N$, we have $N_q(n, N) \leq n/q(1 + o(1))$. We show the existence of a noninteger root for $n \geq 4$ even when N is odd, and for $n \geq 344$ even when N is even. We also prove a conjecture of Krasikov and Litsyn.

I Introduction

The q-ary Krawtchouk polynomial $K_n(x, q, N)$ may be defined by its generating function

$$\sum_{n=0}^{\infty} K_n(x, q, N) z^n = (1 - z)^x (1 + (q - 1)z)^{N-x}, \tag{1.1}$$

which gives the expression

$$K_n(x, q, N) = \sum_{j=0}^{n} (-1)^j (q - 1)^{n-j} \binom{N - x}{n - j} \binom{x}{j}, \tag{1.2}$$

where

$$\binom{x}{j} = \frac{x(x - 1) \cdots (x - j + 1)}{j!}$$

for $j \geq 1$ and $\binom{x}{0} = 1$.

We now give several useful properties of the Krawtchouk polynomials, without proof. The reader may refer to [**13**] for details. We deduce from (1.1) the recurrence relation

$$(n + 1)K_{n+1} = (n + (q - 1)(N - n) - qx)K_n - (q - 1)(N - n + 1)K_{n-1}, \tag{1.3}$$

with the initial values $K_0(x, q, n) = 1$ and $K_1(x, q, N) = N(q - 1) - qx$. From the recurrence relation (1.3) we deduce that

$$n! K_n(x/q, q, N) \in \mathbb{Z}[x]. \tag{1.4}$$

1991 *Mathematics Subject Classification.* Primary 33C45; Secondary 11D99.
Key words and phrases. Krawtchouk polynomials, coding theory, diophantine equations.

From the expression (1.2), we find the symmetry property

$$(q-1)^k \binom{N}{k} K_n(k,q,N) = (q-1)^n \binom{N}{n} K_k(n,q,N), \qquad (1.5)$$

for $0 \leq k, n \leq N$.

In the binary case, an additional property holds:

$$K_n(x,2,N) = (-1)^n K_n(N-x,2,N). \qquad (1.6)$$

Krawtchouk polynomials are of special interest for coding theorists. They occur in many problems and we shall restrict ourselves to those involving their integer roots.

The first problem deals with perfect codes. Let $F_{q,N}$ be the q-ary Hamming space of length N. A perfect code of radius R is a subset C of $F_{q,N}$ with the property that the Hamming spheres of radius R centered at the points of C cover $F_{q,N}$ without intersections. Does such a code exist ?

The second problem generalizes the first one to multiple coverings. A multiple perfect μ-covering of radius R is a multiset C whose elements belong to $F_{q,N}$ with the property that Hamming spheres of radius R centered at the points of C cover $F_{q,N}$ exactly μ times. Does such a code exist ?

Krawtchouk polynomials provide partial answers to these problems. The statements below are q-analogues of the ones given in the binary case by Krasikov and Litsyn [**11**] (see also [**4, 12**] for more details). However the proofs work the same in the q-case [**14**].

THEOREM 1.1. *If $K_R(x,q,N)$ has at least one noninteger root then there is no perfect code of radius R in $F_{q,N}$.*
If $K_R(x,q,N)$ has less than Z integer roots then there is no multiple perfect μ-covering in $F_{q,N}$, where Z is the minimum integer such that

$$\sum_{i=0}^{R} \binom{N}{i} \leq \mu \sum_{j=0}^{Z} \binom{R}{j}.$$

Many authors studied the roots of Krawtchouk polynomials. In the binary case, Clayton [**2**] found all the zeros in the cases $n = 1, 2, 3, N/2$. Diaconis and Graham [**5**] gave a list of zeros of $K_4(x,2,N)$ and conjectured it to be complete. This was proved independently by Mignotte and Pethő [**16**] and by Stroeker and de Weger [**22**]. Hanrot [**8**] solved the case $n = 5$ in his thesis. Recently, Stroeker and de Weger [**23**] completed the cases $n = 6, 7$. Habsieger and Stanton [**7**] also gave complete lists of solutions for $n = (N-t)/2$ with $t = 0, 1, 2, 3, 4, 5, 6$ or with $t = 8$ and x odd.

In the nonbinary cases, fewer results are known. Clayton [**2**] found all the zeros in the cases $n = 1, 2$. Habsieger and Stanton [**7**] gave three infinite families (in q) that are solutions of $K_3(x,q,N)$. Hanrot [**9**] solved completely the case $n = 4$ for $q \leq 20$.

Krasikov and Litsyn [**11**] gave numerous upper bounds for the number of zeros of the binary Krawtchouk polynomials. We shall present in section 2 two new approaches. The first one relies on a new identity and provides results closed to those of Krasikov and Litsyn. The second one gives a general upper bound for q a prime ; in the binary case, it improves on Krasikov and Litsyn's bound for $n \gg \log N$.

Hong [**10**] proved the existence of a noninteger root for nonbinary Krawtchouk polynomials of degree greater than 2. Moreover we know all the possible parameters for perfect binary codes [**24**]. However the problem of the existence of a noninteger root in the binary case remains open (see [**11, conjecture 2**]). The case n and N odd is trivial ($N/2$ is a noninteger root). We shall focus in section 3 on the case n even. Krasikov and Litsyn [**11**] proved that there is always a noninteger root for n large enough. Their proof relies on a result of Shorey and Tijdeman [**21**], which is effective but not explicit. Theoretically it would lead to a bound looking like expexp(C) with C large. Thus the number of cases left would be huge. In section 3, we use other approaches than Shorey and Tijdeman's one to get reasonable explicit results: the case n even and N odd is completely solved, the case n and N even is treated for $n \geq 344$. The cases n odd and N even is the most difficult one, and remains a mystery.

II Number of integral roots

Let $N_q(n,N)$ denote the number of integral roots of the polynomial $K_n(x,q,N)$. Krasikov and Listyn [**11**] proved the next theorem.

THEOREM 2.1. *The following inequalities hold:*
i) for n and N even

$$N_2(n,N) \leq \frac{5n-13}{8} + \frac{\log N}{\log 2} + \frac{\log n}{4\log 2};$$

ii) for n odd and N even

$$N_2(n,N) \leq \frac{5n-10}{8} + \frac{\log N}{\log 2} + \frac{\log n}{4\log 2};$$

i) for n even and N odd

$$N_2(n,N) \leq \frac{4n-8}{7} + \frac{6\log N}{7\log 2} + \frac{2\log n}{7\log 2};$$

i) for n and N odd

$$N_2(n,N) \leq \frac{4n+6}{7} + \frac{6\log N}{7\log 2} + \frac{2\log n}{7\log 2}.$$

We give here alternative approaches which lead to similar results.

LEMMA 2.2. *We have the identities*

$$\sum_k \binom{N}{k}\binom{N}{k+l} K_n(k,2,N)K_n(k+l,2,N) = (-1)^n \binom{N}{n}^2 K_{N+l}(2n,2,2N)$$

and

$$\sum_k \binom{N}{k}\binom{N}{k+1} K_n(k,2,N)K_n(k+1,2,N) = \frac{\binom{2N}{N-1}\binom{N}{n}^3}{\binom{2N}{2n}}\left(1-\frac{2n}{N}\right).$$

PROOF. Let S denote the left-hand side of the first identity. By (1.5) we also have

$$S = \binom{N}{n}^2 \sum_k K_k(n,2,N)K_{k+l}(n,2,N).$$

This last sum represents the coefficient of z^l in the product

$$(1-z)^n(1+z)^{N-n}(1-z^{-1})^n(1+z^{-1})^{N-n} = (-1)^n z^{-N}(1-z)^{2n}(1+z)^{2N},$$

and the first identity follows. The second one corresponds to the case $l = 1$. We use (1.5) to get

$$\sum_{k=0}^{N-1}\binom{N}{k}\binom{N}{k+1}K_n(k,2,N)K_n(k+1,2,N) = (-1)^n \frac{\binom{N}{n}^2\binom{2N}{N-1}}{\binom{2N}{2n}} K_{2n}(N-1,2,2N).$$

We now use (1.3) and (1.6) to get

$$K_{2n}(N-1,2,2n) = \left(1-\frac{2n}{N}\right)K_{2n}(N,2,2n).$$

The generating function (1.1) easily gives $K_{2n}(N,2,2n) = (-1)^n\binom{N}{n}$ and the second identity is proved. □

We introduce now some notations. For p a prime and n a nonnegative integer, the highest power of p in n will be denoted as $\mathrm{ord}_p(n)$. We shall make an extensive use of the following lemma, which can be easily proved by induction.

LEMMA 2.3. *For p a prime and n a nonnegative integer, we have*

$$\mathrm{ord}_p(n!) = \frac{n-\chi_p(n)}{p-1}.$$

where $\chi_p(n)$ denotes the sum of the digits of the p-adic expansion of n.

THEOREM 2.4. $N_2(n,N) \le \frac{2n-1}{3} + \frac{\log N}{\log 2}$.

PROOF. Put $s = N_2(n,N)$ and write $K_n(X,2,N) = P(X)Q(X)$, where P is the monic polynomial whose roots are the integral roots of $K_n(X,2,N)$, and where Q is a polynomial with noninteger roots. By (1.4), $n!K_n(X/2,2,N)$ is a polynomial with integer coefficients. It follows that $2^{-s}n!Q(X/2)$ is also a polynomial with integer coefficients; thus $2^{-2s}n!^2Q(k)Q(k+1)$ is an integer. Since every integer is either odd or even, we also have $P(k)P(k+1) \in 2^s\mathbb{Z}$. Lemma 2.3 then gives

$$\mathrm{ord}_2\left(K_n(k,2,N)K_n(k+1,2,N)\right) \ge s + 2(s-\mathrm{ord}_2(n!)) = 3s-2n+2\chi_2(n).$$

We use the notation from the proof of Lemma 2.2. Since

$$\begin{aligned}\mathrm{ord}_2(S) &= \mathrm{ord}_2\left(\frac{\binom{2N}{N}\binom{N}{n}^3}{\binom{2N}{2n}}\frac{N-2n}{N+1}\right)\\ &= -\chi_2(N)+2\chi_2(n)+2\chi_2(N-n)+\mathrm{ord}_2\left(\frac{N-2n}{N+1}\right)\\ &\le 2\chi_2(n)+3\frac{\log N}{\log 2}-1,\end{aligned}$$

we get $3s-2n \le +3\frac{\log N}{\log 2}-1$, which proves the theorem. □

We now give a similar result in the q-case, for q a prime.

THEOREM 2.5. *Let q be a prime. Then*

$$N_q(n, N) \leq \frac{n}{q}\left(1 + 3\sqrt{\frac{q(q-1)\log N}{2n\log q}}\right)$$

for $n \geq q(q-1)\log N/\log q$.

PROOF. Let l be a nonnegative integer to be chosen later. Put $L = q^l$ and let k be an integer, $0 \leq k < \min(n, L)$. We have

$$|K_k(n, q, N)| \leq \sum_{j=0}^{k}(q-1)^k\binom{n}{j}\binom{N-n}{k-j} = (q-1)^k\binom{N}{k}.$$

From (1.5) we deduce

$$\begin{aligned}
\operatorname{ord}_q(K_n(k, q, N)) &= \operatorname{ord}_q\left((q-1)^{n-k}\frac{\binom{N}{n}}{\binom{N}{k}}K_k(n, q, N)\right)\\
&= \operatorname{ord}_q\left(\frac{k!(N-k)!}{n!(N-n)!}\right) + \operatorname{ord}_q(K_k(n, q, N))\\
&\leq -\operatorname{ord}_q(n!) + \operatorname{ord}_q\left(\frac{k!(N-k)!}{(N-n)!}\right) + \frac{\log K_k(n, q, N)}{\log q}\\
&\leq -\operatorname{ord}_q(n!) + \frac{n}{q-1} + \frac{\log N + k\log(q-1) + \log\binom{N}{k}}{\log q},
\end{aligned}$$

since

$$\begin{aligned}
\operatorname{ord}_q\left(\frac{k!(N-k)!}{(N-n)!}\right) &= \sum_{r=1}^{\left\lfloor\frac{\log N}{\log q}\right\rfloor}\left(\left\lfloor\frac{k}{q^r}\right\rfloor + \left\lfloor\frac{N-k}{q^r}\right\rfloor - \left\lfloor\frac{N-n}{q^r}\right\rfloor\right)\\
&\leq \left\lfloor\frac{\log N}{\log q}\right\rfloor + \sum_{r=1}^{\infty}\frac{n}{q^r} \leq \frac{\log N}{\log q} + \frac{n}{q-1}.
\end{aligned}$$

Put $\omega = \sum_{k=0}^{L-1}\operatorname{ord}_q(K_n(k, q, N))$. This gives the inequality

$$\begin{aligned}
\omega &\leq -L\operatorname{ord}_q(n!) + L\left(\frac{n}{q-1} + \frac{\log N}{\log q}\right) + \sum_{k=1}^{L-1}\frac{k\log(q-1) + \log\binom{N}{k}}{\log q}\\
&\leq -L\operatorname{ord}_q(n!) + L\left(\frac{n}{q-1} + \frac{\log N}{\log q}\right) + \binom{L}{2}\frac{\log(q-1)}{\log q} + (L-1)\frac{\log\binom{N}{L/2}}{\log q}
\end{aligned}$$

by using the log-concavity of the binomial coefficients. One easily checks that $(L/2)! \geq (q-1)^{L/2}$ for $l \geq 1$, and thus

$$\sum_{k=0}^{L-1}\operatorname{ord}_q(K_n(k, q, N)) \leq -L\operatorname{ord}_q(n!) + L\frac{n}{q-1} + \binom{L+1}{2}\frac{\log N}{\log q}. \tag{2.6}$$

As in the proof of Theorem 2.4, we write $K_n(X, q, N) = P(X)Q(X)$, where P is the monic polynomial whose roots are the integral roots of $K_n(X, q, N)$, and where Q is a polynomial with noninteger roots. Note that the degree p of P is exactly $N_q(n, N)$.

By (1.4), $n!K_n(X/q,q,N)$ is a polynomial with integer coefficients, and it still follows that $q^{-p}n!Q(X/q)$ is also a polynomial with integer coefficients. Moreover

$$\operatorname{ord}_q\left(\prod_{k=0}^{L-1}P(k)\right)\geq \operatorname{ord}_q(L!)=\frac{L-1}{q-1}.$$

We get this way

$$\sum_{k=0}^{L-1}\operatorname{ord}_q(K_n(k,q,N))\geq p\frac{L-1}{q-1}-L\operatorname{ord}_q(q^{-p}n!)=-L\operatorname{ord}_q(n!)+p\left(L+\frac{L-1}{q-1}\right).$$

Combining this last inequality with (2.6) leads to the upper bound

$$\begin{aligned}N_q(n,N)&\leq\left(\frac{n}{q-1}+\frac{L+1}{2}\frac{\log N}{\log q}\right)\left(1+\frac{1-1/L}{q-1}\right)^{-1}\\&=\frac{n}{q}\left(1+\frac{L+1}{2\alpha}\right)\left(1-\frac{1}{qL}\right)^{-1}=\varphi(L),\end{aligned}$$

with $\alpha=\frac{n\log q}{(q-1)\log N}$.

Assume now $\alpha\geq q^3/2$ and put $l=\lfloor\frac{\log(2\alpha/q)}{2\log q}\rfloor$, so that $l\geq 1$. We also find $\sqrt{2\alpha/q^3}<L\leq\sqrt{2\alpha/q}$. Since φ is decreasing on this interval, we find

$$\varphi(L)\leq\varphi\left(\sqrt{\frac{2\alpha}{q^3}}\right)\leq 1+3\sqrt{\frac{q}{2\alpha}},$$

after tedious calculations. For $q\leq\alpha\leq q^3/2$, we take $L=q$, and the same inequality holds for $\varphi(L)$. □

III Existence of a noninteger root in the binary case

In [**11**], Krasikov and Litsyn showed that $K_n(x,2,N)$ has a noninteger root for n even large enough. Their result is effective but not explicit; they use a result by Shorey and Tijdeman [**21**] that would lead to a huge bound looking like $\exp\exp(C)$ with C big. In this section we use underlying ideas to their paper and new ones to get a reasonably explicit bound.

The starting point is, as in [**11**], the following result.

THEOREM 3.1. *Assume that n is even and that $K_n(x,2,N)$ has only integer roots. Then*

i) for N odd, the product

$$\prod_{i=0}^{n/2-1}(2i+1)\prod_{i=0}^{n/2-1}(N-2i)$$

is a perfect square;

ii) for N even, the product

$$2^{n/2}\prod_{i=0}^{n/2-1}(2i+1)\prod_{i=0}^{n/2-1}(N/2-i)$$

is a perfect square.

Let us put

$$\Delta(m,k,d)=\prod_{i=0}^{k-1}(m+id)\,.$$

Theorem 3.1 may be written as:

i) For N odd and n even, $\Delta(N-n+2,n/2,2)n!/(n/2)!$ is a perfect square;

ii) For N and n even, $\Delta(N-n/2+1,n/2,1)2^{n/2}n!/(n/2)!$ is a perfect square.

Therefore we shall now study the diophantine equation

$$\Delta(m,k,d)=b_k y^2 \tag{3.2}$$

where $d=1,2$, $m\in d\mathbb{N}+1$ and b is a squarefree integer such that:

i) $b(2k)!/k!$ is a perfect square for $d=1$,

ii) $b2^k(2k)!/k!$ is a perfect square for $d=2$.

For $0\le i<k$ we write

$$m+id=a_i x_i^2\,,$$

where a_i is a squarefree integer and x_i an integer. We start with a general lemma. We use an idea of Erdős and Selfridge [**6**].

LEMMA 3.3. *For $m\ge(k-1)^2d/2$, the a_i's are pairwise distinct.*

PROOF. Assume that there exists $0\le i<j<k$ with $a_i=a_j$. Since $a_i(x_j^2-x_i^2)=d(j-i)>0$, we have $x_j\ge x_i+1$. This implies that

$$(k-1)d\ge(j-i)d>2a_ix_i\ge2\sqrt{a_ix_i^2}\ge2\sqrt{m}\ge(k-1)\sqrt{2d}\,.$$

Since $d\le2$, there is a contradiction and the lemma is proved. □

We now follow Marszalek's approach [**15**] to get an upper bound for the product of the a_i's.

LEMMA 3.4. *Let p be a prime and $k\ge1$. We have*

$$\operatorname{ord}_p\left(\prod_{i=0}^{k-1}a_i\right)-\operatorname{ord}_p\left(\frac{(2k-1)!}{k!}\right)\le\left\lfloor\frac{\log k}{\log p}\right\rfloor+2-\frac{2(k-1)}{p^2-1}\,.$$

Moreover, for $p>\sqrt{k}$

$$\operatorname{ord}_p\left(\prod_{i=0}^{k-1}a_i\right)-\operatorname{ord}_p\left(\frac{(2k-1)!}{k!}\right)\le\begin{cases}0 & \text{if } p \text{ does not divide } 2k,\\ 1 & \text{if } p \text{ divides } 2k.\end{cases}$$

PROOF. When $p=d=2$, the integers $m+id$ are odd. This shows that

$$\operatorname{ord}_2\left(\prod_{i=0}^{k-1}a_i\right)-\operatorname{ord}_2\left(\frac{(2k)!}{k!}\right)=0-k$$

and the lemma follows in this case.

We now assume that p is a prime which does not divide d. We put $\alpha=\lfloor\log k/\log p\rfloor$, so that $p^\alpha\le k<p^{\alpha+1}$. Let j be a nonnegative integer and $i_j=$

$\min\{0 \le i \le k-1 \ : \ p^j \mid n+id\}$, with $i_j = k$ if $\{0 \le i \le k-1 \ : \ p^j \mid n+id\} = \emptyset$. We have

$$|\{1 \le i \le k \ : \ p^m \mid n+id\}| = \left\lfloor \frac{k-1-i_j}{p^j} \right\rfloor + 1$$

and

$$\begin{aligned}\operatorname{ord}_p\left(\prod_{i=0}^{k-1} a_i\right) &= |\{0 \le i \le k \ : \ \operatorname{ord}_p(n+id) \text{ odd}\}| \\ &= \sum_{r=1}^{\infty}(-1)^{r-1}\left(\left\lfloor \frac{k-1-i_r}{p^r} \right\rfloor + 1\right).\end{aligned}$$

For $r > \alpha$, note that

$$\left\lfloor \frac{k-1-i_r}{p^r} \right\rfloor = \begin{cases} 0 & \text{if } 0 \le i_r \le k-1, \\ -1 & \text{if } i_r = k, \end{cases}$$

which implies that

$$\sum_{r=\alpha+1}^{\infty}(-1)^{r-1}\left(\left\lfloor \frac{k-1-i_r}{p^r} \right\rfloor + 1\right) \in \{(-1)^\alpha, 0\}.$$

Since

$$\operatorname{ord}_p\left(\frac{(2k-1)!}{k!}\right) = \sum_{r=1}^{\infty}\left(\left\lfloor \frac{2k-1}{p^r} \right\rfloor - \left\lfloor \frac{k}{p^r} \right\rfloor\right) = \sum_{r=1}^{\alpha}\left(\left\lfloor \frac{2k-1}{p^r} \right\rfloor - \left\lfloor \frac{k}{p^r} \right\rfloor\right),$$

we get

$$\begin{aligned}&\operatorname{ord}_p\left(\prod_{i=0}^{k-1} a_i\right) - \operatorname{ord}_p\left(\frac{(2k-1)!}{k!}\right) \\ &\le \sum_{r=1}^{\alpha}(-1)^{r-1}\left(\left\lfloor \frac{k-1-i_r}{p^r} \right\rfloor + 1\right) + \max((-1)^\alpha, 0) + \sum_{r=1}^{\alpha}\left(\left\lfloor \frac{k}{p^r} \right\rfloor - \left\lfloor \frac{2k-1}{p^r} \right\rfloor\right).\end{aligned}$$

The property $0 \le i_r \le p^r - 1$ gives

$$\begin{aligned}\left\lfloor \frac{k}{p^{2r}} \right\rfloor - \left\lfloor \frac{2k-1}{p^{2r}} \right\rfloor - \left\lfloor \frac{k-1-i_{2r}}{p^{2r}} \right\rfloor - 1 &\le \left\lfloor \frac{k}{p^{2r}} \right\rfloor - \left\lfloor \frac{2k-1}{p^{2r}} \right\rfloor - \left\lfloor \frac{k-p^{2r}}{p^{2r}} \right\rfloor - 1 \\ &\le 1 - \frac{2k}{p^{2r}}\end{aligned}$$

for $1 \le 2r \le \alpha$, and

$$\left\lfloor \frac{k-1-i_{2r-1}}{p^{2r-1}} \right\rfloor + 1 + \left\lfloor \frac{k}{p^{2r-1}} \right\rfloor - \left\lfloor \frac{2k-1}{p^{2r-1}} \right\rfloor \le 1 + \left\lfloor \frac{k-1}{p^{2r-1}} \right\rfloor + \left\lfloor \frac{k}{p^{2r-1}} \right\rfloor - \left\lfloor \frac{2k-1}{p^{2r-1}} \right\rfloor \le 1$$

for $1 \le 2r-1 \le \alpha$. We obtain

$$\operatorname{ord}_p\left(\prod_{i=0}^{k-1} a_i\right) - \operatorname{ord}_p\left(\frac{(2k-1)!}{k!}\right) \le \alpha + \max((-1)^\alpha, 0) - \sum_{1 \le 2r \le \alpha} \frac{2k}{p^{2r}}.$$

Define $\epsilon = 0$ if α is odd and $\epsilon = 1$ otherwise. We have

$$\max((-1)^{\alpha+1}, 0) + \sum_{2r > \alpha} \frac{2k}{p^{2r}} < \epsilon + \frac{2p^{\alpha+1}}{p^{\alpha+1+\epsilon}}\left(1 - \frac{1}{p^2}\right)^{-1} \le 2 + \frac{2}{p^2-1},$$

from which we deduce the inequality

$$\operatorname{ord}_p\left(\prod_{i=0}^{k-1} a_i\right) - \operatorname{ord}_p\left(\frac{(2k-1)!}{k!}\right) < \alpha + 2 + \frac{2}{p^2-1} - \sum_{r\geq 1}\frac{2k}{p^{2r}}.$$

This shows the first part of the lemma. For $p > \sqrt{k}$, we still have

$$\operatorname{ord}_p\left(\prod_{i=0}^{k-1} a_i\right) - \operatorname{ord}_p\left(\frac{(2k-1)!}{k!}\right) \leq 1 + \left\lfloor\frac{k-1}{p}\right\rfloor + \left\lfloor\frac{k}{p}\right\rfloor - \left\lfloor\frac{2k-1}{p}\right\rfloor \leq 1.$$

Moreover, if p does not divide $2k$, we deduce from the condition that $b(2k)!/k!$ is a perfect square or twice a perfect square:

$$\operatorname{ord}_p\left(\prod_{i=0}^{k-1} a_i\right) - \operatorname{ord}_p\left(\frac{(2k-1)!}{k!}\right) = \operatorname{ord}_p\left(\prod_{i=0}^{k-1} a_i\right) - \operatorname{ord}_p\left(\frac{(2k)!}{k!}\right) \equiv 0 \mod 2$$

and the second part of the lemma follows. □

LEMMA 3.5. *For $k \geq 289$, we have*

$$\prod_{i=0}^{k-1} a_i \leq \frac{(2k)!}{k!}\exp\left(-1.139921986189k + 7.70497\sqrt{k} + 3.64146\right).$$

PROOF. By Lemma 3.4 we know that

$$\begin{aligned}\prod_{i=0}^{k-1} a_i &\leq \frac{(2k-1)!}{k!}\exp\left(\sum_{\substack{p\leq\sqrt{k}\\ p \text{ prime}}}\left(\log k + 2\log p - \frac{2(k-1)\log p}{p^2-1}\right) + \sum_{\substack{\sqrt{k}<p\mid 2k\\ p\text{ prime}}}\log p\right)\\ &\leq \frac{(2k-1)!}{k!}\exp\left(\pi(\sqrt{k})\log k + 2\theta(\sqrt{k}) - 2(k-1)\sum_{\substack{p\leq\sqrt{k}\\ p\text{ prime}}}\frac{\log p}{p^2-1} + \log k\right),\end{aligned}$$

where

$$\pi(x) = \sum_{\substack{p\leq x\\ p\text{ prime}}} 1 \quad\text{and}\quad \theta(x) = \sum_{\substack{p\leq x\\ p\text{ prime}}}\log p.$$

We also need Riemann's zeta function

$$\zeta(s) = \sum_{n\geq 1}\frac{1}{n^s} = \prod_{p\text{ prime}}\left(1-\frac{1}{p^s}\right)^{-1},$$

so that

$$-\frac{\zeta'(s)}{\zeta(s)} = \frac{\mathrm{d}}{\mathrm{d}s}\left(\log\left(\prod_{p\text{ prime}}\left(1-\frac{1}{p^s}\right)\right)\right) = \sum_{p\text{ prime}}\frac{\log p}{p^s-1}.$$

Computations on PARI then give

$$\sum_{p\text{ prime}}\frac{\log p}{p^2-1} = -\frac{\zeta'(2)}{\zeta(2)} = 0.569960993094532806399864360019 73\ldots$$

and we get

$$\prod_{i=0}^{k-1} a_i \le \frac{(2k)!}{k!}\exp\left(-1.139921986189k + R(k)\right) ,$$

with

$$R(k) = \pi(\sqrt{k})\log k + 2\theta(\sqrt{k}) + 2(k-1)\sum_{\substack{p>\sqrt{k}\\ p \text{ prime}}} \frac{\log p}{p^2-1} + 1.139921986189 - \log 2 .$$

We use Abel's summation to evaluate the sum in R:

$$\begin{aligned}\sum_{\substack{p>\sqrt{k}\\ p \text{ prime}}} \frac{\log p}{p^2-1} &= \sum_{n>\sqrt{k}} \frac{\log n}{n^2-1}(\pi(n)-\pi(n-1)) \\ &= \sum_{n>\sqrt{k}} \pi(n)(f(n)-f(n+1)) - \pi(\lfloor\sqrt{k}\rfloor)f(\lfloor\sqrt{k}\rfloor+1) ,\end{aligned}$$

with $f(x) = \log x/(x^2-1)$. One easily checks that f'' is nonnegative and

$$f(n) - f(n+1) \le -f'(n) = \frac{2n\log n}{(n^2-1)^2} - \frac{1}{n(n^2-1)} \le \frac{2\log n}{n(n^2-1)} .$$

We now need the estimate from [**18**]:

$$\begin{aligned}\pi(x) &< 1.25506\frac{x}{\log x} \quad \text{for} \quad x>1 ,\\ \pi(x) &> \frac{x}{\log x} \quad \text{for} \quad x \ge 17 .\end{aligned}$$

We obtain

$$\begin{aligned}\sum_{\substack{p>\sqrt{k}\\ p \text{ prime}}} \frac{\log p}{p^2-1} &< 1.25506\sum_{n>\sqrt{k}} \frac{2}{n^2-1} - \frac{\lfloor\sqrt{k}\rfloor}{\log(\lfloor\sqrt{k}\rfloor)} f(\lfloor\sqrt{k}\rfloor+1) \\ &< \frac{2.51012}{\lfloor\sqrt{k}\rfloor} - \frac{0.91278}{\lfloor\sqrt{k}\rfloor} < \frac{1.59734}{\sqrt{k}-1}\end{aligned}$$

for $k \ge 289$. From [**19**], we know that $\theta(x) < 1.000081x$ for $x > 0$. This gives an upper bound for R:

$$\begin{aligned}R(k) &< 2.51012\sqrt{k} + 2.000162\sqrt{k} + 3.19468(\sqrt{k}+1) + 1.139923 - \log 2 \\ &< 7.70497\sqrt{k} + 3.64146\end{aligned}$$

for $k \ge 289$. □

We also need a lower bound for the product of the a_i's.

LEMMA 3.6. *For $k \ge 1$, we have*

$$\prod_{i=0}^{k-1} a_i \ge \exp(-3.285)\left(\frac{25}{16}\right)^k k! .$$

PROOF. We proceed as in [20]. By definition and by Lemma 3.3, the a_i's are distinct squarefree numbers. Therefore

$$\prod_{i=0}^{k-1} a_i \geq \prod_{i=1}^{k} s_i \,,$$

where s_i denotes the i-th squarefree number. Write now $s_i = 900t + r$ with $0 \leq r < 900$. Since every squarefree number has only 576 possible residues modulo 900, we have $i \leq 900t + \min(r, 576)$. As a matter of fact, the integers divisible by $25, 49\dots$ should not appear in the list of possible squarefree integers. There are 207 of them less than or equal to 6929. We get therefore the refinement

$$i + 207 \leq 576t + \min(r, 576) \leq \frac{16}{25}s_i + \min\left(\frac{9}{25}r, 576 - \frac{16}{25}r\right) \leq \frac{16}{25}s_i + \frac{207.36}{25} \,,$$

for $s_i \geq 6929$. This implies that $s_i \geq 25/16i$ for $s_i \geq 6929$ and that the function $\prod_{i=1}^{s} s_i (16/25)^s / s!$ attains its minimum for $s \leq s_{4211} = 6929$. A systematic study shows that this minimum occurs at $s_{31} = 47$ and is greater than $\exp(-3.285)$, which proves the lower bound for the product of the a_i's. □

We can now deduce a nonexistence result.

THEOREM 3.7. *For $m \geq (k-1)^2/d$, equation (3.2) has no solution with*

$$k \geq \begin{cases} 172 & \text{if } d = 1, \\ 5 & \text{if } d = 2. \end{cases}$$

PROOF. We know [**3, p. 33**] that

$$\binom{2k}{k} \leq \frac{4^k}{\sqrt{\pi k}} \,.$$

With Lemma 3.5 and Lemma 3.6, this gives

$$\exp(-3.285)\left(\frac{25}{16}\right)^k \leq \frac{1}{k!}\prod_{i=0}^{k-1} a_i \leq \frac{4^k \exp\left(-1.13992198k + 7.70497\sqrt{k} + 3.64146\right)}{\sqrt{\pi k}} \,,$$

for $k \geq 289$. This implies the inequality

$$(1.13992198 + \log(25/64))k \leq 7.70497\sqrt{k} + 6.92646 - \log(\pi k)/2 \,,$$

which is impossible for $k \geq 1596$.

Let us now assume $4 \leq k \leq 1595$. We use the proof of Lemma 3.4 and Lemma 3.6 to get the inequality

$$\prod_{i=1}^{k} s_i \leq \prod_{\substack{p \leq 2k \\ p \text{ prime}}} p^{v_p}$$

with

$$v_p = \begin{cases} \max((-1)^\alpha, 0) + \sum_{1 \leq 2r-1 \leq \alpha}\left(\left\lfloor \frac{k-1}{p^{2r-1}} \right\rfloor + 1\right) - \sum_{1 \leq 2r \leq \alpha} \left\lfloor \frac{k}{p^{2r}} \right\rfloor & \text{if } p \leq \sqrt{k}, \\ \left\lfloor \frac{2k}{p} \right\rfloor + \left\lfloor \frac{2k}{p^2} \right\rfloor - \left\lfloor \frac{k}{p} \right\rfloor & \text{if } p > \sqrt{k}, \end{cases}$$

where $\alpha = \lfloor \frac{\log k}{\log p} \rfloor$. This condition leads to a contradiction for $212 \le k \le 1595$. As a matter of fact the remaining cases at this stage are $k \le 175$ and $k \in \{177, 178, 181, 184, 187, 191, 199, 211\}$. Define I_k to be the set of odd primes p such that $v_p \equiv \mathrm{ord}_p((2k)!/k!) + 1 \mod 2$. We still have the stronger condition

$$\prod_{i=1}^{k} s_i \le \prod_{\substack{p \le 2k \\ p \text{ prime}}} p^{v_p} \prod_{p \in I_k} p^{-1} .$$

This condition eliminates the cases $k = 172, 173, 174, 175$ and the exceptional cases mentioned above. We thus get $k \le 171$.

We now turn to the case $d = 2$. Since m is odd, so are all the a_i's. Let s'_i denote the i-th odd squarefree integer. If equation (3.2) has a solution, we should have

$$\prod_{i=1}^{k} s'_i \le \prod_{i=0}^{k-1} a_i \le \prod_{\substack{2 < p \le 2k \\ p \text{ prime}}} p^{v_p} .$$

This condition leaves open the cases $k = 1, 2, 3, 4, 9$. The case $k = 9$ is eliminated as before by using $I_9 = \{3\}$. □

This result enables us to get explicit results about the existence of a noninteger root for $K_n(x, 2, N)$.

THEOREM 3.8. *Let n be an even integer greater than 3.*

i) For N even and $n \ge 344$, the Krawtchouk polynomial $K_n(x, 2, N)$ has a noninteger root.

ii) For N odd and $n \ge 4$, the Krawtchouk polynomial $K_n(x, 2, N)$ has a noninteger root.

PROOF. In order to apply Theorem 3.7, we have to check the condition $md \ge (k-1)^2$. As a corollary to Theorem 2.1, Krasikov and Litsyn [**11**] proved that, for $N \ge 23$ and $n > 3 \log N$, the Krawtchouk polynomial $K_n(x, 2, N)$ has a noninteger root. The theorem is easily checked for $N < 23$. Thus we get

$$md = (N - n + 2)/2 \ge (N - n + 2)/2 \ge (\exp(n/3) - n + 2)/2 \ge (n/2 - 1)^2$$

for $d = 1$ and $n \ge 13$, and

$$md = 2(N - n + 2) \ge 2(N - n + 2) \ge 2(\exp(n/3) - n + 2) \ge (n/2 - 1)^2$$

for $d = 2$ and $n \ge 2$. Therefore Theorem 3.7 holds and the first item is proved. The second one is also true for $n \ge 10$. Since this second item is also true for $n = 4, 6$ [**16, 22, 23**], the only remaining case is $n = 8$ and N odd.

Let a'_i be the integer from $\{1, 3, 5, 7\}$ such that $N - 2i \equiv a'_i \mod 8$. We also have $a_i \equiv a'_i \mod 8$, from which it follows that $a_i \ge a'_i$ and $105 \prod_{i=0}^{3} a_i \ge \prod_{i=0}^{3} a'_i = 105$. This in turn implies $a_i = a'_i$. We now distinguish four cases, according to the congruence class of N modulo 8. We shall solve Pell equations [**17, chapter 8**].

1) $N \equiv 1 \mod 8$. We have $(N-1)^2 - 1 = N(N-2) = 7(x_0 x_1)^2$, from which we deduce $N - 1 = ((8 + 3\sqrt{7})^j + (8 - 3\sqrt{7})^j)/2$ for some nonnegative integer j. The only possible residues classes for $N - 1$ modulo 15 are thus $1, 7, 8, 14$, which contradicts the divisibility properties $3 \mid N - 2$ and $5 \mid N - 4$.
2) $N \equiv 3 \mod 8$. We have $(N-1)^2 - 1 = N(N-2) = 3(x_0 x_1)^2$, from which we deduce $N - 1 = ((2 + \sqrt{3})^j + (2 - \sqrt{3})^j)/2$ for some nonnegative integer

j. The only possible residues classes for $N-1$ modulo 5 are thus $1, 2$, which contradicts the divisibility property $5 \mid N-6$.

3) $N \equiv 5 \mod 8$. We have $(N-1)^2 - 1 = N(N-2) = 15(x_0x_1)^2$, from which we deduce $N - 1 = ((4+\sqrt{15})^j + (4-\sqrt{15})^j)/2$ for some nonnegative integer j. The only possible residues classes for $N-1$ modulo 7 are thus $1, 3, 4, 6$, which contradicts the divisibility property $7 \mid N-6$.
4) $N \equiv 7 \mod 8$. We have $N = 7x_0^2$, $N-2 = 5x_1^2$, $N-4 = 3x_2^2$, $N-6 = x_3^2$. By the next lemma, there is no solution with $N \geq n = 8$. □

LEMMA 3.9. *The only positive integer solutions to the system $3b^2 - a^2 = 2$ and $5c^2 - a^2 = 4$ are $(a,b,c) = (1,1,1)$.*

PROOF. We again solve Pell equations [17] to get

$$a = \frac{(1+\sqrt{3})(2+\sqrt{3})^j + (1-\sqrt{3})(2-\sqrt{3})^j}{2} = \left(\frac{1+\sqrt{5}}{2}\right)^{2m+1} + \left(\frac{1-\sqrt{5}}{2}\right)^{2m+1},$$

for some nonnegative integer j and m. Assume $(a,b,c) \neq (1,1,1)$, so that j and m are positive. This gives

$$\left|\left(\frac{1+\sqrt{3}}{2}\right)(2+\sqrt{3})^j \left(\frac{-1+\sqrt{5}}{2}\right)^{2m+1} - 1\right| \leq \left(\frac{-1+\sqrt{5}}{2}\right)^2 < 0.056\,,$$

from which we deduce $|\Lambda| < 0.056$, where

$$\Lambda = \log\left(\frac{1+\sqrt{3}}{2}\right) + j\log(2+\sqrt{3}) + (2m+1)\log\left(\frac{-1+\sqrt{5}}{2}\right).$$

We obtain this way $0.111 + 0.7307m < j < 0.7308m + 0.224$ and therefore $m = \max(j, m, 1)$. As a matter of fact, we also have the more precise estimate

$$|\Lambda| \leq \left(\frac{\sqrt{5}-1}{2}\right)^{4m+2} < 0.382\exp(-1.924m)\,. \tag{3.10}$$

We now apply an estimate of Baker and Wüstholz [1] to get for $m \geq 3$

$$\log|\Lambda| \geq -1.37510^{11}\log m\,.$$

(with their notations, $n=3$, $d=4$, $\alpha_1 = (1+\sqrt{3})(\sqrt{5}-1)/4$, $\alpha_2 = 2+\sqrt{3}$, $\alpha_3 = (\sqrt{5}+1)^2/4$, $b_1 = 1$, $b_2 = j$, $b_3 = -m$, $h'(\alpha_1) = 1/4$, $h'(\alpha_2) = \log(2+\sqrt{3})/4$, $h'(\alpha_3) = 1/4$, $h'(L) = \max(\log m, 1)/4 = \log m$). This last inequality combined with (3.10) gives $m \leq 2.0310^{12}$.

The last cases are treated by continued fractions. Put

$$\delta = \frac{2\log(\frac{\sqrt{5}+1}{2})}{\log(2+\sqrt{3})} \quad \text{and} \quad \lambda = \frac{\log(\frac{\sqrt{5}-1}{2})(\frac{\sqrt{3}+1}{2})}{\log(2+\sqrt{3})}.$$

The inequality (3.10) becomes $|\lambda + j - m\delta| \leq 0.291\exp(-1.924m)$. For a a real number, let $\|a\|$ denote the distance from a to the set of integers. For any positive integer q we have

$$|\lambda + j - m\delta| \geq \frac{\|q\lambda + jq - mq\delta\|}{q} \geq \frac{1}{q}(\|q\lambda\| - m\|q\delta\|)\,.$$

We choose q to be the denominator in one the convergents of δ. The choice $q = 21986236770515$ gives a contradiction for $17 \leq m \leq 2.0310^{12}$: we get

$$0.1546 \leq 0.1795 - 1.22410^{-14}m \leq q|\lambda + j - m\delta|6.410^{12}\exp(-1.924m) < 0.04\,.$$

The range $1 \leq m \leq 16$ is treated similarly by choosing $q = 16$. □

Appendix: proof of a conjecture of Krasikov and Litsyn

In [**11**], Krasikov and Litsyn introduced the polynomial

$$H_k(y) = N!K_k((N-y)/2, 2, N)$$

and conjectured the congruence property $H_{ka+r}(y) \equiv (H_a(y))^k H_r(y) \mod a$. By (1.3), these polynomials satisfy to the recurrence property

$$H_{n+1}(y) = H_n(y) - n(N-n+1)H_{n-1}(y)\,.$$

The congruence property is true for $k = 0$. Let us prove it by induction on k. Assume it is true up to $k-1$. We have

$$H_{ka}(y) = H_{(k-1)a+a}(y) \equiv (H_a(y))^{k-1}H_a(y) \equiv H_a(y)^k \equiv a$$

and

$$\begin{aligned} H_{ka+1}(y) &\equiv (H_a(y))^{k-1}H_{a+1}(y) \equiv (H_a(y))^{k-1}(yH_a(y) - a(N-a+1)H_{a-1}(y)) \\ &\equiv (H_a(y))^k H_1(y) \mod a\,. \end{aligned}$$

Thus the congruence property is true at step k when $r = 0, 1$. We use again the recurrence relation to prove it for larger values of r:

$$\begin{aligned} H_{ka+r+1}(y) &= yH_{ka+r}(y) - (ka+r)(N-ka-r+1)H_{ka+r-1}(y) \\ &\equiv (H_a(y))^k(yH_r(y) - (ka+r)(N-ka-r+1)H_{r-1}(y)) \\ &\equiv (H_a(y))^k H_{r+1}(y) \mod a\,. \end{aligned}$$

This completes the proof of the conjecture.

References

[1] A. Baker and G. Wüstholz, *Logarithmic forms and group varieties*, J. reine angew. Math. **442** (1993), 19–62.

[2] R. Clayton, *Multiple packings and coverings in algebraic coding theory*, Ph. D. thesis, UCLA, 1987.

[3] G. Cohen, I. Honkala, S. Litsyn and A. Lobstein, *Covering codes*, North Holland Mathematical Library vol. 54, Elsevier, 1997.

[4] G. Cohen, I. Honkala, S. Litsyn and H. F. Mattson, *Weighted coverings and packings*, IEEE Trans. Inform. Theory **41** (1995), 1601–1612.

[5] P. Diaconis and R. L. Graham, *The Radon transform on* $\mathbb{Z}_2^k$, Pac. J. Math. **118** (1985), 323–345.

[6] P. Erdős and J. L. Selfridge, *The product of consecutive integers is never a power*, Illinois J. Math. **19** (1975), 292–301.

[7] L. Habsieger and D. Stanton, *More zeros of Krawtchouk polynomials*, Graph and Combin. **9** (1993), 163–172.

[8] G. Hanrot, *Résolution effective d'équations diophantiennes: algorithmes et applications*, Thèse, Université Bordeaux 1, 1997.

[9] G. Hanrot, *On integral zeroes of some Krawtchouk polynomials*, in preparation.

[10] Y. Hong, *On the nonexistence of nontrivial perfect e-codes and tight 2e-designs in Hamming schemes $H(n,q)$ with $e \geq 3$ and $q \geq 3$*, Graph and Combin. **2** (1986), 145–164.
[11] I. Krasikov and S. Litsyn, *On integral zeros of Krawtchouk polynomials*, J. Comb. Theory Ser. A **74** (1996), 71–99.
[12] J. H. van Lint, *A survey on perfect codes*, Rocky Mountain J. Math **5** (1975), 199–224.
[13] J. H. van Lint, *Introduction to coding theory*, Graduate texts in Mathematics **86**, Springer Verlag, second edition, 1992.
[14] S. Litsyn, *personal communication.*
[15] R. Marszalek, *On the product of consecutive elements of an arithmetic progression*, Mh. Math. **100** (1985), 215–222.
[16] M. Mignotte and A. Pethő, *On the system of diophantine equations $x^2 - 6y^2 = -5$ and $x = 2z^2 - 1$*, Math. Scand. **76** (1995), 50–60.
[17] L. J. Mordell, *Diophantine equations*, Pure and Applied Mathematics volume 30, Academic Press, 1969.
[18] J. B. Rosser and L. Schoenfeld, *Approximate formulas for some functions of prime numbers*, Illinois J. Math. **6** (1962), 64–94.
[19] L. Schoenfeld, *Sharper bounds for the Chebyshev functions $\theta(x)$ and $\psi(x)$*. II, Math. Comp. **30** (1976), 337–360.
[20] N. Saradha, *On perfect powers in products with terms from arithmetic progressions*, Acta Arithmetica **82** (1997), 147–172.
[21] Shorey and Tijdeman, *Perfect powers in products of terms in an arithmetic progression* III, Acta Arithmetica **61** (1992), 391–398.
[22] R. J. Stroeker and B. M. M. de Weger, *On a quartic diophantine equation*, Proc. edinburgh Math. Soc. **39** (1996), 97–114.
[23] R. J. Stroeker and B. M. M. de Weger, *On integral zeroes of binary Krawtchouk polynomials*, preprint.
[24] Tietäväinen and A. Perko, *There are no unknown perfect binary codes*, Ann. Univ. Turku., Ser. A **148** (1971), 3–10.

Laboratoire d'Algorithmique Arithmétique, CNRS UMR 5465, Université Bordeaux 1, 351 cours de la Libération, F-33405 Talence Cedex

E-mail address: Laurent.Habsieger@ math.u-bordeaux.fr

DIMACS Series in Discrete Mathematics
and Theoretical Computer Science
Volume **56**, 2001

Some algebra related to P-and Q-polynomial association schemes

Tatsuro Ito, Kenichiro Tanabe, and Paul Terwilliger

ABSTRACT. Inspired by the theory of P-and Q-polynomial association schemes we consider the following situation in linear algebra. Let $\mathbb{K}$ denote a field, and let V denote a vector space over $\mathbb{K}$ with finite positive dimension. We consider a pair of linear transformations $A : V \to V$ and $A^* : V \to V$ satisfying the following four conditions.

(i) A and A^* are both diagonalizable on V.

(ii) There exists an ordering $V_0, V_1, \ldots, V_d$ of the eigenspaces of A such that
$$A^* V_i \subseteq V_{i-1} + V_i + V_{i+1} \qquad (0 \leq i \leq d),$$
where $V_{-1} = 0$, $V_{d+1} = 0$.

(iii) There exists an ordering $V_0^*, V_1^*, \ldots, V_\delta^*$ of the eigenspaces of A^* such that
$$AV_i^* \subseteq V_{i-1}^* + V_i^* + V_{i+1}^* \qquad (0 \leq i \leq \delta),$$
where $V_{-1}^* = 0$, $V_{\delta+1}^* = 0$.

(iv) There is no subspace W of V such that both $AW \subseteq W$, $A^*W \subseteq W$, other than $W = 0$ and $W = V$.

We call such a pair a *Tridiagonal pair*. Referring to the above Tridiagonal pair, we show $d = \delta$. We show that for $0 \leq i \leq d$, the eigenspaces V_i and V_i^* have the same dimension. Denoting this common dimension by ρ_i, we show the sequence $\rho_0, \rho_1, \ldots, \rho_d$ is symmetric and unimodal, i.e. $\rho_i = \rho_{d-i}$ for $0 \leq i \leq d$ and $\rho_{i-1} \leq \rho_i$ for $1 \leq i \leq d/2$. We show that there exists a sequence of scalars $\beta, \gamma, \gamma^*, \varrho, \varrho^*$ taken from $\mathbb{K}$ such that both
$$\begin{aligned} 0 &= [A, A^2A^* - \beta AA^*A + A^*A^2 - \gamma(AA^* + A^*A) - \varrho A^*], \\ 0 &= [A^*, A^{*2}A - \beta A^*AA^* + AA^{*2} - \gamma^*(A^*A + AA^*) - \varrho^* A], \end{aligned}$$
where $[r, s]$ means $rs - sr$. The sequence is unique if $d \geq 3$. Let θ_i (resp. θ_i^*) denote the eigenvalue of A (resp. A^*) associated with V_i (resp. V_i^*), for $0 \leq i \leq d$. We show the expressions
$$\frac{\theta_{i-2} - \theta_{i+1}}{\theta_{i-1} - \theta_i}, \qquad \frac{\theta_{i-2}^* - \theta_{i+1}^*}{\theta_{i-1}^* - \theta_i^*}$$
both equal $\beta + 1$, for $2 \leq i \leq d-1$. We hope these results will ultimately lead to a classification of the Tridiagonal pairs.

2000 *Mathematics Subject Classification.* Primary 05E30, 05E35; Secondary 17B37, 33C45.

Key words and phrases. q-Racah polynomial, Askey scheme, Terwilliger algebra, Askey-Wilson algebra, Dolan-Grady relations, Serre relations.

The first author was supported by a Grant-in-Aid for Scientific Research (No. 10440003), Ministry of Education, Science and Culture, Japan.

1. Introduction

We begin with the following situation in linear algebra. Let $\mathbb{K}$ denote a field, and let V denote a vector space over $\mathbb{K}$ with finite positive dimension. By a *linear transformation on* V, we mean a $\mathbb{K}$-linear map from V to V. Let A denote a linear transformation on V. By an *eigenspace* of A, we mean a nonzero subspace of V of the form

$$\{v \in V \mid Av = \theta v\}, \tag{1.1}$$

where $\theta \in \mathbb{K}$. We say A is *diagonalizable* on V whenever V is spanned by the eigenspaces of A.

DEFINITION 1.1. Let $\mathbb{K}$ denote a field, and let V denote a vector space over $\mathbb{K}$ with finite positive dimension. By a *Tridiagonal pair* (or *TD pair*) on V, we mean an ordered pair (A, A^*), where A and A^* are linear transformations on V that satisfy the following four conditions.

(i) A and A^* are both diagonalizable on V.

(ii) There exists an ordering $V_0, V_1, \ldots, V_d$ of the eigenspaces of A such that

$$A^* V_i \subseteq V_{i-1} + V_i + V_{i+1} \qquad (0 \le i \le d), \tag{1.2}$$

where $V_{-1} = 0$, $V_{d+1} = 0$.

(iii) There exists an ordering $V_0^*, V_1^*, \ldots, V_\delta^*$ of the eigenspaces of A^* such that

$$A V_i^* \subseteq V_{i-1}^* + V_i^* + V_{i+1}^* \qquad (0 \le i \le \delta), \tag{1.3}$$

where $V_{-1}^* = 0$, $V_{\delta+1}^* = 0$.

(iv) There is no subspace W of V such that both $AW \subseteq W$, $A^*W \subseteq W$, other than $W = 0$ and $W = V$.

NOTE 1.2. According to a common notational convention, for a linear transformation A the conjugate-transpose of A is denoted A^*. We emphasize we are *not* using this convention. In a TD pair (A, A^*), the linear transformations A and A^* are arbitrary subject to (i)–(iv) above.

DEFINITION 1.3. Let (A, A^*) denote a TD pair on V. Then (A^*, A) is a TD pair on V, which we refer to as the *dual* of (A, A^*).

We now consider some examples of TD pairs. Our first example is from the theory of association schemes.

EXAMPLE 1.4. Referring to [**26**], [**27**], [**28**], let Y denote a symmetric association scheme, with vertex set X. Assume Y is P-polynomial with respect to an associate matrix A, and Q-polynomial with respect to a primitive idempotent E. Let $\mathrm{Mat}_X(\mathbb{C})$ denote the $\mathbb{C}$-algebra consisting of all matrices with rows and columns indexed by X, and entries in $\mathbb{C}$. Fix a vertex $x \in X$, and let $A^* = A^*(x)$ denote the diagonal matrix in $\mathrm{Mat}_X(\mathbb{C})$ with diagonal entries

$$A^*_{yy} = |X| E_{xy} \qquad (\forall y \in X). \tag{1.4}$$

Let T denote the subalgebra of $\mathrm{Mat}_X(\mathbb{C})$ generated by A and A^*, and let V denote an irreducible T-module. Then the restrictions $A|_V$, $A^*|_V$ form TD pair on V.

PROOF. This is immediate from [**26**, lines (73), (80)]. □

The above example inspired our definition of a TD pair, and motivates all that we do in the present paper.

Our next example of a TD pair comes from Lie algebra.

EXAMPLE 1.5. Let $\mathbb{K}$ denote an algebraically closed field with characteristic 0, and let L denote the Lie algebra $sl_2(\mathbb{K})$. Let A and A^* denote semi-simple elements in L, and assume L is generated by these elements. Let V denote a finite dimensional irreducible L-module. Then A and A^* act on V as a TD pair.

PROOF. We claim L has a basis e, f, h such that A is a nonzero scalar multiple of h, and such that

$$[h, e] = 2e, \qquad [h, f] = -2f, \qquad [e, f] = h, \tag{1.5}$$

where $[\,,\,]$ denotes the Lie bracket. To see this, we identify L with the Lie algebra $sl(W)$, where W is a two dimensional vector space over $\mathbb{K}$. Recall $sl(W)$ is the vector space over $\mathbb{K}$ consisting of all linear transformations on W that have trace 0, together with Lie bracket $[x, y] = xy - yx$. Since A is semisimple, W has a basis u, v consisting of eigenvectors for A. Let r denote the eigenvalue of A associated with u. Since A has trace 0 on W, we find the eigenvalue of A associated with v is $-r$. Observe $r \neq 0$; otherwise $A = 0$, contradicting our assumption that A and A^* generate L. Set $h = r^{-1}A$. Let e (resp. f) denote the linear transformation on W satisfying $ev = u$ and $eu = 0$ (resp. $fu = v$ and $fv = 0$). Observe e and f have trace 0, so they are contained in L. One readily finds e, f, h are linearly independent, and that L has dimension 3, so e, f, h form a basis for L. From the construction, we readily obtain (1.5), and our claim is proved. Using (1.5), we routinely find the L-module V has a basis $v_0, v_1, \ldots, v_d$ satisfying

$$hv_i = (d-2i)v_i \qquad (0 \leq i \leq d), \tag{1.6}$$

$$fv_i = (i+1)v_{i+1} \qquad (0 \leq i < d), \quad fv_d = 0, \tag{1.7}$$

$$ev_i = (d-i+1)v_{i-1} \qquad (0 < i \leq d), \quad ev_0 = 0. \tag{1.8}$$

From (1.6) we find h (and A) is diagonalizable on V. Let V_i denote the span of v_i, for $0 \leq i \leq d$. Then $V_0, V_1, \ldots, V_d$ is an ordering of the eigenspaces of A. By (1.6)–(1.8), and since A^* is a linear combination of e, f, h, we obtain (1.2). We now have Definition 1.1(ii). Interchanging the roles of A and A^* in the above argument, we find A^* is diagonalizable on V, and that Definition 1.1(iii) holds. Since A, A^* generate L, and since V is irreducible as an L module, we obtain Definition 1.1(iv). We have now shown (A, A^*) satisfies the conditions of Definition 1.1, so (A, A^*) is a TD pair on V. □

Inspired by the previous example, we make the following definition. Let (A, A^*) denote a TD pair. We say (A, A^*) is *thin* whenever the eigenspaces of both A and A^* all have dimension 1. In **[30]**, the thin TD pairs are called Leonard pairs, and a complete classification of these objects is given.

Here is another example of a TD pair.

EXAMPLE 1.6. Let $\mathbb{K}$ denote an algebraically closed field of characteristic 0, and let V denote a vector space over $\mathbb{K}$ with finite positive dimension. Let A and A^* denote linear transformations on V satisfying (i), (ii) below.

(i) There exists nonzero scalars $b, b^* \in \mathbb{K}$ such that

$$[A,[A,[A,A^*]]] = b^2[A,A^*], \tag{1.9}$$

$$[A^*,[A^*,[A^*,A]]] = b^{*2}[A^*,A], \tag{1.10}$$

where $[r,s]$ means $rs - sr$.

(ii) There is no subspace W of V such that both $AW \subseteq W$, $A^*W \subseteq W$, other than $W = 0$ and $W = V$.

Then (A, A^*) is a TD pair on V.

PROOF. For all $\theta \in \mathbb{K}$, we define

$$V(\theta) = \{v \in V \mid Av = \theta v\}.$$

Observe $V(\theta)$ is nonzero if and only if θ is an eigenvalue of A, and in this case $V(\theta)$ is the corresponding eigenspace. We claim

$$A^*V(\theta) \subseteq V(\theta - b) + V(\theta) + V(\theta + b) \qquad (\forall \theta \in \mathbb{K}). \tag{1.11}$$

To see (1.11), let θ be given, and pick any $v \in V(\theta)$. Expanding (1.9), we obtain

$$0 = A^3A^* - 3A^2A^*A + 3AA^*A^2 - A^*A^3 - b^2(AA^* - A^*A). \tag{1.12}$$

Applying (1.12) to v, and using $Av = \theta v$, we obtain

$$0 = A^3A^*v - 3\theta A^2A^*v + 3\theta^2 AA^*v - \theta^3 A^*v - b^2AA^*v + b^2\theta A^*v \tag{1.13}$$

$$= (A - \theta + b)(A - \theta)(A - \theta - b)A^*v. \tag{1.14}$$

The scalars $\theta - b, \theta, \theta + b$ are distinct, so apparently

$$A^*v \subseteq V(\theta - b) + V(\theta) + V(\theta + b),$$

and (1.11) follows. Since $\mathbb{K}$ is algebraically closed, all the eigenvalues of A are contained in $\mathbb{K}$. Since V has finite positive dimension, A has at least one eigenvalue θ. Since $\mathbb{K}$ has characteristic zero, the scalars $\theta, \theta-b, \theta-2b, \ldots$ are mutually distinct, so they cannot all be eigenvalues of A; consequently there exists an eigenvalue a for A such that $a - b$ is not an eigenvalue of A. Similarily the scalars $a, a+b, a+2b, \ldots$ are mutually distinct, so they cannot all be eigenvalues of A; consequently there exists a nonnegative integer d such that $a + bi$ is an eigenvalue of A for $0 \leq i \leq d$, but not for $i = d + 1$. Let us abbreviate $V_i = V(a + bi)$ for $0 \leq i \leq d$. Then

$$V_0 + V_1 + \cdots + V_d \tag{1.15}$$

is nonzero and A-invariant by construction. From (1.11) and the construction, we find

$$A^*V_i \subseteq V_{i-1} + V_i + V_{i+1} \qquad (0 \leq i \leq d),$$

where $V_{-1} = 0$, $V_{d+1} = 0$. In particular, (1.15) is A^*-invariant. Applying assumption (ii), we find the sum (1.15) equals V. Now apparently A is diagonalizable on V, and Definition 1.1(ii) holds. Interchanging the roles of A and A^* in the above argument, we find A^* is diagonalizable on V, and Definition 1.1(iii) holds. Observe Definition 1.1(iv) is just assumption (ii) above. We have now shown (A, A^*) satisfy the conditions of Definition 1.1, so it is a TD pair on V. □

The equations (1.9), (1.10) are known as the *Dolan-Grady* relations **[12]**, **[13]**, **[14]**, **[22]**, **[31]** and are important in Statistical Mechanics. The Lie algebra over $\mathbb{C}$ generated by two symbols A, A^* subject to (1.9), (1.10) (where we interpret [,] as the Lie bracket) is infinite dimensional and is known as the Onsager algebra [**1**], [**11**], [**12**], [**13**], [**21**], [**22**], [**31**].

The next example is a kind of q-analog of the previous one.

EXAMPLE 1.7. Let $\mathbb{K}$ denote an algebraically closed field, and let q denote a nonzero element in $\mathbb{K}$ that is not a root of unity. Let V denote a vector space over $\mathbb{K}$ with finite positive dimension, and let A and A^* denote linear transformations on V satisfying (i), (ii) below.

(i) A and A^* are not nilpotent, and

$$0 = [A, A^2A^* - (q+q^{-1})AA^*A + A^*A^2], \tag{1.16}$$

$$0 = [A^*, A^{*2}A - (q+q^{-1})A^*AA^* + AA^{*2}], \tag{1.17}$$

where $[r, s]$ means $rs - sr$.

(ii) There is no subspace W of V such that both $AW \subseteq W$, $A^*W \subseteq W$, other than $W = 0$ and $W = V$.

Then (A, A^*) is a TD pair on V.

PROOF. The proof is similar to that of Example 1.6. For all $\theta \in \mathbb{K}$, we define

$$V(\theta) = \{v \in V \mid Av = \theta v\}.$$

We claim that for all nonzero $\theta \in \mathbb{K}$,

$$A^*V(\theta) \subseteq V(q^{-1}\theta) + V(\theta) + V(q\theta). \tag{1.18}$$

To see (1.18), let θ be given, and pick any $v \in V(\theta)$. Expanding the relation (1.16), we find

$$0 = A^3A^* - (q+1+q^{-1})A^2A^*A + (q+1+q^{-1})AA^*A^2 - A^*A^3. \tag{1.19}$$

Applying each term in (1.19) to v, and using $Av = \theta v$, we routinely find

$$0 = A^3A^*v - (q+1+q^{-1})\theta A^2A^*v + (q+1+q^{-1})\theta^2 AA^*v - \theta^3 A^*v \tag{1.20}$$

$$= (A - q^{-1}\theta)(A-\theta)(A-q\theta)A^*v. \tag{1.21}$$

We assume $\theta \neq 0$ and that q is not a root of unity, so $q^{-1}\theta, \theta, q\theta$ are distinct. Now apparently

$$A^*v \subseteq V(q^{-1}\theta) + V(\theta) + V(q\theta),$$

and (1.18) follows. Since $\mathbb{K}$ is algebraically closed, all the eigenvalues of A are contained in $\mathbb{K}$. Since V has finite positive dimension, and since A is not nilpotent, A has at least one nonzero eigenvalue θ. Since q is not a root of unity, the scalars $\theta, q^{-1}\theta, q^{-2}\theta, \ldots$ are mutually distinct, so they cannot all be eigenvalues of A; consequently there exists an eigenvalue a for A such that $q^{-1}a$ is not an eigenvalue of A. Similarly the scalars $a, aq, aq^2, \ldots$ are mutually distinct, so they cannot all be eigenvalues of A; consequently there exists a nonnegative integer d such that aq^i is an eigenvalue of A for $0 \leq i \leq d$, but not for $i = d+1$. Let us abbreviate $V_i = V(aq^i)$ for $0 \leq i \leq d$. Then

$$V_0 + V_1 + \cdots + V_d \tag{1.22}$$

is nonzero and A-invariant by construction. By (1.18) and the construction, we find

$$A^*V_i \subseteq V_{i-1} + V_i + V_{i+1} \qquad (0 \leq i \leq d),$$

where $V_{-1} = 0$, $V_{d+1} = 0$. In particular, (1.22) is A^*-invariant. Applying assumption (ii) above, we find the sum (1.22) equals V. Now apparently A is diagonalizable on V, and Definition 1.1(ii) holds. Interchanging the roles of A and A^* in the above argument, we find A^* is diagonalizable on V, and Definition 1.1(iii) holds. Observe Definition 1.1(iv) is just assumption (ii) above. We have now shown (A, A^*) satisfy the conditions of Definition 1.1, so it is a TD pair on V. □

The relations (1.16), (1.17) are known as *quantum Serre relations.* They are among the defining relations for the quantum affine algebra $U_r(\widehat{sl}_2)$, where $r^2 = q$. See [**5**], [**6**] for more information about this algebra.

2. Tridiagonal systems

When working with a Tridiagonal pair, it is often convenient to consider a closely related but somewhat more abstract concept, called a Tridiagonal system. To define it, we recall a few concepts from linear algebra. Let V denote a vector space over $\mathbb{K}$ with finite positive dimension, and let $\mathrm{End}(V)$ denote the $\mathbb{K}$-algebra consisting of all linear transformations on V. Let A denote a diagonalizable element in $\mathrm{End}(V)$. Let $\theta_0, \theta_1, \dots, \theta_d$ denote an ordering of the distinct eigenvalues of A, and put

$$E_i = \prod_{\substack{0 \le j \le d \\ j \ne i}} \frac{A - \theta_j I}{\theta_i - \theta_j}$$

for $0 \le i \le d$. By elementary linear algebra,

$$AE_i = E_i A = \theta_i E_i \qquad (0 \le i \le d), \tag{2.1}$$

$$E_i E_j = \delta_{ij} E_i \qquad (0 \le i, j \le d), \tag{2.2}$$

$$\sum_{i=0}^{d} E_i = I. \tag{2.3}$$

From this, one finds $E_0, E_1, \dots, E_d$ is a basis for the subalgebra of $\mathrm{End}(V)$ generated by A. We refer to E_i as the *primitive idempotent* of A associated with θ_i. It is helpful to think of these primitive idempotents as follows. From (2.2), (2.3) one readily finds

$$V = E_0V + E_1V + \cdots + E_dV \qquad \text{(direct sum)}. \tag{2.4}$$

For $0 \le i \le d$, E_iV is the eigenspace of A associated with the eigenvalue θ_i, and E_i is the projection of V onto this eigenspace.

DEFINITION 2.1. Let $\mathbb{K}$ denote a field, and let V denote a vector space over $\mathbb{K}$ with finite positive dimension. By a *Tridiagonal system* (or *TD system*) on V, we mean a sequence

$$\Phi = (A;\ E_0,\ E_1,\ \dots,\ E_d;\ A^*;\ E_0^*,\ E_1^*,\ \dots,\ E_\delta^*) \tag{2.5}$$

that satisfies (i)–(vi) below.

(i) A, A^* are both diagonalizable linear transformations on V.
(ii) $E_0, E_1, \dots, E_d$ is an ordering of the primitive idempotents of A.
(iii) $E_0^*, E_1^*, \dots, E_\delta^*$ is an ordering of the primitive idempotents of A^*.
(iv) $E_i A^* E_j = 0$ if $|i - j| > 1$, $\quad (0 \le i, j \le d)$.
(v) $E_i^* A E_j^* = 0$ if $|i - j| > 1$, $\quad (0 \le i, j \le \delta)$.

(vi) There is no subspace W of V such that both $AW \subseteq W$, $A^*W \subseteq W$, other than $W = 0$ and $W = V$.

We refer to d as the *diameter* of Φ, and say Φ is *over* $\mathbb{K}$. For notational convenience, we set $E_{-1} = 0$, $E_{d+1} = 0$, $E^*_{-1} = 0$, $E^*_{\delta+1} = 0$.

Referring to Definition 2.1, we do not assume the primitive idempotents of A and A^* all have rank 1. A TD system for which these primitive idempotents all have rank 1 is called a Leonard system [**30**]. The Leonard systems are completely classified by Terwilliger [**30**].

In the next two lemmas, we give the connection between TD pairs and TD systems.

LEMMA 2.2. *Let $\mathbb{K}$ denote a field, and let V denote a vector space over $\mathbb{K}$ with finite positive dimension. Let (A, A^*) denote a TD pair on V. Let $V_0, V_1, \ldots, V_d$ denote an ordering of the eigenspaces of A satisfying (1.2), and for $0 \leq i \leq d$ let E_i denote the projection of V onto V_i. Let $V_0^*, V_1^*, \ldots, V_\delta^*$ denote an ordering of the eigenspaces of A^* satisfying (1.3), and for $0 \leq i \leq \delta$ let E_i^* denote the projection of V onto V_i^*. Then*

$$(A; E_0, E_1, \ldots, E_d; A^*; E_0^*, E_1^*, \ldots, E_\delta^*)$$

is a TD system on V.

PROOF. We verify the conditions (i)–(vi) of Definition 2.1. Condition (i) is just Definition 1.1(i). Conditions (ii) and (iii) follow from the comment after (2.4). To obtain condition (iv), pick any integers i, j $(0 \leq i, j \leq d)$, and assume $|i - j| > 1$. For all $v \in V$,

$$\begin{aligned} E_i A^* E_j v &\in E_i A^* V_j \\ &\subseteq E_i(V_{j-1} + V_j + V_{j+1}) \\ &= 0, \end{aligned}$$

so $E_i A^* E_j = 0$. We have now verified condition (iv), and condition (v) is similarily obtained. Condition (vi) is just Definition 1.1(iv). $\square$

LEMMA 2.3. *Let $\mathbb{K}$ denote a field, let V denote a vector space over $\mathbb{K}$ with finite positive dimension, and let*

$$\Phi = (A; E_0, E_1, \ldots, E_d; A^*; E_0^*, E_1^*, \ldots, E_\delta^*)$$

denote a TD sytem on V. Then

$$A^* E_i V \subseteq E_{i-1}V + E_iV + E_{i+1}V \qquad (0 \leq i \leq d), \tag{2.6}$$

$$A E_i^* V \subseteq E_{i-1}^*V + E_i^*V + E_{i+1}^*V \qquad (0 \leq i \leq \delta). \tag{2.7}$$

Moreover, (A, A^*) is a TD pair on V. We refer to (A, A^*) as the TD pair associated with Φ.

PROOF. To obtain (2.6), observe by (2.3) and Definition 2.1(iv) that

$$\begin{aligned} A^* E_i V &= (E_0 + E_1 + \cdots + E_d) A^* E_i V \\ &= (E_{i-1} + E_i + E_{i+1}) A^* E_i V \\ &\subseteq E_{i-1}V + E_iV + E_{i+1}V. \end{aligned}$$

Line (2.7) is similarly obtained, and the last assertion follows. $\square$

We finish this section with an observation.

LEMMA 2.4. *Referring to the TD system (2.5), we have*
(i) $E_i A^* E_j \neq 0$ *if* $|i-j| = 1$, $(0 \leq i, j \leq d)$,
(ii) $E_i^* A E_j^* \neq 0$ *if* $|i-j| = 1$, $(0 \leq i, j \leq \delta)$.

PROOF. (i) Pick any integer i $(1 \leq i \leq d)$, and first assume $E_i A^* E_{i-1} = 0$. Put $W := E_0V + E_1V + \cdots + E_{i-1}V$. By (2.4), and since each of $E_0V, E_1V, \ldots, E_dV$ is nonzero, we find $W \neq 0$, $W \neq V$. Observe $AW \subseteq W$ by the construction. Combining (2.6) and our assumption, we find $A^*W \subseteq W$, and now condition (vi) of Definition 2.1 is contradicted. Next suppose $E_{i-1}A^*E_i = 0$. In this case we routinely find $W := E_iV + E_{i+1}V + \cdots + E_dV$ provides a contradiction to condition (vi) of Definition 2.1.
(ii) Similar to the proof of (i) above. □

3. The relatives of a Tridiagonal system

A given TD system can be modified in several ways to get a new TD system. For instance, let Φ denote the TD system in (2.5), and let α, α^*, β, β^* denote scalars in $\mathbb{K}$ such that $\alpha \neq 0$, $\alpha^* \neq 0$. Then

$$(\alpha A + \beta I; E_0, E_1, \ldots, E_d; \alpha^* A^* + \beta^* I; E_0^*, E_1^*, \ldots, E_\delta^*)$$

is a TD system on V. Also,

$$\Phi^* := (A^*; E_0^*, E_1^*, \ldots, E_\delta^*; A; E_0, E_1, \ldots, E_d), \tag{3.1}$$

$$\Phi^\downarrow := (A; E_0, E_1, \ldots, E_d; A^*; E_\delta^*, E_{\delta-1}^*, \ldots, E_0^*), \tag{3.2}$$

$$\Phi^\Downarrow := (A; E_d, E_{d-1}, \ldots, E_0; A^*; E_0^*, E_1^*, \ldots, E_\delta^*) \tag{3.3}$$

are TD systems on V. We refer to Φ^* (resp. $\Phi^\downarrow$) (resp. $\Phi^\Downarrow$) as the *dual* (resp. *first inversion*) (resp. *second inversion*) of Φ. Viewing $*, \downarrow, \Downarrow$ as permutations on the set of all TD systems,

$$*^2 = \downarrow^2 = \Downarrow^2 = 1, \tag{3.4}$$

$$\Downarrow * = * \downarrow, \qquad \downarrow * = * \Downarrow, \qquad \downarrow\Downarrow = \Downarrow\downarrow. \tag{3.5}$$

The group generated by symbols $*, \downarrow, \Downarrow$ subject to the relations (3.4), (3.5) is the dihedral group D_4. We recall D_4 is the group of symmetries of a square, and has 8 elements. Apparently $*, \downarrow, \Downarrow$ induce an action of D_4 on the set of all TD systems. Two TD systems will be called *relatives* whenever they are in the same orbit of this D_4 action. The relatives of Φ are as follows:

name	relative
Φ	$(A; E_0, E_1, \ldots, E_d; A^*; E_0^*, E_1^*, \ldots, E_\delta^*)$
$\Phi^\downarrow$	$(A; E_0, E_1, \ldots, E_d; A^*; E_\delta^*, E_{\delta-1}^*, \ldots, E_0^*)$
$\Phi^\Downarrow$	$(A; E_d, E_{d-1}, \ldots, E_0; A^*; E_0^*, E_1^*, \ldots, E_\delta^*)$
$\Phi^{\downarrow\Downarrow}$	$(A; E_d, E_{d-1}, \ldots, E_0; A^*; E_\delta^*, E_{\delta-1}^*, \ldots, E_0^*)$
Φ^*	$(A^*; E_0^*, E_1^*, \ldots, E_\delta^*; A; E_0, E_1, \ldots, E_d)$
$\Phi^{\downarrow *}$	$(A^*; E_\delta^*, E_{\delta-1}^*, \ldots, E_0^*; A; E_0, E_1, \ldots, E_d)$
$\Phi^{\Downarrow *}$	$(A^*; E_0^*, E_1^*, \ldots, E_\delta^*; A; E_d, E_{d-1}, \ldots, E_0)$
$\Phi^{\downarrow\Downarrow *}$	$(A^*; E_\delta^*, E_{\delta-1}^*, \ldots, E_0^*; A; E_d, E_{d-1}, \ldots, E_0)$

With reference to Definition 1.3 and Lemma 2.3, it is routine to show two TD systems are relatives if and only if their associated TD pairs are equal or dual.

We now introduce two sequences of parameters that we will use to describe a given TD system.

DEFINITION 3.1. Let Φ denote the TD system in (2.5). For $0 \le i \le d$, we let θ_i denote the eigenvalue of A associated with E_i. We refer to $\theta_0, \theta_1, \ldots, \theta_d$ as the *eigenvalue sequence* of Φ. For $0 \le i \le \delta$, we let θ_i^* denote the eigenvalue of A^* associated with E_i^*. We refer to $\theta_0^*, \theta_1^*, \ldots, \theta_\delta^*$ as the *dual eigenvalue sequence* of Φ. We remark $\theta_0, \theta_1, \ldots, \theta_d$ are mutually distinct, and $\theta_0^*, \theta_1^*, \ldots, \theta_\delta^*$ are mutually distinct.

4. The split decomposition

DEFINITION 4.1. In this section, $\mathbb{K}$ will denote a field, and V will denote a vector space over $\mathbb{K}$ with finite positive dimension. We let

$$\Phi = (A; E_0, E_1, \ldots, E_d; A^*; E_0^*, E_1^*, \ldots, E_\delta^*)$$

denote a TD system on V, with eigenvalue sequence $\theta_0, \theta_1, \ldots, \theta_d$ and dual eigenvalue sequence $\theta_0^*, \theta_1^*, \ldots, \theta_\delta^*$.

Referring to the above definition, we will show $\delta = d$. We will then show there exists a unique sequence $U_0, U_1, \ldots, U_d$ of subspaces of V such that

$$V = U_0 + U_1 + \cdots + U_d \qquad \text{(direct sum)}, \tag{4.1}$$

$$(A - \theta_i I)U_i \subseteq U_{i+1} \quad (0 \le i < d), \qquad (A - \theta_d I)U_d = 0, \tag{4.2}$$

$$(A^* - \theta_i^* I)U_i \subseteq U_{i-1} \quad (0 < i \le d), \qquad (A^* - \theta_0^* I)U_0 = 0. \tag{4.3}$$

The following notation will be useful.

DEFINITION 4.2. With reference to Definition 4.1, we set

$$V_{ij} = \left(\sum_{h=0}^{i} E_h^* V\right) \cap \left(\sum_{k=j}^{d} E_k V\right) \tag{4.4}$$

for all integers i, j. We interpret the sum on the left in (4.4) to be 0 (resp. V) if $i < 0$ (resp. $i > \delta$). Similarily, we interpret the sum on the right in (4.4) to be V (resp. 0) if $j < 0$ (resp. $j > d$).

LEMMA 4.3. *With reference to Definition 4.1 and Definition 4.2, we have*

(i) $V_{i0} = E_0^*V + E_1^*V + \cdots + E_i^*V \qquad (0 \le i \le \delta)$,
(ii) $V_{\delta j} = E_jV + E_{j+1}V + \cdots + E_dV \qquad (0 \le j \le d)$.

PROOF. To get (i), set $j = 0$ in (4.4), and apply (2.4). Line (ii) is similarily obtained. □

LEMMA 4.4. *With reference to Definition 4.1 and Definition 4.2, the following (i)–(iv) hold for $0 \le i \le \delta$ and $0 \le j \le d$.*

(i) $(A - \theta_j I)V_{ij} \subseteq V_{i+1,j+1}$,
(ii) $AV_{ij} \subseteq V_{ij} + V_{i+1,j+1}$,
(iii) $(A^* - \theta_i^* I)V_{ij} \subseteq V_{i-1,j-1}$,
(iv) $A^*V_{ij} \subseteq V_{ij} + V_{i-1,j-1}$.

PROOF. (i) Using (2.7),

$$(A - \theta_j I)\sum_{h=0}^{i} E_h^*V \subseteq \sum_{h=0}^{i+1} E_h^*V, \tag{4.5}$$

and using (2.1),

$$(A - \theta_j I)\sum_{k=j}^{d} E_k V = \sum_{k=j+1}^{d} E_k V. \tag{4.6}$$

Evaluating $(A - \theta_j I)V_{ij}$ using (4.4), (4.5), (4.6), we routinely find it is contained in $V_{i+1,j+1}$.
(ii) Immediate from (i) above.
(iii) Using (2.1),

$$(A^* - \theta^*_i I)\sum_{h=0}^{i} E^*_h V = \sum_{h=0}^{i-1} E^*_h V, \tag{4.7}$$

and using (2.6),

$$(A^* - \theta^*_i I)\sum_{k=j}^{d} E_k V \subseteq \sum_{k=j-1}^{d} E_k V. \tag{4.8}$$

Evaluating $(A^* - \theta^*_i I)V_{ij}$ using (4.4), (4.7), (4.8), we routinely find it is contained in $V_{i-1,j-1}$.
(iv) Immediate from (iii) above. □

LEMMA 4.5. *The scalars d and δ from Definition 4.1 are equal. Moreover, with reference to Definition 4.2,*

$$V_{ij} = 0 \quad \textit{if} \quad i < j, \qquad (0 \le i, j \le d). \tag{4.9}$$

PROOF. Switching A and A^* if necessary, we may assume $\delta \le d$. We first show (4.9). To do this, we show the sum

$$V_{0r} + V_{1,r+1} + \cdots + V_{d-r,d} \tag{4.10}$$

is zero for $0 < r \le d$. Let r be given, and let W denote the sum in (4.10). Applying Lemma 4.4(ii),(iv), we find $AW \subseteq W$ and $A^*W \subseteq W$. By the definition of a TD system, we find $W = 0$ or $W = V$. By Definition 4.2, each term in (4.10) is contained in

$$E_r V + E_{r+1} V + \cdots + E_d V, \tag{4.11}$$

so W is contained in (4.11). The sum (4.11) is properly contained in V by (2.4), and since $r > 0$. Apparently $W \ne V$, so $W = 0$. We have now shown (4.10) is zero for $0 < r \le d$, and (4.9) follows. It remains to show $\delta = d$. Suppose $\delta \ne d$, so that $\delta < d$ by our initial assumption. On one hand, setting $i = \delta$, $j = d$ in (4.9), we find $V_{\delta d} = 0$. On the other hand $V_{\delta d} = E_d V$ by Lemma 4.3(ii), so $V_{\delta d} \ne 0$. We now have a contradiction, so $\delta = d$. □

THEOREM 4.6. *With reference to Definition 4.1, let $U_0, U_1, \ldots, U_d$ denote any subspaces of V. Then the following are equivalent.*

(i) $U_i = (E^*_0 V + E^*_1 V + \cdots + E^*_i V) \cap (E_i V + E_{i+1} V + \cdots + E_d V) \qquad (0 \le i \le d)$.
(ii) *The sequence $U_0, U_1, \ldots, U_d$ satisfies (4.1), (4.2), (4.3).*
(iii) *For $0 \le i \le d$, both*

$$U_i + U_{i+1} + \cdots + U_d = E_i V + E_{i+1} V + \cdots + E_d V, \tag{4.12}$$

$$U_0 + U_1 + \cdots + U_i = E^*_0 V + E^*_1 V + \cdots + E^*_i V. \tag{4.13}$$

PROOF. $(i) \rightarrow (ii)$ To get (4.2) and (4.3), set $j = i$ in Lemma 4.4(i),(iii), and observe $U_i = V_{ii}$. To obtain (4.1), let W denote the sum on the right in that line. Then $AW \subseteq W$ by (4.2), and $A^*W \subseteq W$ by (4.3). By the definition of a TD system, we find $W = 0$ or $W = V$. W contains U_0, and $U_0 = E_0^*V$ is nonzero, so $W \neq 0$. It follows $W = V$, and in other words

$$V = U_0 + U_1 + \cdots + U_d. \tag{4.14}$$

We show the sum (4.14) is direct. To do this, we show

$$(U_0 + U_1 + \cdots + U_{i-1}) \cap U_i = 0$$

for $1 \leq i \leq d$. Let the integer i be given. From the construction

$$U_j \subseteq E_0^*V + E_1^*V + \cdots + E_{i-1}^*V$$

for $0 \leq j \leq i-1$, and

$$U_i \subseteq E_iV + E_{i+1}V + \cdots + E_dV.$$

It follows

$$\begin{aligned} &(U_0 + U_1 + \cdots + U_{i-1}) \cap U_i \\ &\quad \subseteq (E_0^*V + E_1^*V + \cdots + E_{i-1}^*V) \cap (E_iV + E_{i+1}V + \cdots + E_dV) \\ &\quad = V_{i-1,i} \\ &\quad = 0 \end{aligned}$$

in view of Lemma 4.5. We have now shown the sum (4.14) is direct, so (4.1) holds.
$(ii) \rightarrow (iii)$ First consider (4.12). Let i be given, and abbreviate

$$Z \;=\; E_iV + E_{i+1}V + \cdots + E_dV, \qquad W = U_i + U_{i+1} + \cdots + U_d.$$

We show $Z = W$. To obtain $Z \subseteq W$, set $X = \prod_{h=0}^{i-1}(A - \theta_h I)$, and observe $Z = XV$ by (2.1)–(2.3). Using (4.2), we find $XU_j \subseteq W$ for $0 \leq j \leq d$, so $XV \subseteq W$ in view of (4.1). We now have $Z \subseteq W$. To obtain $W \subseteq Z$, set $Y = \prod_{h=i}^{d}(A - \theta_h I)$, and observe

$$Z \;=\; \{v \in V \mid Yv = 0\}. \tag{4.15}$$

Using (4.2), we find $YU_j = 0$ for $i \leq j \leq d$, so $YW = 0$. Combining this with (4.15), we find $W \subseteq Z$. We now have $Z = W$ and hence (4.12). Line (4.13) is similarily obtained.
$(iii) \rightarrow (i)$ We first show the sum $U_0 + U_1 + \cdots + U_d$ is direct. To do this, we show

$$(U_0 + U_1 + \cdots + U_{i-1}) \cap U_i \tag{4.16}$$

is zero for $1 \leq i \leq d$. Let i be given. From (4.12), (4.13), we find (4.16) is contained in

$$(E_0^*V + E_1^*V + \cdots + E_{i-1}^*V) \cap (E_iV + E_{i+1}V + \cdots + E_dV). \tag{4.17}$$

The expression (4.17) equals $V_{i-1,i}$, and is hence zero by Lemma 4.5. It follows (4.16) is zero, and we have now shown the sum $U_0 + U_1 + \cdots + U_d$ is direct. Combining this with (4.12), (4.13), we find

$$\begin{aligned} U_i &= (U_0 + U_1 + \cdots + U_i) \cap (U_i + U_{i+1} + \cdots + U_d) \\ &= (E_0^*V + E_1^*V + \cdots + E_i^*V) \cap (E_iV + E_{i+1}V + \cdots + E_dV), \end{aligned}$$

as desired. □

5. Some projections

DEFINITION 5.1. In this section, $\mathbb{K}$ will denote a field, and V will denote a vector space over $\mathbb{K}$ with finite positive dimension. We let

$$\Phi = (A; E_0, E_1, \ldots, E_d; A^*; E_0^*, E_1^*, \ldots, E_d^*) \tag{5.1}$$

denote a TD system on V, with eigenvalue sequence $\theta_0, \theta_1, \ldots, \theta_d$ and dual eigenvalue sequence $\theta_0^*, \theta_1^*, \ldots, \theta_d^*$. We let $U_0, U_1, \ldots, U_d$ denote the subspaces of V that satisfy (i)–(iii) of Theorem 4.6.

DEFINITION 5.2. With reference to Definition 5.1, for $0 \le i \le d$, we let F_i denote the linear transformation on V satisfying both

$$(F_i - I)U_i = 0, \tag{5.2}$$

$$F_i U_j = 0 \quad \text{if} \quad j \neq i, \qquad (0 \le j \le d). \tag{5.3}$$

In other words, F_i is the projection map from V onto U_i. For notational convenience, we define $F_{-1} = 0$ and $F_{d+1} = 0$.

LEMMA 5.3. *With reference to Definition 5.1 and Definition 5.2,*

$$F_i F_j = \delta_{ij} F_i \qquad (0 \le i, j \le d), \tag{5.4}$$

$$F_0 + F_1 + \cdots + F_d = I, \tag{5.5}$$

$$F_i V = U_i \qquad (0 \le i \le d). \tag{5.6}$$

PROOF. Immediate from Definition 5.2 and (4.1). □

LEMMA 5.4. *With reference to Definition 5.1 and Definition 5.2, the products* $E_i F_j$, $F_i E_j$, $E_j^* F_i$, $F_j E_i^*$ *are all zero for* $0 \le i < j \le d$.

PROOF. Let the integers i, j be given. Combining (2.2), (4.12), (5.6), we find

$$\begin{aligned} E_i F_j V &= E_i U_j \\ &\subseteq E_i(E_j V + E_{j+1} V + \cdots + E_d V) \\ &= 0, \end{aligned}$$

so $E_i F_j = 0$. Similarily, using (4.12) and (5.3),

$$\begin{aligned} F_i E_j V &\subseteq F_i(E_j V + E_{j+1} V + \cdots + E_d V) \\ &= F_i(U_j + U_{j+1} + \cdots + U_d) \\ &= 0, \end{aligned}$$

so $F_i E_j = 0$. The remaining assertions are similarily proved. □

LEMMA 5.5. *With reference to Definition 5.1 and Definition 5.2, we have*

$$F_i E_i F_i = F_i, \qquad E_i F_i E_i = E_i, \tag{5.7}$$

$$F_i E_i^* F_i = F_i, \qquad E_i^* F_i E_i^* = E_i^*, \tag{5.8}$$

for $0 \le i \le d$.

PROOF. To obtain the equation on the left in (5.7), we evaluate the expression

$$F_i(E_0 + E_1 + \cdots + E_d)F_i \tag{5.9}$$

in two ways. First, the expression (5.9) equals F_i, since the sum in the middle is the identity, and since $F_i^2 = F_i$. Pick any integer j $(0 \le j \le d)$. By Lemma 5.4, we have $F_i E_j = 0$ if $j > i$, and $E_j F_i = 0$ if $j < i$, so $F_i E_j F_i = 0$ if $j \neq i$.

Apparently (5.9) equals $F_iE_iF_i$, and we now have the equation on the left in (5.7). The remaining equations are proved in a similar manner. □

We interpret the above lemma as follows.

LEMMA 5.6. *With reference to Definition 5.1 and Definition 5.2, the following (i), (ii) hold for* $0 \le i \le d$.

(i) *The linear transformations*

$$\begin{array}{ccc} U_i & \to & E_iV \\ v & \to & E_iv \end{array} \qquad\qquad \begin{array}{ccc} E_iV & \to & U_i \\ v & \to & F_iv \end{array}$$

are bijections, and moreover, they are inverses.

(ii) *The linear transformations*

$$\begin{array}{ccc} U_i & \to & E_i^*V \\ v & \to & E_i^*v \end{array} \qquad\qquad \begin{array}{ccc} E_i^*V & \to & U_i \\ v & \to & F_iv \end{array}$$

are bijections, and moreover, they are inverses.

PROOF. (i) The maps are inverses in view of (5.7). It follows they are bijections. (ii) Similar to the proof of (i). □

COROLLARY 5.7. *With reference to Definition 5.1, for* $0 \le i \le d$, *the dimensions of* E_iV, U_i, *and* E_i^*V *are equal. Denoting this common dimension by* ρ_i, *we have* $\rho_i = \rho_{d-i}$.

PROOF. It is immediate from Lemma 5.6 that the dimensions of E_iV, U_i, E_i^*V are equal. Denote this common dimension by ρ_i. To show $\rho_i = \rho_{d-i}$, we show E_iV and E_{d-i}^*V have the same dimension. We just showed E_iV and E_i^*V have the same dimension. Applying this result to $\Phi^\downarrow$, we find E_iV and E_{d-i}^*V have the same dimension. □

We finish this section with a remark.

LEMMA 5.8. *Let the TD system* Φ *be as in (5.1). Then for* $0 \le i \le d$, *the space* U_i *from Definition 5.1, the projection* F_i *from Definition 5.2, and the scalar* ρ_i *from Corollary 5.7 are all nonzero.*

PROOF. Recall ρ_i is the dimension of E_iV, and this space is not zero, so $\rho_i \ne 0$. U_i has dimension ρ_i, so $U_i \ne 0$. Recall $F_iV = U_i$, so $F_i \ne 0$. □

6. The raising and lowering maps

In this section, we continue to study the situation of Definition 5.1. We introduce the raising and lowering maps, and use these to show the scalars ρ_i from Corollary 5.7 form a unimodal sequence.

DEFINITION 6.1. With reference to Definition 5.1, we define

$$R = A - \sum_{h=0}^{d} \theta_h F_h, \tag{6.1}$$

$$L = A^* - \sum_{h=0}^{d} \theta_h^* F_h, \tag{6.2}$$

where $F_0, F_1, \ldots, F_d$ are from Definition 5.2. We refer to R (resp. L) as the *raising map* (resp. *lowering map*).

LEMMA 6.2. *With reference to Definition 5.1 and Definition 6.1, for all integers* i $(0 \leq i \leq d)$, *and for all* $v \in U_i$,

$$Rv = (A - \theta_i I)v, \qquad Lv = (A^* - \theta^*_i I)v. \tag{6.3}$$

PROOF. To get the equation on the left, apply both sides of (6.1) to v, and evaluate the result using (5.2), (5.3). The equation on the right is similarily obtained. □

COROLLARY 6.3. *With reference to Definition 5.1 and Definition 6.1,*

(i) $RU_i \subseteq U_{i+1}$ $(0 \leq i < d)$, $RU_d = 0$.
(ii) $LU_i \subseteq U_{i-1}$ $(0 < i \leq d)$, $LU_0 = 0$.

PROOF. Combine Lemma 6.2 with (4.2) and (4.3). □

LEMMA 6.4. *With reference to Definition 5.1, Definition 5.2, and Definition 6.1,*

(i) $RF_i = F_{i+1}R$ $(0 \leq i < d)$, $RF_d = 0$, $F_0R = 0$.
(ii) $LF_i = F_{i-1}L$ $(0 < i \leq d)$, $LF_0 = 0$, $F_dL = 0$.
(iii) $R^{d+1} = 0$.
(iv) $L^{d+1} = 0$.

PROOF. (i) Pick any integer i $(0 \leq i < d)$. Using Definition 5.2 and Corollary 6.3(i), we find the expression $RF_i - F_{i+1}R$ vanishes on each of $U_0, U_1, \ldots, U_d$, so this expression is zero in view of (4.1). Similarily, RF_d and F_0R vanish on each of $U_0, U_1, \ldots, U_d$, so they are zero.
(ii) Similar to the proof of (i).
(iii) Immediate from Corollary 6.3(i) and (4.1).
(iv) Immediate from Corollary 6.3(ii) and (4.1). □

LEMMA 6.5. *With reference to Definition 5.1 and Definition 6.1, pick any integers* i, j $(0 \leq i \leq j \leq d)$. *Then the linear transformation*

$$\begin{array}{ccc} U_i & \to & U_j \\ v & \to & R^{j-i}v \end{array} \tag{6.4}$$

is an injection if $i + j \leq d$, *a bijection if* $i + j = d$, *and a surjection if* $i + j \geq d$. *The linear transformation*

$$\begin{array}{ccc} U_j & \to & U_i \\ v & \to & L^{j-i}v \end{array} \tag{6.5}$$

is an injection if $i + j \geq d$, *a bijection if* $i + j = d$, *and a surjection if* $i + j \leq d$. *(Caution: the maps (6.4), (6.5) are not inverses in general, even in the case* $i+j = d$*).*

PROOF. Concerning the map (6.4), first assume $i + j \leq d$. To show (6.4) is an injection, we pick any vector $v \in U_i$ such that $R^{j-i}v = 0$, and show $v = 0$. Using Lemma 6.2 and Corollary 6.3(i),

$$\begin{aligned} 0 &= R^{j-i}v \\ &= (A - \theta_i I)(A - \theta_{i+1} I) \cdots (A - \theta_{j-1} I)v, \end{aligned}$$

so

$$v \in E_iV + E_{i+1}V + \cdots + E_{j-1}V. \tag{6.6}$$

By (4.13), and since $v \in U_i$,

$$v \in E_0^*V + E_1^*V + \cdots + E_i^*V. \tag{6.7}$$

Applying Lemma 4.5 to $\Phi^{\Downarrow}$, and using $j \leq d - i$, we find

$$(E_0^*V + E_1^*V + \cdots + E_i^*V) \cap (E_0V + E_1V + \cdots + E_{j-1}V) = 0. \tag{6.8}$$

Combining (6.6), (6.7), (6.8), we find $v = 0$, and it follows (6.4) is an injection. Next suppose $i + j = d$. In this case U_i and U_j have the same dimension by Corollary 5.7, so the injection (6.4) is actually a bijection. We now assume $i + j \geq d$, and show (6.4) is a surjection. To do this, we pick any $w \in U_j$, and obtain an element $v \in U_i$ such that $R^{j-i}v = w$. From our comments above, the restriction of R^{2j-d} to U_{d-j} is a bijection onto U_j, so there exists $u \in U_{d-j}$ such that $R^{2j-d}u = w$. Set $v = R^{i+j-d}u$. Applying Corollary 6.3(i), we find $v \in U_i$. Also

$$\begin{aligned} R^{j-i}v &= R^{2j-d}u \\ &= w, \end{aligned}$$

and it follows (6.4) is a surjection. We have now proved all our claims about (6.4). To obtain our claims about (6.5), apply the first part of the lemma to $\Phi^{\downarrow\Downarrow *}$. □

COROLLARY 6.6. *The scalars $\rho_0, \rho_1, \ldots \rho_d$ from Corollary 5.7 satisfy $\rho_{i-1} \leq \rho_i$ for $1 \leq i \leq d/2$.*

PROOF. Let the spaces $U_0, U_1, \ldots, U_d$ be as in Definition 5.1 and recall ρ_i is the dimension of U_i for $0 \leq i \leq d$. Pick any integer i $(1 \leq i \leq d/2)$. By Lemma 6.5, we find the restriction of the raising map R to U_{i-1} is an injection into U_i, so $\rho_{i-1} \leq \rho_i$. □

We end this section with a result we will use later in the paper.

COROLLARY 6.7. *Let the TD system Φ be as in Definition 5.1, let the projections $F_0, F_1, \ldots, F_d$ be as in Definition 5.2, and let the maps R, L be as in Definition 6.1. Then for $0 \leq i \leq j \leq d$, we have $R^{j-i}F_i \neq 0$ and $L^{j-i}F_j \neq 0$.*

PROOF. These maps are nonzero by Lemma 5.8 and Lemma 6.5. □

7. Writing products involving A and A^* in terms of R and L

When working with a TD system such as (5.1), instead of working directly with the elements A and A^*, it is often easier to work with the corresponding raising map R and lowering map L from Definition 6.1. In this section, we consider how to write products involving A and A^* in terms of R and L. We begin with some simple observations.

LEMMA 7.1. *With reference to Definition 5.1 and Definition 5.2,*

(i) $F_iAF_i = \theta_iF_i \qquad (0 \leq i \leq d)$,
(ii) $F_{i+1}AF_i = RF_i \qquad (0 \leq i < d)$,
(iii) $F_jAF_i = 0$ *if* $j - i \notin \{0, 1\}$, $\qquad (0 \leq i, j \leq d)$.

PROOF. To verify these equations, in each case eliminate A using (6.1), and evaluate the result using Lemma 5.3 and Lemma 6.4. □

LEMMA 7.2. *With reference to Definition 5.1 and Definition 5.2,*

(i) $F_iA^*F_i = \theta_i^*F_i \qquad (0 \leq i \leq d)$,
(ii) $F_{i-1}A^*F_i = LF_i \qquad (0 < i \leq d)$,

(iii) $F_j A^* F_i = 0 \quad if \quad j-i \notin \{0,-1\}, \qquad (0 \le i,j \le d).$

PROOF. Similar to the proof of Lemma 7.1. □

LEMMA 7.3. *With reference to Definition 5.1 and Definition 5.2, let* n *denote a positive integer, and let* $B_1, B_2, \ldots, B_n$ *denote a sequence of elements taken from the set* $\{A, A^*\}$. *Then for* $0 \le r, s \le d$, *the product*

$$F_r B_1 B_2 \cdots B_n F_s = \sum_{\sigma} W_\sigma F_s \tag{7.1}$$

with the sum interpreted as follows. The sum is over all sequences of integers

$$\sigma = (i_0, i_1, \ldots, i_n) \tag{7.2}$$

such that $0 \le i_j \le d$ *for* $0 \le j \le n$, *and such that*

$$i_0 = r, \qquad i_n = s,$$
$$i_j - i_{j-1} \in [B_j] \qquad (1 \le j \le n),$$

where we define $[A] = \{0,-1\}$ *and* $[A^*] = \{0,1\}$. *For each sequence* σ *in (7.2), the corresponding* W_σ *is the product* $W_\sigma = W_1 W_2 \cdots W_n$, *where the* W_i *are given by*

$$W_j = \begin{cases} R, & \text{if } i_j = i_{j-1} - 1\ , \\ L & \text{if } i_j = i_{j-1} + 1, \\ \theta_{i_j}, & \text{if } i_j = i_{j-1} \text{ and } B_j = A, \\ \theta^*_{i_j}, & \text{if } i_j = i_{j-1} \text{ and } B_j = A^* \end{cases} \qquad (1 \le j \le n),$$

and where R *and* L *are from Definition 6.1.*

PROOF. The left side of (7.1) equals $F_r B_1 I B_2 I \cdots I B_n F_s$. Eliminating each copy of I in this product using (5.5), and evaluating the result using Lemma 7.1 and Lemma 7.2, we obtain the sum on the right side of (7.1). □

EXAMPLE 7.4. For $0 \le i \le d-1$, we have

$$F_{i+1} A A^* A F_i = RLRF_i + \theta_{i+1}\theta^*_{i+1} RF_i + \theta_i \theta^*_i RF_i.$$

PROOF. Routine application of Lemma 7.3. □

8. Recurrent sequences

It is going to turn out that the eigenvalue sequence and dual eigenvalue sequence of a TD system each satisfy a certain recurrence. In this section, we set the stage by considering this recurrence from several points of view.

DEFINITION 8.1. In this section, $\mathbb{K}$ will denote a field, d will denote a nonnegative integer, and $\theta_0, \theta_1, \ldots, \theta_d$ will denote a sequence of scalars taken from $\mathbb{K}$.

DEFINITION 8.2. With reference to Definition 8.1, let β, γ, ϱ denote scalars in $\mathbb{K}$.

(i) The sequence $\theta_0, \theta_1, \ldots, \theta_d$ is said to be recurrent whenever $\theta_{i-1} \ne \theta_i$ for $2 \le i \le d-1$, and

$$\frac{\theta_{i-2} - \theta_{i+1}}{\theta_{i-1} - \theta_i} \tag{8.1}$$

is independent of i, for $2 \le i \le d-1$.

(ii) The sequence $\theta_0, \theta_1, \dots, \theta_d$ is said to be β-recurrent whenever

$$\theta_{i-2} - (\beta+1)\theta_{i-1} + (\beta+1)\theta_i - \theta_{i+1} \tag{8.2}$$

is zero for $2 \le i \le d-1$.

(iii) The sequence $\theta_0, \theta_1, \dots, \theta_d$ is said to be (β,γ)-recurrent whenever

$$\theta_{i-1} - \beta\theta_i + \theta_{i+1} = \gamma \tag{8.3}$$

for $1 \le i \le d-1$.

(iv) The sequence $\theta_0, \theta_1, \dots, \theta_d$ is said to be (β,γ,ϱ)-recurrent whenever

$$\theta_{i-1}^2 - \beta\theta_{i-1}\theta_i + \theta_i^2 - \gamma(\theta_{i-1}+\theta_i) = \varrho \tag{8.4}$$

for $1 \le i \le d$.

LEMMA 8.3. *With reference to Definition 8.1, the following are equivalent.*

(i) *The sequence $\theta_0, \theta_1, \dots, \theta_d$ is recurrent.*

(ii) *There exists $\beta \in \mathbb{K}$ such that $\theta_0, \theta_1, \dots, \theta_d$ is β-recurrent, and $\theta_{i-1} \ne \theta_i$ for $2 \le i \le d-1$.*

Suppose (i), (ii), and that $d \ge 3$. Then the common value of (8.1) equals $\beta+1$.

PROOF. Routine. □

LEMMA 8.4. *With reference to Definition 8.1, the following are equivalent for all $\beta \in \mathbb{K}$.*

(i) *The sequence $\theta_0, \theta_1, \dots, \theta_d$ is β-recurrent.*

(ii) *There exists $\gamma \in \mathbb{K}$ such that $\theta_0, \theta_1, \dots, \theta_d$ is (β,γ)-recurrent.*

PROOF. *(i)* $\to$ *(ii)* For $2 \le i \le d-1$, the expression (8.2) is zero by assumption, so

$$\theta_{i-2} - \beta\theta_{i-1} + \theta_i = \theta_{i-1} - \beta\theta_i + \theta_{i+1}.$$

Apparently the left side of (8.3) is independent of i, and the result follows.
(ii) $\to$ *(i)* Subtracting the equation (8.3) at i from the corresponding equation obtained by replacing i by $i-1$, we find (8.2) is zero for $2 \le i \le d-1$. □

LEMMA 8.5. *With reference to Definition 8.1, the following (i),(ii) hold for all $\beta, \gamma \in \mathbb{K}$.*

(i) *Suppose $\theta_0, \theta_1, \dots, \theta_d$ is (β,γ)-recurrent. Then there exists $\varrho \in \mathbb{K}$ such that $\theta_0, \theta_1, \dots, \theta_d$ is (β,γ,ϱ)-recurrent.*

(ii) *Suppose $\theta_0, \theta_1, \dots, \theta_d$ is (β,γ,ϱ)-recurrent, and that $\theta_{i-1} \ne \theta_{i+1}$ for $1 \le i \le d-1$. Then $\theta_0, \theta_1, \dots, \theta_d$ is (β,γ)-recurrent.*

PROOF. Let p_i denote the expression on the left in (8.4), and observe

$$p_i - p_{i+1} = (\theta_{i-1} - \theta_{i+1})(\theta_{i-1} - \beta\theta_i + \theta_{i+1} - \gamma)$$

for $1 \le i \le d-1$. Assertions (i), (ii) are both routine consequences of this. □

We mention the standard parametric expressions for recurrent sequences. To state the result, we let $\mathbb{K}^{cl}$ denote the algebraic closure of $\mathbb{K}$.

LEMMA 8.6. *With reference to Definition 8.1, pick any $\beta \in \mathbb{K}$, and assume $\theta_0, \theta_1, \dots, \theta_d$ is β-recurrent. Then the following (i)–(iv) hold.*

(i) *Suppose $\beta \neq 2$, $\beta \neq -2$, and pick $q \in \mathbb{K}^{cl}$ such that $q + q^{-1} = \beta$. Then there exists scalars $\alpha_1, \alpha_2, \alpha_3$ in $\mathbb{K}^{cl}$ such that*

$$\theta_i = \alpha_1 + \alpha_2 q^i + \alpha_3 q^{-i} \qquad (0 \leq i \leq d). \tag{8.5}$$

(ii) *Suppose $\beta = 2$ and char($\mathbb{K}$) $\neq 2$. Then there exists $\alpha_1, \alpha_2, \alpha_3$ in $\mathbb{K}$ such that*

$$\theta_i = \alpha_1 + \alpha_2 i + \alpha_3 i^2 \qquad (0 \leq i \leq d). \tag{8.6}$$

(iii) *Suppose $\beta = -2$ and char($\mathbb{K}$) $\neq 2$. Then there exists $\alpha_1, \alpha_2, \alpha_3$ in $\mathbb{K}$ such that*

$$\theta_i = \alpha_1 + \alpha_2(-1)^i + \alpha_3 i(-1)^i \qquad (0 \leq i \leq d). \tag{8.7}$$

(iv) *Suppose $\beta = 0$ and char($\mathbb{K}$) $= 2$. Then there exists $\alpha_1, \alpha_2, \alpha_3$ in $\mathbb{K}$ such that*

$$\theta_i = \alpha_1 + \alpha_2 i + \alpha_3 \binom{i}{2} \qquad (0 \leq i \leq d), \tag{8.8}$$

where we interpret the binomial coefficient as

$$\binom{i}{2} = \begin{cases} 0 & \text{if } i = 0 \text{ or } i = 1 \pmod 4, \\ 1 & \text{if } i = 2 \text{ or } i = 3 \pmod 4. \end{cases}$$

9. A few comments

In the next section, we are going to show the elements A and A^* in a TD pair satisfy two cubic polynomial equations. In this section, we obtain a few preliminary results that we will need.

DEFINITION 9.1. Throughout this section, $\mathbb{K}$ will denote a field, and V will denote a vector space over $\mathbb{K}$ with finite positive dimension. We let A and A^* denote linear transformations on V, with A diagonalizable. We let $E_0, E_1, \ldots, E_d$ denote an ordering of the primitive idempotents of A. We let θ_i denote the eigenvalue of A associated with E_i, for $0 \leq i \leq d$. We remark $\theta_0, \theta_1, \ldots, \theta_d$ are mutually distinct.

LEMMA 9.2. *With reference to Definition 9.1, suppose*

$$E_i A^* E_j = 0 \quad \text{if} \quad |i - j| > 1, \qquad (0 \leq i, j \leq d), \tag{9.1}$$

and let $\mathcal{D}$ denote the subalgebra of End(V) generated by A. Then

$$\operatorname{Span}\{XA^*Y - YA^*X \mid X, Y \in \mathcal{D}\} = \{XA^* - A^*X \mid X \in \mathcal{D}\}. \tag{9.2}$$

PROOF. For notational convenience set $E_{-1} = 0$, $E_{d+1} = 0$. Observe by (2.3) and (9.1) that for $0 \leq j \leq d$, both

$$A^* E_j = E_{j-1}A^*E_j + E_jA^*E_j + E_{j+1}A^*E_j, \tag{9.3}$$

$$E_j A^* = E_jA^*E_{j-1} + E_jA^*E_j + E_jA^*E_{j+1}. \tag{9.4}$$

Pick any integer i $(0 \leq i \leq d)$. Summing (9.3) over $j = 0, 1, \ldots, i$, summing (9.4) over $j = 0, 1, \ldots, i$, and taking the difference between the two sums, we obtain

$$E_iA^*E_{i+1} - E_{i+1}A^*E_i = L_iA^* - A^*L_i, \tag{9.5}$$

where $L_i := E_0 + E_1 + \cdots + E_i$. Observe $\mathcal{D}$ is spanned by both $E_0, E_1, \ldots, E_d$ and $L_0, L_1, \ldots, L_d$, so

$$\begin{aligned}
&\operatorname{Span}\{XA^*Y - YA^*X \mid X, Y \in \mathcal{D}\} \\
&\quad = \operatorname{Span}\{E_iA^*E_j - E_jA^*E_i \mid 0 \le i, j \le d\} \\
&\quad = \operatorname{Span}\{E_iA^*E_{i+1} - E_{i+1}A^*E_i \mid 0 \le i \le d\} \\
&\quad = \operatorname{Span}\{L_iA^* - A^*L_i \mid 0 \le i \le d\} \\
&\quad = \{XA^* - A^*X \mid X \in \mathcal{D}\},
\end{aligned}$$

and we have (9.2). □

LEMMA 9.3. *With reference to Definition 9.1, suppose*

$$E_iA^*E_j = \begin{cases} 0, & \text{if } |i-j| > 1, \\ \neq 0, & \text{if } |i-j| = 1 \end{cases} \qquad (0 \le i, j \le d), \tag{9.6}$$

and pick any scalars β, γ, ϱ in $\mathbb{K}$. Then

$$0 = [A, A^2A^* - \beta AA^*A + A^*A^2 - \gamma(AA^* + A^*A) - \varrho A^*] \tag{9.7}$$

if and only if the sequence $\theta_0, \theta_1, \ldots, \theta_d$ is (β, γ, ϱ)-recurrent.

PROOF. For notational convenience, we define a two variable polynomial $p \in \mathbb{K}[\lambda, \mu]$ by

$$p(\lambda, \mu) = \lambda^2 - \beta\lambda\mu + \mu^2 - \gamma(\lambda + \mu) - \varrho. \tag{9.8}$$

Let C denote the expression on the right in (9.7), and observe

$$C = \sum_{i=0}^{d}\sum_{j=0}^{d} E_iCE_j \tag{9.9}$$

in view of (2.3). Using (2.1) and the definition of C, we find that for $0 \le i, j \le d$,

$$E_iCE_j = E_iA^*E_j(\theta_i - \theta_j)p(\theta_i, \theta_j). \tag{9.10}$$

Suppose (9.7) holds, so that $C = 0$. Then for $1 \le i \le d$, we have $E_{i-1}CE_i = 0$, and also $E_{i-1}A^*E_i \neq 0$ by (9.6), so $p(\theta_{i-1}, \theta_i) = 0$ by (9.10). Evaluating this using (9.8) and Definition 8.2(iv), we find $\theta_0, \theta_1, \ldots, \theta_d$ is (β, γ, ϱ)-recurrent. Conversely, suppose $\theta_0, \theta_1, \ldots, \theta_d$ is (β, γ, ϱ)-recurrent. By Definition 8.2(iv) and (9.8), we find $p(\theta_{i-1}, \theta_i) = p(\theta_i, \theta_{i-1}) = 0$ for $1 \le i \le d$. Now for $0 \le i, j \le d$, at least one of the factors on the right in (9.10) equals zero, so $E_iCE_j = 0$. Apparently each term on the right in (9.9) is zero, so $C = 0$, and we have (9.7). □

10. Two polynomial equations for A and A^*

THEOREM 10.1. *Let $\mathbb{K}$ denote a field, and let (A, A^*) denote a TD pair over $\mathbb{K}$. Then there exists a sequence of scalars $\beta, \gamma, \gamma^*, \varrho, \varrho^*$ taken from $\mathbb{K}$ such that both*

$$0 = [A, A^2A^* - \beta AA^*A + A^*A^2 - \gamma(AA^* + A^*A) - \varrho A^*], \tag{10.1}$$

$$0 = [A^*, A^{*2}A - \beta A^*AA^* + AA^{*2} - \gamma^*(A^*A + AA^*) - \varrho^* A]. \tag{10.2}$$

The sequence is unique if the diameter of the pair is at least 3.

PROOF. Let $\Phi = (A; E_0, E_1, \ldots, E_d; A^*; E_0^*, E_1^*, \ldots, E_d^*)$ denote a TD system associated with (A, A^*). Let $\theta_0, \theta_1, \ldots, \theta_d$ denote the eigenvalue sequence of Φ, and let $\theta_0^*, \theta_1^*, \ldots, \theta_d^*$ denote the dual eigenvalue sequence of Φ. Let $F_0, F_1, \ldots, F_d$ denote the projection maps for Φ, from Definition 5.2. Let R and L denote the raising and lowering maps for Φ, from Definition 6.1. First assume $d \geq 3$. Observe Φ satisfies (9.1) by Definition 2.1(iv), so Lemma 9.2 applies. Taking $X = A^2, Y = A$ on the left in (9.2), we find there exists scalars $\alpha_1, \alpha_2, \ldots, \alpha_d$ in $\mathbb{K}$ such that

$$A^2A^*A - AA^*A^2 = \sum_{i=1}^{d} \alpha_i(A^iA^* - A^*A^i). \tag{10.3}$$

We show $\alpha_i = 0$ for $4 \leq i \leq d$. Suppose not, and set

$$t := \max\{i \mid 4 \leq i \leq d, \quad \alpha_i \neq 0\}.$$

Multiplying each term in (10.3) on the left by F_t and on the right by F_0, and evaluating the result using Lemma 7.3, we routinely find R^tF_0 times $(\theta_0^* - \theta_t^*)\alpha_t$ is zero. However $R^tF_0 \neq 0$ by Corollary 6.7, $\theta_0^* \neq \theta_t^*$ by construction, and we assumed $\alpha_t \neq 0$, so we have a contradiction. We now have $\alpha_i = 0$ for $4 \leq i \leq d$, so (10.3) becomes

$$\begin{aligned} A^2A^*A - AA^*A^2 = \alpha_1(AA^* - A^*A) &+ \alpha_2(A^2A^* - A^*A^2) \\ &+ \alpha_3(A^3A^* - A^*A^3). \end{aligned} \tag{10.4}$$

We show $\alpha_3 \neq 0$. Suppose $\alpha_3 = 0$. Multiplying each term in (10.4) on the left by F_3 and on the right by F_0, and evaluating the result using Lemma 7.3, we routinely find R^3F_0 times $\theta_1^* - \theta_2^*$ is zero. Observe $R^3F_0 \neq 0$ by Corollary 6.7, and $\theta_1^* \neq \theta_2^*$, so we have a contradiction. We have now shown $\alpha_3 \neq 0$. Set

$$\begin{aligned} C := \alpha_1A^* + \alpha_2(AA^* + A^*A) &+ \alpha_3(A^2A^* + A^*A^2) \\ &+ (\alpha_3 - 1)AA^*A. \end{aligned} \tag{10.5}$$

Observe $AC - CA$ equals

$$\begin{aligned} \alpha_1(AA^* - A^*A) + \alpha_2(A^2A^* - A^*A^2) &+ \alpha_3(A^3A^* - A^*A^3) \\ &+ AA^*A^2 - A^2A^*A, \end{aligned}$$

and this equals zero in view of (10.4). Hence A and C commute. Dividing C by α_3 and using (10.5), we find A commutes with

$$A^2A^* - \beta AA^*A + A^*A^2 - \gamma(AA^* + A^*A) - \varrho A^*,$$

where

$$\beta := \alpha_3^{-1} - 1, \qquad \gamma := -\alpha_2\alpha_3^{-1}, \qquad \varrho := -\alpha_1\alpha_3^{-1}. \tag{10.6}$$

We now have (10.1). Pick any integer i $(2 \leq i \leq d-1)$. Multiplying each term in (10.1) on the right by F_{i-2} and on the left by F_{i+1}, and evaluating the result using Lemma 7.3, we routinely find R^3F_{i-2} times

$$\theta_{i-2}^* - (\beta+1)\theta_{i-1}^* + (\beta+1)\theta_i^* - \theta_{i+1}^* \tag{10.7}$$

is zero. Observe $R^3F_{i-2} \neq 0$ by Corollary 6.7, so (10.7) is zero, and it follows $\theta_0^*, \theta_1^*, \ldots, \theta_d^*$ is β-recurrent in view of Definition 8.2(ii). Applying Lemma 8.4 and then Lemma 8.5 to the sequence $\theta_0^*, \theta_1^*, \ldots, \theta_d^*$, we find there exists scalars γ^*, ϱ^* in $\mathbb{K}$ such that this sequence is $(\beta, \gamma^*, \varrho^*)$-recurrent. Applying Lemma 9.3 to Φ^*, we obtain (10.2). Concerning uniqueness, let $\beta, \gamma, \gamma^*, \varrho, \varrho^*$ denote any scalars in

$\mathbb{K}$ satisfying (10.1), (10.2). Applying Lemma 9.3, we find $\theta_0, \theta_1, \ldots, \theta_d$ is (β, γ, ϱ)-recurrent. From this and our assumption $d \geq 3$, one routinely finds β, γ, ϱ are uniquely determined. Similarily γ^* and ϱ^* are uniquely determined, and we have proved the theorem for the case $d \geq 3$. Next assume $d \leq 2$, and let β denote any scalar in $\mathbb{K}$. If $d = 2$ set $\gamma = \theta_0 - \beta\theta_1 + \theta_2$, and if $d \leq 1$ let γ denote any scalar in $\mathbb{K}$. If $d \geq 1$ set

$$\varrho = \theta_0^2 - \beta\theta_0\theta_1 + \theta_1^2 - \gamma(\theta_0 + \theta_1),$$

and if $d = 0$ let ϱ denote any scalar in $\mathbb{K}$. Applying Definition 8.2(iv), we find the sequence $\theta_0, \theta_1, \ldots, \theta_d$ is (β, γ, ϱ)-recurrent. Applying Lemma 9.3, we find (10.1) holds. Applying the above argument to Φ^*, we find there exists scalars γ^*, ϱ^* for which (10.2) holds. □

REMARK 10.2. The Dolan-Grady relations (1.9), (1.10) are the same thing as (10.1), (10.2) with $\beta = 2$, $\gamma = \gamma^* = 0$, $\varrho = b^2$, $\varrho^* = b^{*2}$. The quantum Serre relations (1.16), (1.17) are the same thing as (10.1), (10.2) with $\beta = q + q^{-1}$, $\gamma = \gamma^* = 0$, $\varrho = \varrho^* = 0$.

11. The eigenvalues and dual eigenvalues of a TD system

In the proof of Theorem 10.1, we got an indication that the eigenvalue sequence and the dual eigenvalue sequence of a TD system are recurrent. In this section, we give detailed results along this line.

THEOREM 11.1. *Let $\mathbb{K}$ denote a field, let (A, A^*) denote a TD pair over $\mathbb{K}$, and let $\beta, \gamma, \gamma^*, \varrho, \varrho^*$ denote scalars in $\mathbb{K}$ that satisfy (10.1), (10.2). Let $\theta_0, \theta_1, \ldots, \theta_d$ (resp. $\theta^*_0, \theta^*_1, \ldots, \theta^*_d$) denote the eigenvalue sequence (resp. dual eigenvalue sequence) for an associated TD system. Then the expressions*

$$(11.1) \qquad \frac{\theta_{i-2} - \theta_{i+1}}{\theta_{i-1} - \theta_i}, \qquad \frac{\theta^*_{i-2} - \theta^*_{i+1}}{\theta^*_{i-1} - \theta^*_i}$$

both equal $\beta + 1$, for $2 \leq i \leq d - 1$. Moreover,

$$\begin{aligned}
(11.2) \quad \gamma &= \theta_{i-1} - \beta\theta_i + \theta_{i+1} && (1 \leq i \leq d-1),\\
(11.3) \quad \gamma^* &= \theta^*_{i-1} - \beta\theta^*_i + \theta^*_{i+1} && (1 \leq i \leq d-1),\\
(11.4) \quad \varrho &= \theta^2_{i-1} - \beta\theta_{i-1}\theta_i + \theta^2_i - \gamma(\theta_{i-1} + \theta_i) && (1 \leq i \leq d),\\
(11.5) \quad \varrho^* &= \theta^{*2}_{i-1} - \beta\theta^*_{i-1}\theta^*_i + \theta^{*2}_i - \gamma^*(\theta^*_{i-1} + \theta^*_i) && (1 \leq i \leq d).
\end{aligned}$$

PROOF. Let Φ denote the TD system referred to in the statement of the theorem. Applying Lemma 9.3 to Φ, we find the sequence $\theta_0, \theta_1, \ldots, \theta_d$ is (β, γ, ϱ)-recurrent, giving (11.4). Applying Lemma 8.5, and since the θ_i are distinct, we find the sequence $\theta_0, \theta_1, \ldots, \theta_d$ is (β, γ)-recurrent, giving (11.2). Applying Lemma 8.4, we find $\theta_0, \theta_1, \ldots, \theta_d$ is β-recurrent, so the expressions on the left in (11.1) all equal $\beta + 1$. Applying our above comments to Φ^*, we obtain the remaining assertions. □

THEOREM 11.2. *Let $\mathbb{K}$ denote a field. Let Φ denote a TD system over $\mathbb{K}$, with eigenvalue sequence $\theta_0, \theta_1, \ldots, \theta_d$ and dual eigenvalue sequence $\theta^*_0, \theta^*_1, \ldots, \theta^*_d$. Let β denote the scalar from Theorem 10.1. Then (i)–(iv) hold below.*

(i) *Suppose* $\beta \neq 2, \beta \neq -2$, *and pick* $q \in \mathbb{K}^{cl}$ *such that* $q+q^{-1}=\beta$. *Then there exists scalars* $\alpha_1, \alpha_2, \alpha_3, \alpha_1^*, \alpha_2^*, \alpha_3^*$ *in* $\mathbb{K}^{cl}$ *such that*

$$\theta_i = \alpha_1 + \alpha_2 q^i + \alpha_3 q^{-i}, \tag{11.6}$$

$$\theta^*_i = \alpha^*_1 + \alpha^*_2 q^i + \alpha^*_3 q^{-i}, \tag{11.7}$$

for $0 \leq i \leq d$. *Moreover* $q^i \neq 1$ *for* $1 \leq i \leq d$.

(ii) *Suppose* $\beta = 2$ *and char*$(\mathbb{K}) \neq 2$. *Then there exists* $\alpha_1, \alpha_2, \alpha_3, \alpha_1^*, \alpha_2^*, \alpha_3^*$ *in* $\mathbb{K}$ *such that*

$$\theta_i = \alpha_1 + \alpha_2 i + \alpha_3 i^2, \tag{11.8}$$

$$\theta^*_i = \alpha^*_1 + \alpha^*_2 i + \alpha^*_3 i^2, \tag{11.9}$$

for $0 \leq i \leq d$. *Moreover char*$(\mathbb{K}) = 0$ *or char*$(\mathbb{K}) > d$.

(iii) *Suppose* $\beta = -2$ *and char*$(\mathbb{K}) \neq 2$. *Then there exists* $\alpha_1, \alpha_2, \alpha_3, \alpha_1^*, \alpha_2^*, \alpha_3^*$ *in* $\mathbb{K}$ *such that*

$$\theta_i = \alpha_1 + \alpha_2 (-1)^i + \alpha_3 i(-1)^i, \tag{11.10}$$

$$\theta^*_i = \alpha^*_1 + \alpha^*_2 (-1)^i + \alpha^*_3 i(-1)^i, \tag{11.11}$$

for $0 \leq i \leq d$. *Moreover char*$(\mathbb{K}) = 0$ *or char*$(\mathbb{K}) > d/2$.

(iv) *Suppose* $\beta = 0$ *and char*$(\mathbb{K}) = 2$. *Then* $d \leq 3$.

PROOF. Most of the assertions are immediate from Lemma 8.6, but a few points need explanation. Suppose we are in the situation of (i). Then using (8.5), we find that for $1 \leq i \leq d$, $q^i = 1$ implies $\theta_i = \theta_0$, a contradiction. Suppose we are in the situation of (ii). Then using (8.6), we find that for $1 \leq i \leq d$, $i = 0$ (in $\mathbb{K}$) implies $\theta_i = \theta_0$, a contradiction. Hence char$(\mathbb{K}) = 0$ or char$(\mathbb{K}) > d$. Suppose we are in the situation of (iii). Then using (8.7), we find that for $1 \leq i \leq d$, $i = 0$ (in $\mathbb{K}$) and i even implies $\theta_i = \theta_0$. Hence char$(\mathbb{K}) = 0$ or char$(\mathbb{K}) > d/2$. Suppose we are in the situation of (iv), and that $d \geq 4$. Then applying (8.8), we find $\theta_0 = \theta_4$, a contradiction. Hence $d \leq 3$. □

12. The raising and lowering maps, revisited

In this section we consider the implications of Theorem 10.1 for the raising map R and the lowering map L.

THEOREM 12.1. *Let* Φ *denote a TD system. Let* $\theta_0, \theta_1, \ldots, \theta_d$ *denote the eigenvalue sequence and let* $\theta^*_0, \theta^*_1, \ldots, \theta^*_d$ *denote the dual eigenvalue sequence. Let* $U_0, U_1, \ldots, U_d$ *denote the spaces from Definition 5.1. Let* R *and* L *denote the raising and lowering maps for* Φ, *from Definition 6.1. Then for* $0 \leq i \leq d-2$,

$$R^3L - (\beta+1)R^2LR + (\beta+1)RLR^2 - LR^3 + (\beta+1)\varepsilon_i R^2 \tag{12.1}$$

vanishes on U_i, *and*

$$RL^3 - (\beta+1)LRL^2 + (\beta+1)L^2RL - L^3R + (\beta+1)\varepsilon_i L^2 \tag{12.2}$$

vanishes on U_{i+2}, *where* β *is from Theorem 10.1, and where*

$$\varepsilon_i = (\theta_i - \theta_{i+2})(\theta^*_{i+1} - \theta^*_{i+2}) - (\theta^*_{i+2} - \theta^*_i)(\theta_{i+1} - \theta_i). \tag{12.3}$$

PROOF. Let the projections $F_0, F_1, \ldots, F_d$ be as in Definition 5.2. To get (12.1), multiply each term in (10.1) on the right by F_i, and on the left by F_{i+2}, and simplify using Lemma 7.3. To get (12.2), multiply each term in (10.2) on the left by F_i, and on the right by F_{i+2}, and simplify using Lemma 7.3. □

Referring to the above theorem, it is natural to consider when ε_i is independent of i. Below, we give one instance where this occurs.

THEOREM 12.2. *Let $\mathbb{K}$ denote a field. Let Φ denote a TD system over $\mathbb{K}$, with eigenvalue sequence $\theta_0, \theta_1, \dots, \theta_d$ and dual eigenvalue sequence $\theta_0^*, \theta_1^*, \dots, \theta_d^*$. Suppose there exists scalars q, a, b, a^*, c^* in $\mathbb{K}$ such that*

$$\theta_i = a + bq^i, \qquad \theta_i^* = a^* + c^* q^{-i} \qquad (0 \leq i \leq d). \tag{12.4}$$

Then the raising map R and the lowering map L from Definition 6.1 satisfy the quantum Serre relations

$$0 = [R, R^2L - (q + q^{-1})RLR + LR^2], \tag{12.5}$$

$$0 = [L, L^2R - (q + q^{-1})LRL + RL^2], \tag{12.6}$$

where $[r, s]$ means $rs - sr$.

PROOF. Let Δ denote the expression on the right in (12.5). Evaluating (12.3) using (12.4), we find $\varepsilon_i = 0$ for $0 \leq i \leq d-2$. Evaluating (11.1) using (12.4), we find $\beta = q + q^{-1}$. Using this information, we find (12.1) equals Δ. Applying Theorem 12.1, we find Δ vanishes on U_i for $0 \leq i \leq d-2$. Using Corollary 6.3, we find Δ vanishes on U_i for $i = d-1$ and $i = d$. Apparently Δ vanishes on each of $U_0, U_1, \dots, U_d$, and these span V, so $\Delta = 0$, and we have (12.5). Applying a similar argument, we obtain (12.6). □

13. Remarks and conjectures

In this section, we give some open problems and ideas for future research.

As we have indicated earlier, the main problem that we would like to see solved is the following.

PROBLEM 13.1. Classify all the TD pairs.

The above problem might be difficult, but the fact that there is a classification in the thin case [**30**] gives us hope. To make headway on the non-thin case, we suggest one of the following two problems.

PROBLEM 13.2. Classify all the TD pairs for which the sequence $\rho_0, \rho_1, \dots, \rho_d$ from Corollary 5.7 is of the form $1, 2, 2, 2, \dots, 2, 1$. We remark these TD pairs play an important role in the theory of P-and Q-polynomial association schemes [**19**], [**24**].

PROBLEM 13.3. Classify all the TD systems that satisfy the assumption of Theorem 12.2, where we assume the scalar q in that theorem is not a root of unity. Given the results of that theorem, it is natural to guess these TD sytems are related to representations of the quantum affine algebra $U_r(\widehat{sl}_2)$, where $r^2 = q$. See [**5**], [**6**] for information on this algebra.

In light of Theorem 10.1, we consider the following generalization of a TD pair. Let $\mathbb{K}$ denote a field, and let V denote a vector space over $\mathbb{K}$ with finite positive dimension. By a *generalized TD pair* on V, we mean an ordered pair (A, A^*), where A and A^* are linear transformations on V such that (i), (ii) hold below.

(i) There exists scalars $\beta, \gamma, \gamma^*, \varrho, \varrho^*$ taken from $\mathbb{K}$ such that

$$0 = [A, A^2A^* - \beta AA^*A + A^*A^2 - \gamma(AA^* + A^*A) - \varrho A^*], \tag{13.1}$$

$$0 = [A^*, A^{*2}A - \beta A^*AA^* + AA^{*2} - \gamma^*(A^*A + AA^*) - \varrho^* A], \tag{13.2}$$

where $[r, s]$ means $rs - sr$.

(ii) There is no subspace W of V such that both $AW \subseteq W$, $A^*W \subseteq W$, other than $W = 0$ and $W = V$.

By Theorem 10.1, any TD pair is a generalized TD pair. However, we should not expect every generalized TD pair to be a TD pair. We saw some indications of this in Example 1.5, Example 1.6, and Example 1.7. In each of these examples, to get a TD pair, we needed to make an assumption about the ground field $\mathbb{K}$. Also, in Example 1.7, we made an assumption about the parameter q involved, which implies a restriction on $\beta = q + q^{-1}$. In addition, we assumed A and A^* are not nilpotent. With these comments in mind, we pose the following problem.

PROBLEM 13.4. Classify all the generalized TD pairs that are not TD pairs.

Given scalars $\beta, \gamma, \gamma^*, \varrho, \varrho^*$ in a field $\mathbb{K}$, let $T = T(\beta, \gamma, \gamma^*, \varrho, \varrho^*)$ denote the associative $\mathbb{K}$-algebra with identity generated by two symbols A, A^* subject to the relations (13.1), (13.2). Let us call T a *Tridiagonal algebra* (or *TD algebra*). The generalized TD pairs are essentially the same thing as irreducible finite dimensional modules for TD algebras. Given the results of this paper, we would not be suprised if the TD algebras had some interesting properties, and we encourage the reader to look into this.

We mention the relations (13.1), (13.2) previously appeared in [**28**], in the context of P- and Q-polynomial association schemes [**2**], [**3**], [**20**], [**25**], [**29**]. In [**28**], it is shown that for these schemes the matrices A and A^* in Example 1.4 satisfy (13.1), (13.2). In this context the algebra generated by A and A^* is known as the subconstituent algebra or the Terwilliger algebra [**4**], [**7**], [**8**], [**9**], [**10**], [**15**], [**19**], [**24**], [**26**], [**27**].

We mention the relations (13.1), (13.2) are satisfied by the generators of both the classical and quantum "Quadratic Askey-Wilson algebra" introduced by Granovskii, Lutzenko, and Zhedanov [**16**]. See [**17**], [**18**], [**23**], [**32**], [**33**], [**34**] for more information on this algebra.

Here are a few conjectures about TD systems.

CONJECTURE 13.5. *Let Φ denote a TD system with diameter d, and let the scalars $\rho_0, \rho_1, \dots, \rho_d$ be as in Corollary 5.7. We conjecture*

$$\rho_i \leq \binom{d}{i} \qquad (0 \leq i \leq d). \tag{13.3}$$

The above conjecture has been proven for $d \leq 3$ by Tanabe.

CONJECTURE 13.6. *Let $\mathbb{K}$ denote a field, and let V denote a vector space over $\mathbb{K}$ with finite positive dimension. Let Φ denote a TD system on V, and let d denote the diameter. Let the maps R and L be as in Definition 6.1. Let the space U_0 be as in Definition 5.1, and pick any nonzero vector $v \in U_0$. We conjecture V is spanned by the vectors of the form*

$$L^{i_1}R^{i_2}L^{i_3}R^{i_4}\cdots R^{i_n}v,$$

where $i_1, i_2, \ldots, i_n$ ranges over all sequences such that n is nonnegative and even, and such that $i_1, i_2, \ldots i_n$ are integers satisfying $0 \leq i_1 < i_2 < \cdots < i_n \leq d$. For example, for $d = 3$, we are asserting V is spanned by

$$v, \quad Rv, \quad R^2v, \quad R^3v, \quad LR^2v, \quad LR^3v, \quad L^2R^3v, \quad RL^2R^3v.$$

Conjecture 13.6 has been proved for $d \leq 3$ by Tanabe. We remark that Conjecture 13.6 implies Conjecture 13.5.

CONJECTURE 13.7. *Let Φ denote a TD system with diameter d, and let the scalars $\rho_0, \rho_1, \ldots, \rho_d$ be as in Corollary 5.7. Consider the polynomial*

$$\sum_{i=0}^{d} \rho_i t^i \tag{13.4}$$

in a variable t. We conjecture (13.4) equals

$$(1 + t + t^2 + \cdots + t^{d_1})(1 + t + t^2 + \cdots + t^{d_2}) \cdots (1 + t + t^2 + \cdots + t^{d_n}), \tag{13.5}$$

where $d_1, d_2, \ldots, d_n$ are positive integers whose sum is d.

We remark that Conjecture 13.7 implies Conjecture 13.5.

References

[1] C. Ahn and K. Shigemoto, *Onsager algebra and integrable lattice models*, Modern Phys. Lett. A **6** (1991), no. 38, 3509–3515.

[2] E. Bannai and T. Ito, *Algebraic combinatorics I: Association schemes*, Benjamin/Cummings, London, 1984.

[3] A. E. Brouwer, A. M. Cohen, and A. Neumaier, *Distance-regular graphs*, Springer-Verlag, Berlin, 1989.

[4] J. S. Caughman IV, *The Terwilliger algebras of bipartite P- and Q-polynomial schemes*, Discrete Math. **196** (1999), no. 1-3, 65–95.

[5] Vyjayanthi Chari and Andrew Pressley, *Quantum affine algebras*, Comm. Math. Phys. **142** (1991), no. 2, 261–283.

[6] ———, *Quantum affine algebras and their representations*, Representations of groups (Banff, AB, 1994), Amer. Math. Soc., Providence, RI, 1995, pp. 59–78.

[7] B. V. C. Collins, *The girth of a thin distance-regular graph*, Graphs Combin. **13** (1997), no. 1, 21–30.

[8] B. Curtin, *Bipartite distance-regular graphs. I*, Graphs Combin. **15** (1999), no. 2, 143–158.

[9] ———, *Bipartite distance-regular graphs. II*, Graphs Combin. **15** (1999), no. 4, 377–391.

[10] B. Curtin and K. Nomura, *Distance-regular graphs related to the quantum enveloping algebra of $sl(2)$*, J. Algebraic Combin. (to appear).

[11] E. Date and S.S. Roan, *The algebraic structure of the Onsager algebra*, Czech. J. Phys. **50** (2000), no. 1, 37–44.

[12] ———, *The structure of quotients of the Onsager algebra by closed ideals*, J. Phys. A: Math. Gen. **33** (2000), 3275–3296.

[13] B. Davies, *Onsager's algebra and the Dolan-Grady condition in the non-self-dual case*, J. Math. Phys. **32** (1991), no. 11, 2945–2950.

[14] L. Dolan and M. Grady, *Conserved charges from self-duality*, Phys. Rev. D (3) **25** (1982), no. 6, 1587–1604.

[15] J. Go, *The Terwilliger algebra of the Hypercube Q_D*, Graphs Combin. (submitted).

[16] Ya. I. Granovskiĭ, I. M. Lutzenko, and A. S. Zhedanov, *Mutual integrability, quadratic algebras, and dynamical symmetry*, Ann. Physics **217** (1992), no. 1, 1–20.

[17] Ya. I. Granovskiĭ and A. S. Zhedanov, *"Twisted" Clebsch-Gordan coefficients for* $\mathrm{su}_q(2)$, J. Phys. A **25** (1992), no. 17, L1029–L1032.

[18] Ya. I. Granovskiĭ, A. S. Zhedanov, and I. M. Lutsenko, *Quadratic algebras and dynamical symmetry of the Schrödinger equation*, Soviet Phys. JETP **72** (1991), no. 2, 205–209.

[19] S. Hobart and T. Ito, *The structure of nonthin irreducible T-modules of endpoint 1: ladder bases and classical parameters*, J. Algebraic Combin. **7** (1998), no. 1, 53–75.

[20] D. A. Leonard, *Parameters of association schemes that are both P- and Q-polynomial*, J. Combin. Theory Ser. A **36** (1984), no. 3, 355–363.

[21] J. H. H. Perk, *Star-triangle equations, quantum Lax pairs, and higher genus curves*, Theta functions—Bowdoin 1987, Part 1 (Brunswick, ME, 1987), Amer. Math. Soc., Providence, RI, 1989, pp. 341–354.

[22] S.S. Roan, *Onsager's algebra, loop algebra and chiral Potts model*, MPI 91–70, Max-Planck Inst. fur Mathematik (1991).

[23] V. Spiridonov, L. Vinet, and A. Zhedanov, *Periodic reduction of the factorization chain and the Hahn polynomials*, J. Phys. A **27** (1994), no. 18, L669–L675.

[24] K. Tanabe, *The irreducible modules of the Terwilliger algebras of Doob schemes*, J. Algebraic Combin. **6** (1997), no. 2, 173–195.

[25] P. Terwilliger, *A characterization of P- and Q-polynomial association schemes*, J. Combin. Theory Ser. A **45** (1987), no. 1, 8–26.

[26] ———, *The subconstituent algebra of an association scheme. I*, J. Algebraic Combin. **1** (1992), no. 4, 363–388.

[27] ———, *The subconstituent algebra of an association scheme. II*, J. Algebraic Combin. **2** (1993), no. 1, 73–103.

[28] ———, *The subconstituent algebra of an association scheme. III*, J. Algebraic Combin. **2** (1993), no. 2, 177–210.

[29] ———, *A new inequality for distance-regular graphs*, Discrete Math. **137** (1995), no. 1-3, 319–332.

[30] ———, *Two linear transformations each tridiagonal with respect to an eigenbasis of the other*, Linear Algebra Appl. (submitted).

[31] D. B. Uglov and I. T. Ivanov, sl(N) *Onsager's algebra and integrability*, J. Statist. Phys. **82** (1996), no. 1-2, 87–113.

[32] A. S. Zhedanov, *"Hidden symmetry" of Askey-Wilson polynomials*, Teoret. Mat. Fiz. **89** (1991), no. 2, 190–204.

[33] ———, *The "Higgs algebra" as a "quantum" deformation of* su(2), Modern Phys. Lett. A **7** (1992), no. 6, 507–512.

[34] ———, *Quantum* $\mathrm{su}_q(2)$ *algebra: "Cartesian" version and overlaps*, Modern Phys. Lett. A **7** (1992), no. 18, 1589–1593.

Department of Computational Science, Faculty of Science, Kanazawa University, Kakuma-machi, Kanazawa 920–1192, Japan
E-mail address: ito@kappa.s.kanazawa-u.ac.jp

Combinatorial and Computational Mathematics Center, Pohang University of Science and Technology, San 31 Hyoja Dong, Nam-Gu, Pohang, 790-784, Korea
E-mail address: tanabe@com2mac.postech.ac.kr

Department of Mathematics, University of Wisconsin, 480 Lincoln drive, Madison, Wisconsin, 53706, USA
E-mail address: terwilli@math.wisc.edu

DIMACS Series in Discrete Mathematics
and Theoretical Computer Science
Volume **56**, 2001

Bounds for the Christoffel-Darboux kernel of the binary Krawtchouk Polynomials

Ilia Krasikov

ABSTRACT. We give a lower bound on the Christoffel-Darboux kernel for $x = y$ of the binary Krawtchouk polynomials.

1. Introduction

The binary Krawtchouk polynomials $K_k^n(x) = K_k(x)$ are a family of discrete orthogonal polynomials defined by the following recurrence

$$(k+1)K_{k+1}(x) = (n-2x)K_k(x) - (n-k+1)K_{k-1}(x), \tag{1.1}$$

and $K_{-1}(x) = 0$, $K_0(x) = 1$. The orthogonality relation is given by

$$\sum_{i=0}^{n} \binom{n}{x} K_i(x)K_j(x) = 2^n \binom{n}{k} \delta_{ij}. \tag{1.2}$$

Throughout the paper n is supposed to be fixed and dependence on n is omitted in the notation where it does not lead to confusion. We also put $p_k = 2\sqrt{k(n-k)}$. Behaviour and properties of Krawtchouk polynomials play a crucial role in the linear programming method of the coding theory and in some other combinatorial problems (see e.g. [**4**],[**7**],[**10**]). In particular it is important to know lower bounds on the corresponding Christoffel-Darboux kernel

$$W_k(x,y) = \frac{K_{k+1}(x)K_k(y) - K_k(x)K_{k+1}(y)}{y-x} = \frac{2\binom{n}{k}}{k+1} \sum_{i=0}^{k} \frac{K_i(x)K_i(y)}{\binom{n}{i}}, \tag{1.3}$$

especially in the limiting case, $x \to y$, i.e.

$$W_k(x,x) = W_k(x) = K_{k+1}(x)K_k'(x) - K_k(x)K_{k+1}'(x) = \frac{2\binom{n}{k}}{k+1} \sum_{i=0}^{k} \frac{(K_i(x))^2}{\binom{n}{i}}. \tag{1.4}$$

It has been shown in [**5**] that the following function

$$U_k(x) = (K_k(x))^2 - K_k(x-1)K_k(x+1)$$

provides a good approximation for $W_k(x)$ for $\frac{n}{2} - \frac{n-2k}{2n}p_k < x < \frac{n}{2} + \frac{n-2k}{2n}p_k$, that is on a part of the oscillatory region as all the roots of $K_k(x)$ are in the interval

1991 *Mathematics Subject Classification.* Primary 33C45, 33C47; Secondary 26D05.

$(\frac{n-p_k}{2}, \frac{n+p_k}{2})$. The goal of this paper is to establish a lower bound on $W_k(x)$ in the whole oscillatory region for $k < \frac{n}{2}$. Namely we prove the following

THEOREM 1.1. *Let* $3 \le k < \frac{n}{2}$, $\mu_k = \max\{\frac{n-2k+5}{2(n-k+2)}, \binom{n-3}{k-3}2^{3-n}\}$, *and define for* k *even*

$$G_k(x) = \frac{2k(2x+p_k-n)(n-k)^2\Gamma(x)\Gamma(n-x)}{n^3(p_k+2)\Gamma(\frac{n}{2}+1)\Gamma(\frac{n}{2}-1)}\binom{n/2}{k/2}^2.$$

Then

$$W_k(x) \ge \mu_k G_k(x)\frac{n-k+1}{k+1}, \quad k = 0(mod 2),$$
$$W_k(x) \ge \mu_k G_{k+1}(x), \quad k = 1(mod 2).$$

On the other hand it is easy to find a rough upper bound on $W_k(x)$. For instance, multiplying (1.4) by $\binom{n}{x}$ and summing up over x, by (1.2) we get $\sum_{x=0}^{n}\binom{n}{x}W_k(x) = 2^{n+1}\binom{n}{k}$. This simple identity shows that our lower bounds is quite accurate, at least for integer x. More precise upper bounds may be obtained by the same method as the lower bound. We make some necessary remarks at the end of the paper.

Note also that there are bounds on the extreme roots of Krawtchouk polynomials sharper than $\frac{n\pm p_k}{2}$ (see e.g. [**4**]). Moreover, it has been shown recently [**3**] that for n and k growing in such a way that $\frac{k}{n} \to c$, $0 < c < \frac{1}{2}$, the smallest zero x_1 of $K_k(x)$ satisfies

$$x_1 \ge \frac{n-p_k}{2} + \frac{3(n-2k)^{2/3}}{8k^{1/6}(n-k)^{1/6}}.$$

Thus the above result readily gives upper bounds on the corresponding Christoffel numbers. We omit the details.

2. Proofs

The proof of the theorem will be split into a number of lemmas. First we list some properties of the Krawtchouk polynomials needed in the sequel. Proofs can be found in [**4**],[**7**],[**10**].

$$(n-x)K_k(x+1) = (n-2k)K_k(x) - xK_k(x-1); \tag{2.1}$$

$$K_{k+1}(x+1) = K_{k+1}(x) - K_k(x+1) - K_k(x); \tag{2.2}$$

$$K_k(n-x) = (-1)^k K_k(x); \tag{2.3}$$

We start with the following simply inequality:

LEMMA 2.1. *For* $i \le k$,

$$W_i(x) \le \frac{(n-k)!(k+1)!}{(n-i)!(i+1)!}W_k(x), \tag{2.4}$$

PROOF. Follows from

$$W_k(x) = \frac{2\binom{n}{k}}{k+1}\sum_{i=0}^{k}\frac{(K_i(x))^2}{\binom{n}{i}} = \frac{2(n-k+1)\binom{n}{k-1}}{k(k+1)}\sum_{i=0}^{k}\frac{(K_i(x))^2}{\binom{n}{i}} \ge$$
$$\frac{n-k+1}{k+1}W_{k-1}(x).$$

□

The following two lemmas have been proved in [**2**], [**5**]. Since they play a key role we include their proofs for completeness.

LEMMA 2.2. *$U_k(x) > 0$ for $1 \le k \le \frac{n}{2}$.*

PROOF. Put $K_k(x) = a\prod_{i=0}^{k}(x-x_i)$, $x_1 < x_2 < ... < x_k$, and $d_i = 1 - \frac{1}{(x-x_i)^2}$. It is known [**1**] that for $k < \frac{n}{2}$ the distance between consecutive roots of $K_k(x)$ is greater than 2. It is enough to show that $F(x) = \frac{K_k(x-1)K_k(x+1)}{(K_k(x))^2} = \prod_{i=0}^{k} d_i < 1$. Note that $|x - x_i| > \sqrt{2}$ implies $0 < d_i < 1$. Hence there are at most two d_i such that $|d_i| > 1$. Since $|d_i| > 1$ only if $d_i < 0$ then for $x < x_1$ or $x > x_k$ we get that either all d_i are less than one or $F(x) < 0$. Hence we may assume that there is an index j such that $x_j < x < x_{j+1}$. By $x_{i+1} - x_i > 2$ we have $0 < d_i < 1, i \ne j, j+1$. Now the following inequalities complete the proof:

$$d_j d_{j+1} \le \max_{x_j < x < x_{j+1}} d_j d_{j+1} = (1 - \frac{4}{(x_{j+1}-x_j)^2})^2 < 1.$$

□

LEMMA 2.3.

$$W_k(x) = \frac{1}{2}(U_{k+1}(x) - U_k(x)) + \frac{2\binom{n}{k}}{k+1}\sum_{i=0}^{k}\frac{U_i(x)}{\binom{n}{i}}.$$

PROOF. Using $U_k(x) = (K_k(x))^2 - K_k(x-1)K_k(x+1)$ and (1.3), (1.4) we get

$$\frac{4\binom{n}{k}}{k+1}\sum_{i=0}^{k}\frac{U_i(x)}{\binom{n}{i}} = W_k(x) - K_k(x+1)K_{k+1}(x-1) + K_{k+1}(x+1)K_k(x-1).$$

Applying (2.2) to $K_{k+1}(x+1)K_k(x-1)$, first to $K_{k+1}(x+1)$ and then to $K_{k+1}(x)$ we get the claim. □

Now we can reduce a problem of bounding $W_k(x)$ to the corresponding problem for $U_k(x)$.

LEMMA 2.4. *For $3 \le k \le \frac{n}{2}$,*

$$W_k(x) \ge \frac{\binom{n-3}{k-3}}{2\sum_{i=0}^{k-3}\binom{n-3}{i}}U_{k+1}(x).$$

In particular

$$W_k(x) \ge \max\{\frac{n-2k+5}{2(n-k+2)}, \binom{n-3}{k-3}2^{3-n}\}U_{k+1}(x).$$

PROOF. From the previous lemma we have

$$2W_k(x) \ge U_{k+1}(x) - \frac{k-3}{k+1}U_k(x).$$

Since $U_0(x) = 0$, it yields

$$2\sum_{i=0}^{k}\binom{i+1}{4}W_i(x) \ge \sum_{i=0}^{k}\binom{i+1}{4}(U_{i+1}(x) - \frac{i-3}{i+1}U_i(x)) = \binom{k+1}{4}U_{k+1}(x).$$

Applying inequality (2.4) we get

$$\sum_{i=0}^{k}\binom{i+1}{4}W_i(x) \le W_k(x)\sum_{i=0}^{k}\frac{(n-k)!(k+1)!}{(n-i)!(i+1)!}\binom{i+1}{4} =$$

$$W_k(x)\frac{(n-k)!(k+1)!}{4!(n-3)!}\sum_{i=3}^{k}\binom{n-3}{i-3},$$

giving

$$W_k(x) \geq \frac{\binom{n-3}{k-3}}{2\sum_{i=0}^{k-3}\binom{n-3}{i}}U_{k+1}(x) > \binom{n-3}{k-3}2^{3-n}.$$

Also, by induction on k one gets that for $k \leq \frac{n}{2}$,

$$\frac{\binom{n}{k}}{\sum_{i=0}^{k}\binom{n}{i}} \geq \frac{n-2k+5}{n-k+2}.$$

and the result follows. □

LEMMA 2.5. *Let* $x > \frac{n-p_k}{2}$, *then*

$$\frac{U_k(x)}{U_k(x+1)} \geq \frac{(n-x-1)(p_k+2x-n)}{x(p_k+2x+2-n)}. \tag{2.5}$$

PROOF. Follows by $U_k(x) > 0$ and the following identity

$$U_k(x+1) - \frac{x(2x+p_k+2-n)}{(n-x-1)(2x+p_k-n)}U_k(x) =$$

$$-\frac{(\sqrt{n+p_k}K_k(x+1) - \sqrt{n-p_k}K_k(x))^2}{(n-x-1)(2x+p_k-n)}.$$

□

Probably for k even $U_k(x)$ has the only minimum at $n/2$. We don't know how to prove this, but the following lemma provides an estimates on $U_k(x)$ near to $x = \frac{n}{2}$.

LEMMA 2.6. *For* $[\frac{n}{2}] \leq x \leq [\frac{n}{2}] + 1$ *and* k *even*

$$U_k(x) \geq \frac{4k(n-k)^2}{n^3}\binom{n/2}{k/2}^2.$$

PROOF. By recurrence (1.1) we get

$$(-1)^{k/2}K_k(\frac{n}{2}) = A = \binom{n/2}{k/2},\ (-1)^{k/2}K_k(\frac{n}{2}+1) = B = \frac{n-2k}{n}\binom{n/2}{k/2}. \tag{2.6}$$

Since $K_k(x)$ has only real zeros at distances greater than 2 apart, from (2.3) it follows that for k even a graph of $K_k(x)$ is symmetric with respect $\frac{n}{2}$. Moreover interlacing of zeros of $K_k(x)$ and its derivative implies that $|K_k(x)|$ increasing on $\frac{n}{2} - 1 \leq x < \frac{n}{2}$, decreasing on $\frac{n}{2} < x \leq \frac{n}{2} + 1$, and is concave for $|x - \frac{n}{2}| \leq 1$.

Assume first that $x = \frac{n}{2} + \lambda$, $0 \leq \lambda \leq 1$, and put $v = K_k(x-1)$, $u = K_k(x)$. Then $|v| \geq \lambda A + (1-\lambda)B$, $|u| \geq \lambda B + (1-\lambda)A$. Using recurrence (2.1) we may rewrite $U_k(x)$ in the following form

$$U_k(x) = u^2 - \frac{n-2k}{n-x}vu + \frac{x}{n-x}v^2. \tag{2.7}$$

Using (2.6) we have

$$\frac{\partial}{\partial x}U_k(x) \geq 0,\ \frac{\partial}{\partial v}U_k(x) \geq 0,\ \frac{\partial}{\partial u}U_k(x) \geq 0.$$

Hence

$$U_k(x) \geq u^2 - \frac{2n-4k}{n}vu + v^2 \geq$$

$$\min_{\lambda}\{(\lambda B+(1-\lambda)A)^2-\frac{2(n-2k)}{n}(\lambda A+(1-\lambda)B)(\lambda B+(1-\lambda)A)+(\lambda A+(1-\lambda)B)^2\}$$

$$=\frac{4k}{n}(\frac{A+B}{2})^2=\frac{4k(n-k)^2}{n^3}\binom{n/2}{k/2}^2.$$

For $\frac{n-1}{2}\le x\le\frac{n}{2}$ the proof is similar. We put $w=K_k(x+1)$ and use

$$U_k(x)=u^2-\frac{n-2k}{x}uw+\frac{n-x}{x}w^2,$$

instead of (2.7). □

LEMMA 2.7. *Let* $\frac{n-p_k}{2}<x<\frac{n-1}{2}$, *then for* $k<\frac{n}{2}$ *even*

$$U_k(x)\ge\frac{4k(2x+p_k-n)(n-2k)^2\Gamma(x)\Gamma(n-x)}{n^3(p_k+2)\Gamma(\frac{n}{2}+1)\Gamma(\frac{n}{2}-1)}\binom{n/2}{k/2}^2. \tag{2.8}$$

PROOF. Let $y=\lfloor x\rfloor$, $\theta=x-y$ then multiplying (2.5) we get

$$\prod_{j=y}^{\lfloor n/2\rfloor-1}U_k(j+\theta)\ge\prod_{j=y}^{\lfloor n/2\rfloor-1}\frac{(j+\theta)(2j+2\theta+p_k+2-n)}{(n-j-\theta-1)(2j+2\theta+p_k-n)}U_k(j+1+\theta),$$

yielding

$$U_k(x)\ge\frac{(2x+p_k-n)\Gamma(x)\Gamma(n-x)}{(p_k+2\theta)\Gamma(\frac{n}{2}+\theta)\Gamma(\frac{n}{2}-\theta)}U_k(\frac{n}{2}+\theta),$$

for n even, and

$$U_k(x)\ge\frac{(2x+p_k-n)\Gamma(x)\Gamma(n-x)}{(p_k-1+2\theta)\Gamma(\frac{n-1}{2}+\theta)\Gamma(\frac{n+1}{2}-\theta)}U_k(\frac{n-1}{2}+\theta),$$

for n odd. Now the result follows from the previous two lemmas and log-convexity of $\Gamma(x)$. □

Now Theorem 1.1 follows from (2.4), the above bound on $U_k(x)$ and (2.4) if k is even.

Finally, let us make a remark concerning an upper bound on $W_k(x)$. It can be easily deduced from Lemma 2.3 and the following counterpart of Lemma 2.5:

For $x<\frac{n+p_k}{2}$,

$$\frac{U_k(x)}{U_k(x+1)}\le\frac{x(n+p_k-2x-2)}{(n-x-1)(n+p_k-2x)}.$$

The last inequality follows from the identity:

$$U_k(x+1)-\frac{x(n+p_k-2x-2)}{(n-x-1)(n+p_k-2x)}U_k(x)=$$

$$\frac{(\sqrt{n-p_k}K_k(x+1)-\sqrt{n+p_k}K_k(x))^2}{(n-x-1)(n+p_k-2x)}.$$

One can check that as in the case of continuous orthogonal polynomials, $W_k(x)\binom{n}{x}$ is close to a constant function in x, see **[8]** for excellent discussion.

References

[1] L. Chihara and D. Stanton, *Zeros of generalized Krawtchouk polynomials*, J. Approx. Theory 60, No.1 1990, 43-57.
[2] I. Krasikov, *Nonnegative quadratic forms and bounds on orthogonal polynomials*, submitted.
[3] W.H. Foster and I. Krasikov, *Bounds on extreme roots of orthogonal polynomials*, submitted.
[4] I. Krasikov and S. Litsyn, *On integral zeros of Krawtchouk polynomials*, J. Combin. Theory, Ser. A, Vol.74,1 1996, pp.71-99.
[5] I. Krasikov and S. Litsyn, *On the distance distributions of BCH codes and their duals,* Designs, Codes and Cryptography, to appear.
[6] I. Krasikov and S. Litsyn, *On the distance distribution of duals of BCH codes*, IEEE Trans. Inform. Theory, Vol.45, 1, (1999), pp. 247-250.
[7] V.I. Levenshtein, *Universal bounds for codes and designs*, In: Handbook of coding theory, Vol.I, North-Holland, 1998, pp.499-648.
[8] A.L. Levin, D.S. Lubinsky, *Bounds for orthogonal polynomials for exponential weights on (-1,1) and* $(-\infty, \infty)$, in Supplemento ai Rendiconti del Circolo Matematico di Palermo, Ser.II, 33. Proceedings of the second international conference in functional analysis and approximation theory, 1993, pp.75-96.
[9] J.H. van Lint, *Introduction to Coding Theory*, Springer-Verlag, 1992.
[10] F.J. MacWilliams and N.J.A. Sloane, *The Theory of Error-Correcting Codes*, New York: North-Holland, 1977.

Department of Math. Sci., Brunel University, Uxbridge, Middlesex UB8 3PH, U.K.
E-mail address: `mastiik@brunel.ac.uk, ilia@eng.tau.ac.il`

DIMACS Series in Discrete Mathematics
and Theoretical Computer Science
Volume **56**, 2001

Survey of binary Krawtchouk polynomials

Ilia Krasikov and Simon Litsyn

ABSTRACT. Krawtchouk polynomials play a crucial role in many combinatorial problems related to the Hamming association scheme. We survey some old and new results concerning binary Krawtchouk polynomials.

Binary Krawtchouk polynomials play an important role in combinatorial problems related to the binary Hamming association schemes. In the paper we attempt to review what is known about such polynomials. Most of the properties which are mentioned without reference are standard for orthogonal polynomials and their proofs can be found in [**7, 23, 33, 36, 38, 39, 45**]. Among other results we also present several new properties of Krawtchouk polynomials.

1. Definitions and Basic Properties

The **Krawtchouk polynomials** is a family of polynomials orthogonal with respect to a discrete measure μ supported on the integer points $0, 1, ..., n$, and defined by $\mu(i) = \binom{n}{i} p^i q^{n-i}$. The binary Krawtchouk polynomials $K_k^n(x)$ are a particular case of these with $p = q = 1/2$. They are defined by any of the following expressions:

- Via generating function:

$$\sum_{k=0}^{\infty} K_k^n(x) z^k = (1-z)^x (1+z)^{n-x} \tag{1}$$

- Explicit expressions:

$$K_k^n(x) = \sum_{j=0}^{k} (-1)^j \binom{x}{j} \binom{n-x}{k-j} \tag{2}$$

$$= \sum_{j=0}^{k} (-2)^j \binom{x}{j} \binom{n-j}{k-j} \tag{3}$$

$$= \sum_{j=0}^{k} (-1)^j 2^{k-j} \binom{n-x}{k-j} \binom{n-k+j}{j} \tag{4}$$

1991 *Mathematics Subject Classification.* 33C45.

Supported in part by Binational Science Foundation (BSF) under grant 1999099.

- Cauchy's integral formula:

(5) $$K_k^n(x) = \frac{1}{2\pi i}\oint \frac{(1+z)^{n-x}(1-z)^x}{z^{k+1}}dz$$

- Integral formula:

(6) $$K_k^n(x) = \frac{2^{n-1}(-i)^x}{\pi}\int_0^{2\pi} \exp[i(n/2-k)\theta]\cos^{n-x}(\theta/2)\sin^x(\theta/2)d\theta$$

- Hypergeometric representation (see for definition e.g. [**4**, Chapter 2]):

(7) $$K_k^n(x) = \binom{n}{k}\ {}_2\mathrm{F}_1[-k,-x,-n,2]$$

- A value of Jacobi polynomials:

(8) $$K_k^n(x) = (-2)^k P_k^{(x-k,n-x-k)}(0)$$

This formula follows from comparison of (2) with an explicit expression [**45**, (4.3.2)] for Jacobi polynomials:

(9) $$P_k^{a,b} = \sum_{i=1}^{k}\binom{k+a}{k-i}\binom{k+b}{i}\left(\frac{x-1}{2}\right)^i\left(\frac{x+1}{2}\right)^{k-i}$$

- A limiting case of Hahn polynomials [**18**]:

(10) $$K_k^n(x) = \binom{n}{k}\lim_{t\to\infty} H_k^n(x,t/2,t/2)$$

here

(11) $$H_k^n(x,\alpha,\beta) = \sum_{j=0}^{k}(-1)^j\frac{\binom{k}{j}\binom{k+\alpha+\beta+j}{j}}{\binom{j+\alpha}{j}\binom{n}{j}}\binom{x}{j}$$

- Recurrence relations:

The following initial conditions are standard:

(12) $$K_0^n(x) = 1, \quad K_1^n(x) = n-2x,$$

(13) $$K_k^n(0) = \binom{n}{k}, \quad K_k^n(1) = \frac{n-2k}{n}\binom{n}{k}$$

Using the hypergeometric representation of $K_k^n(x)$ and fifteen Gauss relations between contiguous functions (see [**4**, §2.8]) one can derive many recurrent relations for Krawtchouk polynomials. Some of them are listed below. The first two are the most used, being the recurrence and difference relations, (16) is proved in [**23**], the others appear in [**6**].

(14) $$(k+1)K_{k+1}^n(x) = (n-2x)K_k^n(x) - (n-k+1)K_{k-1}^n(x)$$

(15) $$(n-x)K_k^n(x+1) = (n-2k)K_k^n(x) - xK_k^n(x-1)$$

(16) $$(n-k+1)K_k^{n+1}(x) = (3n-2k-2x+1)K_k^n(x) - 2(n-x)K_k^{n-1}(x)$$

$$K_k^n(x) = K_k^n(x-1) - K_{k-1}^n(x) - K_{k-1}^n(x-1) \tag{17}$$

$$K_k^n(x) = K_k^{n-2}(x-1) - K_{k-2}^{n-2}(x-1) \tag{18}$$

$$K_k^n(x) = K_k^{n-1}(x) + K_{k-1}^{n-1}(x) \tag{19}$$

$$K_k^n(x) = K_k^{n-1}(x-1) - K_{k-1}^{n-1}(x-1) \tag{20}$$

$$K_k^n(x) = K_k^n(x-1) - 2K_{k-1}^{n-1}(x-1) \tag{21}$$

$$K_k^n(x) = K_k^n(x-2) - 4K_{k-1}^{n-2}(x-2) \tag{22}$$

The **moment sums** of Krawtchouk polynomials (can be easily obtained from the generating function):

$$\sum_{i=0}^{n} K_i^n(x) = 2^n \delta_{k,0} \tag{23}$$

$$\sum_{i=0}^{n} iK_i^n(x) = 2^{n-1}(n\delta_{k,0} - \delta_{k,1}) \tag{24}$$

$$\sum_{i=0}^{n} i^2 K_i^n(x) = 2^{n-2}(n(n+1)\delta_{k,0} - 2n\delta_{k,1} + 2\delta_{k,2}) \tag{25}$$

Coefficients of Krawtchouk polynomials can be directly calculated. If $K_k^n(x) = c_k x^k + \ldots + c_0$, then

$$c_k = (-2)^k/k! \tag{26}$$

$$c_{k-1} = (-2)^{k-1}n/(k-1)! \tag{27}$$

$$c_{k-2} = \frac{(-2)^{k-2}}{6}(3n^2 - 3n + 2k - 4)/(k-2)! \tag{28}$$

$$c_0 = \binom{n}{k} \tag{29}$$

The following **symmetry relations** hold for Krawtchouk polynomials:

$$\binom{n}{x} K_k^n(x) = \binom{n}{k} K_x^n(k), \quad \text{(for nonnegative integer } x\text{);} \tag{30}$$

$$K_k^n(x) = (-1)^k K_k^n(n-x) \tag{31}$$

$$K_k^n(x) = (-1)^x K_{n-k}^n(x), \quad \text{(for integer } x,\ 0 \le x \le n\text{)} \tag{32}$$

The **orthogonality relations** are:

$$\sum_{i=0}^{n} \binom{n}{i} K_k^n(i) K_l^n(i) = \delta_{kl} \binom{n}{k} 2^n \tag{33}$$

$$\sum_{i=0}^{n} K_l^n(i) K_i^n(k) = \delta_{lk} 2^n \tag{34}$$

The following **identities** are valid for Krawtchouk polynomials [**13**]

$$\sum_{i=0}^{n} \binom{n}{i} \binom{n}{i+m} K_k^n(i) K_k^n(i+m) = (-1)^k \binom{n}{k}^2 K_{n+m}^{2n}(2k) \tag{35}$$

$$\sum_{i=0}^{n} \binom{n}{i} \binom{n}{i+1} K_k^n(i) K_k^n(i+1) = \frac{\binom{2n}{n-1}\binom{n}{k}^3}{\binom{2n}{2k}} (1 - 2k/n) \tag{36}$$

2. Expansions

We start from the **multiplication formula** [**7**]:

$$K_k^n(x) K_i^n(x) = \sum_{j=\max(0,k+i-n)}^{\min(k,i)} a(k,i,j) K_{k+i-2j}(x) \tag{37}$$

where

$$a(k,i,j) = \binom{k+i-2j}{k-j} \binom{n-k-i+2j}{j} \tag{38}$$

In particular,

$$(K_k^n(x))^2 = \sum_{j=0}^{k} \binom{2j}{j} \binom{n-2j}{k-j} K_{2j}^n(x) \tag{39}$$

As orthogonal polynomials Krawtchouk polynomials satisfy the **Christoffel-Darboux formula**:

$$W_k^n(x,y) = \frac{K_{k+1}^n(x) K_k^n(y) - K_k^n(x) K_{k+1}^n(y)}{y - x} = \frac{2}{k+1} \binom{n}{k} \sum_{i=0}^{k} \frac{K_i^n(x) K_i^n(y)}{\binom{n}{i}} \tag{40}$$

The limiting case $x = y$, is

$$W_k^n(x,x) = K_{k+1}^n(x)(K_k^n(x))' - K_k^n(x)(K_{k+1}^n(x))' = \frac{2\binom{n}{k}}{k+1} \sum_{i=0}^{k} \frac{(K_i^n(x))^2}{\binom{n}{i}} \tag{41}$$

For any polynomial $f(x)$ of degree d there is the unique Krawtchouk expansion:

$$f(x) = \sum_{k=0}^{d} f_k K_k^n(x) \tag{42}$$

where the coefficients are

$$f_k = 2^{-n} \sum_{i=0}^{n} f(i) K_i^n(k) \tag{43}$$

Notice that the Krawtchouk expansion of the polynomials symmetric (resp. antisymmetric) with respect to $n/2$ contains only Krawtchouk polynomials of even (resp. odd) degrees.

For the **derivative** of Krawtchouk polynomials the following formula easily follows from recurrence relations:

$$(K_k^n(x))' = -2 \sum_{i=0}^{\lfloor (k-1)/2 \rfloor} \frac{K_{k-1-2i}^n(x)}{2i+1} \tag{44}$$

The following (probably new) expansions are valid and can be derived using recurrences:

$$K_{2k+1}^n(x) = (n-2x) \sum_{j=0}^{k} (-1)^j \frac{K_{2k-2j}^n(x)}{n-2k-1} \prod_{i=0}^{j} \frac{n-2k-1+2i}{2k+1-2i} \tag{45}$$

$$K_k^m(x) = \sum_{i=0}^{\lfloor k/2 \rfloor} \binom{\frac{n-m}{2}+i-1}{i} K_{k-2i}^n(x) \tag{46}$$

$$(K_k^n(x))^2 - \frac{k+1}{k} K_{k+1}^n(x) K_{k-1}^n(x) = \sum_{i=1}^{k} \frac{(n-2k+2i)\binom{n-k+i}{i-1}}{k^2 \binom{k-1}{i}} (K_{k-i}^n(x))^2 \tag{47}$$

We continue with expansions of **annihilating polynomials**. Expression (48) is due to [**7**], (49)-(52) are derived in [**25**], (53) and (55) appear in [**27**].

$$\frac{2^{n-k+1}}{(n-k+1)!} \prod_{j=k}^{n} (j-x) = \sum_{i=0}^{n} \binom{n-i}{k-1} K_i^n(x) \tag{48}$$

(zeros of this polynomial are $x = k, k+1, \ldots, n$)

$$\frac{1}{(2k)!} \prod_{i=0}^{k-1} ((n-2x)^2 - 4i^2) = \sum_{i=0}^{k} \binom{n/2-i}{k-i} K_{2i}^n(x) \tag{49}$$

(zeros of this polynomial are $x = n/2 \pm i$, $i = 0, 1, \ldots, k-1$; $x = n/2$ is a multiple zero).

$$\frac{1}{(2k)!} \prod_{i=0}^{k-1} ((n-2x)^2 - (2i+1)^2) = \sum_{i=0}^{k} \binom{(n-1)/2-i}{k-i} K_{2i}^n(x) \tag{50}$$

(zeros of this polynomial are $x = (n-1)/2 \pm i$, $i = 0, 1, \ldots, k-1$; $x = (n-1)/2$ is a multiple zero).

$$\frac{1}{(2k)!} \prod_{i=0}^{k-1} ((n-2x)^2 - 16i^2) = \sum_{i=0}^{k} \frac{n-4i}{n-2k-2i} \binom{n/2-k-i}{k-i} K_{2i}^n(x) \tag{51}$$

(zeros of this polynomial are $x = n/2 \pm 2i$, $i = 0, 1, \ldots, k-1$; $x = n/2$ is a multiple zero).

$$\frac{1}{(2k)!}\prod_{i=0}^{k-1}((n-2x)^2 - 4(2i+1)^2) = \sum_{i=0}^{k}\binom{n/2-k-i-1}{k-i}K_{2i}^n(x) \tag{52}$$

(zeros of this polynomial are $x = n/2 \pm (2i+1)$, $i = 0, 1, \ldots, k-1$; $x = n/2$ is a multiple zero).

In general,

$$\mathcal{B}_h^n(x,k) = \frac{1}{(2k)!}\prod_{i=0}^{k-1}((n-2x)^2 - h^2 i^2) = \sum_{j=0}^{k} A_{2j}(n,h,k)K_{2j}^n(x) \tag{53}$$

where

$$A_{2j}(n,h,k) = \frac{(-1)^j h^{2k}}{2^k k!}\left(\frac{d^k}{dz^k}\sin^{2j}\Psi_h \cos^{n-2j}\Psi_h\right)\Big|_{z=1} \tag{54}$$

and

$$\Psi_h = \frac{\arccos z}{h}$$

These polynomials are even in respect to $n/2$. The zeros of $\mathcal{B}_h^n(x,k)$ are $\frac{n}{2} \pm \frac{hi}{2}, i = 0, \ldots, k-1$. Notice that $n/2$ is a root of multiplicity 2.

In particular, for $h = 8$ this gives for $j \leq k-1$,

$$A_{2j}(n,8,k) = \frac{1}{(2k-2j)2^{(n-k)/2}}\sum_{i=0}^{n/2-2j}(i-2j)\binom{n/2-2j}{i}\binom{i/2-k-1}{k-j-1} \tag{55}$$

and

$$A_{2k}(n,8,k) = 1$$

3. Connection to other polynomials

The **Lloyd polynomials** $L_k^n(x)$ are defined as sums of Krawtchouk polynomials:

$$L_k^n(x) := K_k^{n-1}(x-1) = \sum_{i=0}^{k} K_i^n(x) \tag{56}$$

There are intimate relations that connect Krawtchouk polynomials to other systems of orthogonal polynomials. They can be verified using recurrences.

Discrete **Hahn polynomials** (see (11), for $\beta = \alpha = -n-1$ have the following representation

$$H_k^n(x,-n-1,-n-1) = \frac{\binom{n}{k} - \binom{n}{k-1}}{2^k}\sum_{i=0}^{k}\frac{\binom{k}{i}}{\binom{n}{i}\binom{n}{k-i}}K_i^n(x)K_{k-i}^n(x) \tag{57}$$

Discrete **Chebyshev polynomials** (proportional to the Hahn polynomials with $\alpha = \beta = 0$) are defined as follows:

$$T_k^n(x) = \sum_{i=0}^{k}(-1)^i\binom{x}{i}\binom{n-x}{k-i}\binom{k}{i} \tag{58}$$

The following expansion in the basis of Krawtchouk polynomials can be easily derived using induction on k and the corresponding three term recurrence relations (the recurrence relations for the discrete Chebyshev polynomials can be found in [**39**]):

$$(59) \qquad T_k^n(x) = \frac{\binom{2k}{k}}{2^k} \sum_{i=0}^{\lfloor k/2 \rfloor} \frac{\binom{k}{i}\binom{n-k+2i}{2i}}{\binom{2k}{2i}} K_{k-2i}^n(x)$$

The **Hermite polynomials** $H_k(x)$ are a limiting case of Krawtchouk polynomials, They are defined by the recurrence relation

$$(60) \qquad H_{k+1}(x) = 2xH_k(x) - 2kH_{k-1}(x),\ H_0(x) = 1,\ H_1(x) = 2x$$

If one compares (60) with (14), the following connection holds:

$$K_k^n\left(\frac{n-\sqrt{2n}y}{2}\right) =$$

(61)

$$\frac{1}{k!2^{k/2}}(n^{k/2}H_k(y) + 4n^{(k-2)/2}\left(\binom{k}{3}H_{k-2}(y) + 2\binom{k}{4}H_{k-4}(y)) + n^{(k-4)/2}R_k(y)\right)$$

where $R_0(y) = R_1(y) = 0$, and

$$R_{k+1}(y) =$$

$$(62) \quad 2yR_k(y) - \frac{2k(n-k+1)}{n}R_{k-1}(y) + 8(k-1)\binom{k}{4}(3H_{k-3}(y) + 2yH_{k-4}(y))$$

In particular [**29**], for fixed k and y,

$$(63) \qquad K_k^n\left(\frac{n-\sqrt{2n}y}{2}\right) = \frac{n^{k/2}}{k!2^{k/2}}H_k(y) + O(n^{k/2-1})$$

4. Modular properties

For the following modular properties see [**23**].

Let $n \geq k + 2^m$. Then

$$K_k^n(x) \equiv K_k^n(x + 2^m) \pmod{2^{m+1}}$$

For any integer $s \geq 0$ and $q = p^s$, p is a prime

$$(64) \qquad K_k^n(x) \equiv K_k^{n-q}(x) + K_{k-q}^{n-q}(x) \pmod{p}$$

$$(65) \qquad K_k^n(x) \equiv K_k^{n-q}(x-q) - K_{k-q}^{n-q}(x-q) \pmod{p}$$

$$(66) \qquad K_k^n(x) \equiv -K_k^n(x-q) + 2K_k^{n-q}(x-q) \pmod{p}$$

For $p = 3$:

$$(67) \qquad K_k^n(x) \equiv K_k^{n-q}(x-2q) + K_{k-2q}^{n-2q}(x-2q) \pmod{3}$$

For arbitrary p prime we have got the following weak analogue of the Lucas theorem. Let $n = n_l p^l + \ldots + n_0$, $k = k_s p^s + \ldots + k_0$, $x = x_l p^l + \ldots + x_0$, and $n_i \geq x_i$ for $i = 0, \ldots, s$. Then

$$K_k^n(x) \equiv \prod_{i=0}^{s} K_{k_i}^{n_i}(x_i) \pmod{p} \tag{68}$$

Define $\mathcal{H}_k^n(y) = k! K_k^n(\frac{n-y}{2})$. Then [**13**]

$$\mathcal{H}_{ka+r}^n(y) \equiv (\mathcal{H}_a^n(y))^k \, \mathcal{H}_r^n(y) \pmod{a}$$

5. Zeros of Krawtchouk polynomials

The Krawtchouk polynomial $K_k^n(x)$ has k simple roots $0 < r_{1,n}(k) < r_{2,n}(k) < \ldots < r_{k,n}(k) < n$. The roots are symmetric with respect to $n/2$, that is

$$r_{i,n}(k) + r_{k+1-i,n}(k) = n, \; i = 1, \ldots, k.$$

More information on location of the smallest root can be easily derived from the following result [**33**]:

$$r_{1,n}(k) = n/2 - \max\left(\sum_{i=0}^{k-2} x_i x_{i+1} \sqrt{(i+1)(n-i)}\right) \tag{69}$$

where maximum is taken over all x_i subjected to $\sum_{i=0}^{k-1} x_i^2 = 1$.
In particular, we get for $k \leq n/2$

$$r_{1,n}(k) \geq n/2 - \sqrt{(k-1)(n-k+2)} \tag{70}$$

Another estimate for $k \leq n/2$ is

$$r_{1,n}(k) \geq n/2 - \sqrt{k(n-k)} \tag{71}$$

A more accurate estimate [**11**] provided k is growing linearly with n, $0 < k/n < 1/2$ is given by

$$r_{1,n}(k) \geq n/2 - \sqrt{k(n-k)} + \frac{3(n-2k)^{2/3}}{8k^{1/6}(n-k)^{1/6}} + O(1) \tag{72}$$

In [**34**] the following upper bound is given:

$$r_{1,n}(k) \leq n/2 - \sqrt{\frac{n-k+2}{2}}\, h_k(k) \tag{73}$$

where $h_1(k) < \ldots < h_k(k)$ are the roots of the Hermite polynomial $H_k(z)$, and [**40**]

$$h_k(k) = \sqrt{2k+1} - c(2k+1)^{-1/6} + O(k^{-5/6}) \quad \text{where} \quad c = 1.85575\ldots$$

The following upper bound is valid [**34**, (6.25)]:

$$r_{1,n}(k) \leq n/2 - \sqrt{k(n-k)} + k^{1/6}\sqrt{n-k} \quad \text{for} \quad k \leq [n/2] \tag{74}$$

The roots of the Krawtchouk polynomials satisfy some interlacing properties:

$$r_{i,n+1}(k) < r_{i,n}(k-1) < r_{i+1,n+1}(k), \quad i = 1, \ldots, k-1 \tag{75}$$

$$r_{i,n}(k) < r_{i,n+1}(k) < r_{i+1,n}(k) < r_{i+1,n+1}(k), \quad i = 1, \ldots, k-1 \tag{76}$$

Moreover [**6**], for $i = 1, \dots, n$,

$$r_{i,n}(k) - r_{i,n+1}(k) < 1 \tag{77}$$

For fixed n and $1 \le k < n/2$ (see [**6**]):

$$r_{i+1,n}(k) - r_{i,n}(k) > 2 \tag{78}$$

for $k < n$,

$$r_{i+1,n}(k) - r_{i,n}(k) > 1 \tag{79}$$

Let $t = x/k$, $\gamma = n/k$, $r = \frac{\gamma}{2} - \sqrt{\gamma - 1}$. Provided $k \to \infty$, the distribution of zeros of $K_k^n(x)$ is given by the measure with density [**9, 10, 16, 32**]:

$$v(t) = 1 + \frac{1}{\pi} arccot\left(\frac{\gamma - 2}{2\sqrt{(t-r)(\gamma - t - r)}}\right), \qquad t \in (r, \gamma - r), \tag{80}$$

$$v(t) = \chi(2 - \gamma), \quad t \in (0, r) \cup (\gamma - r, \gamma), \tag{81}$$

where χ is the characteristic function of the interval $(0, \infty)$.

Significant attention has been devoted to integer roots (see [**6, 8, 13, 14, 23, 29, 30**]). Knowledge of existence of non-integer roots is important in proving non-existence of perfect codes with given parameters and reconstruction problems in graphs (see [**23**] for a survey).

The integer roots are completely classified for $k \le 7$, and for $k = (n-t)/2$ with $t \le 6$, and $t = 8$ when the root is odd. For k odd and n even a trivial integral root at $n/2$ exists. In general for every fixed $k \ge 4$, $K_k^n(x)$ can have nontrivial integer root only for finitely many n.

Let $t > 6$ be either an odd prime, a power of 2 or of the form $2pq$, where p is an odd prime, q is odd, and p does not divide q. Then for $k = (n-t)/2$, $K_k^n(x)$ can possess nontrivial even roots only for finitely many n.

Let digits of p-adic expansion of a number $a = \sum_i a_p(i)p^i$ will be denoted by $a_p(i)$ If $a_p(i) \ge b_p(i)$ for all i then we will write it as $a \succ_p b$. Then the following facts are known.

There are no integer roots if $n \succ_2 k$, or else if $n \succ_3 k$ and all digits in the ternary expansions of n and k are confined to $\{0, 2\}$.

At least one noninteger root exists for k and n even, $k \ge 4$, or for k even, n odd, $k \ge 344$. For k and n odd the trivial root is nonintegral. The case k odd and n even is still open. However, some particular cases were resolved. Namely, there is a noninteger root when $n \equiv 0 \pmod 4$ and $k \equiv 3 \pmod 4$, $n \equiv 0 \pmod 8$ and $k \equiv 5 \pmod 8$.

The total number of integer roots (conjectured to be at most 4) does not exceed

$$\frac{k}{2} + \frac{3}{2}\sqrt{k \log_2 n}\,. \tag{82}$$

For growing k the coefficient $\frac{3}{2}$ can be replaced by $\frac{1}{2} + o(1)$.

For large k's the following upper bound on the number of integer roots is better:

$$\frac{3}{2}\sqrt[3]{2(k-1)(n-k+2)}\,. \tag{83}$$

In particular, for arbitrary k, the number of integer roots is at most $(9n/4)^{2/3}$.

6. Bounds on Krawtchouk polynomials

By the symmetry with respect to $n/2$, when it is convenient, we can consider only values of Krawtchouk polynomials for $x \le n/2$.

Let a^* be the greatest even integer $\le a$, then (6) yields for $2 \le x \le n-2$,

$$(84)\qquad |K_k^n(x)| \le 2^{n-x^*-(n-x)^*} \frac{\binom{x^*+(n-x)^*}{(x^*+(n-x)^*)/2}\binom{(x^*+(n-x)^*)/2}{x^*/2}}{\binom{x^*+(n-x)^*}{x^*}}$$

For n and x even, $0 \le x \le n$, this gives [**22, 35**],

$$(85)\qquad |K_k^n(x)| \le |K_{n/2}^n(x)|$$

$$(86)\qquad |K_k^n(x)| \le \frac{\binom{n}{n/2}\binom{n/2}{k/2}}{\binom{n}{x}}$$

For $x \in [0, n]$ integer, from the orthogonality relation we deduce

$$(87)\qquad |K_k^n(x)| \le 2^{n/2}\sqrt{\binom{n}{k}\binom{n}{x}^{-1}}$$

It is natural to separate two cases: bounds on the oscillatory region (i.e. where the roots exist) and the rest of the interval $[0, n]$. Let also $k < n/2$.

Define $A_k(x) = \frac{x!(n-x)!}{(\lfloor \frac{k}{2} \rfloor)!^2(\lfloor \frac{k}{2} \rfloor)!^2}$ and

$$\tau(n,k) = \begin{cases} \frac{n}{2} + k(n-k), & n, k \text{ even} \\ 4, & n \text{ even, } k \text{ odd} \\ 2k+1, & n \text{ odd, } k \text{ even} \\ 2n-2k+1, & n, k \text{ odd} \end{cases}$$

and $h(x) = 4x(2k+1) - (n-2k-2x)^2$. Then [**21**] for

$$\frac{n-1-\sqrt{(2k+1)(2n-2k+1)}}{2} < x \le \frac{n}{2},$$

$$(88)\qquad |K_k^n(x))| \le 2\sqrt{\frac{A_k(x)\tau(n,k)}{h(x+1)}}$$

Outside the oscillatory region the following bounds are valid. Let $p = 2\sqrt{k(n-k)}$, $s = \sqrt{(n-2k)^2 - 4x(n-x)}$. Let $k < n/2$ and $x < \frac{n-p}{2}$, then for some $0 < \theta < 1$,

$$(89)\qquad \frac{K_k(x+1)}{K_k(x)} = \frac{n-2k+s}{2(n-x)} + \theta R(x)$$

where

$$R(x) = \frac{x(n-2x+p)}{s^3}.$$

This leads to the following bound ([**21**], see also [**17**, (19)] and [**16**, (3.20)] for alternative forms of the expression for the main term) For $k < n/2$ and $x < \frac{n-p-1}{2}$, integer,

$$\ln K_k(x+1) = \ln\binom{n}{k} +$$

$$n\ln\frac{3n-2k-2x-s}{2n}+k\ln\frac{n-2x+s}{2(n-k)}+x\ln\frac{n-2k+s}{2(n-x)}+\psi(x) \tag{90}$$

where

$$\ln\frac{n-2k+s}{2(n-x)}<\psi(x)<\frac{4nx(n-x)}{ps(n-2k)}.$$

The preceding results are based on the following Turan type inequality [**21**]:

$$U_k^n(x)=(K_k^n(x))^2-K_k^n(x-1)K_k^n(x+1)>0 \tag{91}$$

provided $k\le n/2$. The function $U_k^n(x)$ is intimately connected to the Christoffel-Darboux formula (41). Namely,

$$W_k^n(x,x)=\frac{U_{k+1}^n(x)-U_k^n(x)}{2}+\frac{2\binom{n}{k}}{k+1}\sum_{i=0}^{k}\frac{U_i^n(x)}{\binom{n}{i}} \tag{92}$$

and for $3\le k\le n/2$ [**20**],

$$W_k^n(x,x)\ge\max\left\{\frac{n-2k+5}{2(n-k+2)},\binom{n-3}{k-3}2^{3-n}\right\}U_{k+1}^n(x) \tag{93}$$

Notice that in a certain sense $(W_k^n(x,y))^2$ is the optimal nonnegative polynomial for the linear programming method (see [**26**] and [**41**]). For bounds on $W_k^n(x)$ and $U_k^n(x)$ see [**20, 26, 28**].

The strong asymptotics for values of Krawtchouk polynomials in the oscillatory region was obtained in [**16**]. Let

$$t=x/k \qquad \gamma=n/k,$$
$$p(z)=\ln z-t\ln(1+z)-(\gamma-t)\ln(1-z),$$
$$z_1=\frac{-\gamma+2t+\sqrt{(\gamma-2t)^2-4(\gamma-1)}}{2(\gamma-1)},$$
$$\alpha(t)=\arg\left(e^{-p(z_1)}\right),\qquad \beta(t)=\arg\left(z_1\sqrt{p''(z_1)}\right).$$

For $n\to\infty$ and $x\in\left(\frac{n}{2}-\sqrt{k(n-k)},\frac{n}{2}\right)$,

$$K_k^n(x)=\frac{1}{2^k\pi}e^{-k\mathcal{R}(p(z_1))}\sin(k\alpha(t)-\beta(t)+o(1))\times \tag{94}$$
$$\times\left|\frac{\sqrt{\pi}n^{-1/2}}{z_1\sqrt{2p''(z_1)}}+O\left(n^{-1}\right)\right|$$

References

[1] E. Bannai, *On perfect codes in the Hamming schemes $H(n,q)$ with q arbitrary*, J. Comb. Theory, Ser.A, 23, 1977, pp.52-67.

[2] E. Bannai, *Orthogonal polynomials in coding theory and algebraic combinatorics*, In book: Orthogonal Polynomials, P. Nevai (ed.), Kluwer, 1990, pp.25-53.

[3] E. Bannai and T. Ito, *Algebraic Combinatorics I: Association Schemes*, Benjamin/Cummings, Menlo Park, 1984.

[4] H. Bateman and A.Erdélyi, *Higher Transcendental Functions*, Vol.1, McGraw-Hill, 1953.

[5] M. Best, *A contribution to the nonexistence of perfect codes*, Ph.D. Thesis, University of Amsterdam, 1982.

[6] L. Chihara and D. Stanton, *Zeros of generalized Krawtchouk polynomials*, J. Approx. Theory, 60, 1, 1990, pp.43–57.

[7] P. Delsarte, *Four Fundamental Parameters of a Code*, MBLE Research Lab., Report R 184, January 1972, Information and Control, vol.23, 1973, pp.407-438.

[8] P. Diaconis and R. L. Graham, *The Radon transform on Z_2^k*, Pac. J. Math., 118, 1985, pp.323-345.
[9] P. D. Dragnev and E. B. Saff, *Constrained energy problems with applications to orthogonal polynomials of a discrete variable*, Journal d'Analyse Mathematique (Jerusalem), 72, 1997, pp. 223-259.
[10] P. D. Dragnev and E. B. Saff, *A problem in potential theory and zero asymptotics of Krawtchouk polynomials*, Journal of Approximation Theory, 102, 1999, pp 120-140.
[11] W. H. Foster and I. Krasikov, *Bounds for the extreme roots of orthogonal polynomials*, submitted.
[12] G. Gasper, *Positivity and special functions*, Theory and Applications of Special Functions, ed. by R. Askey, Academic Press, New York (1975), pp. 375–433.
[13] L. Habsieger, *Integral zeroes of Krawtchouk polynomials*, this issue.
[14] L. Habsieger and D. Stanton, *More zeros of Krawtchouk polynomials*, Graphs and Combinatorics, 2, 1993, pp.163-172.
[15] Y. Hong, *On the nonexistence of nontrivial perfect e-codes and tight 2e-designs in Hamming schemes $H(N,q)$ with $e \geq 3$ and $q \geq 3$*, Graphs and Combinatorics, 2, 1986, pp.145-164.
[16] M. Ismail and P. Simeonov, *Strong asymptotics for Krawtchouk polynomials*, J. Comput. Appl. Math., 100, 2, 1998, pp. 121–144.
[17] G. Kalai and N. Linial, *On the distance distribution of codes,* IEEE Trans. Inform. Theory, 41, 5, 1995, pp.1467–1472.
[18] S. Karlin and J. McGregor, *The Hahn polynomials, formulas and an application*, Scripta Math. 26, 1961, pp.33-46.
[19] T. Kasami, T. Fujiwara and S. Lin, *An approximation to the weight distribution of binary linear codes*, IEEE Trans. Inform. Theory, 31, 6, 1985, pp.769–780.
[20] I. Krasikov, *Bounds for the Christoffel-Darboux kernel of the binary Krawtchouk polynomials*, this issue.
[21] I. Krasikov, *Nonnegative quadratic forms and bounds on orthogonal polynomials*, submitted.
[22] I.Krasikov, and S.Litsyn, *On spectra of BCH codes*, IEEE Transactions on Information Theory, 41, 3, 1995, pp.786-788.
[23] I.Krasikov, and S.Litsyn, *On integral zeros of Krawtchouk polynomials*, Journal of Combinatorial Theory, Ser.A, 150, 1996, pp.441–447.
[24] I.Krasikov, and S.Litsyn, *Linear programming bounds for doubly-even self-dual codes*, IEEE Transactions on Information Theory, 43, 4, 1997, pp.1238–1244.
[25] I.Krasikov, and S.Litsyn, *Bounds on spectra of codes with known dual distance*, Designs, Codes and Cryptography, 13, 3, 1998, pp.285–298.
[26] I.Krasikov, and S.Litsyn, *On the distance distribution of duals of BCH codes*, IEEE Transactions on Information Theory, 45, 1, 1999, pp.247–250.
[27] I.Krasikov, and S.Litsyn, *An improved upper bound on the minimum distance of doubly-even self-dual codes*, IEEE Trans. Inform. Theory, 46, 1, 2000, pp.274–278.
[28] I.Krasikov, and S.Litsyn, *On the distance distribution of BCH codes and their duals*, Designs, Codes and Cryptography, to appear.
[29] I.Krasikov, and Y.Roditty, *Switching reconstruction and Diophantine equations*, JCT-B, 54, 1992, pp.189–195.
[30] I.Krasikov, and Y.Roditty, *More on vertex switching reconstruction*, JCT-B, 60, 1994, pp.40–55.
[31] M. Krawtchouk, *Sur une généralisation des polynomes d'Hermite*, Comptes Rendus, 189, 1929, pp.620–622.
[32] A.B.J. Kuijlaars, W. Van Assche, *The asymptotic zero distribution of orthogonal polynomials with varying recurrence coefficients*, J. Approx. Theory, 99, 1999, pp.167-197
[33] V. Levenshtein, *Krawtchouk polynomials and universal bounds for codes and designs in Hamming spaces*, IEEE Trans. Info. Theory, 41, 1995, pp.1303–1321.
[34] V. Levenshtein, *Bounds for packings of metric spaces and some their applications*, In: Problemy Kibernetiki, 40, Nauka, Moscow, 1983, pp.43-110 (in Russian).
[35] F. Levy-dit-Vehel and S.Litsyn, *Parameters of Goppa codes revisited*, IEEE Transactions on Information Theory, 43, 1997, pp.1811–1819.
[36] J. H. van Lint, *Introduction to Coding Theory*, Springer-Verlag, 1992.

[37] R. J. McEliece, E. R. Rodemich, H. C. Rumsey and L. R. Welch, *New upper bounds on the rate of a code via the Delsarte-MacWilliams inequalities*, IEEE Trans. Inform. Theory, 23, 1977, pp.157-166.

[38] F.J. MacWilliams and N.J.A. Sloane, *The Theory of Error-Correcting Codes*, New York: North-Holland, 1977.

[39] A. F. Nikiforov, S. K. Suslov and V. B. Uvarov, *Classical Orthogonal Polynomials of a Discrete Variable*, (Translated from the Russian) Springer-Verlag, Berlin, 1991.

[40] P. E. Ricci, *Improving the asymptotics for the greatest zeros of Hermite polynomials*, Computers, Math. Applic., 30, 3–6, 1995, pp.409–416.

[41] A. Samorodnitsky *On the optimum of Delsarte's linear program*, preprint, 1998.

[42] I. I. Sharapudinov, *Asymptotic properties of Krawtchouk polynomials*, Matematicheskie Zametki, 44, 5, 1988, pp.682–693; English translation: Mathematical Notes, pp.855-862.

[43] V. M. Sidel'nikov, *Weight spectrum of binary Bose-Chaudhuri-Hoquingham codes*, Probl. Peredachi Inform., 7, 1, 1971, pp.14–22.

[44] P. Solé, *A limit law on the distance distribution of binary codes*, IEEE Trans. Inform.Theory, 36, 1990, pp.229–232.

[45] G. Szegö, *Orthogonal Polynomials*, Amer. Math. Soc. Colloq. Publ., v.23, Providence, RI, 1975.

[46] H. C. A, van Tilborg, *Uniformly packed codes*, Ph. D. Thesis, Eindhoven, 1976.

DEPARTMENT OF MATHEMATICAL SCIENCES, BRUNEL UNIVERSITY, UXBRIDGE UB8 3PH UNITED KINGDOM

E-mail address: Ilia.Krasikov@brunel.ac.uk

EE-SYSTEMS, TEL AVIV UNIVERSITY, 69987 RAMAT AVIV, ISRAEL

E-mail address: litsyn@eng.tau.ac.il

DIMACS Series in Discrete Mathematics
and Theoretical Computer Science
Volume **56**, 2001

On an Algebraic Method for Bounding the Covering Radius

Tero Laihonen

ABSTRACT. A combinatorial parameter of a block code called the covering radius gives significant information on geometric properties of the code. In this paper a new bound on the covering radius is derived with a given dual distance. This bound improves on the known estimates on covering radius in a certain interval. The used method is also somewhat simpler than the approach leading to the best previously known bound in that range.

1. Introduction

One of the basic geometric parameters of a block code is the covering radius and it is defined for the code $C \subseteq F_2^n$ (here F_2^n is the n-dimensional binary Hamming space) by

$$R = \max_{x \in F_2^n} \min_{c \in C} d(x, c)$$

where d denotes the Hamming distance. If the covering radius of a code is smaller than the minimum distance, then the code is maximal. Other applications of this parameter include, for example, data compression, write-once memories, decoding [**3**, Chapter 1] and identifying codes [**11**].

The problem of determining the covering radius of a code has attracted, as covering problems in Hamming spaces generally, a great deal of attention lately [**3**]. It has turned out to be a very hard task to find the covering radius of a code.

As intuition suggests [**23**], the covering radius of a code can be estimated with the aid of its dual distance. For linear codes the dual distance is the minimum distance of the dual code. Its exact value or at least some bounds are often known from the construction of the code (see, e.g., BCH codes, Reed-Muller codes, etc). Numerous results concerning this relationship have already been established (see, e.g., [**5, 7, 18, 6, 9, 8, 4, 14, 12, 15, 17, 22, 21**]). Most of these results are, however, significant only for relatively large values of the normalized (with respect to the length) dual distance. The aim of this paper is to derive a better upper bound on covering radius than the known ones when the value of the normalized dual distance is not too large. There are only two known results in this situation (see [**22, 2**]). The bound given in the next theorem is proved in [**2**].

1991 *Mathematics Subject Classification.* Primary 94B75, 94B65; Secondary 05B15,05B40.
This work was supported by the Academy of Finland under grant #46186.

Let $H_2(x) = -x\log_2 x - (1-x)\log_2(1-x)$ be the binary entropy function and denote

$$\Phi(\tau,\eta) = H_2(\tau) + \int_0^\eta \log_2\left(\frac{1-2\tau+\sqrt{(1-2\tau)^2-4z(1-z)}}{2-2z}\right)dz.$$

THEOREM 1.1. *Let $(C_n)_{n=1}^\infty$ be a family of binary linear codes of length n, dual distance $d' = d'(n)$ and covering radius $R = R(n)$ where $\lim_{n\to\infty}(d'/n, R/n) = (\delta', \rho)$. Here $0 < \delta' < 1/2$. Let p and q be positive integers, $\hat{\imath}(\xi,\tau,l) = \max\{0,\xi-\tau\} + (\tau - \max\{0,\xi-\tau\})l/p$, $\alpha_i = \delta' + (\min\{1/2, 2\tau\} - \delta')i/q$ and choose β_i $(i = 1,...,q)$ in such a way that $(\alpha_i(1-\alpha_i) - \beta_i(1-\beta_i))/(1+2\sqrt{\beta_i(1-\beta_i)}) < \delta'/2$. Let further $0 < \beta_i < \alpha_{i-1}$ and $\beta_1 < 1/2(1-\sqrt{1-2\delta'})$. If $\delta'/2 < \tau < 1/2$ and*

$$\Phi(\tau, 1/2-\sqrt{\tau(1-\tau)}) > \max_{v=0,\dots,q-1}\{H_2(\beta_{v+1}) + \max_{l=0,..,p-1}\{\Phi(\hat{\imath}(\alpha_v,\tau,l),$$
$$1/2-\sqrt{\tau(1-\tau)}) - H_2(\hat{\imath}(\alpha_v,\tau,l)) + \alpha_v + (1-\alpha_v)H_2\left(\tfrac{\tau+\hat{\imath}(\alpha_v,\tau,l+1)-\alpha_v}{2(1-\alpha_v)}\right)\}\}$$

then

$$\rho \le 1/2 - \sqrt{\tau(1-\tau)}. \tag{1.1}$$

This theorem gives the best known bound when the normalized dual distance is not close to 1/2 — notice that [**18**] since $\rho = 0$ when $\delta' > 1/2$, we are only interested in the interval $\delta' \in [0, 1/2]$. The bound that is superior to the theorem above for δ' close to 1/2 can be found from [**12, 13**]. We remark that there exist [**23**] families of binary linear codes giving the *lower* bound $\rho \ge H_2^{-1}(1-H_2(\delta'))$ for all $0 < \delta' < 1/2$. When $0 < \delta' < 1/2$ there is, however, a gap between the known lower and upper estimates (see [**2**, Figure 1]).

In Theorem 3.10 we get a bound which improves on the estimate of the theorem above in a quite wide range of normalized dual distances. The approach of this paper is a refinement of the method discovered in [**2**] and the current method is also somewhat simpler. The improvement is based on a better estimate for the cardinality of a constant weight code (see, e.g., [**20, 13**]).

2. Preliminaries

Although the covering radius is a combinatorial parameter, we have a powerful algebraic method for estimating it. Let us briefly describe it. Let C be a code and assume that $c \in C$ and $x \in S_t = \{y \in F_2^n | w(y) = t\}$ where $w(\cdot)$ denotes the Hamming weight. We denote by $N(b,t)$ the number of solutions $(c,x) \in C \times S_t$ to the equation $c + x = b$. Fix a natural number R'. If for all $b \in F_2^n$ we have $\sum_{t=0}^n \alpha_t N(b,t) > 0$ where $\alpha_t \le 0$ for $t = R'+1,\dots,n$, then the covering radius is at most R'. It is easy to derive the following theorem from this argument (for details consult, e.g., [**3**, p. 230]). Denote by $M(n,d,w)$ the largest cardinality of a binary code of length n, minimum distance at least d and fixed weight w. Let $K_t(x)$ be the Krawtchouk polynomial of degree t, i.e.,

$$K_t(x) = \sum_{j=0}^t (-1)^j \binom{x}{j}\binom{n-x}{t-j}.$$

THEOREM 2.1. *Suppose that a binary code C is linear and denote by d' its dual distance. Let R' be a natural number and $F(x) = \sum_{j=0}^{n} F_j K_j(x)$ a polynomial such that*

$$F_0 > \sum_{j=d'}^{n} |F_j| M(n, d', j) \tag{2.1}$$

and for $i = R' + 1, \dots, n$

$$F(i) \leq 0. \tag{2.2}$$

Then the covering radius of C is at most R'.

Let $x_1^{(t)}$ denote the smallest zero of the Krawtchouk polynomial $K_t(x)$. We apply to the previous theorem the polynomial

$$f(x) = \frac{K_t(a)^2}{a - x} \left(K_{t+1}(x) + K_t(x)\right)^2.$$

where a $(= a_n = a_{n,t})$ is such that $x_1^{(t+1)} < a < x_1^{(t)}$ and $K_t(a) = -K_{t+1}(a)$. By [**19**] this choice is possible. This polynomial is originally designed for the linear programming method for error-correcting codes [**19**]. Since we choose later $R' = \lceil a_{n,t} \rceil$, we are interested in optimizing on the degree of $f(x)$ for a given dual distance in Theorem 2.1.

The (Krawtchouk) coefficients of the polynomial $f(x)$ needed in the theorem can easily be calculated [**19**].

PROPOSITION 2.2. *We can write the polynomial $f(x)$ defined above as $f(x) = \sum_{j=0}^{n} f_j K_j(x)$ where*

(2.3)

$$f_j = \frac{2K_t(a)}{t+1} \binom{n}{t} \sum_{i=0}^{t} \frac{K_i(a)}{\binom{n}{i}} \left\{ \binom{n-j}{\frac{t+1+i-j}{2}} \binom{j}{\frac{t+1-i+j}{2}} + \binom{n-j}{\frac{t+i-j}{2}} \binom{j}{\frac{t-i+j}{2}} \right\}.$$

Here a binomial coefficient with fractional or negative lower index is to be interpreted as zero.

3. A bound on covering radii of linear codes

Evidently, we can deduce from (2.3) that

$$f_0 = \frac{2K_t(a)^2}{t+1} \binom{n}{t}. \tag{3.1}$$

The value of $K_t(a)$ can be asymptotically estimated according to the lemma below. For the proof consult [**2**].

Recall that the function $\Phi(\tau, \eta)$ stems naturally from the fact [**10**] that

$$\frac{1}{n} \log_2 K_t(x) = \Phi(\tau, \eta) + o(1)$$

where $\lim_{n\to\infty}(t/n, x/n) = (\tau, \eta)$ and $0 \leq \eta < 1/2 - \sqrt{\tau(1-\tau)}$.

LEMMA 3.1. *Let $\lim_{n\to\infty} t/n = \tau$. For the sequence $(a_n)_{n=1}^{\infty}$ defined above*

$$\liminf_{n\to\infty} \frac{1}{n} \log_2 K_t(a_n) \geq \Phi(\tau, \psi)$$

where $\psi = \lim_{n\to\infty} a_n/n = \frac{1}{2} - \sqrt{\tau(1-\tau)}$.

This lemma is useful in evaluating the left-hand side of (2.1) when $F = f$. In order to deal with the right-hand side we concentrate now on upper bounds for f_j ($j \neq 0$). We assume that $t < (n-1)/2$. Observe that the natural requirement $\deg f(x) = 2t+1 \leq n$ is then fulfilled. Recalling [**3**, p. 33] that

$$\binom{n}{k} \leq 2^{nH_2(k/n)},$$

we get by Proposition 2.2

$$\begin{aligned} f_j &\leq \frac{2K_t(a)}{t+1}\binom{n}{t}\sum_{i=0}^{t}\frac{K_i(a)}{\binom{n}{i}}2^j\left\{\binom{n-j}{\frac{t+1+i-j}{2}}+\binom{n-j}{\frac{t+i-j}{2}}\right\} \\ &\leq \frac{2K_t(a)}{t+1}\binom{n}{t}\sum_{i=\max\{0,j-t-1\}}^{t}\frac{K_i(a)}{\binom{n}{i}}2^{j+(n-j)H_2\left(\frac{t+1+i-j}{2n-2j}\right)}. \end{aligned} \tag{3.2}$$

In [**2**] it is shown that we may estimate f_j asymptotically as follows. Recall that $\hat{\imath}(\xi,\tau,l) = \hat{\imath}(\xi,\tau,l,p) = \max\{0,\xi-\tau\} + (\tau - \max\{0,\xi-\tau\})l/p$.

LEMMA 3.2. *Let p be a positive integer. Let further $\lim_{n\to\infty} t/n = \tau$ and $\lim_{n\to\infty} j/n = \xi$. Assume that $\xi < 2\tau$. Then for sufficiently large n*

$$f_j \leq \frac{2K_t(a)}{t+1}\binom{n}{t}p\left\lceil\frac{t-\max\{0,j-t-1\}}{p}\right\rceil 2^{n(M(\xi,\tau)+\epsilon)} \tag{3.3}$$

where $\epsilon > 0$ is arbitrary and

$$\begin{aligned} M(\xi,\tau) &= \max_{l=0,\ldots,p-1}\left\{\Phi(\hat{\imath}(\xi,\tau,l),\frac{1}{2}-\sqrt{\tau(1-\tau)}) - H_2(\hat{\imath}(\xi,\tau,l))\right. \\ &\quad \left. +\xi + (1-\xi)H_2\left(\frac{\tau+\hat{\imath}(\xi,\tau,l+1)-\xi}{2(1-\xi)}\right)\right\}. \end{aligned}$$

3.1. Modifications of the first condition. Let us now return to the condition (2.1):

$$f_0 > \sum_{w=d'}^{n}|f_w|M(n,d',w).$$

By virtue of (3.2) and (3.1) this is fulfilled if

$$\begin{aligned} \frac{2K_t(a)^2}{t+1}\binom{n}{t} &> \sum_{w=d'}^{2t+1}M(n,d',w)\frac{2K_t(a)}{t+1}\binom{n}{t} \\ &\quad \cdot \sum_{i=\max\{0,w-t-1\}}^{t}\frac{K_i(a)}{\binom{n}{i}}2^{w+(n-w)H_2\left(\frac{t+1+i-w}{2n-2w}\right)}. \end{aligned} \tag{3.4}$$

Dividing both sides by $K_t(a)\binom{n}{t}\frac{2}{t+1}$ and denoting

$$\hat{f}(w) = \sum_{i=\max\{0,w-t-1\}}^{t}\frac{K_i(a)}{\binom{n}{i}}2^{w+(n-w)H_2\left(\frac{t+1+i-w}{2n-2w}\right)}$$

we obtain

$$K_t(a) > \sum_{w=d'}^{2t+1}M(n,d',w)\hat{f}(w) \tag{3.5}$$

Next we try to get a simpler condition to verity (2.1) for f. To that end we bound above the right-hand side of (3.5). Let us first give an obvious auxiliary lemma.

LEMMA 3.3. *Let $w \in \{d', ..., 2t+1\}$. When w grows, $\hat{f}(w)$ does not increase.*

In order to estimate $M(n, d', w)$ in (3.5) we utilize the following two results.

THEOREM 3.4 ([**16**]). *Let r and s be integers such that $r \equiv s \pmod 2$. Then if $0 \le r - s \le 2(n-w)$ and $0 \le r + s \le 2w$, it follows that*

$$M(n,d,w) \le \frac{\binom{n}{w-s}}{\binom{w}{(r+s)/2}\binom{n-w}{(r-s)/2}} M(n,d,r).$$

Denote the asymptotic rate function by

$$R(\delta,\omega) = \sup_{(d_n),(w_n)} \{\limsup_{n\to\infty} \frac{1}{n} \log_2 M(n, d_n, w_n)\}$$

where the supremum is taken over all sequences (d_n) and (w_n) with

$$\lim_{n\to\infty} (d_n/n, w_n/n) = (\delta, \omega).$$

THEOREM 3.5 ([**19**]). *We have*

$$R(\delta,\omega) \le \begin{cases} 0 & \text{if } 0 \le \omega \le \zeta^* \\ m(\delta,\omega) := H_2((1-\sqrt{1-u^2})/2) & \text{if } \zeta^* \le \omega \le 1/2 \end{cases} \tag{3.6}$$

where $\zeta^* = (1-\sqrt{1-2\delta})/2$ and $u = -\delta + (\delta^2 - 2\delta + 4\omega(1-\omega))^{1/2}$.

We choose the parameters r and s of Theorem 3.4 in the following way. Optimality of these choices is examined in [**16**] and [**20**]. We take

$$\begin{cases} r = \lfloor \eta n \rfloor, \\ \frac{(n-2w)r}{n-2r} - 2 < s \le \frac{(n-2w)r}{n-2r}, \\ s \equiv r \pmod 2 \end{cases} \tag{3.7}$$

where η is the point at which the function

$$\Lambda(x) := m(\delta, x) + 1 - H_2(x) \tag{3.8}$$

obtains its minimum.

In order to use Theorem 3.4 in (3.5) we have to make sure that the parameters r and s given above fulfill the requirements of the theorem for *every* weight $w = d', ..., 2t+1$.

LEMMA 3.6. *The parameter $s = s(w)$ chosen by (3.7) is not increasing as w grows.*

PROOF. Notice that r is fixed for a given n. Let s satisfy (3.7). If w grows, then the upper bound $(n-2w)r/(n-2r)$ becomes smaller. Since the parity of r is obviously the same for all $w = d', .., 2t+1$, s cannot increase with respect to w. □

Assume from now on that $\lim_{n\to\infty} d'/n = \delta'$ and $\lim_{n\to\infty} t/n = \tau$.

LEMMA 3.7. *Let $\eta < \delta' < 2\tau < 1/2$. Then for sufficiently large n the requirements of Theorem 3.4*

$$\begin{cases} 0 \le r+s \le 2w \\ 0 \le r-s \le 2(n-w) \end{cases}$$

are satisfied for all $w = d', ..., 2t+1$.

PROOF. Let $\lim_{n\to\infty} w/n = \omega$. Then

$$\lim_{n\to\infty} \frac{r+s}{n} = \frac{2\eta(1-\eta-\omega)}{1-2\eta} =: B_1(\omega)$$

and

$$\lim_{n\to\infty} \frac{r-s}{n} = \frac{2\eta(\omega-\eta)}{1-2\eta} =: B_2(\omega).$$

Recall that $\lim_{n\to\infty}(d'/n, (2t+1)/n) = (\delta', 2\tau)$. One can check that $B_1(2\tau) > 0$, $B_1(\delta') < 2\delta'$, $B_2(\delta') > 0$ and $B_2(2\tau) < 2(1-2\tau)$. Therefore, by the previous lemma we can conclude the claim for sufficiently large n. □

3.2. Applying the bound on $M(n, d', w)$ to the condition. According to the preceding result we may rewrite (3.5) for sufficiently large n in the form

$$K_t(a) > M(n,d',r) \sum_{w=d'}^{2t+1} \frac{\binom{w}{w-s}}{\binom{w}{(r+s)/2}\binom{n-w}{(r-s)/2}} \hat{f}(w). \tag{3.9}$$

In handling the binomial coefficients in the denominator it is convenient to rely on the next estimate [**3**, p. 33]: Suppose that $0 < \lambda < 1$ and λn is an integer, then

$$\frac{2^{nH_2(\lambda)}}{\sqrt{8n\lambda(1-\lambda)}} \le \binom{n}{\lambda n}.$$

Since $\lambda(1-\lambda)$ achieves its maximum at $\lambda = 1/2$, we get

$$\frac{1}{\sqrt{2n}} 2^{nH_2(\lambda)} \le \binom{n}{\lambda n}.$$

Therefore we can bound above the right-hand side of (3.9) by

$$2nM(n,d',r) \sum_{w=d'}^{2t+1} \frac{\binom{n}{w-s}\hat{f}(w)}{2^{wH_2\left(\frac{r+s}{2w}\right)+(n-w)H_2\left(\frac{r-s}{2(n-w)}\right)}}.$$

Let us denote

$$y(w) = \frac{r + \frac{(n-2w)r}{n-2r}}{2w} = \frac{r(n-r-w)}{w(n-2r)}$$

and

$$z(w) = \frac{r - \frac{(n-2w)r}{n-2r}}{2(n-w)} = \frac{r(w-r)}{(n-w)(n-2r)}.$$

LEMMA 3.8. *Let $w = d', \ldots, 2t+1$. Suppose that $\eta(1-\eta-2\tau) > \tau(1-2\eta)$ and $2\eta(2\tau-\eta) < (1-2\tau)(1-2\eta)$. Then we have for sufficiently large n that*

$$H_2\left(\frac{r+s}{2w}\right) \ge H_2(y(w))$$

and

$$H_2\left(\frac{r-s}{2(n-w)}\right) \ge H_2(z(w)).$$

PROOF. Because $(r+s)/(2w)$ achieves its minimum at $w = 2t+1$ and $\eta(1-\eta-2\tau) > \tau(1-2\eta)$, we know that $(r+s)/(2w) > 1/2$ for all $w = d', ..., 2t+1$ when n is sufficiently large. Therefore, by (3.7)

$$H_2\left(\frac{r+s}{2w}\right) \geq H_2(y(w))$$

and so the first part of the lemma is verified. The second part goes in the same way using the assumption $2\eta(2\tau-\eta) < (1-2\tau)(1-2\eta)$. □

Essentially the same result can be proved without the constraints, but this special case is enough for our purposes and it also simplifies the proof.

Consequently, we obtain the following upper estimate for the right-hand side of (3.9) for sufficiently large n

$$2nM(n,d',r)\sum_{w=d'}^{2t+1}\frac{\binom{n}{w-s}\hat{f}(w)}{2^{n\left(\frac{w}{n}H_2(y(w))+\frac{n-w}{n}H_2(z(w))\right)}}.$$

The next lemma is easily verified.

LEMMA 3.9. *Let $w = d', ..., 2t+1$ and $2t+1 < n/2$. The function*

$$c(w) = \frac{w}{n}H_2(y(w)) + \frac{n-w}{n}H_2(z(w))$$

increases with respect to w.

Now we are ready to state the theorem which was our goal in this paper. Let us denote $\alpha_i := \delta' + (2\tau-\delta')i/q$ $(q \in \mathbb{N})$ and

$$\begin{aligned} L_q(\delta',\tau) \quad := \max_{v=0,\ldots,q-1}&\left\{H_2\left(\alpha_{v+1}-\frac{(1-2\alpha_{v+1})\eta}{1-2\eta}\right)-\alpha_v H_2\left(\frac{\eta(1-\eta-\alpha_v)}{\alpha_v(1-2\eta)}\right)\right. \\ &\left. -(1-\alpha_v)H_2\left(\frac{\eta(\alpha_v-\eta)}{(1-\alpha_v)(1-2\eta)}\right)+M(\alpha_v,\tau)\right\}. \end{aligned}$$

THEOREM 3.10. *Let $(C_n)_{n=1}^{\infty}$ be a sequence of binary linear codes of length n, dual distance d' and covering radius R where $\lim_{n\to\infty}(R/n, d'/n) = (\rho,\delta')$ and $q \in \mathbb{N}$. Choose η $(\eta < \delta')$ from 3.8 and τ such that $\delta' < 2\tau < 1/2$, $\eta(1-\eta-2\tau) > \tau(1-2\eta)$ and $2\eta(2\tau-\eta) < (1-2\tau)(1-2\eta)$. If*

$$\Phi(\tau, 1/2-\sqrt{\tau(1-\tau)}) > m(\delta',\eta) + L_q(\delta',\tau),$$

then

$$\rho \leq 1/2 - \sqrt{\tau(1-\tau)}.$$

PROOF. Let us divide the interval $[d', 2t+1]$ of the sum in (3.9) into q pieces, so that we have the sub-intervals $[w_{i-1}, w_i]$ $(i = 1, ..., q)$ where

$$w_i = \lfloor d' + \frac{2t+1-d'}{q}i\rfloor \quad (i = 0, ..., q).$$

Thus $\alpha_i = \lim_{n\to\infty} w_i/n$.

Since $\binom{n}{w-s}$ is increasing on w (by Lemma 3.6), we obtain using Lemmas 3.3 and 3.9 that the condition (3.9) can be rewritten in a stronger form (provided that n is sufficiently large)

$$K_t(a) > 2n^2M(n,d',r)\sum_{v=0}^{q-1}\frac{\binom{n}{w_{v+1}-s(w_{v+1})}\hat{f}(w_v)}{2^{n\left(\frac{w_v}{n}H_2(y(w_v))+\frac{n-w_v}{n}H_2(z(w_v))\right)}}.$$

Thus for sufficiently large n the condition is fulfilled (by Lemma 3.2) if

$$K_t(a) > M(n, d', r)2^{n(L_q(\delta', \tau)+\epsilon)}. \tag{3.10}$$

Pick then $R' = \lceil a \rceil$ in the condition (2.2). Combining (3.10) with Theorem 3.5 and Lemma 3.1 we get that for sufficiently large n the requirement (2.1) is satisfied if

$$\Phi(\tau, 1/2 - \sqrt{\tau(1-\tau)}) > m(\delta', \eta) + L_q(\delta', \tau).$$

Thus the assertion follows from Theorem 2.1. □

Observe that the theorem above is a general result for linear binary codes with no additional constraints on the codes. The theorem does not need the β_i's, contrary to (1.1). Although it contains several conditions, the theorem is easy to use — we just fix some δ' and increase τ as long as possible without violating the constraints.

This result improves on the best known bound (1.1) on the interval $\delta' \in (0.04, 0.20)$ (recall [**18**] that for $\delta' \geq 0.5$ we have $\rho = 0$). For instance, at $\delta' = 0.12$ the bound (1.1) yields $\rho \leq 0.252$ and Theorem 3.10 gives $\rho \leq 0.249$. For the values $\delta' \leq 0.04$ we get the same bound as Tietäväinen's bound [**22**] and (1.1).

As mentioned by a referee, it is worth noticing [**1**] that $\Phi(\tau, 1/2 - \sqrt{\tau(1-\tau)}) = 1/2(H_2(\tau) - H_2(1/2 - \sqrt{\tau(1-\tau)}) + 1)$.

The bound of Theorem 3.10 is especially designed for small values of the normalized dual distance (this is due to the fact that the improvement on the estimate of the cardinality of a constant weight code [**20, 13**] is valid only for small values of the normalized distance), so for other values one should rely on (1.1) and the result of [**12**].

Acknowledgements

The author wishes to thank A. Ashikhmin, I. Honkala, S. Litsyn and A. Tietäväinen for inspiring discussions. The author is also grateful to the anonymous referee for comments and suggestions which helped to improve the paper.

References

[1] A. Ashikhmin and A. Barg, *Binomial moments of the distance distribution: bounds and applications,* IEEE Trans. Inform. Theory **45** (1999), 438–452.

[2] A. Ashikhmin, I. Honkala, T. Laihonen and S. Litsyn, *On relations between covering radius and dual distance,* IEEE Trans. Inform. Theory **45** (1999), 1808–1816.

[3] G. Cohen, I. Honkala, S. Litsyn and A. Lobstein, *Covering Codes*, Amsterdam: Elsevier, 1997.

[4] C. Delorme and P. Solé, *Diameter, covering index, covering radius and eigenvalues*, European J. Combin. **12** (1991), 95–108.

[5] P. Delsarte, *Four fundamental parameters of a code and their combinatorial significance,* Information and Control **23** (1973), 407–438.

[6] G. Fazekas and V. I. Levenshtein, *On upper bounds for code distance and covering radius of designs in polynomial metric spaces*, J. Combin. Theory Ser. A **70** (1995), 267–288.

[7] T. Helleseth, T. Kløve, J. Mykkeltveit, *On the covering radius of binary codes,* IEEE Trans. Inform. Theory **24** (1978), 627–628.

[8] I. Honkala, T. Laihonen and S. Litsyn, *On covering radius and discrete Chebyshev polynomials*, Applicable Algebra in Engineering, Communication and Computing **8** (1997), 395–402.

[9] I. Honkala, S. Litsyn and A. Tietäväinen, *On algebraic methods in covering radius problems,* In: Applied Algebra, Algebraic Algorithms and Error-Correcting Codes, Lecture Notes in Comput. Sci., 948, eds., G. Cohen, M. Giusti and T. Mora, Berlin, Heidelberg, New York: Springer (1995), pp. 21–32.

[10] G. Kalai and N. Linial, *On the distance distribution of codes,* IEEE Trans. Inform. Theory **41** (1995), 1467–1472.

[11] M. G. Karpovsky, K. Chakrabarty and L. B. Levitin, *On a new class of codes for identifying vertices in graphs,* IEEE Trans. Inform. Theory **44** (1998), 599–611.

[12] T. Laihonen, *Covering radius of self-complementary codes and BCH codes*, Proc. of the IEEE International Symposium on Information Theory (1998), p. 434.

[13] T. Laihonen, *Estimates on the Covering Radius When the Dual Distance is Known*, doctoral dissertation, Turku Centre for Computer Science TUCS Dissertations, no. 11, University of Turku, Finland, 1998.

[14] T. Laihonen and S. Litsyn, *On upper bounds for minimum distance and covering radius of non-binary codes*, Designs, Codes and Cryptography **14** (1998), 71–80.

[15] T. Laihonen and S. Litsyn, *New bounds on covering radius as a function of dual distance*, SIAM Journal on Discrete Mathematics **12** (1999), 243–251.

[16] V. I. Levenshtein, *Upper-bound estimates for fixed-weight codes*, Problems of Info. Trans., **7** (1974), 281–287, (in Russian: Problemy Peredachi Informatsii, **7** (1971)).

[17] S. Litsyn, P. Solé and R. Struik, *On the covering radius of unrestricted code as a function of the rate and dual distance,* Discrete Appl. Math. **82** (1998), 177–191.

[18] S. Litsyn and A. Tietäväinen, *Upper bounds on the covering radius of a code with a given dual distance,* European J. Combin. **17** (1996), 265–270.

[19] R. J. McEliece, E. R. Rodemich, H. C. Rumsey and L. R. Welch, *New upper bounds on the rate of the code via the Delsarte-MacWilliams inequalities*, IEEE Trans. Inform. Theory **23** (1977), 157–166.

[20] A. Samorodnitsky, *On the optimum of Delsarte's linear program*, preprint.

[21] P. Solé and P. Stokes, *Covering radius, codimension, and dual-distance width,* IEEE Trans. Inform. Theory **39** (1993), 1195–1203.

[22] A. Tietäväinen, *An upper bound on the covering radius as a function of the dual distance,* IEEE Trans. Inform. Theory **36** (1990), 1472–1474.

[23] A. Tietäväinen, *Covering radius and dual distance,* Designs, Codes and Cryptography **1** (1991), 31-46.

Department of Mathematics and Turku Centre for Computer Science TUCS, University of Turku, FIN-20014 Turku, Finland

E-mail address: terolai@utu.fi

DIMACS Series in Discrete Mathematics
and Theoretical Computer Science
Volume **56**, 2001

Design systems: combinatorial characterizations of Delsarte $\mathcal{T}$-designs via partially ordered sets

William J. Martin

ABSTRACT. A number of important design-theoretic structures, such as combinatorial block designs and orthogonal arrays, can be characterized as Delsarte $\mathcal{T}$-designs in cometric association schemes. Several recent papers extend this theory to more exotic types of designs; here, the relevant association schemes fail to be cometric. In all cases, however, the language of partially ordered sets can be used to describe the designs and to establish the connection to association schemes. In this paper, we introduce "design systems" as a tool for describing all of the known examples in a uniform fashion. A design system consists of an association scheme with a partial order on its eigenspaces together with a second partially ordered set having the vertices of the scheme as its maximal elements. The key axiom for these systems ties the incidence matrices of this partial order to the eigenspaces of the corresponding association scheme.

The results of the paper are as follows. First, given a design system, we prove an equivalence between certain families of Delsarte $\mathcal{T}$-designs and "designs" in the attached partially ordered set. Next, we give several ways of obtaining new design systems from known ones. Two general bounds, one on the size of a $\mathcal{T}$-design and the other on the degree of a $\mathcal{T}$-design, are established. A tight design in a design system is shown to induce a new design system of its own. We conclude by exploring some of the examples, including ordered orthogonal arrays and λ-transitive sets of permutations.

1. Association schemes

We refer the reader to [**2**, Chapter 2] for background material on association schemes. We essentially follow the notation of [**2**], but also suggest [**4**] and [**7**] for background material on codes and designs. A *(symmetric) association scheme* $(X, \mathcal{A})$ consists of a finite set X, of size N say, together with a collection $\mathcal{A} = \{A_i : i \in \mathcal{I}\}$ of symmetric 01-matrices each having rows and columns indexed by X and satisfying the following three conditions:

- there is a distinguished element $0 \in \mathcal{I}$ such that $A_0 = I$;
- $\sum_i A_i = J$, the matrix of all ones;
- for any choice of indices i and j from $\mathcal{I}$, the product $A_i A_j$ lies in the linear span of the matrices in $\mathcal{A}$.

2000 *Mathematics Subject Classification.* Primary 05E30, 05B30; Secondary 06A11.

If $|\mathcal{I}| = d+1$, we say that $(X, \mathcal{A})$ has *d classes*. The matrices A_i will be referred to as *Schur idempotents* of the scheme. We sometimes view each A_i as the adjacency matrix of a symmetric binary relation R_i on X and, for $x, y \in X$, we say x is *i-related* to y to indicate that the (x, y) entry of A_i is equal to one.

The *Bose-Mesner algebra* of $(X, \mathcal{A})$ is the linear span of $\mathcal{A}$ and is denoted by $\mathbb{A}$. The primitive idempotents for this algebra (under ordinary matrix multiplication) are denoted $\{E_j : j \in \mathcal{J}\}$ for some index set $\mathcal{J}$ of size $d+1$ with the convention that $\mathcal{J}$ contains a distinguished element 0 satisfying $E_0 = \frac{1}{N}J$. Schur multiplication is denoted $\circ$ and the *Krein parameters* of the scheme, q_{ij}^k $(0 \leq i, j, k \leq d)$, are defined by the equations

$$E_i \circ E_j = \frac{1}{N} \sum_{k \in \mathcal{J}} q_{ij}^k E_k.$$

The *Krein conditions* state that $q_{ij}^k \geq 0$ and, in several important classes of association schemes, many Krein parameters vanish.

The change-of-basis matrices P and Q are defined by the equations

$$A_i = \sum_{j \in \mathcal{J}} P_{ji} E_j \qquad (i \in \mathcal{I})$$

and

$$E_j = \frac{1}{N} \sum_{i \in \mathcal{I}} Q_{ij} A_i \qquad (j \in \mathcal{J}).$$

We will refer to P and Q as the *first* and *second eigenmatrix* of the scheme, respectively.

Let $(X, \mathcal{A})$ be an association scheme and let $\mathcal{T} \subseteq \mathcal{J}$. Consider a subset $D \subseteq X$ of the vertices of the scheme. If $\chi = \chi_D$ denotes the characteristic vector of D as a subset of X, then D is a *Delsarte $\mathcal{T}$-design* in $(X, \mathcal{A})$ provided $E_j \chi = 0$ for all $j \in \mathcal{T} - \{0\}$. Alternatively, this may be viewed geometrically. The *standard module* $\mathbb{R}^N$ decomposes into $d+1$ eigenspaces V_j $(j \in \mathcal{J})$ where V_j is the column space of E_j. A subset D is a Delsarte $\mathcal{T}$-design precisely when its characteristic vector is orthogonal to all eigenspaces V_j $(j \in \mathcal{T})$ except, of course, V_0 which is the span of the all-ones vector.

2. Design systems

The definition we are about to give is motivated by Delsarte's definition of a regular semilattice, though the axioms here are much weaker than those found in [**5**]. In [**5**], we find $\mathcal{J} = \{0, \ldots, d\}$ linearly ordered and a graded semilattice $(\mathcal{P}, \preceq)$ with X as its set of maximal elements. The rank function maps elements of $\mathcal{P}$ to eigenspaces in such a way that the incidence matrix W_j of X versus elements of rank j is associated to the jth eigenspace in the following way: $W_j W_j^T$ is a linear combination of $E_0, \ldots, E_j$ with this last coefficient non-zero. Examples include the Hamming and Johnson schemes as well as their q-analogues.

A design system consists of an association scheme and two partial orders related to it. On the one hand, we suppose there is a natural partial order on the eigenspace indices $\mathcal{J}$ of the association scheme. The larger partially ordered set ("poset"), $(\mathcal{P}, \preceq)$, has the vertices of the scheme as its maximal elements. The remaining elements are ostensible in general, but each corresponds to a particular

Roos antidesign [**18**] (see also [**16**]) and therefore these are often intrinsic to the association scheme. We hypothesize a generalized height function from the first poset to the second. This function associates an eigenspace index $j \in \mathcal{J}$ to each element of the larger poset in an order-preserving fashion. For each $j \in \mathcal{J}$, we require the incidence relation between X and the pre-image of j to relate to the corresponding eigenspace V_j.

We now give the precise definition. Assume the existence of the following objects.

1. A symmetric association scheme $(X, \mathcal{A})$ with primitive idempotents $\{E_j : j \in \mathcal{J}\}$;
2. A partially ordered set $(\mathcal{P}, \preceq)$ with $X \subseteq \mathcal{P}$;
3. A partial order $\trianglelefteq$ on the index set $\mathcal{J}$ for the primitive idempotents of the scheme;
4. An order-preserving surjection

$$\mathsf{eig} : (\mathcal{P}, \preceq) \to (\mathcal{J}, \trianglelefteq).$$

In our definition of a design system, we will impose three conditions on the ordered four-tuple

$$\Delta = (X, \mathcal{A};\ \mathcal{P}, \preceq;\ \mathcal{J}, \trianglelefteq;\ \mathsf{eig}).$$

For $z \in \mathcal{P}$, define $\phi(z) = \{x \in X : z \preceq x\}$. Observe that, for any $w, z \in \mathcal{P}$,

$$w \preceq z \qquad \Rightarrow \qquad \phi(z) \subseteq \phi(w).$$

Condition (i): For all $z \in \mathcal{P}$, we require that $\phi(z)$ uniquely determines z.

For each $j \in \mathcal{J}$, we consider the jth *fibre*, $\mathcal{P}^j = \mathsf{eig}^{-1}(j)$, of $(\mathcal{P}, \preceq)$. Thus $\mathcal{P}$ can be expressed as a disjoint union of fibres. Our second condition is

Condition (ii): For each $j \in \mathcal{J}$, there is an integer ρ_j such that each $z \in \mathcal{P}^j$ is dominated by exactly ρ_j elements of X:

$$|\phi(z)| = \rho_j, \qquad (z \in \mathcal{P}^j).$$

(The next condition will ensure that each ρ_j is strictly positive.)

For $j \in \mathcal{J}$, define the *incidence matrix*, W_j, to be the 01-matrix with rows indexed by elements of X and columns indexed by elements of $\mathcal{P}^j$ having

$$(W_j)_{x,z} = \begin{cases} 1, & \text{if } z \preceq x; \\ 0, & \text{otherwise.} \end{cases}$$

Condition *(ii)* is equivalent to the requirement that each W_j has constant column sum.

Condition (iii): We assume, for each j,

$$V_j \subseteq \mathsf{colsp} W_j \subseteq \bigoplus_{k \trianglelefteq j} V_k.$$

If conditions *(i)*, *(ii)*, and *(iii)* above are satisfied, we call the ordered quadruple

$$\Delta = (X, \mathcal{A};\ \mathcal{P}, \preceq;\ \mathcal{J}, \trianglelefteq;\ \mathsf{eig})$$

a *design system.* The following lemma gives some elementary properties of design systems.

LEMMA 2.1. *Let* $(X, \mathcal{A}; \mathcal{P}, \preceq; \mathcal{J}, \trianglelefteq; \mathsf{eig})$ *be a design system. Then:*

(a): *each fibre* $\mathcal{P}^j$ *is an antichain in* $(\mathcal{P}, \preceq)$*;*
(b): *the set of maximal elements of* $(\mathcal{P}, \preceq)$ *is equal to* X*;*
(c): *the poset* $(\mathcal{J}, \trianglelefteq)$ *has a unique maximal element,* 1 *say. Moreover,* $\mathcal{P}^1 = X$*;*
(d): *the poset* $(\mathcal{J}, \trianglelefteq)$ *has* 0 *as its unique minimal element;*
(e): *the poset* $(\mathcal{P}, \preceq)$ *has a unique minimal element,* $\mathbf{0}$ *say, and* $\mathcal{P}^0 = \{\mathbf{0}\}$*;*
(f): *the association scheme* $(X, \mathcal{A})$ *is rational; i.e., the first eigenmatrix* P *is an integer matrix.*

PROOF. **(a):** Suppose $\mathsf{eig}(z) = \mathsf{eig}(w) = j$ with $z \preceq w$. Then the column of W_j indexed by w is dominated entry-wise by the column indexed by z. On the other hand, W_j has constant column sum (condition *(ii)*), so the two columns must be equal. Now condition *(i)* gives $w = z$.

(b): If $z \in \mathcal{P} - X$ is maximal with $\mathsf{eig}(z) = j$, then the column of W_j indexed by z consists of all zeros. Since W_j has constant column sum, it must therefore be a zero matrix. But this contradicts *(iii)* which stipulates that the column space of W_j contain V_j. Now let y be a maximal element in $\mathcal{P}$. Then $y \in X$. Let $j = \mathsf{eig}(y)$. By *(ii)*, W_j has constant column sum equal to one. We know that $V_j \subseteq \mathsf{colsp} W_j$ and that, for each $x \in X$, the elementary basis vector $\mathbf{e}_x$ is not orthogonal to V_j, hence $\mathbf{e}_x \not\perp \mathsf{colsp} W_j$. So W_j cannot have any zero row. Thus $\mathcal{P}^j = X$ and each $x \in X$ is maximal.

(c): Define $1 \in \mathcal{J}$ by $\mathcal{P}^1 = X$. Since eig is an order-preserving map, this is the unique maximal element of $(\mathcal{J}, \trianglelefteq)$.

(d): Let $j \in \mathcal{J}$. As W_j is a non-zero 01-matrix, its column space is not orthogonal to V_0. Hence $0 \trianglelefteq j$ by condition *(iii)*.

(e): From **(d)**, $j \trianglelefteq 0$ if and only if $j = 0$. Hence, by *(iii)*, it follows that W_0 has column space V_0, so it is an all-ones matrix. Therefore, if $\mathsf{eig}(z) = 0$, then $z \preceq x$ for all $x \in X$. By condition *(i)*, this element z is unique.

(f): For each j, we trivially have a rational basis for $\mathsf{colsp} W_j$. Of course, V_0 admits a rational basis. Now fix $j \in \mathcal{J}$ and assume by way of induction that, for each $i \lhd j$, V_i admits a rational basis. Now it is clear that

$$V_j = \mathsf{colsp} W_j \cap \left(\bigoplus_{i \lhd j} V_i \right)^{\perp}$$

admits a rational basis as well. Now if $\theta = P_{ji}$, we have $A_i x = \theta x$ for $x \in V_j$. As A_i is rational and x can be taken to be rational and non-zero, we see that θx is rational, so θ is a rational integer. □

REMARK 2.2. We observe that, with 1 denoting the unique maximal element of $\mathcal{J}$, W_1 is an identity matrix and its column space is $\mathbb{R}^N$, the direct sum of all $d+1$ eigenspaces.

Let $(\mathcal{Q}, \leq)$ be a partially ordered set. A *lower ideal* in $(\mathcal{Q}, \leq)$ is a subset $\mathcal{T} \subseteq \mathcal{Q}$ having the property: if $j \in \mathcal{T}$ and $i \leq j$, then $i \in \mathcal{T}$.

Consider a design system $\Delta = (X, \mathcal{A}; \mathcal{P}, \preceq; \mathcal{J}, \trianglelefteq; \mathsf{eig})$. Suppose $\mathcal{T} \subseteq \mathcal{J}$. For $D \subseteq X$, we say that D is a *poset* $\mathcal{T}$*-design* in Δ if, for each $j \in \mathcal{T}$, there exists a

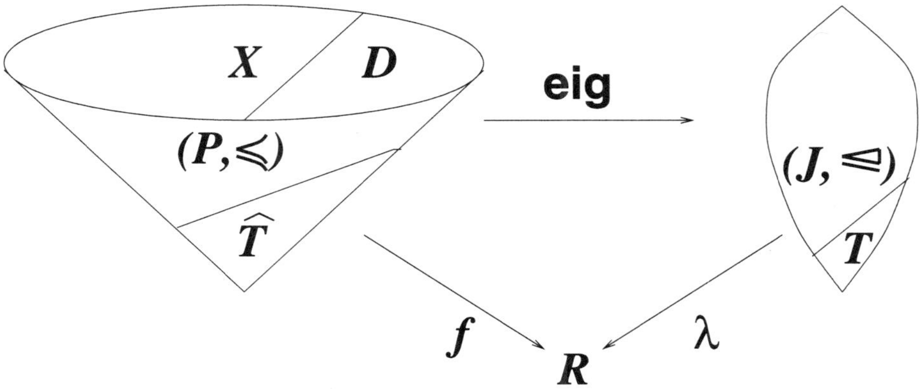

FIGURE 1. Subset D is a poset $\mathcal{T}$-design if and only if $f|_{\hat{\mathcal{T}}} = \lambda \circ \mathsf{eig}$ for some function λ.

constant λ_j such that

$$|\{x \in D : z \preceq x\}| = \lambda_j$$

for all $z \in \mathcal{P}^j$. If χ denotes the characteristic vector of D as a subset of X, then, for a fixed j, this is equivalent to the condition that $\chi^T W_j = \lambda_j \mathbf{1}$ where $\mathbf{1}$ denotes the all-ones vector of appropriate dimension.

Alternatively, we may view any subset $D \subseteq X$ as a function on $\mathcal{P}$ as follows: for $z \in \mathcal{P}$, define

$$f(z) = |\{x \in D : z \preceq x\}|.$$

(This definition extends to weighted designs in the obvious way.) Let $\widehat{\mathcal{T}} = \mathsf{eig}^{-1}(\mathcal{T})$ and let $f|_{\hat{\mathcal{T}}}$ denote the restriction of f to $\widehat{\mathcal{T}}$. Then D is a poset $\mathcal{T}$-design if and only if there exists a function $\lambda : \mathcal{T} \to \mathcal{R}$ such that

$$f|_{\hat{\mathcal{T}}} = \lambda \circ \mathsf{eig}.$$

This equation is depicted in Figure 1. Condition *(ii)* guarantees that X itself is a poset $\mathcal{J}$-design in the system. The only other $\mathcal{J}$-design is the empty set.

We are now prepared to present the first main result.

THEOREM 2.3. *Let $\Delta = (X, \mathcal{A};\ \mathcal{P}, \preceq;\ \mathcal{J}, \trianglelefteq;\ \mathsf{eig})$ be a design system, let $\mathcal{T} \subseteq \mathcal{J}$ and let $D \subseteq X$.*

(a): *If D is a poset $\mathcal{T}$-design in Δ, then D is a Delsarte $\mathcal{T}$-design in $(X, \mathcal{A})$;*
(b): *if $\mathcal{T}$ is a lower ideal in $(\mathcal{J}, \trianglelefteq)$ and D is a Delsarte $\mathcal{T}$-design in $(X, \mathcal{A})$, then D is a poset $\mathcal{T}$-design in Δ.*

PROOF. **(a)** Suppose D is a poset $\mathcal{T}$-design in Δ. Then there exist integers $\{\lambda_j : j \in \mathcal{T}\}$ such that $\chi^T W_j = \lambda_j \mathbf{1}^T$ for all $j \in \mathcal{T}$. Let $j \in \mathcal{T} - \{0\}$. Let $\mathbf{e}$ be any column of E_j. By condition *(iii)* of the definition, there exists a vector $\mathbf{r}$ such that $\mathbf{e} = W_j \mathbf{r}$. We have $\mathbf{1}^T W_j = \rho_j \mathbf{1}^T$. Since $j \neq 0$,

$$0 = \langle \mathbf{1}, \mathbf{e} \rangle = \mathbf{1}^T W_j \mathbf{r} = \rho_j \mathbf{1}^T \mathbf{r}.$$

Thus we find

$$\langle \chi, \mathbf{e} \rangle = \chi^T W_j \mathbf{r} = \lambda_j \mathbf{1}^T \mathbf{r} = 0.$$

This shows that $E_j \chi = 0$ for all non-zero $j \in \mathcal{T}$.

We will use a result of Roos to prove **(b)**. Let D and A be subsets of X with characteristic vectors χ and ψ, respectively. If for all $j \in \mathcal{J} - \{0\}$, either $E_j\chi = 0$ or $E_j\psi = 0$, then Corollary 3.3 in **[18]** tells us that

$$|A \cap D| = \frac{|A| \cdot |D|}{|X|}. \tag{2.1}$$

(If D is a $\mathcal{T}$-design in $(X, \mathcal{A})$, then such a set A is called a $\mathcal{T}$*-antidesign.*)

Suppose now that D is a Delsarte $\mathcal{T}$-design in $(X, \mathcal{A})$ and that $\mathcal{T}$ is a lower ideal in $(\mathcal{J}, \trianglelefteq)$. Condition *(iii)* of the definition implies that, for each $z \in \widehat{\mathcal{T}}$, $\phi(z)$ is a $\mathcal{T}$-antidesign. Thus, by Roos's result and condition *(ii)*, we quickly see that D is a poset $\mathcal{T}$-design in Δ. □

REMARK 2.4. The above proof gives us the following number-theoretic restrictions for the existence of a poset $\mathcal{T}$-design D. Using (2.1), we see that N divides $\rho_j|D|$ for all $j \in \mathcal{T}$. Since $\lambda_j = \rho_j|D|/N$ for $j \in \mathcal{T}$, we can also write $\lambda_k = \frac{\rho_k}{\rho_j}\lambda_j$ for $j, k \in \mathcal{T}$.

3. Products and extensions of design systems

The first theorem in this section generalizes Theorem 2.3 in **[10]**. Let

$$\Delta = (X, \mathcal{A}; \mathcal{P}, \preceq; \mathcal{I}, \trianglelefteq; \mathsf{eig})$$

and

$$\Xi = (Y, \mathcal{B}; \mathcal{Q}, \preceq'; \mathcal{J}, \trianglelefteq'; \mathsf{eig}')$$

be two design systems. The *product* of these two systems is the ordered quadruple

$$\Delta \otimes \Xi = (X \times Y, \mathcal{A} \otimes \mathcal{B}; \mathcal{P} \times \mathcal{Q}, \preceq; \mathcal{I} \times \mathcal{J}, \trianglelefteq; \mathsf{Eig})$$

where

- $(X \times Y, \mathcal{A} \otimes \mathcal{B})$ is the product association scheme with Schur idempotents
$$\{A_h \otimes B_i : A_h \in \mathcal{A},\ B_i \in \mathcal{B}\};$$
- $(\mathcal{P} \times \mathcal{Q}, \preceq)$ is the product of posets $(\mathcal{P}, \preceq)$ and $(\mathcal{Q}, \preceq')$:
$$(p_1, q_1) \preceq (p_2, q_2) \qquad \text{iff} \qquad (p_1 \preceq p_2 \text{ and } q_1 \preceq' q_2);$$
- $(\mathcal{I} \times \mathcal{J}, \trianglelefteq)$ is the product of posets $(\mathcal{I}, \trianglelefteq)$ and $(\mathcal{J}, \trianglelefteq')$;
- the new height function is defined by
$$\mathsf{Eig}(p, q) = \big(\mathsf{eig}(p), \mathsf{eig}'(q)\big).$$

It is easy to check that, up to equivalence, this product is associative, commutative, and admits an identity. The details are left to the reader.

THEOREM 3.1. *The product of two design systems is also a design system.*

PROOF. Clearly, Eig is an order-preserving surjection from $(\mathcal{P} \times \mathcal{Q}, \preceq)$ to $(\mathcal{I} \times \mathcal{J}, \trianglelefteq)$. For (w, w') in $\mathcal{P} \times \mathcal{Q}$, we have

$$\{(x, y) \in X \times Y : (w, w') \preceq (x, y)\} = \{x \in X : w \preceq x\} \times \{y \in Y : w' \preceq' y\},$$

and it follows that condition *(i)* holds for the product. If W_i denotes the incidence matrix of X versus $\mathcal{P}^i$ in the first system and W_j' denotes the incidence matrix of Y versus $\mathcal{Q}^j$ in the second system, then the incidence matrix of $X \times Y$ versus the fibre $(\mathcal{P} \times \mathcal{Q})^{(i,j)}$ in the product is given by the Kronecker product of these two matrices:

$$W_{(i,j)} = W_i \otimes W'_j.$$

From this equation, it is easy to see that $W_{(i,j)}$ has constant column sum. Hence condition *(ii)* holds for the product.

The primitive idempotents for the product association scheme are of the form

$$E_{(i,j)} = E_i \otimes E'_j$$

where E_i is any primitive idempotent for $(X, \mathcal{A})$ and E'_j is any primitive idempotent for $(Y, \mathcal{B})$. Thus the eigenspace $V_{(i,j)}$ of the product scheme is spanned by vectors $\mathbf{v} \otimes \mathbf{v}'$ where $\mathbf{v} \in V_i$ and $\mathbf{v}' \in V'_j$. Now since the two component systems satisfy condition *(iii)* of the definition, there exist vectors $\mathbf{r}$ and $\mathbf{r}'$ such that

$$\mathbf{v} = W_i \mathbf{r} \qquad \text{and} \qquad \mathbf{v}' = W'_j \mathbf{r}'.$$

Thus

$$\mathbf{v} \otimes \mathbf{v}' = \left(W_i \otimes W'_j\right)\left(\mathbf{r} \otimes \mathbf{r}'\right)$$

and it follows that $V_{(i,j)}$ is contained in the column space of $W_{(i,j)}$. On the other hand, if $(k, \ell) \not\trianglelefteq (i,j)$ in $(\mathcal{I} \times \mathcal{J}, \trianglelefteq)$, we may assume without loss that $k \not\trianglelefteq i$. Then, for every $\mathbf{e} \in V_k$, the kth eigenspace for $(X, \mathcal{A})$, we have $\mathbf{e}^T W_i = 0$. Now let $\mathbf{e}'$ be any eigenvector in the eigenspace V'_ℓ of the second component system. We have

$$(\mathbf{e} \otimes \mathbf{e}')^T W_{(i,j)} = (\mathbf{e}^T W_i) \otimes (\mathbf{e}'^T W'_j) = 0$$

showing that

$$\mathsf{colsp} W_{(i,j)} \subseteq \bigoplus_{(k,\ell) \trianglelefteq (i,j)} V_{(k,\ell)}.$$

Thus condition *(iii)* holds and our proof is complete. □

Consider a design system

$$\Delta = (X, \mathcal{A}; \mathcal{P}, \preceq; \mathcal{J}, \trianglelefteq; \mathsf{eig}).$$

The m-fold product of this system with itself is again a design system by the above theorem. Each element of this larger system admits a natural action by the symmetric group $\mathsf{Sym}(m)$. For example, if $\underline{j} = (j_1, \dots, j_m) \in \mathcal{J}^m$ and $\sigma \in \mathsf{Sym}(m)$, then we write

$$\underline{j}^\sigma = (j_{\sigma(1)}, \dots, j_{\sigma(m)}).$$

Clearly, $\underline{i} \trianglelefteq \underline{j}$ if and only if $\underline{i}^\sigma \trianglelefteq \underline{j}^\sigma$. Thus we have a well-defined partial order on the orbits of $\mathsf{Sym}(m)$ acting on the m-fold product poset $(\mathcal{J}^m, \trianglelefteq)$. Let $(\mathcal{J}^{(m)}, \trianglelefteq)$ denote this "symmetrization" of $(\mathcal{J}^m, \trianglelefteq)$. Let orb denote the function from $\mathcal{J}^m$ to $\mathcal{J}^{(m)}$ which sends a tuple to its corresponding orbit. Then $\mathsf{orb} \circ \mathsf{eig}$ surjectively maps elements of the product poset $(\mathcal{P}^m, \preceq)$ to elements of $(\mathcal{J}^{(m)}, \trianglelefteq)$ in an order-preserving fashion.

In [**4**, Sec. 2.5], Delsarte proves that the m-fold product scheme $(X, \mathcal{A})^{\otimes m}$ contains a fusion scheme which is invariant under the action of $\mathsf{Sym}(m)$; this latter scheme is called the m-fold *extension* of $(X, \mathcal{A})$. For an m-tuple of indices

$$\underline{j} = (j_1, \dots, j_m) \in \mathcal{J}^m,$$

let $[\underline{j}]$ denote the orbit containing it under the action of $\mathsf{Sym}(m)$. Then the primitive idempotents of the extension scheme are indexed by $\mathcal{J}^{(m)}$ and take the form

$$E_{[\underline{j}]} = \sum_{\underline{h} \in [\underline{j}]} E_{\underline{h}}$$

where, for $\underline{h} = (h_1, \ldots, h_m)$, the product scheme $(X, \mathcal{A})^m$ contains the primitive idempotent

$$E_{\underline{h}} = \bigotimes_{\ell=1}^{m} E_{h_\ell}.$$

(The Schur idempotents $A_{\underline{i}}$ for the extension scheme are obtained in an analogous manner.) We will denote the m-fold extension of $(X, \mathcal{A})$ by $\mathsf{Ext}^m(X, \mathcal{A})$.

We now aim to show that the ordered quadruple

$$\mathsf{Ext}^m(\Delta) = (\mathsf{Ext}^m(X, \mathcal{A});\ \mathcal{P}^m, \preceq;\ \mathcal{J}^{(m)}, \trianglelefteq;\ \mathsf{orb} \circ \mathsf{eig})$$

is a design system. Note that the vertex set is the same as that of the m-fold product. Moreover, the partial order $(\mathcal{P}^m, \preceq)$ is simply the m-fold product of the original partial order $(\mathcal{P}, \preceq)$. We will refer to this structure as the *m-fold extension* of the original design system Δ.

THEOREM 3.2. *For any positive integer m, the m-fold extension of a design system is again a design system.*

PROOF. Condition *(i)* is immediately satisfied, by Theorem 3.1. We use the group symmetry to prove condition *(ii)*. Suppose $\underline{i}, \underline{j} \in \mathcal{J}^m$ with $\mathsf{orb}(\underline{i}) = \mathsf{orb}(\underline{j})$. Then, for $z = (z_1, \ldots, z_m)$ in $\mathcal{P}^m$ with $\mathsf{eig}(z) = \underline{i}$, there exists $\sigma \in \mathsf{Sym}(m)$ such that $\mathsf{eig}(z^\sigma) = \underline{j}$. Obviously

$$|\{x \in X^m : z \preceq x\}| = |\{x \in X^m : z^\sigma \preceq x\}|.$$

Thus, in the m-fold product system, we have $\rho_{\underline{i}} = \rho_{\underline{j}}$. So, upon collapsing fibres via the function orb, we find that there is a well-defined value $\rho_{[\underline{j}]}$ such that

$$|\{x \in X^m : z \preceq x\}| = \rho_{[\underline{j}]}$$

whenever $\mathsf{orb}(\mathsf{eig}(z)) = [\underline{j}]$.

For $[\underline{j}] \in \mathcal{J}^{(m)}$, the eigenspace $V_{[\underline{j}]}$ is the direct sum of eigenspaces $V_{\underline{i}}$ of the m-fold product scheme as $\underline{i}$ ranges over the orbit $[\underline{j}]$. By the same token, the incidence matrix $W_{[\underline{j}]}$ for the m-fold extension is obtained by simply stacking the matrices $\{W_{\underline{i}} : \underline{i} \in [\underline{j}]\}$ one alongside the other. Thus the column space of $W_{[\underline{j}]}$ is the join of the column spaces of these various submatrices. Since condition *(iii)* of our definition holds for the m-fold product scheme, the column space of $W_{\underline{i}}$ contains $V_{\underline{i}}$ for each $\underline{i} \in [\underline{j}]$. Hence the column space of $W_{[\underline{j}]}$ contains $V_{[\underline{j}]}$. On the other hand, the column space of $W_{\underline{i}}$ is orthogonal to an eigenspace $V_{\underline{k}}$ unless $\underline{k} \trianglelefteq \underline{i}$ in $\mathcal{J}^m$. Thus the join, $\mathsf{colsp}W_{[\underline{j}]}$ is orthogonal to $V_{\underline{k}}$ unless $\underline{i} \trianglelefteq \underline{k}$ for some $\underline{i} \in [\underline{j}]$. But orb is an order-preserving map. So $V_{[\underline{j}]}$ is orthogonal to $V_{\underline{k}}$ — and therefore also to $V_{[\underline{k}]}$ — unless $[\underline{k}] \trianglelefteq [\underline{j}]$ in $\mathcal{J}^{(m)}$. This completes the proof of condition *(iii)*. □

4. The Delsarte bounds

In this section, we extend two well-known bounds of Delsarte to the setting of association schemes with many vanishing Krein parameters. Originally, these results were proved for cometric association schemes in [**4**]. Here, we replace the cometric property with certain vanishing conditions for Krein parameters with reference to a partial order $\trianglelefteq$ on the set $\mathcal{J}$ of eigenspaces of the association scheme under consideration.

For $\mathcal{E}$ and $\mathcal{F}$, subsets of $\mathcal{J}$, define

$$\mathcal{E} \star \mathcal{F} = \{k \in \mathcal{J} : \sum_{i\in\mathcal{E}} \sum_{j\in\mathcal{F}} q_{ij}^k > 0\}.$$

In view of the Krein conditions, an index k lies in $\mathcal{E} \star \mathcal{F}$ whenever $q_{ij}^k \neq 0$ for some $i \in \mathcal{E}$ and some $j \in \mathcal{F}$. As an example, observe that, in a cometric scheme, if we take $\mathcal{E} = \{0, \dots, e\}$ and $\mathcal{F} = \{0, \dots, f\}$, then

$$\mathcal{E} \star \mathcal{F} \subseteq \{0, \dots, e+f\}.$$

For $j \in \mathcal{J}$, let m_j denote the jth *multiplicity*, i.e., the rank of the primitive idempotent E_j of our association scheme $(X, \mathcal{A})$. We now use a slightly modified version of Delsarte's linear programming approach to give a generalization of the Ray-Chaudhuri/Wilson and Rao bounds, which were proved for designs in Johnson and Hamming schemes, respectively.

THEOREM 4.1. *Let $(X, \mathcal{A})$ be an association scheme and let $\mathcal{T} \subseteq \mathcal{J}$. Assume $\mathcal{E} \subseteq \mathcal{J}$ satisfies $\mathcal{E} \star \mathcal{E} \subseteq \mathcal{T}$. Then, for any Delsarte $\mathcal{T}$-design $D \subseteq X$, we have*

$$|D| \geq \sum_{j\in\mathcal{E}} m_j.$$

Moreover, if equality holds, then, for $\ell \neq 0$ in $\mathcal{J}$,

$$\sum_{j\in\mathcal{E}} Q_{\ell j} = 0$$

whenever D contains a pair of ℓ-related elements.

PROOF. Any matrix $M \in \mathbb{A}$ can be expanded in the form

$$M = {\scriptstyle N} \sum_{j\in\mathcal{J}} \beta_j E_j$$

and also as

$$M = \sum_{i\in\mathcal{I}} \alpha_i A_i$$

where $\alpha_i = \sum_j Q_{ij}\beta_j$ for each $i \in \mathcal{I}$. Suppose we restrict attention to non-negative matrices $M \in \mathbb{A}$ which satisfy the following two conditions:

(a): $\beta_j \leq 0$ for all $j \notin \mathcal{T}$; and
(b): $\beta_0 = 1$.

Without loss of generality, assume $0 \in \mathcal{T}$. Let $D \subseteq X$ be a Delsarte $\mathcal{T}$-design with characteristic vector χ. We expand $\chi^T M \chi$ in two ways to obtain the string of inequalities

$$\begin{aligned} |D|\alpha_0 &= \alpha_0 \chi^T A_0 \chi \\ &\leq \sum_{\mathcal{I}} \alpha_i \chi^T A_i \chi = N \sum_{\mathcal{J}} \beta_j \chi^T E_j \chi \\ &= N \chi^T E_0 \chi + N \sum_{\mathcal{T}-\{0\}} \beta_j \chi^T E_j \chi + N \sum_{j \notin \mathcal{T}} \beta_j \chi^T E_j \chi \\ &\leq N \chi^T E_0 \chi = |D|^2. \end{aligned}$$

This gives us the bound $|D| \geq \alpha_0$. (This is nothing more than an alternative form of Delsarte's linear programming bound, as described in [**10**].) We now proceed to obtain an explicit dual solution which is motivated by Delsarte's "Wilson polynomial" [**4**, p74].

Let

$$F = \sum_{j \in \mathcal{E}} E_j.$$

Then $F \circ F$ is a non-negative matrix with spectral decomposition

$$F \circ F = \sum_{k \in \mathcal{J}} \left(\frac{1}{N} \sum_{i \in \mathcal{E}} \sum_{j \in \mathcal{E}} q_{ij}^k \right) E_k. \tag{4.1}$$

Here, we have used the definition of the Krein parameters given in Section 1. Now, by choice of $\mathcal{E}$, we have $q_{ij}^k = 0$ whenever $i, j \in \mathcal{E}$ and $k \notin \mathcal{T}$. So condition **(a)** is satisfied by any non-negative multiple of $F \circ F$. We scale by

$$\gamma = \frac{N^2}{\sum_{j \in \mathcal{E}} m_j}$$

to obtain a non-negative matrix $M = \gamma(F \circ F)$ which satisfies conditions **(a)** and **(b)**, It is straightforward to check that the diagonal entries of M are all equal to

$$\alpha_0 = \sum_{j \in \mathcal{E}} m_j.$$

Now if $|D| = \alpha_0$, we return to the above string of equations and inequalities to discover that, for each $\ell \neq 0$,

$$\alpha_\ell \left(\chi^T A_\ell \chi \right) = 0$$

must hold. Thus, if D contains a pair of ℓ-related elements, we are forced to have

$$\alpha_\ell = \sum_{k \in \mathcal{J}} \beta_k Q_{\ell k} = 0.$$

Now, from Equation (4.1), we find

$$\beta_k = \frac{\gamma}{N} \sum_{i \in \mathcal{E}} \sum_{j \in \mathcal{E}} q_{ij}^k$$

so that

$$\alpha_\ell = \frac{\gamma}{N} \sum_{i \in \mathcal{E}} \sum_{j \in \mathcal{E}} \sum_{k \in \mathcal{J}} q_{ij}^k Q_{\ell k}$$

which gives us

$$\alpha_\ell = \gamma \left(\sum_{j\in\mathcal{E}} Q_{\ell j} \right)^2 = 0$$

as desired. □

Let $D \subseteq X$. For each matrix M in the Bose-Mesner algebra $\mathbb{A}$ of $(X, \mathcal{A})$, let $\bar{M}$ denote the submatrix of M with rows and columns indexed by the elements of D. The vector space $\{\bar{M} : M \in \mathbb{A}\}$ is occasionally the Bose-Mesner algebra of another association scheme. We express this by saying that D is a *schematic design* or that D *induces* an association scheme inside $(X, \mathcal{A})$. Let us refer to the new scheme as the *induced scheme* on vertex set D.

For $D \subseteq X$, we define the *degree* of D to the the number of non-zero indices $i \in \mathcal{I} - \{0\}$ for which $\bar{A}_i$ is not the zero matrix. Call D a *tight design* if, for some $\mathcal{E}$, equality holds in the Delsarte bound of Theorem 4.1.

THEOREM 4.2. *Let $(X, \mathcal{A})$ be an association scheme. Let $\mathcal{T} \subseteq \mathcal{J}$ and assume $\mathcal{E} \subseteq \mathcal{J}$ satisfies $\mathcal{E} \star \mathcal{E} \subseteq \mathcal{T}$.*

(a): *if D is any Delsarte $\mathcal{T}$-design in $(X, \mathcal{A})$ with degree s, then $s + 1 \geq |\mathcal{E}|$;*
(b): *if $|\mathcal{E}| = s + 1$, then D is a tight design and D is a schematic design in $(X, \mathcal{A})$;*
(c): *if $|\mathcal{E}| = s$, then either D is a tight design or D is a schematic design in $(X, \mathcal{A})$.*

The proof, based on that of Theorem 5.25 in [**4**], is only a slight variation of the proof of Theorem 3.3 in [**10**] and is therefore omitted.

COROLLARY 4.3. *Let $\Delta = (X, \mathcal{A}; \mathcal{P}, \preceq; \mathcal{J}, \trianglelefteq; \mathsf{eig})$ be a Delsarte system. Let $\mathcal{T}$ be a lower ideal in $\mathcal{J}$ and let $\mathcal{E} \subseteq \mathcal{J}$ satisfying $0 \in \mathcal{E}$ and $\mathcal{E} \star \mathcal{E} \subseteq \mathcal{T}$. Suppose D is a Delsarte $\mathcal{T}$-design in $(X, \mathcal{A})$ satisfying one of the following two sets of conditions:*

(A): *D has degree $s = |\mathcal{E}| - 1$;*
(B): *D has degree $s = |\mathcal{E}|$ and satisfies $|D| > \sum_{j\in\mathcal{E}} m_j$.*

In case **(A)**, *define $\mathcal{K} = \mathcal{E} - \{j\} \cup \{1\}$ where j is any maximal element of $\mathcal{E}$ in $(\mathcal{J}, \trianglelefteq)$; in case* **(B)**, *define $\mathcal{K} = \mathcal{E} \cup \{1\}$. Define*

$$\mathcal{B} = \{\bar{A}_i : i \in \mathcal{I},\ \bar{A}_i \neq 0\}.$$

Then there is a subset $\mathcal{Q}$ of $\mathsf{eig}^{-1}(\mathcal{K})$ such that

$$(D, \mathcal{B};\ \mathcal{Q}, \preceq;\ \mathcal{K}, \trianglelefteq;\ \mathsf{eig})$$

is a design system, where $\preceq$, $\trianglelefteq$ and eig are obtained by restriction.

PROOF. We will prove the result for condition **(B)**; the proof for condition **(A)** is similar.

Part (c) of Theorem 4.2 tells us that D is a schematic design. So $(D, \mathcal{B})$ is indeed an association scheme.

By Theorem 2.3, D is also a poset $\mathcal{T}$-design in Δ. Since $0 \in \mathcal{E}$, we must have $\mathcal{E} \subseteq \mathcal{T}$. Therefore the incidence matrices W_k ($k \in \mathcal{K}$) of the new system all have constant column sum, regardless of which subset of $\mathcal{P}^k$ we choose for inclusion in our new poset $(\mathcal{Q}, \preceq)$. So condition *(ii)* will be satisfied.

We next establish condition *(iii)* under the assumption that $\mathcal{Q}^k = \mathcal{P}^k$ for all $k \in \mathcal{E}$. (We will obviously choose $\mathcal{Q}^1 = D$.) Later, we will delete elements if necessary in order to establish condition *(i)*, but we will show that this can be achieved in such a way that condition *(iii)* remains true.

If M is a matrix with rows indexed by X, let $\pi(M)$ denote the submatrix obtained by restricting to those rows indexed by elements of D. Similarly, if U is a subspace of $\mathbb{R}^X$, let $\pi(U)$ denote the vector subspace of $\mathbb{R}^D$ obtained by restricting each vector in U to the coordinates indexed by elements of D. The key to establishing condition *(iii)* is that the process of taking the column space of a matrix commutes with this projection operator π.

It is easy to see from the proof of Theorem 4.2 (see [**4**] or [**10**]) that the primitive idempotents of the induced association scheme are the following: for each $j \in \mathcal{E}$, we take an appropriate scalar multiple — F_j, say — of the submatrix $\bar{E}_j$ and , for $j = 1$, our primitive idempotent is

$$F_1 = I - \sum_{j \in \mathcal{E}} F_j.$$

Let U_j denote the column space of F_j $(j \in \mathcal{K})$; this is the jth eigenspace of the induced scheme. Since $\bar{E}_j$ has the same rank as E_j itself, we know that $U_j = \pi(V_j)$ for $j \in \mathcal{E}$.

Consider $j \in \mathcal{E}$. Since the incidence matrix W_j contains V_j in its column space, it follows that the submatrix $W_j' = \pi(W_j)$ contains U_j in its column space. On the other hand, we have

$$\mathsf{colsp} W_j \subseteq \bigoplus_{i \trianglelefteq j} V_i$$

in the original system. Observe that, for all i, $i \trianglelefteq j$ implies $i \in \mathcal{E}$. Now if we restrict each vector in the column space to (coordinates indexed by elements in) D, we obtain the column space of the incidence matrix of $\mathcal{Q}^j$ versus D. Thus we have shown that

$$\mathsf{colsp} W_j' \subseteq \bigoplus_{i \trianglelefteq j} U_i$$

which establishes condition *(iii)* for the system

$$(D, \mathcal{B}; \mathsf{eig}^{-1}(\mathcal{K}), \preceq; \mathcal{K}, \trianglelefteq; \mathsf{eig}).$$

(Note that, for $j = 1$, W_j' is the identity matrix and there is nothing to prove.)

We now obtain $(\mathcal{Q}, \preceq)$ from $(\mathsf{eig}^{-1}(\mathcal{K}), \preceq)$ by deleting points according to the following two rules, applied in turn:

- if $z \in \mathsf{eig}^{-1}(j)$ and the column of W_j' indexed by z is orthogonal to U_j, then delete z;
- if W_j' contains repeated columns, then delete all but one of the corresponding points of $\mathsf{eig}^{-1}(j)$.

Clearly, the resulting system will satisfy condition *(i)* of the definition and conditions *(ii)* and *(iii)* will continue to hold. □

REMARK 4.4. In [**4**, Theorem 5.21], it is shown that case **(A)** cannot occur in the familiar cometric setting where $\mathcal{T} = \{0, \ldots, t\}$ and $\mathcal{E} = \{0, \ldots, \lfloor \frac{t}{2} \rfloor\}$.

EXAMPLE 4.5. As a sample application of this corollary, consider a design D of strength t in the Johnson scheme $J(v,k)$ which falls under case **(B)** of the corollary. Then we obtain a cometric d-class association scheme where $d = \lfloor \frac{t}{2} \rfloor + 1$ and the eigenspaces are indexed by $\mathcal{K} = \{0, \ldots, \lfloor \frac{t}{2} \rfloor + 1\}$ with a linear order. The corollary gives us a design system with $\mathcal{P} = \{z \subseteq \{1, \ldots, v\} : |z| \leq d-1\} \cup D$. Thus a subset D' of the blocks of D is a Delsarte $\{0, \ldots, t'\}$-design in this new association scheme ($1 \leq t' \leq d$) if and only if the blocks from D' form an ordinary t'-design in $J(v,k)$.

5. Applications

In this final section, we discuss some of the applications of the above combinatorial theory. In many cases, we give a binary operation on $\mathcal{J}$ (with an enlarged codomain) which sheds light on the operation $\star$ of the previous section and thus helps us choose a good set $\mathcal{E}$ for use in Theorem 4.1.

5.1. Cometric schemes and their products. When $(X, \mathcal{A})$ is cometric, there is a linear ordering on the set $\mathcal{J} = \{0, \ldots, d\}$ indexing the eigenspaces. It is natural to choose this linear order for $\trianglelefteq$ and seek a ranked poset $(\mathcal{P}, \preceq)$ which yields a design system for $(X, \mathcal{A})$. In addition to Delsarte's regular semilattices and a construction of Stanton [**22**] for certain families of metric and cometric association schemes with classical parameters, there are a number of sporadic examples. (See also Munemasa [**17**] and Terwilliger [**23**].) For the Shrikhande graph, Song [**21**] found the following poset. There are exactly twelve 8-element subsets of the vertices which induce a 4-regular subgraph. Denote these subsets by Z_i ($1 \leq i \leq 12$). Let

$$\mathcal{P} = \{\{x\} : x \in X\} \cup \{Z_i : 1 \leq i \leq 12\} \cup \{X\}$$

and let $\preceq$ denote reverse inclusion. With $\mathcal{J} = \{0, 1, 2\}$ linearly ordered, this gives a design system for the Shrikhande graph.

The incidence graph G of points versus hyperplanes in the finite projective geometry $PG(n,q)$ yields a three-class metric and cometric association scheme with two Q-polynomial orderings on its eigenspaces. We will order the set $\{0,1,2,3\}$ naturally with $\theta_0 > \theta_1 > \theta_2 > \theta_3$ denoting the distinct eigenvalues of G. If we wish to construct a design system for this graph using reverse inclusion as our partial order $\preceq$, we will clearly take $\mathcal{P}^3 = X$ and $\mathcal{P}^0 = X$. A natural choice for $\mathcal{P}^2$ is the set $\{\{p, H\} : p \in H\}$ of "flags" of the geometry. The only non-trivial part is the construction of $\mathcal{P}^1$. In view of Lemma 2.1 **(f)**, if $\theta_1 = q^{(n-1)/2}$ is not an integer, it is impossible to find a design system in this case. On the other hand, if q is a perfect square or if n is odd, it is possible to construct design systems as we now show.

Suppose q is a perfect square, say $q = s^2$. A *Baer subgeometry* is a subconfiguration of $(s^{n+1}-1)/(s-1)$ points and $(s^{n+1}-1)/(s-1)$ hyperplanes of $PG(n,q)$ inducing an incidence structure isomorphic to $PG(n,s)$. In our notation, a Baer subgeometry $\mathcal{S}$ is a subset of X consisting of the points and the hyperplanes belonging to that subgeometry.

PROPOSITION 5.1. *Suppose $n \geq 2$ and $q = s^2$ is a prime power. Consider the association scheme $(X, \mathcal{A})$ coming from the incidence graph G of $PG(n,q)$. Let the (linear) ordering on eigenspaces be determined by the ordering*

$$\theta_0 = \frac{q^n - 1}{q-1}, \quad \theta_1 = q^{\frac{n-1}{2}}, \quad \theta_2 = -q^{\frac{n-1}{2}}, \quad \theta_3 = -\frac{q^n-1}{q-1}$$

of the eigenvalues of G. Let $\mathcal{P}$ consist of $\mathcal{P}^0$, $\mathcal{P}^2$, $\mathcal{P}^3$ as above and $\mathcal{P}^1$ being the set of all Baer subgeometries of X. Let $\preceq$ denote reverse inclusion. Then the ordered *quadruple*

$$(X, \mathcal{A};\ \mathcal{P}, \preceq;\ \{0,1,2,3\}, \leq; \mathsf{eig})$$

is a design system, where eig *maps an element of* $\mathcal{P}^j$ *to the integer* j $(0 \leq j \leq 3)$.

PROOF. It is not hard to see that the characteristic vectors of flags span the space $V_0+V_1+V_2$ since the one-dimensional space V_3 is spanned by the vector which takes value $+1$ on every point and value -1 on every hyperplane. The partition of X into a Baer subgeometry and its complement is an equitable partition [**7**, Chapter 5] with parameter matrix

$$\begin{pmatrix} \frac{s^n-1}{s-1} & \frac{q^n-1}{q-1} - \frac{s^n-1}{s-1} \\ \frac{s^{n-1}-1}{s-1} & \frac{q^n-1}{q-1} - \frac{s^{n-1}-1}{s-1} \end{pmatrix}.$$

Thus the characteristic vector of any Baer subgeometry lies in $V_0 \oplus V_1$. Now the automorphism group of $PG(n,q)$ acts 2-transitively on points. So the point sets of all Baer subgeometries form a combinatorial 2-design on the points of $PG(n,q)$. By the proof of Fisher's inequality, the rank of the incidence matrix of points versus Baer subgeometries is equal to $\frac{q^n-1}{q-1}$. This shows that the incidence vectors of all Baer subgeometries (each now viewed as a set consisting of both points and hyperplanes) span the sum $V_0 \oplus V_1$ of eigenspaces of the association scheme. It now follows quickly that we have a design system. □

As a direct consequence, we can say that a subset $D \subseteq X$ of the vertices of this association scheme is a Delsarte $\{1\}$-design precisely when D meets every Baer subgeometry in a set of size ν for some ν (though we cannot determine how many of these ν objects are points and how many are blocks).

Recall that a point of $PG(n,q)$ is a 1-dimensional subspace, P, of $\mathbb{F}_q^{n+1}$ and a block of $PG(n,q)$ is an n-dimensional subspace, H, of $\mathbb{F}_q^{n+1}$. For a k-dimensional subspace K, denote by Y_K the following subset of X:

$$Y_K = \{P :\ P \text{ is a point of } PG(n,q),\ P \subseteq K\} \bigcup \{H :\ H \text{ is a block of } PG(n,q),\ K \subseteq H\}.$$

PROPOSITION 5.2. *Suppose* $n \geq 3$ *is odd and* q *is a prime power. Consider the association scheme* $(X, \mathcal{A})$ *coming from the incidence graph* G *of* $PG(n,q)$. *Let the (linear) ordering on eigenspaces be determined by the ordering*

$$\theta_0 = \frac{q^n-1}{q-1}, \quad \theta_1 = q^{\frac{n-1}{2}}, \quad \theta_2 = -q^{\frac{n-1}{2}}, \quad \theta_3 = -\frac{q^n-1}{q-1}$$

of the eigenvalues of G. *Let* $\mathcal{P}$ *consist of* $\mathcal{P}^0$, $\mathcal{P}^2$, $\mathcal{P}^3$ *as above and*

$$\mathcal{P}^1 = \{Y_K : \dim(Y) = (n+1)/2\}.$$

Let $\preceq$ *denote reverse inclusion. Then the ordered quadruple*

$$(X, \mathcal{A};\ \mathcal{P}, \preceq;\ \{0,1,2,3\}, \leq; \mathsf{eig})$$

is a design system, where eig *maps an element of* $\mathcal{P}^j$ *to the integer* j $(0 \leq j \leq 3)$. □

In [**10**], the author studies products of association schemes in which each scheme has an attached graded poset as in the previous examples. For the bounds (parallel to Section 4), we require each scheme to be cometric. In these examples, the poset $(\mathcal{J}, \trianglelefteq)$ is a product $\prod_{i=1}^{m}\{0,\ldots,d_i\}$ of chains and we have a binary operation

$$+ : \mathcal{J} \times \mathcal{J} \to (\mathbb{N} \cup \{0\})^m$$

given by coordinate-wise sum of integers such that $q_{ij}^k = 0$ unless $k \trianglelefteq i + j$ where i, j and k are m-tuples of non-negative integers and $a \trianglelefteq b$ if and only if each entry of a is less than or equal to the corresponding entry of b. Thus, for $\mathcal{T} = \mathcal{J}_t := \{(t_1,\ldots,t_m) : \sum t_i \le t\}$, we may choose $\mathcal{E} = \mathcal{J}_{\lfloor t/2 \rfloor}$ for use in Theorem 4.1. Some interesting special cases are studied in [**8**], [**9**] and [**20**] with connections to cryptography, the Assmus-Mattson theorem, mixed-level orthogonal arrays and mixed-alphabet error-correcting codes.

5.2. The ordered Hamming scheme. Let F be a finite alphabet of size $q \ge 2$. Let $\hat{F} = F \cup \{*\}$ and

$$\mathcal{P} = \left\{ a = (a_1,\ldots,a_\ell) \in \hat{F}^\ell : \forall i,j\, (a_i = *, j > i) \Rightarrow\ a_j = * \right\}$$

where ℓ is a positive integer. We define an association scheme on $X = F^\ell$ as follows. The ith associate relation is defined as follows: for $a = (a_1,\ldots,a_\ell)$ and $b = (b_1,\ldots,b_\ell)$, $(a,b) \in R_i$ if $a_j = b_j$ for all $j \le \ell - i$ but (provided $i \neq 0$) $a_i \neq b_i$. This is a special case of a large family of association schemes discovered by Yamamoto, Fujii and Hamada. This association scheme is self-dual and satisfies $q_{ij}^k = 0$ whenever $k > \max(i,j)$. The set $\mathcal{J} = \{0,\ldots,\ell\}$ of eigenspaces is linearly ordered. This design system is very similar to one arising from a cometric scheme. But this association scheme is not cometric: we sometimes have $q_{ij}^{i+j} = 0$.

The ordered Hamming scheme, studied in [**15**], can be embedded in a design system which is simply the s-fold extension of the above system. Eigenspaces are now indexed by ferrers diagrams which fit inside a rectangle having s rows and ℓ columns. The partial order $\trianglelefteq$ is containment. Suppose we view this partially ordered set $(\mathcal{J}, \trianglelefteq)$ as a lower ideal in Young's lattice $(\mathcal{Y}, \trianglelefteq)$. Then it is possible to define a binary operation — namely $+ : \mathcal{J} \times \mathcal{J} \to \mathcal{Y}$ given by concatenation of ferrers diagrams (with re-shuffling, if necessary, to maintain monotonicity) — so that $q_{ij}^k = 0$ unless $k \trianglelefteq i + j$ where q_{ij}^k denotes the Krein parameter of the ordered Hamming scheme.

Consider an array M with entries from F having m rows and $s\ell$ columns, these being indexed by $\{(i,j) : 1 \le i \le s,\ 1 \le j \le \ell\}$. We say that M is an ordered orthogonal array $OOA_\nu(t,s,\ell,q)$ is provided, for every choice of $t_1,\ldots,t_s$ satisfying $0 \le t_i \le \ell$ for all i and $\sum t_i = t$, the subarray obtained by restricting to columns $\{(i,j) : 1 \le i \le s,\ 1 \le j \le t_i\}$ has each t-tuple $a \in F^t$ appearing exactly ν times as a row. Let

$$\mathcal{T} = \{(j_1,\ldots,j_s) : 0 \le j_i \le \ell,\ \sum_{i=1}^{s} j_i \le t\}.$$

It is not hard to see that $D \subseteq X = F^{s\ell}$ constitutes the set of rows of an ordered orthogonal array $OOA_\nu(t,s,\ell,q)$ (for some value of ν) if and only if D is a poset $\mathcal{T}$-design in Δ. Thus we have a new proof of Theorem 3.4 in [**15**]: an ordered orthogonal array $OOA_\nu(t,s,\ell,q)$ is equivalent to a Delsarte $\mathcal{T}$-design in the ordered

Hamming scheme. If we choose

$$\mathcal{E} = \{(j_1, \ldots, j_s) : 0 \le j_i \le \ell, \sum_{i=1}^{s} j_i \le t/2\},$$

then the Delsarte Bound, Theorem 4.1, specializes to Theorem 4.5 in [**15**]. (But see also Theorem 3.5 in [**14**].)

These ordered orthogonal arrays have received considerable attention recently due to their connection with quasi-Monte Carlo methods for problems such as numerical integration. (See the references in [**15**].) But this association scheme is also a natural tool for the study of a certain type of error-correcting code introduced by Tsfasman and Rosenbloom in [**19**]. These codes can be shown to protect against synchronization errors in a certain channel model. This application will be discussed elsewhere.

5.3. The association scheme of the symmetric group. Finally, consider the association scheme of the symmetric group S_n. We have one relation for each conjugacy class and one eigenspace for each irreducible representation. The set $\mathcal{J}$ can be viewed as the set of all partitions $\mu = (\mu_1, \ldots, \mu_k)$ $(\mu_1 \ge \mu_2 \ge \cdots)$ of n. These partitions are endowed with the reverse dominance order:

$$\mu \trianglelefteq \lambda \quad \Leftrightarrow \quad \forall i \sum_{j=1}^{i} \mu_j \ge \sum_{j=1}^{i} \lambda_j.$$

The elements of $\mathcal{P}$ are all cosets of all Young subgroups of S_n, ordered by reverse inclusion. The map eig sends each coset to the "shape" of the corresponding Young subgroup. This gives a design system for S_n [**13**].

In joint work with Bruce Sagan, the author studies designs in this system. Suppose λ is a partition of n. A set $G \subseteq S_n$ is said to be a *λ-transitive* set of permutations if there exists a constant r such that the following holds. For any two set partitions $P = (P_1, \ldots, P_k)$ and $Q = (Q_1, \ldots, Q_k)$ of $\Omega = \{1, \ldots, n\}$ of shape λ (i.e., for $1 \le i \le k$, $|P_i| = |Q_i| = \lambda_i$), there exist exactly r elements of G which map P to Q:

$$\forall i \, (P_i^g = Q_i) \, .$$

If the set G happens to be a group, then we call G a *λ-transitive permutation group*. For $\lambda = (n - t, t)$, this reduces to the definition of a t-homogeneous permutation group, and for $\lambda = (n - t, 1, 1 \ldots, 1)$, this reduces to the definition of a t-transitive permutation group. In [**13**], we prove

THEOREM 5.3 (Martin and Sagan). *Let G be a subset of S_n with characteristic vector x. Then G is λ-transitive if and only if G is a Delsarte $\mathcal{T}$-design, where*

$$\mathcal{T} = \{\mu \vdash n : \mu \trianglelefteq \lambda\}.$$

See [**13**] for an application of λ-transitive sets of permutations to the construction of perpendicular arrays, which are connected to cryptosystems with perfect t-fold secrecy.

Acknowledgments

I would like to thank Paul Terwilliger, David Wagner and Akihiro Munemasa for helpful comments regarding the material in this paper.

This paper was written while I was visiting the Department of Combinatorics and Optimization at the University of Waterloo. I wish to thank the department for its hospitality and accommodation during this visit. My research is supported by the Canadian government through NSERC grant number OGP0155422. Additional support was provided by MITACS and CITO through the Centre for Applied Cryptographic Research.

References

[1] E. Bannai and T. Ito. Algebraic Combinatorics I: Association Schemes. Benjamin-Cummings Lecture Note Series **58**, The Benjamin-Cummings Publishing Company, Inc., London, 1984.

[2] A. E. Brouwer, A. M. Cohen, and A. Neumaier. Distance-Regular Graphs. Springer-Verlag, Berlin, 1989.

[3] B. A. Davey and H. A. Priestley. Introduction to Lattices and Order. Cambridge Univ. Press, London, 1990.

[4] P. Delsarte. An algebraic approach to the association schemes of coding theory. *Philips Res. Reports Suppl.* **10** (1973).

[5] ———. Association schemes and t-designs in regular semilattices. *J. Combin. Th. (A)* **20** (1976), 230–243.

[6] ———. Pairs of vectors in the space of an association scheme. *Philips Res. Reports* **32** (1977), 373–411.

[7] C. D. Godsil. Algebraic Combinatorics. Chapman and Hall, London, 1993.

[8] V. I. Levenshtein. Split orthogonal arrays and maximum resilient systems of functions. *Designs, Codes and Crypt.* **12** (1997), 31–160.

[9] W. J. Martin. Mixed block designs. *J. Combin. Designs* **6** no. 2 (1998), 151–163.

[10] ———. Designs in product association schemes. *Designs, Codes and Crypt.* **16** no. 3 (1999), 271–289.

[11] ———. Minimum distance bounds for s-regular codes. To appear: Geometric and Algebraic Combinatorics. *Proceedings of conference in honor of Jaap Seidel's 80th birthday,* in: Designs, Codes and Crypt.

[12] ———. Symmetric designs, sets with two intersection numbers and Krein parameters of incidence graphs. To appear: *Proceedings of the Thirteenth Midwestern Conference on Combinatorics, Cryptography and Computing.* J. Combin. Math. Combin. Comput.

[13] ——— and B. E. Sagan. Groups and sets of permutations transitive relative to a partition. *In preparation.*

[14] W. J. Martin and D. R. Stinson. A generalized Rao bound for ordered orthogonal arrays and (t, m, s)-nets. *Canad. Math. Bull.* **42** no. 3 (1999), 359–370.

[15] ———. Association schemes for ordered orthogonal arrays and (T, M, S)-nets. *Canad. J. Math.* **51** no. 2 (1999), 326–346.

[16] A. D. Meyerowitz. Cycle-balanced partitions in distance-regular graphs. To appear, *Preprint.*

[17] A. Munemasa. An analogue of t-designs in the association schemes of alternating bilinear forms. *Graphs Combin.* **2** no. 3 (1986), 259–267.

[18] C. Roos. On antidesigns and designs in an association scheme. *Delft Progress Report* **7** (1982), 98–109.

[19] M. Yu. Rosenbloom and M. A. Tsfasman. Codes for the m-metric. *Problems of Information Transmission* **33** (no. 1) (1997), 45–52.

[20] N. J. A. Sloane and J. Stufken. A linear programming bound for orthogonal arrays with mixed levels. *J. Stat. Plan. Inf.* **56** (1996), 295–306.

[21] S. Y. Song. Posets related to some association schemes. *J. Korean Soc. Math. Educ.* **25** (1987), 57–69.

[22] D. Stanton. t-Designs in classical association schemes. *Graphs Combin.* **2** (1986), 283–286.

[23] P. Terwilliger. The incidence algebra of a uniform poset, pp 193–212 in: Coding Theory and Design Theory, Part I, IMA Vol. Math. Appl., **20**, Springer, New York, 1990.

Mathematics and Statistics, University of Winnipeg, Winnipeg, Manitoba R3B 2E9

Current address: Combinatorics and Optimization, University of Waterloo, Waterloo, Ontario N2L 3G1

E-mail address: `William.Martin@UWinnipeg.ca`

DIMACS Series in Discrete Mathematics
and Theoretical Computer Science
Volume **56**, 2001

The isomorphism problem for circulant graphs via Schur ring theory

Mikhail Muzychuk, Mikhail Klin, and Reinhard Pöschel

ABSTRACT. This paper concerns the applications of Schur ring theory to the isomorphism problem of circulant graphs (Cayley graphs over cyclic groups). A digest of the most important facts about Schur rings presented in the first sections provides the reader with a gentle self-contained introduction to this area. The developed machinery allows us to give proofs of two conjectures (Zibin' 1975 and Toida 1977) about necessary conditions on isomorphisms of the circulants. These new results together with a few new proofs of some known facts show the feasibility of the technique of Schur rings in algebraic combinatorics.

The concluding section contains a short historical and bibliographical survey of various results related to the isomorphism problem for Cayley graphs.

1. Introduction

A circulant graph with n vertices is a Cayley graph over the cyclic group C_n of order n, that is a graph which is invariant with respect to a cyclic rotation of all its vertices.

Circulant graphs form a wide and important class of vertex-transitive (directed or undirected) graphs. Such graphs can be regarded as so-called quasi-random graphs (in the sense of [**16**]), which means that various graph-theoretical properties can be successfully modelled and investigated within the class of circulants.

Though the history of circulants spans more than a century, modern interest in these graphs did not resurface until the 1960s. In 1967 A. Ádám posed his famous conjecture about a necessary and sufficient isomorphism condition for circulant graphs, thus opening the current wave of interest in the circulants. The history of this conjecture over 20 years can be observed in Pálfy's excellent survey [**72**].

One of the most successful solutions of the isomorphism problem for circulant graphs is based on the use of Schur rings (S-rings for brevity).

The notion of an S-ring was introduced by I. Schur in 1933 [**77**]. It was developed and used during a few decades mostly for purely group-theoretical purposes by H. Wielandt and other disciples and followers of I. Schur. In 1972-78 two of the authors (M.K. and R.P.) suggested using S-ring theory in graph theory, namely

---

1991 *Mathematics Subject Classification.* Primary 05C60, 05E99; Secondary 05C25, 20B25 .
The first author was partially supported by grant of the DAAD.

for the investigation of the symmetry properties of circulant graphs. Most of their results were available, for a long time, only in the form of preprints with a very limited distribution. During the last decade the interest in such application of S-rings was supported by the publications of the first author (M.M.) and a few other mathematicians. It became evident that S-rings give a very suitable and natural way to classify circulants.

The present paper may be regarded as a brief survey of the techniques of S-rings and applications in graph theory. The methodology presented in the paper is illustrated by the unified proofs of a number of known and new criteria for the isomorphism of circulant graphs. In particular, we give complete solutions of two related problems which were posed by D.K. Zibin' (1975) and S. Toida (1977).

During the last two decades circulant graphs found various interesting applications. For example, numerous graph and network models were used for the elaboration of reliable architectures of computers and computer networks. For these purposes one needs to select graphs with certain optimal or quasi-optimal combinations of invariants of a graph such as small diameter, fixed degree, uniformity, redundant paths, short wires, simple routing algorithm, *etc.* [**35**]. It turns out that rather good solutions of the above-mentioned problems may be found in the class of circulant graphs, see, for example, [**10, 9, 27, 90, 75**].

We hope that our survey will be of interest not only to the pure experts in the graph isomorphism problem but also to those who are interested in various applications of circulants. This is why we are trying to present here a self-contained gentle introduction to the subject which may be available to a wide mathematical audience.

On the other hand our interest here is limited by the consideration of only one class of Cayley graphs: those over cyclic groups. Other classes of Cayley graphs, as well as purely group-theoretical approaches to the isomorphism problem of Cayley graphs (and circulant graphs in particular) appear only in the context of a general outline of activities, without sufficient details.

The paper consists of seven sections. Section 2 gives a short historical introduction to the isomorphism problem of circulant graphs with a special emphasis on its natural generalization, suggested by L. Babai: the description of Cayley graphs with the Cayley isomorphism property (CI-graphs).

Section 3 presents a self-contained introduction to Schur rings. We start with the main definitions and the characterization of Schur rings as certain subalgebras of group algebras (3.1). Important properties of subgroups appearing as elements of S-rings (first formulated by O. Tamaschke) are given in 3.2. In 3.3 we consider three different sorts of S-ring isomorphisms (algebraic, combinatorial and Cayley), while in 3.4 we explain how isomorphisms of Cayley graphs are related to isomorphisms of S-rings. In 3.5, starting from a classical theorem of Schur, a Galois correspondence between S-rings and permutation groups is established. Example 3.6 provides the reader an opportunity to "play" a little bit with all the above-mentioned notions. We have to admit that Section 3 requires from the reader a certain level of a general algebraic culture.

Section 4 is a short digest of all results about Schur rings over cyclic groups C_n necessary for our presentation. We start from the classical theorem of Schur, claiming that for a non-prime n each non-trivial S-ring over C_n is imprimitive. A number of propositions presented in 4.2 was originally formulated and proved by Muzychuk in [**64**], a paper mostly aimed at experts in algebra. In this text we try to find transparent formulations of all claims used, such that they will be clear to a wider mathematical audience.

Section 5 properly concerns the isomorphism problem for circulant graphs. The machinery developed in the previous sections allows us to give short and rather simple proofs of various known and new results. In 5.1 we present the proof of Zibin's conjecture. As an immediate consequence we get the proof of Toida's conjecture and of the theorem about arc-transitive circulants (which recently was proved by C.H. Li with the aid of the Classification of Finite Simple Groups). It is worth mentioning that during the last two decades there appeared many papers dealing with various particular cases of the graph isomorphism problem for circulant graphs. As a rule these results may be reproved by the standard use of S-rings. In 5.2 we illustrate this claim, giving a new proof of the theorem by X. Fang and M. Xu [**26**] about the so-called double-loop circulants. We conclude Section 5 with a computational example: the solution of the isomorphism problem for 8-vertex circulant graphs. We think that a careful examination of this example will provide the reader with an extra insight, which will help to enter the subject more deeply.

Schur rings are just a particular case of more general structures, commonly treated in Algebraic Combinatorics: association schemes and coherent configurations. A brief outline of such interrelations is given in Section 6, however a deeper consideration of these notions is out of the scope of this paper.

In Sections 1-6 we were tried to avoid a huge amount of information of historical and bibliographical nature. This is why most of such information is concentrated in the final Section 7. Here an interested reader may consider the subject more widely, making a glance on other techniques and other classes of Cayley graphs. Nevertheless we are not pretending to submit in this section a comprehensive survey, being self-restricted by the consideration of only those results which have natural links with the main content of our paper.

2. Genesis of the problem

Let Γ be a (di)graph on n vertices which we identify with the elements of $\mathbb{Z}_n = \{0, 1, ..., n-1\}$. The (di)graph Γ is called ***circulant*** if it has a cyclic symmetry, *i.e.*, if the permutation $(0, 1, 2, ..., n-1)$ is an automorphism of the graph. Each circulant graph is completely determined by its ***connection set*** S which is the set of all points connected to 0 in Γ. Two vertices i and j are Γ-connected if and only if $i - j \in S$ where the difference is taken in the cyclic group $\mathbb{Z}_n$ (additive notation). Thus a circulant graph is a ***Cayley graph*** over the cyclic group $\mathbb{Z}_n$.

In what follows we shall write $\mathsf{Cay}(\mathbb{Z}_n, S)$ for a circulant graph $\Gamma = (V, E)$ with connection set S, *i.e.* $V = \mathbb{Z}_n$, $E = \{(x, y) \mid x - y \in S\}$.

The isomorphism problem for circulant graphs is stated as follows:

> Given two circulant graphs $\mathsf{Cay}(\mathbb{Z}_n, S)$ and $\mathsf{Cay}(\mathbb{Z}_n, T)$.
> Find an efficient algorithm which recognizes an isomorphism between these graphs.

Let $\mathbb{Z}_n^*$ be the set of all residues modulo n which are relatively prime to n. If $S = mT$ for some $m \in \mathbb{Z}_n^*$, then the permutation $x \mapsto mx, x \in \mathbb{Z}_n$ is an isomorphism between the two graphs. In this case we shall say that the connection sets (and the graphs) are ***conjugate by multiplier***. In 1967 A. Ádám conjectured that two circulant graphs are isomorphic if and only if they are conjugate by a multiplier [**1**]. Rather soon it was found that the conjecture is false in general: the minimal counterexample on 8 vertices was constructed by B. Elspas and J. Turner in 1970 [**21**]. In general, it was shown by B. Alspach and T.D. Parsons in 1979 [**2**] and, independently by V.N. Egorov and A.I. Markov [**20**] that Ádám's conjecture fails if n is divisible by 8 or by a square of an odd prime. On the other hand, it was shown that for specific values of n the conjecture is true (see Section 7 for a detailed survey). This led to the "corrected" Ádám's conjecture formulated by P.P. Pálfy in 1987 [**72**]:

> Adam's conjecture holds if and only if 8 does not divide n and p^2 does not divide n for each odd prime number p.

Since then, the topic has attracted the attention of many researchers. Two aspects have been considered:

1. To prove the corrected Ádám's conjecture;
2. To find an algorithm which solves the isomorphism problem for some specific values of n which are not covered by the corrected Ádám's conjecture.

An affirmative answer to the corrected Ádám's conjecture was given in 1995-97 by Muzychuk [**66, 67**]. The proof combines group theory with computations in Schur rings.

There are not many results in the second direction: in 1978, M. Klin and R Pöschel [**46**] and, in 1979, B. Alspach and T.D. Parsons [**2**] gave a complete solution for $n = p^2$. In 1980, Klin and Pöschel solved the isomorphism problem for $n = p^\alpha$, $p > 2$, [**43**] (in both cases p is a prime).

An important role in the recognition of isomorphisms is played by so-called ***layers*** of connection sets.

Let d be an arbitrary divisor of n. Then the d-th layer of $S \subseteq \mathbb{Z}_n$ is defined as $(S)_d := S \cap (\mathbb{Z}_n)_d$ where $(\mathbb{Z}_n)_d := \{x \in \mathbb{Z}_n \mid \gcd(x, n) = d\}$. It was shown by Klin and Pöschel [**43**] that if $n = p^\alpha$, then two circulants $\mathsf{Cay}(\mathbb{Z}_n, S)$ and $\mathsf{Cay}(\mathbb{Z}_n, T)$ are isomorphic if and only if for each $d|n$ there exist multipliers $m_d \in \mathbb{Z}_n^*$, such that $m_d(T)_d = (S)_d$ and which satisfy some compatibility condition.

The layers and multipliers appeared also in a computer enumeration of small circulants which was done in 1975 by D.K. Zibin' who made the following conjecture.

Zibin's conjecture [**88**]. If $\mathsf{Cay}(\mathbb{Z}_n, S)$ and $\mathsf{Cay}(\mathbb{Z}_n, T)$ are two isomorphic circulants, then there exist multipliers $m_d \in \mathbb{Z}_n^*$, $d|n$, such that $m_d(S)_d = (T)_d$.

A particular case of this conjecture was independently formulated by S. Toida in 1977:

Toida's conjecture [**80**]. Let $S \subseteq \mathbb{Z}_n^*$ and $T \subseteq \mathbb{Z}_n$. Then $\mathsf{Cay}(\mathbb{Z}_n, S)$ is isomorphic to $\mathsf{Cay}(\mathbb{Z}_n, T)$ if and only if the connection sets T and S are conjugate by a multiplier.

Inspired by Ádám's conjecture L. Babai started the study of the isomorphism problem of ***Cayley combinatorial objects*** in 1977 [**3**]. He showed that the solution of the isomorphism problem for arbitrary Cayley combinatorial objects may be "found" in the automorphism group of the object. To formulate his results we need a few definitions.

Let H be a finite group in multiplicative notation and let

$$\mathsf{Cay}(H, S) := \{(x, y) \in H^2 \mid xy^{-1} \in S\}$$

be an arbitrary Cayley graph over H ($S \subset H$). We shall say that two Cayley graphs $\mathsf{Cay}(H,S)$ and $\mathsf{Cay}(H,T)$ are ***Cayley isomorphic*** if there exists an automorphism $\varphi \in \mathsf{Aut}(H)$ such that $S^\varphi = T$ (S and T are ***conjugate*** by φ; $S^\varphi := \{s^\varphi \mid s \in S\}$, where we use the notation $s^\varphi := \varphi(s)$). It is a trivial observation that two Cayley isomorphic Cayley graphs are isomorphic as usual graphs. The converse is not true even for circulants. A subset $S \subseteq H$ is called a ***CI-subset*** if for each $T \subseteq H$ the graphs $\mathsf{Cay}(H,S)$, $\mathsf{Cay}(H,T)$ are isomorphic if and only if the sets T and S are conjugate by an element from $\mathsf{Aut}(H)$. A group H is called ***CI-group*** if each subset of H is a CI-subset.

Babai [**3**] gave a group-theoretical criterion whether a combinatorial "Cayley structure" has the CI-property. The restriction of his result to Cayley graphs immediately gives the following theorem.

THEOREM 2.1. *A subset $S \subseteq H$ is a CI-subset if and only if every two H-isomorphic*[1] *regular subgroups of* $\mathsf{Aut}(\mathsf{Cay}(H,S))$ *are conjugate in* $\mathsf{Aut}(\mathsf{Cay}(H,S))$.

Another, purely combinatorial approach, was suggested by Klin and Pöschel in 1978 [**46**]. They proposed to look more carefully at the combinatorial invariants of a graph in order to attack the isomorphism problem for Cayley graphs. They found a very strong combinatorial invariant for a Cayley graph: the Schur ring generated by its connection set. The key fact is that an isomorphism between Cayley graphs is also an isomorphism between S-rings generated by their connection sets. Since S-rings are algebraic objects, an isomorphism testing for S-rings is much easier than an isomorphism testing for Cayley graphs. We shall see that for cyclic graphs the structure properties of S-rings yield rather strong necessary conditions for graph isomorphism. To our knowledge there is no group-theoretical way to obtain these conditions. Finally, it is worth mentioning that in some cases the necessary conditions obtained using Schur rings are also sufficient, which leads to an efficient algorithm for recognizing graph isomorphism.

3. Schur rings.

3.1. Basic definitions. Let H be an arbitrary finite multiplicatively written group with identity e. The group algebra $\mathbb{Q}H$ consists of all formal sums $\sum_{h\in H}\alpha_h h$, $\alpha_h \in \mathbb{Q}$ which are multiplied according to the following rule

$$(\sum_{h\in H}\alpha_h h)(\sum_{h\in H}\beta_h h) = \sum_{h\in H}\sum_{f\in H}(\alpha_h\beta_f)(fh).$$

The ***Schur-Hadamard product*** in $\mathbb{Q}H$ is defined as follows:

$$(\sum_{h\in H}\alpha_h h)\circ(\sum_{h\in H}\beta_h h) = \sum_{h\in H}(\alpha_h\beta_h)h.$$

For each $m \in \mathbb{Z}$ and $x = \sum_{h\in H} x_h h$ let $x^{(m)} := \sum_{h\in H} x_h h^m$. For a nonempty subset $T \subseteq H$ we set $\underline{T} := \sum_{t\in T} t$. Following [**85**] elements of this form will be called ***simple quantities***. A simple quantity $\underline{T}$ is called ***symmetric*** if $T^{(-1)} = T$.

A subalgebra $\mathcal{A}$ of the group algebra $\mathbb{Q}H$ is called a ***Schur ring***, for short ***S-ring***, if it satisfies the following axioms:

(S1) $\mathcal{A}$ has a basis of simple quantities $\underline{T_0}, \ldots, \underline{T_r}$, where $T_0 = \{e\}$,
(S2) $T_i \cap T_j = \emptyset$ for $i \neq j$, and $\bigcup_{j=0}^r T_j = H$,

[1] H-isomorphic means isomorphic to H

(S3) For each $i \in \{0, 1, \ldots, r\}$ there exists $i' \in \{0, 1, \ldots, r\}$, such that $T_{i'} = T_i^{(-1)}$

By (S1) and (S2), the basis $\underline{T_0}, \ldots, \underline{T_r}$ is unique and it is called the ***standard basis*** of $\mathcal{A}$. The sets T_i, $0 \leq i \leq r$, are called the ***basic sets*** of $\mathcal{A}$ and the notation $\mathcal{A} = \langle \underline{T_0}, \ldots, \underline{T_r} \rangle$ will be used if $\mathcal{A}$ is an S-ring over H whose basic sets are $T_0, \ldots, T_r$. The set of all basic sets is also denoted by $\mathsf{Basic}(\mathcal{A})$. By (S2), every $h \in H$ is contained in exactly one basic set, which will be denoted by $T_{(h)}$. There are two ***trivial*** S-rings over an arbitrary group H, namely $\langle \underline{e}, \underline{G \setminus \{e\}} \rangle$ and $\mathbb{Q}H$.

The following characterization of Schur rings is folklore (see Section 7 for more details):

THEOREM 3.1. *Let $\mathcal{A} \subseteq \mathbb{Q}H$ be a subalgebra of the group algebra $\mathbb{Q}H$. Then $\mathcal{A}$ is an S-ring over H if and only if $\underline{e} \in \mathcal{A}$, $\underline{H} \in \mathcal{A}$ and $\mathcal{A}$ is closed with respect to $\circ$ and $^{(-1)}$.*

It follows from this result that the intersection of S-rings over H is an S-ring over the same group. Given $x_1, ..., x_k \in \mathbb{Q}H$, we define $\langle\langle x_1, ..., x_k \rangle\rangle$ to be the least S-ring which contains the elements $x_1, ..., x_k$. If $\underline{S}$ is a simple quantity, then we shall write $\langle\langle S \rangle\rangle$ instead of $\langle\langle \underline{S} \rangle\rangle$.

An S-ring $\mathcal{B} \subseteq \mathbb{Q}H$ is called a ***fusion*** of an S-ring $\mathcal{A} \subseteq \mathbb{Q}H$ (or an ***S-subring*** of $\mathcal{A}$) if $\mathcal{B} \subseteq \mathcal{A}$. An S-ring $\mathcal{A}$ is called a ***fission*** of an S-ring $\mathcal{B}$ if $\mathcal{B}$ is a fusion of $\mathcal{A}$. It is a trivial observation that $\mathcal{B}$ is a fusion of $\mathcal{A}$ if and only if each basic set T of $\mathcal{B}$ is the union of some basic sets of $\mathcal{A}$.

The following operation is a useful tool in computing $\langle\langle x_1, ..., x_k \rangle\rangle$. Let $q \in \mathbb{Q}$ be an arbitrary number. Define the mapping $I_q : \mathbb{Q}H \to \mathbb{Q}H$ by the following formula

$$I_q(\sum_{h \in H} x_h h) := \sum_{h \in H \wedge x_h = q} h\,.$$

PROPOSITION 3.2 (SCHUR-WIELANDT PRINCIPLE [**85**]). *If $\mathcal{A}$ is an S-ring over a group H, then $I_q(x) \in \mathcal{A}$ for each $x \in \mathcal{A}$ and $q \in \mathbb{Q}$.*

We finish this section by a simple example how to calculate $\langle\langle x \rangle\rangle$. Let $x = g^1 + g^2 + g^3$ be a simple quantity in the group ring $\mathbb{Z}C_8$. Then $x^2 = g^2 + 2g^3 + 3g^4 + 2g^5 + g^6$. By the Schur-Wielandt principle $g^2 + g^6, g^3 + g^5, g^4 \in \langle\langle x \rangle\rangle$. Now $g^3 = x \circ (g^3 + g^5) \in \langle\langle x \rangle\rangle$ implies that $\langle\langle x \rangle\rangle = \mathbb{Q}C_8$.

3.2. Subgroup properties of S-rings. Let $\mathcal{A} = \langle \underline{T_0}, ..., \underline{T_d} \rangle$ be an S-ring over a group H. A subgroup $F \leq H$ is called an $\mathcal{A}$-subgroup [**79**] if $\underline{F} \in \mathcal{A}$. There are two ***trivial*** $\mathcal{A}$-subgroups: $\{e\}$ and H. If $\mathcal{A}$ has a non-trivial $\mathcal{A}$-subgroup, then $\mathcal{A}$ is called ***imprimitive***, otherwise it is called ***primitive***. A subgroup $F \leq H$ is an $\mathcal{A}$-subgroup if and only if it is the union of some basic sets of $\mathcal{A}$.

PROPOSITION 3.3 ([**79**]). *Let F, G be two arbitrary $\mathcal{A}$-subgroups. Then $G \cap F$ and*[2] *$\langle F \cup G \rangle$ are $\mathcal{A}$-subgroups.*

In other words, the set of all $\mathcal{A}$-subgroups is a sublattice of the subgroup lattice of H. It is not difficult to show that $\langle T \rangle$ is an $\mathcal{A}$-subgroup for each $T \in \mathsf{Basic}(\mathcal{A})$.

For each $T \in \mathsf{Basic}(\mathcal{A})$ let

$$\begin{array}{lcll} \mathsf{L}(T) & := & \{h \in H \mid hT = T\} & \text{(left stabilizer)}, \\ \mathsf{R}(T) & := & \{h \in H \mid Th = T\} & \text{(right stabilizer)}, \\ \mathsf{Sta}(T) & := & \mathsf{L}(T) \cap \mathsf{R}(T) & \text{(stabilizer)}. \end{array}$$

[2]We recall that $\langle X \rangle$ denotes a subgroup generated by a subset $X \subseteq G$.

PROPOSITION 3.4 ([**85**, Prop. 23.5]). *For every* $T \in \mathsf{Basic}(\mathcal{A})$*, the stabilizers* $\mathsf{L}(T)$, $\mathsf{R}(T)$ *and* $\mathsf{Sta}(T)$ *are* $\mathcal{A}$*-subgroups.*

Let F be an $\mathcal{A}$-subgroup. Define $\mathcal{A}_F := \mathcal{A} \cap \mathbb{Q}F$. It is easy to check that $\mathcal{A}_F$ is an S-ring over the group F and $\mathsf{Basic}(\mathcal{A}_F) = \{T \in \mathsf{Basic}(\mathcal{A}) \mid T \subset F\}$. We shall call $\mathcal{A}_F$ an ***induced S-subring*** of $\mathcal{A}$. If F is an $\mathcal{A}$-subgroup which is normal in H one can define a ***quotient*** S-ring $\mathcal{A}/F$ over the factor-group H/F [**76**], as as shown in the next proposition. Notice that the natural homomorphism $\pi : H \to H/F$ can be canonically extended to an homomorphism $\mathbb{Q}H \to \mathbb{Q}(H/F)$ that also will be denoted by π.

PROPOSITION 3.5 ([**76**]). *Let* $\mathcal{A}$ *be an S-ring over* H *and* $F \trianglelefteq H$ *be a normal* $\mathcal{A}$*-subgroup. Then the algebra* $\mathcal{A}/F := \pi(\mathcal{A})$ *is an S-ring over* H/F *the basic sets of which are given by* $\mathsf{Basic}(\mathcal{A}/F) = \{T/F \mid T \in \mathsf{Basic}(\mathcal{A})\}$.[3]

3.3. Isomorphisms between S-rings. Two S-rings $\mathcal{A}$ and $\mathcal{B}$ are called ***algebraically isomorphic***, notation $\mathcal{A} \cong_{alg} \mathcal{B}$, if there exists a bijection $g : \mathsf{Basic}(\mathcal{A}) \to \mathsf{Basic}(\mathcal{B}) : T \mapsto T^g$ such that the map $\underline{T} \mapsto \underline{T^g}$ induces (via $\sum \alpha_i \underline{T_i} \mapsto \sum \alpha_i \underline{T_i^g}$) an isomorphism between the algebras $\mathcal{A}$ and $\mathcal{B}$.

Two S-rings $\mathcal{A} = \langle \underline{T_0}, ..., \underline{T_r} \rangle \subseteq \mathbb{Q}H$ and $\mathcal{B} = \langle \underline{S_0}, ..., \underline{S_s} \rangle \subseteq \mathbb{Q}K$ will be called ***combinatorially isomorphic***, notation $\mathcal{A} \cong_{com} \mathcal{B}$, if $r = s$ and there exists a bijection $f : H \to K$ such that $\{\mathsf{Cay}(H, T_i)^f \mid i = 0, \ldots, r\} = \{\mathsf{Cay}(K, S_i) \mid i = 0, \ldots, s\}$. (Here we use the notation $R^f := \{(x^f, y^f) \mid (x, y) \in R\}$ for $R \subseteq H \times H$.) The bijection f is called a ***combinatorial isomorphism***. Each combinatorial isomorphism f induces a bijection $f^* : \mathsf{Basic}(\mathcal{A}) \to \mathsf{Basic}(\mathcal{B})$ given by $\mathsf{Cay}(H, T)^f = \mathsf{Cay}(K, T^{f^*})$ for $T \in \mathsf{Basic}(\mathcal{A})$. A combinatorial isomorphism $f : H \to K$ is called ***normalized***[4] if $f(e) = e$. If f is a normalized combinatorial isomorphism, then $T^f = T^{f^*}$ holds for each $T \in \mathsf{Basic}(\mathcal{A})$.

PROPOSITION 3.6 ([**69**]). *If* f *is a combinatorial isomorphism between two S-rings* $\mathcal{A}$ *and* $\mathcal{B}$*, then* f^* *is an algebraic isomorphism between them.*

Two S-rings are called ***Cayley isomorphic***, notation $\mathcal{A} \cong_{Cay} \mathcal{B}$, if there exists a combinatorial isomorphism $f : H \to K$ which is simultaneously an isomorphism from the group H onto K. The connection between the three types of isomorphisms is given by the following implications:

$$\mathcal{A} \cong_{Cay} \mathcal{B} \quad \Longrightarrow \quad \mathcal{A} \cong_{com} \mathcal{B} \quad \Longrightarrow \quad \mathcal{A} \cong_{alg} \mathcal{B}.$$

But the converse is not true in general:

$$\mathcal{A} \cong_{Cay} \mathcal{B} \quad \not\Longleftarrow \quad \mathcal{A} \cong_{com} \mathcal{B} \quad \not\Longleftarrow \quad \mathcal{A} \cong_{alg} \mathcal{B}.$$

We shall give a simple example which shows that $\mathcal{A} \cong_{Cay} \mathcal{B} \not\Longleftarrow \mathcal{A} \cong_{com} \mathcal{B}$. Let $H = \mathbb{Z}_4$ and $K = \mathbb{Z}_2 \oplus \mathbb{Z}_2$ be two abelian groups of order 4 written additively. Take the following Schur rings $\mathcal{A} = \langle \underline{0}, \underline{2}, \underline{1,3} \rangle \subseteq \mathbb{Q}\mathbb{Z}_4$, $\mathcal{B} = \langle \underline{(0,0)}, \underline{(1,0)}, \underline{(1,1),(0,1)} \rangle \subseteq \mathbb{Q}[\mathbb{Z}_2 \oplus \mathbb{Z}_2]$. These S-rings cannot be Cayley isomorphic, since $\mathbb{Z}_4 \not\cong \mathbb{Z}_2 \oplus \mathbb{Z}_2$. On the other hand the bijection $0 \mapsto (0,0)$, $1 \mapsto (0,1)$, $2 \mapsto (1,0)$, $3 \mapsto (1,1)$ is a combinatorial isomorphism between them.

Let $\mathcal{A} \subseteq \mathbb{Q}H$ and $\mathcal{B} \subseteq \mathbb{Q}K$ be two S-rings over the groups H and K respectively. We shall write $\mathsf{Iso}_{Cay}(\mathcal{A}, \mathcal{B}), \mathsf{Iso}_{com}(\mathcal{A}, \mathcal{B}), \mathsf{Iso}_{alg}(\mathcal{A}, \mathcal{B})$ for the sets of all Cayley,

[3]Here T/F stands for the set of left F-cosets: $T/F := \{tF \mid t \in T\}$.

[4]S-isomorphisms in [**47**]

combinatorial and algebraic isomorphisms, respectively, between $\mathcal{A}$ and $\mathcal{B}$. If $\mathcal{A} = \mathcal{B}$, then

$$\begin{aligned} \mathsf{Aut}_{Cay}(\mathcal{A}) &:= \mathsf{Iso}_{Cay}(\mathcal{A},\mathcal{A}) \leq \mathsf{Sym}(H); \\ \mathsf{Aut}_{com}(\mathcal{A}) &:= \mathsf{Iso}_{com}(\mathcal{A},\mathcal{A}) \leq \mathsf{Sym}(H); \\ \mathsf{Aut}_{alg}(\mathcal{A}) &:= \mathsf{Iso}_{alg}(\mathcal{A},\mathcal{A}) \leq \mathsf{Sym}(\mathsf{Basic}(\mathcal{A})). \end{aligned}$$

The subgroup of $\mathsf{Aut}_{com}(\mathcal{A})$ consisting of all automorphisms f which induce the trivial algebraic automorphism of $\mathcal{A}$ (i.e. f^* is the identity) will be denoted by $\mathsf{Aut}(\mathcal{A})$. We call these automorphisms ***strong automorphisms*** of a Schur ring (cf. with similarly defined notions in [**22**]). If $\mathsf{Basic}(\mathcal{A}) = \{T_0, ..., T_r\}$, then

$$\mathsf{Aut}(\mathcal{A}) = \bigcap_{0\leq i\leq r} \mathsf{Aut}(\mathsf{Cay}(H, T_i)).$$

We have $\mathsf{Aut}(\mathcal{A}) \trianglelefteq \mathsf{Aut}_{com}(\mathcal{A})$.

The symmetric group $\mathsf{Sym}(H)$ contains two natural copies of H, namely H_R and H_L defined as follows:

$$\begin{aligned} H_R &:= \{h_R \mid h \in H\}, \quad x^{h_R} := xh,\ x \in H, \\ H_L &:= \{h_L \mid h \in H\}, \quad x^{h_L} := h^{-1}x,\ x \in H\,. \end{aligned}$$

Since H_R is a subgroup of the automorphism group of an arbitrary Cayley graph $\mathsf{Cay}(H, S)$, we have $H_R \leq \mathsf{Aut}(\mathcal{A})$ for each S-ring $\mathcal{A} \subseteq \mathbb{Q}H$.

The group $\mathsf{Aut}_{Cay}(\mathcal{A})$ is a subgroup of the normalizer of H_R in $\mathsf{Sym}(H)$ which is isomorphic to the holomorph $H \rtimes \mathsf{Aut}(H)$ of H and acts on H by the following rule: $x^{(h,\varphi)} = x^{\varphi}h$. More precisely $\mathsf{Aut}_{Cay}(\mathcal{A}) = H \rtimes \Phi$ where $\Phi \leq \mathsf{Aut}(H)$ consists of all automorphisms which satisfy

$$T^{\varphi} = T \text{ for all } T \in \mathsf{Basic}(\mathcal{A})\,.$$

An S-ring $\mathcal{A}$ over a group H is called ***CI-S-ring*** if for arbitrary S-ring $\mathcal{B}$ over H and any $f \in \mathsf{Iso}_{com}(\mathcal{A},\mathcal{B})$ there exists an automorphism $g \in \mathsf{Aut}(H)$ such that $T^{f^*} = T^g$ for each $T \in \mathsf{Basic}(\mathcal{A})$. If $f \in \mathsf{Iso}_{com}(\mathcal{A},\mathcal{B})$ is a combinatorial isomorphism, then fh_R is also a combinatorial isomorphism between the same Schur rings, and, moreover $f^* = (fh_R)^*$. If $h = f(e)^{-1}$, then fh_R will be a normalized combinatorial isomorphism. Therefore an S-ring $\mathcal{A}$ is a CI-S-ring if and only if for each normalized combinatorial isomorphism $f \in \mathsf{Iso}_{com}(\mathcal{A},\mathcal{B})$ there exists an automorphism $\varphi \in \mathsf{Aut}(H)$ such that $T^f = T^{\varphi}$ for each $T \in \mathsf{Basic}(\mathcal{A})$.

3.4. S-rings and Cayley graphs. Given a Cayley graph $\mathsf{Cay}(H, S), S \subseteq H$, we can associate the S-ring $\langle\langle \underline{S} \rangle\rangle$ generated by its connection set. It turns out that there exists a deep connection between the properties of S and $\langle\langle \underline{S} \rangle\rangle$. In what follows we shall write $\langle\langle S \rangle\rangle$ instead of $\langle\langle \underline{S} \rangle\rangle$.

THEOREM 3.7 ([**69**]). *Let $f : H \to K$ be an isomorphism between the Cayley graphs $\mathsf{Cay}(H, S)$ and $\mathsf{Cay}(K, R)$. Then*

(1) *f is a combinatorial isomorphism between $\langle\langle S \rangle\rangle$ and $\langle\langle R \rangle\rangle$. In particular* $|\mathsf{Basic}(\langle\langle S \rangle\rangle)| = |\mathsf{Basic}(\langle\langle R \rangle\rangle)|$;
(2) $\langle\langle S \rangle\rangle \cong_{alg} \langle\langle R \rangle\rangle$;
(3) *Let $S = \bigcup\{B \mid B \in I\}$ be the decomposition of S into a disjoint union of basic sets of $\langle\langle S \rangle\rangle$ ($I \subseteq \mathsf{Basic}(\mathcal{A})$), then $R = \bigcup\{B^{f^*} \mid B \in I\}$.*

In a particular case, namely when $H = K$ and $S = T$ we have a stronger result which is very useful in computing of the automorphism group of a Cayley graph.

THEOREM 3.8 ([**69**]). *Let $f \in \mathsf{Sym}(H)$ be an automorphism of a Cayley graph* $\mathsf{Cay}(H,S)$. *Then $f \in \mathsf{Aut}(\langle\langle S\rangle\rangle)$ and $f^* \in \mathsf{Aut}_{alg}(\langle\langle S\rangle\rangle)$.*

As a consequence we can easily obtain the following useful statement:

THEOREM 3.9. *A subset $S \subseteq H$ is a CI-subset if and only if $\langle\langle S\rangle\rangle$ is a CI-S-ring over H.*

PROOF. Let S be an CI-subset, $\mathcal{A} := \langle\langle S\rangle\rangle$ and $f \in \mathsf{Iso}_{com}(\mathcal{A},\mathcal{B})$ be an arbitrary combinatorial isomorphism between $\mathcal{A}$ and some S-ring $\mathcal{B}$. W.l.o.g. we may assume that f is normalized. Let $S = \bigcup_{i\in I} S_i$ be a decomposition of S into basic sets of $\mathcal{A}$. Then $\mathsf{Cay}(H,S_i)^f = \mathsf{Cay}(H,S_i^{f^*})$ and $\mathsf{Cay}(H,S)^f = \mathsf{Cay}(H,T)$, where $T = \bigcup_{i\in I} S_i^{f^*}$. Since S is a CI-subset, there exists $g \in \mathsf{Aut}(H)$ such that $T = S^g$. Therefore $fg^{-1} \in \mathsf{Aut}(\mathsf{Cay}(H,S))$. By Theorem 3.7, $fg^{-1} \in \mathsf{Aut}(\mathcal{A})$, implying $R^{(fg^{-1})^*} = R$ for each $R \in \mathsf{Basic}(\mathcal{A})$. Since fg^{-1} is normalized, $R^{fg^{-1}} = R$ implying that $R^{f^*} = R^f = R^g$ for each $R \in \mathsf{Basic}(\mathcal{A})$, as desired.

Assume now that $\mathcal{A} := \langle\langle S\rangle\rangle$ is a CI-S-ring and let $\mathsf{Cay}(H,T)$ be a Cayley graph isomorphic to $\mathsf{Cay}(H,S)$. Let f be an arbitrary isomorphism between $\mathsf{Cay}(H,S)$ and $\mathsf{Cay}(H,T)$, i.e., $\mathsf{Cay}(H,S)^f = \mathsf{Cay}(H,T)$. Then, by Theorem 3.7, f is a combinatorial isomorphism between the S-rings $\mathcal{A}$ and $\mathcal{B} := \langle\langle T\rangle\rangle$. Since $\mathcal{A}$ is a CI-S-ring there exists $g \in \mathsf{Aut}(H)$ such that $fg^{-1} \in \mathsf{Aut}(\mathcal{A})$. Then $R^{f^*} = R^g$ holds for each $R \in \mathsf{Basic}(\mathcal{A})$ and our claim follows easily from part (3) of Theorem 3.7. □

The following result is a direct analogue to Theorem 2.1:

THEOREM 3.10. *An S-ring $\mathcal{A}$ over a group H is a CI-S-ring if and only if any two regular H-isomorphic subgroups of $\mathsf{Aut}(\mathcal{A})$ are conjugate in $\mathsf{Aut}(\mathcal{A})$.*

We shall say that an algebraic isomorphism $f : \mathcal{A} \to \mathcal{B}$ is ***induced by a group automorphism*** if there exists a group automorphism $\varphi \in \mathsf{Aut}(H)$ such that $T^f = T^\varphi$ for each basic set T of $\mathcal{A}$. The following claim is a direct consequence of the definition of an CI-S-ring and gives a sufficient condition for being a CI-S-ring which easily can be verified.

THEOREM 3.11. *Let $\mathcal{A}$ be an S-ring over a group H. Assume that for every S-ring $\mathcal{B} \subseteq \mathbb{Q}H$ algebraically isomorphic to $\mathcal{A}$ each algebraic isomorphism $f \in \mathsf{Iso}_{alg}(\mathcal{A},\mathcal{B})$ is induced by a group automorphism. Then $\mathcal{A}$ is a CI-S-ring.*

3.5. S-rings and permutation groups. As we have seen before, the automorphism group of a Cayley graph $\mathsf{Cay}(H,S)$ contains the subgroup H_R which is a regular subgroup of the symmetric group $\mathsf{Sym}(H)$. Consider an arbitrary group $G \le \mathsf{Sym}(H)$ which contains H_R as a subgroup. Let $\{e\} = T_0, T_1, ..., T_r$ be all the orbits of G_e. The vector space spanned by the simple quantities $\underline{T_i}$ $(i \in \{0,\dots,r\})$ is called the ***transitivity module*** of G and is denoted by $\mathcal{S}(H,G_e)$. The following result was proved by I. Schur in 1933.

THEOREM 3.12 ([**85**, Theorem 24.1]). *The transitivity module $\mathcal{S}(H,G_e)$ is an S-ring over H.*

Not every S-ring is the transitivity module of some permutation group. An S-ring which is the transitivity module of some group G, $H_R \le G \le \mathsf{Sym}(H)$, is called a ***Schurian S-ring*** ([**73**]).

Let $\Phi \le \mathsf{Aut}(H)$ be an arbitrary subgroup. The group $G := H \rtimes \Phi$ acts on H by the following rule: $x^{(h,\varphi)} = x^\varphi h$. Since G contains H_R, the transitivity module

$\mathcal{S}(H, G_e)$ is an S-ring. The basic sets of this S-ring are exactly the orbits of Φ on H. It may be characterized as the subspace of $\mathbb{Q}H$ which consists of all elements fixed by Φ. We denote this S-ring by $\mathcal{S}(H, \Phi)$.

The connection between permutation groups and Schur rings is reflected in the following theorem.

THEOREM 3.13. *Let $\mathcal{A}$ and $\mathcal{B}$ be arbitrary S-rings over a group H and let G and K be arbitrary overgroups of H_R in $\mathsf{Sym}(H)$. Then*

(1) $\mathcal{A} \subseteq \mathcal{B} \implies \mathsf{Aut}(\mathcal{A}) \geq \mathsf{Aut}(\mathcal{B})$,
(2) $G \leq K \implies \mathcal{S}(H, G_e) \supseteq \mathcal{S}(H, K_e)$,
(3) $\mathcal{A} \subseteq \mathcal{S}(H, \mathsf{Aut}(\mathcal{A})_e)$,
(4) $G \leq \mathsf{Aut}(\mathcal{S}(H, G_e))$.

This statement describes a Galois correspondence between S-rings over H and overgroups of the regular permutation group H_R in $\mathsf{Sym}(H)$. It is a particular case of a Galois correspondence between coherent configurations and permutation groups [**23, 25**]. Theorem 3.13 implies that an S-ring $\mathcal{A}$ is Schurian if and only if $\mathcal{A} = \mathcal{S}(H, \mathsf{Aut}(\mathcal{A})_e)$.

3.6. An example. In this subsection we show by example how to use S-ring theory for proving that a certain group is a CI-group. We take the quaternion group Q_8 which is known to be a CI-group (see, for example, [**54**]). We use the standard notation for the elements of Q_8 : $1, \mathsf{i}, \mathsf{j}, \mathsf{k}, -1, -\mathsf{i}, -\mathsf{j}, -\mathsf{k}$ where $\mathsf{i}^2 = \mathsf{j}^2 = \mathsf{k}^2 = \mathsf{ijk} = -1$.

The complete list of all S-rings over Q_8 can be found e.g. by the computer package COCO [**24**]. There are 24 non-trivial S-rings over Q_8 which are spread into 10 Cayley-isomorphism classes. We give a list of representatives of these classes:

$\mathcal{A}_1 = \langle \underline{T_0}, \underline{T_1}, \underline{T_2}, \underline{T_3}, \underline{T_4}, \underline{T_5} \rangle$ with
$T_0 = \{1\},\ T_1 = \{\mathsf{k}\},\ T_2 = \{-1\},\ T_3 = \{-\mathsf{k}\},\ T_4 = \{\mathsf{i}, -\mathsf{i}\},\ T_5 = \{\mathsf{j}, -\mathsf{j}\}$;

$\mathcal{A}_2 = \langle \underline{T_0}, \underline{T_1}, \underline{T_2}, \underline{T_3}, \underline{T_4} \rangle$ with
$T_0 = \{1\},\ T_1 = \{-1\},\ T_2 = \{\mathsf{i}, -\mathsf{i}\},\ T_3 = \{\mathsf{j}, \mathsf{k}\},\ T_4 = \{-\mathsf{j}, -\mathsf{k}\}$;

$\mathcal{A}_3 = \langle \underline{T_0}, \underline{T_1}, \underline{T_2}, \underline{T_3}, \underline{T_4} \rangle$ with
$T_0 = \{1\},\ T_1 = \{-1\},\ T_2 = \{\mathsf{i}, -\mathsf{i}\},\ T_3 = \{\mathsf{j}, -\mathsf{j}\},\ T_4 = \{\mathsf{k}, -\mathsf{k}\}$;

$\mathcal{A}_4 = \langle \underline{T_0}, \underline{T_1}, \underline{T_2}, \underline{T_3}, \underline{T_4} \rangle$ with
$T_0 = \{1\},\ T_1 = \{\mathsf{k}\},\ T_2 = \{-1\},\ T_3 = \{-\mathsf{k}\},\ T_4 = \{\mathsf{i}, -\mathsf{i}, \mathsf{j}, -\mathsf{j}\}$;

$\mathcal{A}_5 = \langle \underline{T_0}, \underline{T_1}, \underline{T_2}, \underline{T_3} \rangle$ with
$T_0 = \{1\},\ T_1 = \{-1\},\ T_2 = \{\mathsf{k}, -\mathsf{k}\},\ T_3 = \{\mathsf{i}, -\mathsf{i}, \mathsf{j}, -\mathsf{j}\}$;

$\mathcal{A}_6 = \langle \underline{T_0}, \underline{T_1}, \underline{T_2}, \underline{T_3} \rangle$ with
$T_0 = \{1\},\ T_1 = \{-1\},\ T_2 = \{\mathsf{i}, \mathsf{j}, \mathsf{k}\},\ T_3 = \{-\mathsf{i}, -\mathsf{j}, -\mathsf{k}\}$;

$\mathcal{A}_7 = \langle \underline{T_0}, \underline{T_1}, \underline{T_2} \rangle$ with $T_0 = \{1\},\ T_1 = \{\mathsf{k}, -1, -\mathsf{k}\},\ T_2 = \{\mathsf{i}, \mathsf{j}, -\mathsf{i}, -\mathsf{j}\}$;

$\mathcal{A}_8 = \langle \underline{T_0}, \underline{T_1}, \underline{T_2} \rangle$ with $T_0 = \{1\},\ T_1 = \{-1\},\ T_2 = \{\mathsf{i}, \mathsf{j}, \mathsf{k}, -\mathsf{i}, -\mathsf{j}, -\mathsf{k}\}$;

$\mathcal{A}_9 = \langle \underline{T_0}, \underline{T_1} \rangle$ with $T_0 = \{1\},\ T_1 = \{-1, \mathsf{i}, \mathsf{j}, \mathsf{k}, -\mathsf{i}, -\mathsf{j}, -\mathsf{k}\}$;

$\mathcal{A}_{10} = \mathbb{Q}Q_8$.

Our goal is to show that each S-ring in the list is a CI-S-ring. By Theorem 3.11 it is sufficient to show that each algebraic isomorphism between two S-rings is induced by a group automorphism. Thus we have to study algebraic isomorphisms between the enumerated S-rings. Two algebraically isomorphic S-rings not only have the same number of basic sets, but also have the same multisets of basic set cardinalities. Using this observation we can easily find that all S-rings $\mathcal{A}_1 - \mathcal{A}_{10}$ are pairwise non-isomorphic algebraically. Thus each algebraic isomorphism between two S-rings is, in fact, an algebraic automorphism of one of the S-rings $\mathcal{A}_i$, $i = 1, ..., 8$. There are S-rings the non-trivial basic sets of which are of distinct cardinalities (S-rings $\mathcal{A}_5, \mathcal{A}_7, \mathcal{A}_8, \mathcal{A}_9$); $\mathcal{A}_{10}$ is a trivial S-ring and therefore it is an CI-S-ring. Therefore their groups of algebraic automorphisms are trivial and the corresponding S-rings are CI. We consider the remaining cases.
S-ring $\mathcal{A}_1$: There is only one basic set X which satisfies $X^2 = T_0$, namely T_2. Therefore $T_2^f = T_2$ for each $f \in \mathsf{Aut}_{alg}\mathcal{A}_1$. Analogously, the equation $X^2 = T_2$ has two solutions: T_1 and T_3. Hence $\{T_1, T_3\}^f = \{T_1, T_3\}$ for every $f \in \mathsf{Aut}_{alg}\mathcal{A}_1$. As we mentioned before, an algebraic automorphism preserves the cardinality of basic sets. Therefore $\{T_4, T_5\}^f = \{T_4, T_5\}$ and, finally, $\mathsf{Aut}_{alg}\mathcal{A}_1 \leq \{1, (13), (45), (13)(45)\}$. On the other hand, a direct check shows $(\mathsf{Aut}_{Cay}\mathcal{A}_1)^* = \{1, (13), (45), (13)(45)\}$. By Theorem 3.11, $\mathcal{A}_1$ is a CI-S-ring.
S-ring $\mathcal{A}_2$: Since T_1 is the unique non-trivial basic set of cardinality 1, $T_1^f = T_1$ for each $f \in \mathsf{Aut}_{alg}\mathcal{A}_1$. T_2 is the unique symmetric basic set of cardinality 2. Therefore $T_2^f = T_2$ for each $f \in \mathsf{Aut}_{alg}\mathcal{A}_1$. Thus $\mathsf{Aut}_{alg}\mathcal{A}_2 \leq \{1, (34)\}$. Together with $(\mathsf{Aut}_{Cay}\mathcal{A}_2)^* = \{1, (34)\}$ we obtain that $\mathcal{A}_2$ is a CI-S-ring.

In the same fashion one can show that every S-ring in our list has the property $(\mathsf{Aut}_{Cay}\mathcal{A}_i)^* = \mathsf{Aut}_{alg}\mathcal{A}_i, i = 1, ..., 10$. Thus, by Theorem 3.11, every S-ring over Q_8 is a CI-S-ring and, therefore, Q_8 is a CI-group.

4. Schur rings over a cyclic group

In order to avoid confusion between addition in the group algebra and addition in $\mathbb{Z}_n$ we shall use multiplicative notation for the cyclic group of order n. Thus C_n will stand for a cyclic group of order n, the identity element will be denoted by e, and by g_0 we denote a generator of the group. The automorphism group of C_n will be identified with $\mathbb{Z}_n^* := \{x \in \mathbb{Z}_n \mid \gcd(x, n) = 1\}$, where the action of $m \in \mathbb{Z}_n^*$ on C_n is defined by $g \mapsto g^m$. If $X = \sum_{g \in C_n} x_g g$ is an arbitrary element of $\mathbb{Q}C_n$, then for an integer m let $X^{(m)}$ denote the element $\sum_{g \in C_n} x_g g^m$. The mapping $X \mapsto X^{(m)}$ is bijective if and only if $\gcd(m, n) = 1$. For $T \subseteq C_n$ and an integer m let $T^{(m)} := \{t^m \mid t \in T\}$. If $\gcd(m, n) = 1$, then $\underline{T}^{(m)} = \underline{T^{(m)}}$. For an arbitrary subset $S \subseteq C_n$ we define two groups: $\mathsf{Sta}_+(S) := \{g \in C_n \mid gS = S\} \leq C_n$ and $\mathsf{Sta}_*(S) := \{m \in \mathbb{Z}_n^* \mid S^{(m)} = S\} \leq \mathbb{Z}_n^*$.

4.1. Classical results.

The first result about the structure of S-rings over a cyclic group was obtained by I. Schur in 1933.

THEOREM 4.1 ([**77**]). *If n is not prime, then each non-trivial S-ring over C_n is imprimitive.*

If n is prime, then each S-ring over C_n will be primitive. The classification of all S-rings over a cyclic group of prime order p can be easily obtained (see the references in Section 7).

THEOREM 4.2. *Let p be a prime number. Then each S-ring over C_p coincides with $\mathcal{S}(C_p, \Phi)$ for some $\Phi \leq \mathbb{Z}_n^*$.*

The following statement is very useful for the study of S-rings over cyclic groups.

PROPOSITION 4.3. [**85**] *Let $\mathcal{A}$ be an S-ring over C_n. Then, for each $m \in \mathbb{Z}_n^*$, we have $T \in \mathsf{Basic}(\mathcal{A}) \iff T^{(m)} \in \mathsf{Basic}(\mathcal{A})$.*

Two basic sets S and T of $\mathcal{A}$ are called ***conjugate*** if there exists an $m \in \mathbb{Z}_n^*$ such that $S^{(m)} = T$. It is easy to see that conjugacy is an equivalence relation on $\mathsf{Basic}(\mathcal{A})$. The set $\mathring{T} := \bigcup\{T^{(m)} \mid m \in \mathbb{Z}_n^*\}$ is called the ***trace*** of T. Let $\mathring{\mathcal{A}}$ denote the vector space spanned by the simple quantities $\underline{\mathring{T}}$, $T \in \mathsf{Basic}(\mathcal{A})$. Since $\mathring{\mathcal{A}} = \mathcal{A} \cap \mathcal{S}(C_n, \mathbb{Z}_n^*)$, $\mathring{\mathcal{A}}$ is an S-ring over C_n which is a fusion of $\mathcal{A}$. An S-ring is called ***rational*** (or an ***S-ring of traces***) if $\mathcal{A} = \mathring{\mathcal{A}}$.

4.2. Algebraic isomorphisms between S-rings over a cyclic group. We start with the following theorem.

THEOREM 4.4 ([**64**, Theorem 1.1′]). *Let $\lambda : \mathcal{A} \to \mathcal{B}$ be an algebraic isomorphism between two S-rings $\mathcal{A}$ and $\mathcal{B}$ over C_n and let $\underline{T}$ be a basic quantity of $\mathcal{A}$. Then $(\underline{T})^\lambda = \underline{T}^{(m)}$ for some $m \in \mathbb{Z}_n^*$.*

REMARK. Notice that the exponent m may depend on the basic set T.

As a consequence of this theorem and Proposition 4.3 we obtain:

THEOREM 4.5. ([**64**, Theorem 1.1]) *Any two algebraically isomorphic S-rings over C_n coincide.*

Theorem 4.4 and Theorem 4.5 imply the following theorem.

THEOREM 4.6 ([**64**]). *Let $\mathcal{A} = \langle \underline{T_0}, \dots, \underline{T_r} \rangle$ be an S-ring over C_n and let $\lambda : \mathsf{Basic}(\mathcal{A}) \to \mathsf{Basic}(\mathcal{A})$ be an algebraic automorphism of $\mathcal{A}$. Then for every $T \in \mathsf{Basic}(T)$ there exists $m_T \in \mathbb{Z}_n^*$ such that $T^\lambda = T^{(m_T)}$.*

Thus each algebraic automorphism of $\mathcal{A}$ is uniquely determined by the data $(m_T)_{T \in \mathsf{Basic}(\mathcal{A})}$ but not vice versa, since each m_T may be replaced by any product $m_T c$ with $c \in \mathsf{Sta}_*(T)$. On the other hand, an arbitrary choice of $(m_T)_{T \in \mathsf{Basic}(\mathcal{A})}$ may not correspond to a combinatorial automorphism of $\mathcal{A}$, since there may be some dependencies between the exponents m_T. To show the simplest connections between the exponents we need the following statement.

LEMMA 4.7 ([**64**, Prop. 3.1]). *Let f be an algebraic automorphism of an S-ring $\mathcal{A}$ over C_n. Then for every simple quantity $\underline{T} \in \mathcal{A}$ and $m \in \mathbb{Z}_n^*$ we have*

(1) $(\underline{T}^{(m)})^f = (\underline{T}^f)^{(m)}$,
(2) $\mathsf{Sta}_*(\underline{T}^f) = \mathsf{Sta}_*(\underline{T})$.

For $d|n$, let $(C_n)_d$ denote the set of all elements in C_n of order n/d. Given $S \subseteq C_n$ we define the ***layer***

$$(S)_d := S \cap (C_n)_d .$$

LEMMA 4.8 ([**69**]). *Let f be an algebraic automorphism of an S-ring $\mathcal{A}$ over C_n. Then for each $d|n$ there exists an $m_d \in \mathbb{Z}_n^*$ such that $T^f = T^{(m_d)}$ for all $T \in \mathsf{Basic}(\mathcal{A})$ with $(T)_d \neq \emptyset$.*

PROOF. Fix an arbitrary $T \in \mathsf{Basic}(\mathcal{A})$ with $T \cap (C_n)_d \neq \emptyset$. By Theorem 4.6 we know that $T^f = T^{(m_T)}$ for some $m_T \in \mathbb{Z}_n^*$. If T and T' are two basic sets of $\mathcal{A}$ which intersect non-trivially $(C_n)_d$, then there exists an $m \in \mathbb{Z}_n^*$ with $T^{(m)} \cap T' \neq \emptyset$. Therefore $T' = T^{(m)}$. By Lemma 4.7, $T'^f = (T^{(m)})^f = (T^f)^{(m)} = (T^{(m_T)})^{(m)} = T^{(m_T m)} = T^{(m m_T)} = T'^{(m_T)}$. Thus we can choose $m_d := m_T$. □

5. Application of Schur rings to the isomorphism problem for circulants

5.1. Proofs of Zibin's and Toida's conjectures.

THEOREM 5.1 (ZIBIN'S CONJECTURE [69]). *Let $\mathsf{Cay}(C_n, S)$ and $\mathsf{Cay}(C_n, R)$ be two isomorphic circulant graphs. Then for each $d|n$ there exists $m_d \in \mathbb{Z}_n^*$ such that $(S)_d^{(m_d)} = (R)_d$.*

PROOF. Let $f : C_n \to C_n$ be an isomorphism between the Cayley graphs $\mathsf{Cay}(C_n, S)$ and $\mathsf{Cay}(C_n, R)$. Then, by Theorem 3.7(1)-(2), f is a combinatorial isomorphism between $\langle\langle S\rangle\rangle$ and $\langle\langle R\rangle\rangle$ which induces an algebraic isomorphism f^* between $\mathsf{B}_S := \mathsf{Basic}(\langle\langle S\rangle\rangle)$ and $\mathsf{B}_R := \mathsf{Basic}(\langle\langle R\rangle\rangle)$. By 3.7(3) we have $R = \bigcup\{T^{f^*} \mid T \in \mathsf{B}_S \text{ and } T \subseteq S\}$. Therefore

$$\begin{aligned}
(R)_d &= \bigcup\{T^{f^*} \cap (C_n)_d \mid T \in \mathsf{B}_S \text{ and } T \subseteq S\} \\
&\overset{(4.6)}{=} \bigcup\{T^{(m_T)} \cap (C_n)_d \mid T \in \mathsf{B}_S \text{ and } T \subseteq S\} \\
&= \bigcup\{(T \cap (C_n)_d)^{(m_T)} \mid T \in \mathsf{B}_S \text{ and } T \subseteq S\} \\
&= \bigcup\{(T \cap (C_n)_d)^{(m_T)} \mid T \in \mathsf{B}_S, T \subseteq S, T \cap (C_n)_d \neq \emptyset\} \\
&\overset{(4.8)}{=} \bigcup\{(T \cap (C_n)_d)^{(m_d)} \mid T \in \mathsf{B}_S, T \subseteq S, T \cap (C_n)_d \neq \emptyset\} \\
&= \left(\bigcup\{(T \cap (C_n)_d) \mid T \in \mathsf{B}_S, T \subseteq S, T \cap (C_n)_d \neq \emptyset\}\right)^{(m_d)} \\
&= (S)_d^{(m_d)}.
\end{aligned}$$

□

Since Toida's conjecture is a partial case of Zibin's conjecture, the above result implies the following:

THEOREM 5.2 (TOIDA'S CONJECTURE). *Every $S \subseteq \mathbb{Z}_n^*$ is a CI-subset of $\mathbb{Z}_n$.*

Another class of CI-subsets of the cyclic group is described in the following statement which is a direct consequence of Theorem 4.6.

THEOREM 5.3. *If a subset $T \subset C_n$ is a basic set of some S-ring over C_n, then T is a CI-subset. In particular, this is true if the circulant $\mathsf{Cay}(C_n, T)$ is arc-transitive.*

REMARK. It was shown by C.H. Li [57] that an arc-transitive circulant graph $\mathsf{Cay}(C_n, T)$ is a CI-graph. His proof is purely group-theoretical and relies on the classification of finite simple groups.

5.2. Isomorphism problem for double-loop circulants. As an illustration of the previous results we shall solve the isomorphism problem for a special class of circulant graphs, so-called double-loop circulants. Following [82] a ***double-loop circulant graph*** is (in the original additive notation) a circulant $\mathsf{Cay}(\mathbb{Z}_n, S)$

with $S = \{a, -a, b, -b\}$, $|S| = 4$. The isomorphism problem for double-loop circulants was considered in several publications [**26, 60, 62**]. It was shown in [**26**] that S is a CI-subset of $\mathbb{Z}_n$. We give a short proof of this result based on S-ring theory.

THEOREM 5.4. *Every 4-element subset $S = \{a, -a, b, -b\}, |S| = 4$ of $\mathbb{Z}_n$ is a CI-subset.*

PROOF. First we switch to the multiplicative notation. Let $H = C_n$, $S = \{a, a^{-1}, b, b^{-1}\}$ and $\mathcal{A} := \langle\langle S \rangle\rangle$. Without loss of generality we may assume that the graph $\mathsf{Cay}(C_n, S)$ is connected. If S is a basic set of $\mathcal{A}$, then S is a CI-subset by Theorem 5.3. Assume the contrary, i.e., let S be a union of at least two basic sets. By Theorem 3.9 we have to show that $\mathcal{A}$ is a CI-S-ring. The simple quantity $\underline{S}$ belongs to the vector space $\mathcal{B} := \mathsf{Span}_{\mathbb{Q}}\{g + g^{-1} \mid g \in C_n\}$. Since $\mathcal{B} = \mathcal{S}(C_n, \Phi)$ for $\Phi = \{1, -1\} \leq \mathbb{Z}_n^*$, it is a Schur ring over C_n. Therefore $\mathcal{A} \subseteq \mathcal{B}$ and each basic set of $\mathcal{A}$ is symmetric. But there is a unique way to split S into a union of symmetric subsets of C_n: $S = \{a, a^{-1}\} \cup \{b, b^{-1}\}$. Therefore $A := \{a, a^{-1}\}$ and $B := \{b, b^{-1}\}$ are basic sets of $\mathcal{A}$. Since the graph $\mathsf{Cay}(C_n, S)$ is connected, $\langle a, b \rangle = C_n$.

To prove that $\mathcal{A}$ is a CI-S-ring it is sufficient to show that each algebraic automorphism of $\mathcal{A}$ is induced by some automorphism of C_n. Let λ be an algebraic automorphism of $\mathcal{A}$. Then by Theorem 4.6 there exist $m_a, m_b \in \mathbb{Z}_n^*$ such that $A^\lambda = A^{(m_a)}, B^\lambda = B^{(m_b)}$. We have to show that there exists an automorphism $m \in \mathbb{Z}_n^*$ such that $A^{(m_a)} = A^{(m)}$ and $B^{(m_b)} = B^{(m)}$. Since the group $\langle a, b \rangle$ is cyclic, the automorphisms $x \mapsto x^{m_a}$, $x \in \langle a \rangle$, and $x \mapsto x^{m_b}$, $x \in \langle b \rangle$, are induced by an automorphism of $\langle a, b \rangle$ if and only if they agree on the intersection $F = \langle a \rangle \cap \langle b \rangle$. Thus we have to show that the automorphisms m_a and m_b may be chosen in such a way that their restrictions to F induce the same automorphism, i.e., $x^{m_a} = x^{m_b}$ for each $x \in F$. Let s, t be natural numbers such that $f := a^s = b^t \in F$ and f generates F.

The elements $A^{(k)}$, $B^{(k)}$ $(k \geq 1)$ satisfy the recurrences $AA^{(k)} = A^{(k-1)} + A^{(k+1)}$, $BB^{(k)} = B^{(k-1)} + B^{(k+1)}$. Therefore, by induction on k, we obtain

$$(A^{(k)})^\lambda = (A^{(k)})^{(m_a)},$$
$$(B^{(k)})^\lambda = (B^{(k)})^{(m_b)}$$

In particular,

$$(a^t + a^{-t})^\lambda = (a^t + a^{-t})^{(m_a)},$$

and

$$(b^s + b^{-s})^\lambda = (b^s + b^{-s})^{(m_b)}.$$

Together with

$$(a^t + a^{-t})^\lambda = (b^s + b^{-s})^\lambda$$

we obtain $b^{sm_b} \in \{a^{tm_a}, a^{-tm_a}\}$. Replacing m_b by $-m_b$, if necessary, we obtain $b^{sm_b} = a^{tm_a}$, or, equivalently, $f^{m_a} = f^{m_b}$. Since m_a and m_b agree on the intersection $\langle a \rangle \cap \langle b \rangle$, there exists an $m \in \mathbb{Z}_n^*$ such that the restrictions of m onto the subgroups $\langle a \rangle$, $\langle b \rangle$ coincide with m_a and m_b respectively. □

5.3. Solution of the isomorphism problem for $\mathbb{Z}_8$. In previous subsections we dealt only with necessary conditions for being a CI-subset. Here we shall apply S-rings to get a solution for the isomorphism problem for a particular cyclic group, namely the cyclic group of order 8. In this subsection, instead of C_8 we use the additive notation $\mathbb{Z}_8$ for the cyclic group.

We start with a complete list of S-rings over $\mathbb{Z}_8$ which was obtained in [**32, 33, 44**]. It contains the following S-rings:

$\mathcal{A}_1 = \langle \underline{T_0}, \underline{T_1}, \underline{T_2} \rangle$ with $T_0 = \{0\}$, $T_1 = \{1,2,3,4,5,6,7\}$;

$\mathcal{A}_2 = \langle \underline{T_0}, \underline{T_1}, \underline{T_2} \rangle$ with $T_0 = \{0\}$, $T_1 = \{1,3,5,7\}$, $T_2 = \{2,6,4\}$;

$\mathcal{A}_3 = \langle \underline{T_0}, \underline{T_1}, \underline{T_2} \rangle$ with $T_0 = \{0\}$, $T_1 = \{1,3,5,7,2,6\}$, $T_2 = \{4\}$;

$\mathcal{A}_4 = \langle \underline{T_0}, \underline{T_1}, \underline{T_2}, \underline{T_3} \rangle$ with $T_0 = \{0\}$, $T_1 = \{1,3,5,7\}$, $T_2 = \{2,6\}$, $T_3 = \{4\}$;

$\mathcal{A}_5 = \langle \underline{T_0}, \underline{T_1}, \underline{T_2}, \underline{T_3}, \underline{T_4} \rangle$ with
$T_0 = \{0\}$, $T_1 = \{1,3,5,7\}$, $T_2 = \{2\}$, $T_3 = \{6\}$, $T_4 = \{4\}$;

$\mathcal{A}_6 = \langle \underline{T_0}, \underline{T_1}, \underline{T_2}, \underline{T_3}, \underline{T_4}, \underline{T_5} \rangle$ with
$T_0 = \{0\}$, $T_1 = \{1,5\}$, $T_2 = \{3,7\}$, $T_3 = \{2\}$, $T_4 = \{6\}$, $T_5 = \{4\}$;

$\mathcal{A}_7 = \langle \underline{T_0}, \underline{T_1}, \underline{T_2}, \underline{T_3}, \underline{T_4} \rangle$ with
$T_0 = \{0\}$, $T_1 = \{1,5\}$, $T_2 = \{3,7\}$, $T_3 = \{2,6\}$, $T_4 = \{4\}$;

$\mathcal{A}_8 = \langle \underline{T_0}, \underline{T_1}, \underline{T_2}, \underline{T_3}, \underline{T_4} \rangle$ with
$T_0 = \{0\}$, $T_1 = \{1,3\}$, $T_2 = \{5,7\}$, $T_3 = \{2,6\}$, $T_4 = \{4\}$;

$\mathcal{A}_9 = \langle \underline{T_0}, \underline{T_1}, \underline{T_2}, \underline{T_3}, \underline{T_4} \rangle$ with
$T_0 = \{0\}$, $T_1 = \{1,7\}$, $T_2 = \{3,5\}$, $T_3 = \{2,6\}$, $T_4 = \{4\}$;

$\mathcal{A}_{10} = \mathbb{Q}\mathbb{Z}_8$.

Our goal is to find all non-CI-S-rings among this list. We shall use the sufficient condition formulated in Theorem 3.11 which means that we have to study algebraic isomorphisms between the enumerated S-rings.

First we note that $\mathcal{A}_1$ and $\mathcal{A}_{10}$ are CI, since they are trivial S-rings. By Theorem 4.5 two distinct S-rings over $\mathbb{Z}_8$ are pairwise non-isomorphic algebraically. Therefore, each algebraic isomorphism between two S-rings is, in fact, an algebraic automorphism of one of the S-rings $\mathcal{A}_i$, $i = 1, ..., 10$.

Let now $\mathcal{A}$ be one of the above S-rings and let f be one of its algebraic automorphisms. Since $T_0^f = T_0$ and f preserves the cardinalities of basic sets, those S-rings where all non-identical basic sets have distinct cardinalities have a trivial group $\mathsf{Aut}_{alg}\mathcal{A}$, and, therefore, are CI-S-rings. This excludes S-rings $\mathcal{A}_1 - \mathcal{A}_4$ from being non-CI-S-rings.

The S-rings $\mathcal{A}_7, \mathcal{A}_8, \mathcal{A}_9$ have a unique non-trivial basic set of cardinality one: $\{4\}$. Therefore $\{4\}^f = \{4\}$ for each $f \in \mathsf{Aut}_{alg}\mathcal{A}$. Furthermore, $\{2,6\}$ is the unique symmetric basic set T of $\mathcal{A}$ which satisfies $T + T = \{4\}$. Therefore $\{2,6\}^f = \{2,6\}$ for each $f \in \mathsf{Aut}_{alg}\mathcal{A}$. Thus each algebraic automorphism of $\mathcal{A}$ fixes T_0, T_3 and T_4. Thus either it is trivial or it swaps T_1 and T_2. In both cases it is induced by a group automorphism of $\mathbb{Z}_8$.

Thus $\mathcal{A}_6$ is the remaining candidate for a non-CI-S-ring. Take an arbitrary $f \in \mathsf{Aut}_{alg}\mathcal{A}$. Since T_3 and T_4 are the only basic sets which satisfy the equation $X + X + X = T_0$, either $T_3^f = T_3$, $T_4^f = T_4$ or $T_3^f = T_4$, $T_4^f = T_3$. As f preserves the cardinalities of basic sets, $\{T_1, T_2\}$ and $\{T_3, T_4, T_5]$ are f-invariant sets. Combining these results we obtain that f must be one of the following four permutations: $id, (3,4), (1,2), (3,4)(1,2)$. A direct check shows that each of this permutations is an algebraic automorphism of $\mathcal{A}_6$. The automorphism group of $\mathbb{Z}_n$ contains four permutations $x \mapsto x$, $x \mapsto 5x$, $x \mapsto 3x$, $x \mapsto 7x$. Each of these permutes the basic sets of $\mathcal{A}_6$, and, therefore, induces an algebraic automorphism of $\mathcal{A}_6$. The first two permutations act trivially on $\{T_0, T_1, ..., T_5\}$. The two others induce the permutation $(3,4)(1,2)$. Therefore the algebraic automorphisms $(3,4)$ and $(1,2)$ are not induced by a group automorphism. If they are not induced by a combinatorial automorphism, then $\mathcal{A}_6$ also would be a CI-S-ring. In fact, it is sufficient to check one of them, say $(3,4)$. So we are faced with the following question:

Does there exist a permutation $g \in \mathsf{Sym}(\mathbb{Z}_8)$ such that

$$(*) \qquad \begin{cases} \mathsf{Cay}(\mathbb{Z}_8, T_1)^g = \mathsf{Cay}(\mathbb{Z}_8, T_1), & \mathsf{Cay}(\mathbb{Z}_8, T_2)^g = \mathsf{Cay}(\mathbb{Z}_8, T_2), \\ \mathsf{Cay}(\mathbb{Z}_8, T_3)^g = \mathsf{Cay}(\mathbb{Z}_8, T_4), & \mathsf{Cay}(\mathbb{Z}_8, T_4)^g = \mathsf{Cay}(\mathbb{Z}_8, T_3), \\ \mathsf{Cay}(\mathbb{Z}_8, T_5)^g = \mathsf{Cay}(\mathbb{Z}_8, T_5)\,? \end{cases}$$

It turns out that the answer to this question is affirmative! Each element $x \in \mathbb{Z}_8$ has a unique presentation in the form $x = x_0 + 2x_1$ where $x_0 \in \{0,1\}$, $x_1 \in \{0,1,2,3\}$ and all arithmetic is done modulo 8. Define the permutation $g \in \mathsf{Sym}(\mathbb{Z}_8)$ as follows $(x_0 + 2x_1)^g = x_0 + 6x_1$. A direct check shows that g satisfies (*) ! Thus $\mathcal{A}_6$ is not a CI-S-ring. To build a non-CI-subset it is sufficient to find a union of basic sets of $\mathcal{A}_6$ which generates the whole S-ring. For this purpose we can examine the list of S-rings over $\mathbb{Z}_8$. The set $S = T_1 \cup T_3 = \{1,5,2\}$ does not appear as a union of basic sets of any S-ring except $\mathcal{A}_6$. Therefore $\langle\langle S \rangle\rangle = \mathcal{A}_6$. Using Theorem 3.9 we obtain that $\{1,5,2\}$ is a non-CI-subset. It should be mentioned that $\{1,5,2\}$ was the first known example of a non-CI-subset of a cyclic group ([**21**]).

Our detailed analysis of S-rings yields the complete solution of the isomorphism problem for $\mathbb{Z}_8$. In fact, any two isomorphic circulants on 8 points are either conjugate by a multiplier or an isomorphism between them may be realized as a product of g (as defined above) and a multiplier from $\mathbb{Z}_8^*$.

6. Schur rings and association schemes

S-rings were introduced by I. Schur in a purely group-theoretical way. Much later it was understood that S-rings are a special case of more general combinatorial objects known as association schemes. The aim of this section is to give a short introduction to the connection between these two objects. We start with the definition of an association scheme.

Let X be a finite set. A set $\mathcal{R} = \{R_0, ..., R_r\}$ of binary relations forms an ***association scheme*** on X ([**6**]) if it satisfies the following conditions

AS1. $R_0 = I_X$, and $X^2 = R_0 \cup ... \cup R_r$ is a disjoint union;
AS2. $\forall 0 \le i \le r \, \exists 0 \le i' \le r : R_i^\top = R_{i'}$;[5]
AS3. For each triple $i,j,k \in \{0,\dots,r\}$ and arbitrary $(u,v) \in R_k$ the cardinality p_{ij}^k of the set $\{y \in X \mid (u,y) \in R_j \text{ and } (y,v) \in R_k\}$ depends on the triple (i,j,k) only and does not depend on the choice of the pair $(u,v) \in R_k$.

The numbers p_{ij}^k are called the ***intersection numbers*** of the association scheme $(X, \mathcal{R})$. The relations R_k are called ***basic relations*** of the association scheme. The (di)graphs $\Gamma_i = (X, R_i)$ are called ***basic graphs*** of the association scheme.

The ***Bose-Mesner algebra*** of an association scheme $(X, \mathcal{R})$ is the matrix algebra $\mathcal{W}$ the basis of which consists of the adjacency matrices A_i of the graphs

[5]Here $R^\top := \{(y,x) \mid (x,y) \in R\}$.

Γ_i, $i \in \{0, ..., r\}$. This basis of adjacency matrices is called the ***standard basis*** of the Bose-Mesner algebra.

The ***automorphism group*** $\mathsf{Aut}(\mathcal{W})$ of an association scheme $\mathcal{W} = (X, \mathcal{R})$ is defined as the intersection of the automorphism groups of all basic graphs Γ_i.

Every S-ring $\mathcal{A} = \langle \underline{T_0}, \dots, \underline{T_r} \rangle$ over a group H yields an association scheme on H the basic relations of which are the Cayley graphs $\mathsf{Cay}(H, T_i)$, $i \in \{0, ..., r\}$. We shall denote it by $\mathsf{As}(\mathcal{A}) = (H; \{\mathsf{Cay}(H, T_i)\}_{i=0}^{r})$. A direct check shows that the structure constants of $\mathsf{As}(\mathcal{A})$ coincide with the structure constants of $\mathcal{A}$. A combinatorial isomorphism between S-rings as introduced in Section 2 is a particular case of an association scheme isomorphism. More precisely, two S-rings $\mathcal{A} = \langle \underline{T_0}, \dots, \underline{T_r} \rangle \subseteq \mathbb{Q}H$ and $\mathcal{B} = \langle \underline{S_0}, \dots, \underline{S_r} \rangle \subseteq \mathbb{Q}F$ are combinatorially isomorphic if and only if the corresponding association schemes are isomorphic in the usual way. It is well-known that the automorphism group $\mathsf{Aut}(\mathcal{A})$ as defined in Section 2 coincides with the automorphism group of the association scheme $\mathsf{As}(\mathcal{A})$. As we have seen before, $\mathsf{Aut}(\mathcal{A})$ contains the subgroup H_R, the right regular representation of the group H. Thus $\mathsf{Aut}(\mathsf{As}(\mathcal{A}))$ contains a regular subgroup isomorphic to H. This property characterizes the S-rings among the association schemes. Let $(X, \mathcal{R})$ $(\mathcal{R} = \{R_0, ..., R_r\})$ be an association scheme on a set X. We say that $(X, \mathcal{R})$ is a ***Cayley association scheme*** over a group H if $\mathsf{Aut}(\mathcal{R})$ contains a regular subgroup isomorphic to H.

Let $(X, \mathcal{R})$ be a Cayley association scheme, let H be a regular subgroup of $\mathsf{Aut}(X, \mathcal{R})$and let $x \in X$ be an arbitrary point. We define the ***Schur projection*** of a basic relation $R_i \in \mathcal{R}$ on H at $x \in X$ by

$$\mathsf{Spr}_{H,x}(R_i) := \{h \in H \mid (x^h, x) \in R_i\}$$

Let $T_i := \mathsf{Spr}_{H,x}(R_i)$ for $i \in \{0, 1, ..., r\}$. Then the following properties are easy to check:

(1) $\{T_0, ..., T_r\}$ is a partition of H with $T_0 = \{e\}$;
(2) If $R_i^T = R_j$ then $T_j = T_i^{(-1)}$;
(3) $\mathsf{Spr}_{H,x^h}(R_i) = h^{-1}\mathsf{Spr}_{H,x}(R_i)h$.

Every $y \in X$ is of the form x^h for some uniquely determined $h \in H$. Thus by (3), if H is abelian, then $\mathsf{Spr}_{H,x}(R_i)$ does not depend on the choice of $x \in X$.

The following theorem follows directly from the definitions.

THEOREM 6.1. *The vector space $\mathcal{A} := \langle \underline{T_0}, ..., \underline{T_r} \rangle$ is an S-ring over H. The association schemes $(X, \mathcal{R})$ and $\mathsf{As}(\mathcal{A})$ are isomorphic (as association schemes).*

Moreover, the above S-ring has the same structure constants as $(X, \mathcal{R})$ does:

$$\underline{T_i}\,\underline{T_j} = \sum_{k=0}^{r} p_{ij}^k \underline{T_k}\,.$$

In many cases the automorphism group of $(X, \mathcal{A})$ may contain more than one regular subgroup. The following result is a direct analogue to Babai's Theorem 2.1.

THEOREM 6.2. *Let $(X, \mathcal{R})$ be a Cayley association scheme and let H,K be two regular subgroups of $\mathsf{Aut}(\mathcal{R})$. Then the Schur projections of $(X, \mathcal{R})$ on H and K are Cayley isomorphic if and only if H and K are conjugate in $\mathsf{Aut}(\mathcal{R})$.*

Finally it should be mentioned that association schemes are a particular case of ***coherent configurations*** introduced by D. Higman [**36, 37**]. The shortest way to define them is to drop the requirement that the diagonal relation is one of the basic relations of a scheme. Equivalent algebraic objects known as ***cellular algebras*** were introduced by B. Weisfeiler and A. Leman [**83, 84**].

7. Concluding remarks

7.1. Circulant matrices. Circulants have a long history. One of the first references is attributed to a paper by E. Catalan in 1846. In the 19th century, the theory of circulants was developed within the origins of matrix theory and especially determinants theory. The book [**17**] by Ph.J. Davis gives a vivid description of the place of circulants in matrix theory from the beginning of the century to the 1970's. It also introduces the reader to the powerful methods of spectral analysis of circulant matrices. These methods may be reformulated in modern terms of the so-called second standard basis in commutative coherent algebras, cf. [**6, 25**].

7.2. Different approaches to Ádám's conjecture. It is worth mentioning that the dates of the publications sometimes do not give exact attribution to the time when the corresponding result became known to experts. For example, [**29**] and [**72**] were available as preprints many years before the corresponding papers appeared, results of [**46, 43**] were reported in [**45**], etc. (We refer to the final note in [**5**], which cites already in August 1977 several of the above-mentioned results.)

7.3. Conjectures of Zibin' and Toida. The authors are much obliged to D.K. Zibin' (Riga) – an expert in technical cybernetics –, who attracted the attention of Klin and Pöschel to the isomorphism problem of circulant graphs. The computer experiments, arranged by him and his colleagues have played a very stimulative role. In fact, his conjecture was formulated on a less rigorous level, as an empirical observation based on the analysis of the experimental results. Toida's conjecture was independently considered and attacked in [**15**] by C.Y. Chao. However, the proof of the conjecture presented there contains a gap, namely: in the proof of the main result the author wrote (page 95, line 2) that, ω denoting the n-th root of unity, $\omega^0, \omega^1, ..., \omega^{n-1}$ are linearly independent over $\mathbb{Q}$ contrary to the well-known identity $\omega^0 + \omega^1 + ... + \omega^{n-1} = 0$. This error makes the whole proof incorrect.

7.4. Schur rings. Schur's method as a tool in group theory was known to experts for a long time due to its detailed presentation in [**85, 50, 78**]. As a combinatorial tool, it is still unpopular and has the reputation of a "heavy technique" (cf. [**41**]). One can mention a number of publications, where Schur's methodology was rediscovered, for example in the framework of the description of strongly regular circulant graphs (partial difference sets over cyclic groups) [**61**]. In the encyclopedical monograph [**12**], objects equivalent to S-rings are called translation association schemes.

Theorem 3.1 was implicitely used by many authors, see for example [**11, 18**]. In explicit form it was formulated in [**63**].

7.5. Circulants with a prime number of vertices. In explicit form all S-rings over $\mathbb{Z}_p$, p a prime, were first described in [**73**]. Much earlier, in 1965, B. Gordon introduced in [**34**] α-partitions of a group H, which are equivalent to S-rings. He also determined all α-partitions over $\mathbb{Z}_p$.

Chao used the method of Schur rings in [**13**] as a tool for the classification of vertex-transitive graphs. J. Turner in [**81**] proved Adam's conjecture for p-vertex circulant graphs (though the conjecture itself is not mentioned in this paper). There is a number of other interesting publications (see for example references in [**42**] which are devoted to the circulants with a prime and a prime-power number of vertices), however their consideration does not fit the framework of our survey.

7.6. Schur rings over cyclic groups. The problem of complete enumeration of S-rings over the cyclic group of order n was first raised in [**73**], see also Chapter 8 of the book [**74**] (this chapter was written jointly with Klin). Although the isomorphism problem of circulants was the main application of S-rings, the enumeration problem was also studied as an interesting combinatorial problem by itself. The first enumeration results were obtained by M. Klin, Ja. Gol'fand, N. Naimark and R. Pöschel during 1974-85 in [**73, 46, 74, 32, 44, 33**]. They gave a complete list of all S-rings over $\mathbb{Z}_n$ if n is a prime power or a product of two distinct primes. Their results were used for a solution of the isomorphism problem for circulants. All S-rings over $\mathbb{Z}_n$ for several small values of n were enumerated by V.A. Zaichenko in [**87**], using a computer. The results of an attempt to enumerate all S-rings over $\mathbb{Z}_n$ where n is a product of three distinct primes were reported in [**28**]. The rational S-rings over the cyclic group of square-free order were classified Ja. Gol'fand in 1985 ([**31**]). This result was generalized by Muzychuk for a cyclic group of arbitrary order, [**63**]. All S-rings over the cyclic group of square-free order were classified by Muzychuk in 1993, see [**65**]. It should be mentioned that although the latter classification was not used directly for proving the corrected Ádám's conjecture, the Schur ring theory was an important ingredient of its proof.

A beautiful survey of some results on enumeration of S-rings is presented in [**51**]. An important classification result about Schur rings over a cyclic group was obtained in [**52, 53**]. It claims that each S-ring over $\mathbb{Z}_n$ is either a tensor or a so-called wedge product of two S-rings over cyclic groups of smaller orders. We are not aware of any attempts to apply these results to the study of the algebraic and combinatorial isomorphisms of Schur rings.

7.7. Isomorphism criteria for circulant graphs. Just a few results are known in the case where Adam's conjecture is not true. Klin and Pöschel presented in [**45, 46, 74, 47**] isomorphism criteria for $n = p^2$, p a prime. The proof is based on the following "complete scheme":

- enumeration of S-rings,
- description of the automorphism groups,
- finding the normalisers of the automorphism groups in the symmetric group,
- formulation of an isomorphism criterion.

In [**43**] necessary and sufficient isomorphism conditions for p^m-vertex circulant graphs, p an odd prime, were obtained on the base of the classification of S-rings in [**73**]. In all these papers the multiplier approach was used.

7.8. Multipliers. It turns out that the multiplier approach can be used as a universal tool for the classification of combinatorial objects with cyclic symmetry. The main idea of the multiplier approach may be formulated as follows. If two cyclic objects are not conjugate by one multiplier, then we may take a set of multipliers and use them to construct an isomorphism between these objects which will be a local multiplier isomorphism, *i.e.* a multiplier isomorphism on each "section" of $\mathbb{Z}_n$

(of course, the notion of section must be rigorously defined simultaneously with the exact definition of a combinatorial object). This idea was successfully implemented for cyclic codes in [**39, 40**]. Concerning circulant graphs, this approach was used in [**2, 43**]. It follows from Theorem 5.1 that the layers of the connection sets of two isomorphic circulants are conjugate by multipliers. The main problem now is to find necessary and sufficient conditions which allow us to reconstruct an isomorphism between the graphs from a list of multipliers. Another approach to the isomorphism problem for cyclic combinatorial objects was suggested in [**68, 70**] in the framework of so-called solving sets.

A very surprising fact is that the origins of this method go back to the beginning of the century (S. Bays, P. Lambossy), when it was used for the isomorphism testing of circulant Steiner systems.

7.9. The isomorphism problem for Cayley graphs. As already mentioned, the isomorphism problem for circulants is a particular case of the general isomorphism problem for Cayley graphs. The main open problem in this area is to classify all CI-groups. During the last decade C.H. Li and his colleagues, using the classification of finite simple groups, made significant progress towards a complete solution of the problem. They found a very narrow list of groups which still remain as candidates for being CI. The main problem now is to check which of these groups are really CI. In general, it is a rather hard problem to show that a concrete group is a CI-group. In almost all known cases the proof of being CI either essentially exploits S-ring theory or may be simplified by using calculations in S-rings. For this reason, we are sure that a combination of group-theoretical techniques with S-ring theory will be helpful in order to make a further progress in the area.

7.10. Coherent algebras. Coherent (cellular) algebras from the beginning were considered by representatives of the Soviet school as a tool for the solution of the isomorphism problem for all graphs. The minimal coherent algebra $W(A)$ which contains an adjacency matrix A of a given graph Γ is a source of various important invariants of Γ.

In [**71**] coherent algebras together with S-rings are used for the solution of the recognition problem for circulant graphs with prime number of vertices. We refer to [**48, 49**] for the description of other applications of coherent algebras in isomorphism testing.

7.11. Towards further progress. One can modify the "complete scheme" (see Subsection 7.7) proposed by Klin and Pöschel in order to make it suitable for a solution of the isomorphism problem for Cayley graphs over an arbitrary group. In the original "complete scheme" approach we enumerate all S-rings over a given group. But for the Cayley graph isomorphism problem we need only Schurian S-rings. This observation together with rather elementary group-theoretical arguments lead sometimes to an essential reduction of the enumeration part. For example such ideas led to a proof that $\mathbb{Z}_p^4$ is a CI-group [**38**].

8. Acknowledgements

We thank a referee and Alyssa Sankey for helpful remarks and linguistical improvements.

References

[1] A. Ádám, *Research problem 2-10*, J. Combin. Theory **2**(1967), 393.

[2] B. Alspach, T.D. Parsons, *Isomorphism of circulant graphs and digraphs*, Discrete Math. **25**(1979), 97-108.

[3] L. Babai, *Isomorphism problem for a class of point-symmetric structures*, Acta Math. Acad. Sci. Hungar. **29**(1977), 329-336.

[4] L. Babai, P. Frankl, *Isomorphism of Cayley graphs I*, in: Colloq. Math. Soc. J. Bolyai **18**. Combinatorics, Keszthely, 1976, North - Holland, Amsterdam, 1978, 35-52.

[5] L. Babai, P. Frankl, *Isomorphism of Cayley graphs II*, Acta Math. Acad. Sci. Hungar. **34**(1979), 177-183.

[6] E. Bannai, T. Ito, *Algebraic Combinatorics I: Association Schemes*, Benjamin/Cummings, Menlo Park, CA, 1984.

[7] E. Bannai, Song-Sung-Yell, *Character tables of fission schemes*, European J. of Combin. **14**(1993), no. 5, 385-396.

[8] J.-C. Bermond, G. Illiades, C. Peyrat, *An optimization problem in distributing loop computer networks*, Annals New York Acad. Sci. **555**(1989), 45-55.

[9] F.T. Boesch, J.-F. Wang, *Reliable circulant networks with minimum transmission delay*, IEEE Trans. on circ. and syst. **C32**(1985), 1286-1291.

[10] J.E. Boilatt, *Fast load balancing in Cayley graphs and in circuits*, Lect. Nites in Comp. Science **790**(1994), 315-326.

[11] W.G. Bridges, R.A. Mena. *Rational G-matrices with rational eigenvalues*, J. of Combin. Theory A **108**(1982), no. 6, 264-280.

[12] A.E. Brouwer, A.E. Cohen, A.Neumaier, *Distance-regular graphs*, Springer Verlag, 1989.

[13] C.Y. Chao, *Some applications of a theorem of Schur to graphs and to a class of endomorphisms*, Portugaliae Math. **26**(1967), 505-523.

[14] C.Y. Chao, *On a type of circulants*, Linear Algebra and Appl. **6**(1973), 241-248.

[15] C.Y. Chao, *On digraphs with circulant adjacency matrices*, Arch. Math. **54**(1990), 93-104.

[16] F.P.K. Chung, R.L. Graham, *Quasi-random subsets of* $\mathbf{Z}_n$, J. Comb. Theory (A) **61**(1992), 64-86.

[17] Ph. J. Davis, *Circulant matrices*, John Wiley & Sons. New York, 1979.

[18] Ph. Delsarte, *Regular schemes over a finite Abelian group*, Geom. Dedicatae **8**(1979), 477-490.

[19] D.Ž. Djoković, *Isomorphism problem for a special class of graphs*, Acta Math. Acad. Sci. Hung. **21**(1970), 267-270.

[20] V.N. Egorov, A.I. Markov, *On Ádam's conjecture for graphs with circulant adjacency matrices*, Soviet Math. Dokl. **20**(1979), 1292-1296. (Russian)

[21] B. Elspas, J. Turner, *Graphs with circulant adjacency matrices*, J. Combin. Theory **9**(1970), 297-307.

[22] S. Evdokimov, I. Ponomarenko, *On highly closed cellular algebras and highly closed isomorphism*, Electron. J. Comb. **6**(1999), # R18.

[23] I.A. Faradžev, A.A. Ivanov, M.H. Klin. *Galois correspondence between permutation groups ang cellular rings (association schemes)*, Graphs and Combinatorics **6**(1990), 303-332.

[24] I.A. Faradžev, M.H. Klin, *Computer package for computations with coherent configurations*, Proc. of the 1991 Intern. Symp. on Symb. and Algebr. Computations, ISSAC'91 (Bonn, July 15-17, 1991), ACM press, 219-223.

[25] I.A. Faradžev, M.H. Klin, M.E.Muzichuk, *Cellular rings and groups of automorphisms of graphs*, In: Investigations on Algebraic Theory of Combinatorial Objects. Mathematics and Its Applications (Soviet Series), I.A. Faradžev, A.A.Ivanov, M.H.Klin, A.J.Woldar (eds.),Kluwer Acad. Publ. **84**(1993), 1-152.

[26] Xingui Fang, Mingyao Xu, *On isomorphisms of Cayley graphs of small valency*, Algebra Colloq. **1:1**(1994), 67-76.

[27] C.M. Fiduccia, R.W. Foreade and G.S. Zito, *Geometry and diameter bounds of directed Cayley graphs on abelian groups*, SIAM Journal of Disc. Math. **11**(1998), 157-167.

[28] A.G. Ganushkin, *Schur rings over cyclic groups of order pqr*, In: Sixth All-Union Conference in group theory. Abstracts, Kiev, 1978, p. 15. (Russian)

[29] C.D. Godsil, *On Cayley graphs isomorphism*, Ars Combinatoria **15**(1983), 231-246.

[30] C. D. Godsil, *On the full automorphism group of Cayley graphs*, Combinatorica **1**(1981) 243-256.

[31] Ja.Ju. Gol'fand, *A description of subrings in* $V(S_{p_1} \times \ldots \times S_{p_m})$, Investigations in the algebraic theory of combinatorial objects, Proc. Seminar Inst. System Studies, Moscow, 1985, 65-76. (Russian); English translation in: Investigations on Algebraic Theory of Combinatorial Objects, Mathematics and Its Applications (Soviet Series), I.A. Faradžev, A.A. Ivanov, M.H. Klin, A.J. Woldar (Eds), Kluwer Acad. Publ. **84**(1994), 209-223.

[32] Ja.Ja. Gol'fand, M.H. Klin, N.L. Naimark, *The structure of S-rings over* $\mathbb{Z}_{2^m}$, In:"Sixteenth All-Union Algebraic Conference", part 2, Leningrad, 1981, 195-196.

[33] Ja.Ju. Gol'fand, N.L. Naimark, R. Pöschel, *Schur rings over* $\mathbb{Z}_{2^m}$, Preprint P-MATH-01/85 Akad. der Wiss. der DDR, ZIMM, Berlin, 1985.

[34] B. Gordon. *A generalization of the coset decomposition of a finite group*, Pacif. J. Math. **15**(1965), 503-509.

[35] W.D. Hillis, *The Connection Machine*, The MIT Press, Cambridge, 1986.

[36] D.G. Higman, *Coherent configurations I*, Rend. Sem. Mat. Univ. Padova **44**(1970), 1-25.

[37] D.G. Higman, *Coherent algebras*, Linear algebra and its applications **93**(1987), 209-239.

[38] M. Hirasaka, M. Muzychuk, *The elementary abelain group of rank* 4 *is a CI-group*, J. of Combin. Theory, accepted.

[39] W.G. Huffman, A. Job, V. Pless, *Multipliers and generalized multipliers of cyclic objects and cyclic codes*, J. of Comb. Th. (A) **62**(1993), no. 2, 183-215.

[40] W.C. Huffman, *The equivalence of two cyclic objects on pq elements*, Discrete Math. **154**(1996), 103-127.

[41] A. Joseph, *The isomorphism problem for Cayley digraphs on graphs of prime-squared order*, Discrete Math. **141**(1995), no. 1-3, 173-183.

[42] M. Klin, V. Liskovets, R. Pöschel, *Analytical enumeration of circulant graphs with prime-squared number of vertices*, Semin. Lothar. Comb. **36**(1996), B36d.

[43] M.Ch. Klin, R. Pöschel, *The isomorphism problem for circulant digraphs with* p^n *vertices*, Preprint P-34/80 Akad. der Wiss. der DDR, ZIMM, Berlin, 1980.

[44] M.Ch. Klin, N.L. Naimark, R. Pöschel, *Schur rings over* $\mathbb{Z}_{2^m}$, Preprint P-MATH-14/81 Akad. der Wiss. der DDR, ZIMM, Berlin, 1981.

[45] M.H. Klin, Pöschel, *The isomorphism problem for cyclic graphs with* p^2 *or pq vertices*, Abstract presented at the A.A.Zykov's Seminar on Graph Theory, Odessa, 1975. (Russian)

[46] M.Ch. Klin, R. Pöschel, *The König problem, the isomorphism problem for cyclic graphs and the characterization of Schur rings*, Preprint AdWd DDR, ZIMM, Berlin, March 1978.

[47] M.H. Klin, R. Pöschel, *The König problem, the isomorphism problem for cyclic graphs and the method of Schur rings*, Colloq. Math. Soc. J. Bolyai **25**, Algebraic methods in graph theory, Szeged, 1978. North-Holland, Amsterdam, 1981; 405-434.

[48] M. Klin, Ch. Rücker, G. Rücker, G. Tinhofer, *Algebraic combinatorics in mathematical chemistry. Methods and algorithms. I. Permutation groups and coherent (cellular) algebras*, MATCH **40**(1999), 7-138.

[49] M. Klin, G. Tinhofer, *Algebraic combinatorics in mathematical chemistry. Methods and algorithms. III. Graph invariants and stabilization methods*, Technical report TUM–M9902. Technische Universität München, März 99, 64 pp.

[50] R. Kochendörfer, *Lehrbuch der Gruppentheorie unter besonder Berücksichtigung der endlichen Gruppen*, Akad. Verlagsgesellschaft, Leipzig, 1966.

[51] K.H. Leung, S.L. Ma, *The structure of Schur rings over cyclic groups*, J. Pure Appl. Algebra **66**(1990), 287-302.

[52] K.H. Leung, S.H. Man, *On Schur rings over cyclic groups*, Israel J. Math. **106**(1998), 251-267.

[53] K.H. Leung, S.H. Man, *On Schur rings over cyclic groups, II*, J.of Algebra **183**(1996), no. 2, 273-285.

[54] C.H. Li, M. Conder, *On the isomorphisms of finite Cayley graphs*, Europ. J. Combin. **19**(1998), no. 8, 911-919.

[55] C.H. Li, C.E. Praeger, M.Y. Xu, *On finite groups with Cayley isomorphism property*, J. of Graph Theory **27**(1998), 1-11.

[56] C.H. Li, *On isomorphism of connected Cayley graphs*, Discrete Math. **178**(1998), 109-122.

[57] C.H. Li, *Permutation groups and circulant graphs*, Manuscript 1999.

[58] C.H. Li, *The finite groups with the 2-DCI-property*, Comm. in Algebra **24**(1996), 1749-1757.

[59] C.H. Li. *On isomorphism of connected Cayley graphs II*, J. Combin. Theory (B) **74**(1998), 28-34.

[60] B. Litow, B. Mans, *On the Ádám property for circulant graphs*, Tech. Report of the Dept. of Comp. Science, James Cook Univ., TR-96-11 (1996), 1-8.

[61] S.L. Ma, *On association schemes, Schur rings, strongly regular graphs and partial difference sets*, Ars Combinatoria **27**(1989), 211-220.

[62] B. Mans, F. Pappalardi, I. Shparlinski, *On Ádám's conjecture on circulant graphs*, Lect. Notes in Comp. Science **1449**(1998), 260-285.

[63] M. Muzychuk, *The structure of rational Schur rings over cyclic groups*, Europ. J. of Combin. **14**(1993), 479-490.

[64] M. Muzychuk, *On the structure of basic sets of Schur rings over cyclic groups*, J. of Algebra **169**(1994), no. 2, 655-678.

[65] M. Muzychuk, *The structure of Schur rings over cyclic groups of square-free order*, Acta Appl. Math. **52**(1998), 163-181.

[66] M. Muzychuk, *Ádám's conjecture is true in the square-free case*, J. of Combin. Th. (A) **72**(1995), no. 1, 118-134.

[67] M. Muzychuk, *On Ádám's conjecture for circulant graphs*, Discrete Math. **176**(1997), 285-298.

[68] M. Muzychuk, *On the number of cyclic combinatorial objects isomorphic to a given one*, Preprint Math-AL-16-1996, Technische Universität Dresden 1996.

[69] M. Muzychuk, R. Pöschel, *Isomorphism criterion for circulant graphs*, Preprint Math-AL-9-1999, Technische Universität Dresden, 1999.

[70] M. Muzychuk,*On the isomorphism problem for cyclic combinatorial objects*, Discrete Math. **197/198**(1999), 589-606.

[71] M. Muzychuk, G. Tinhofer, *Recognizing circulant graphs of prime order in polynomial time*, Electron. J. Comb. **5**(1998), Research paper R25, 28 pp.

[72] P.P. Pálfy, *Isomorphism problem for relational structures with a cyclic automorphism*, Europ. J. Comb. **8**(1987), 35-43.

[73] R. Pöschel, *Untersuchungen von S-Ringen, insbensondere im Gruppenring von p-Gruppen*, Math. Nachr. **60**(1974), 1-27.

[74] R. Pöschel, L.A. Kalužnin, *Funktionen- und Relationenalgebren*, Birkhäuser Verlag, Basel u. Stuttgart, and Deutscher Verlag der Wissenschaften, Berlin, 1979.

[75] S.T. Schibell, R.M. Stafford, *Processor interconnection networks from Cayley graphs*, Disc. Appl. Math. **40**(1992), 333-357.

[76] Wai-Chee Shiu, *Algebraic structure of Schur rings*, Chinese Journal of Math. **21**(1993), no. 1, 55-71.

[77] I. Schur, *Zur Theorie der einfach transitiven Permutationgruppen*, S.-B. Preuss. Akad. Wiss. phys.-math. Kl. **18/20**(1933), 598-623.

[78] W.R. Scott, *Group Theory*, Prentice-Hall, 1964.

[79] O. Tamaschke, *A generalization of conjugacy in groups*, Rendiconti del Seminario Mathematico dell'Universitá di Padova **40**(1968), 408-427.

[80] S. Toida, *A note on Ádám's conjecture*, J.Combin. Theory (B), **23**(1977), 239-246.

[81] J. Turner, *Point symmetric graphs of prime order*, J. Combin. Theory **3**(1967), 136-145.

[82] J.L.A. Vebra, M.A.Fiol, P. Morillo, I. Alegre, *The diameter of undirected graphs associated to plane tescellations*, Ars Combinatoria **20 B**(1985), 159-171.

[83] B.Ju. Weisfeiler, A.A.Leman, *Reduction of a graph to a canonical form and an algebra arising during this reduction*, Nauchno-Technicheskaya Informatsia **9** Seria 2, (1968), 12-16. (Russian)

[84] B. Weisfeiler (editor), *On the construction and identification of graphs*, Lect. Notes in Math. **558**(1976).

[85] H. Wielandt, *Finite Permutation Groups*, Academic Press, 1964, Berlin.

[86] H. Wielandt, *Permutation groups through invariant relations and invariant functions*, Lect. Notes., Dept. Math.m Ohio St. Univ, Colimbus, 1969.

[87] V.A. Zaichenko, *An algorithmic approach to the construction of combinatorial objects and to computations in permutation groups based on the method of ivariant relations*, Ph.D. Thesis, Moscow, MPhTI, 1981. (Russian)

[88] D.K. Zibin', Private communication, August 1975.

[89] P.H. Zieschang, *An Algebraic Approach to Association Schemes*, Springer, LNM 1628.

[90] J. Žerovnik, T. Pisanski, *Computing the diameter in multiple-loop networks*, J. of Algorithms **14**(1993), 226-243.

Department of Mathematics and Computer Science, Netanya Academic College, 16 Kibutz Galuyot St., 42365 Netanya, Israel
E-mail address: muzy@Netanya.ac.il

Department of Mathematics, Ben-Gurion University of the Negev, 84105 Beer-Sheva, Israel,
E-mail address: klin@cs.bgu.ac.il

Institut für Algebra, Technische Universität Dresden, D - 01062 Dresden, Germany
E-mail address: poeschel@math.tu-dresden.de

DIMACS Series in Discrete Mathematics
and Theoretical Computer Science
Volume **56**, 2001

New 1-generator Quasi-Twisted Codes over GF(5)

Irfan Siap, Nuh Aydin, and Dijen Ray-Chaudhuri

ABSTRACT. One of the most important problems of coding theory is to construct codes with best possible minimum distances. Recently, quasi-cyclic (QC) and quasi-twisted (QT) codes have been proven to contain many such codes. In this paper, we make use of the methods introduced in [**5**] to search for new linear codes over $GF(5)$ which are QT and we have been able to find 16 such codes with better minimum distances.

1. Introduction

The classes of quasi-cyclic (QC) and quasi-twisted (QT) codes have been shown to be promising to solve one of the most important problems in coding theory: to construct codes with the best possible parameters. Recently, the structure of QT codes have been investigated and a BCH-type bound on minimum distance is given in the special case of 1-generator QT codes in [**5**]. In this paper we make use of these results to develop an efficient method to search for new linear QT codes over the field $GF(5)$ and we have been able to find many.

We start with recalling basic definitions and notations. A linear code C of length n over F_q, the finite field with q elements, also denoted by $GF(q)$, is a vector subspace of $V := F_q^n$ which is equipped with the metric (called the Hamming distance) d defined by $d(\boldsymbol{u},\boldsymbol{v}) := |\{i : u_i \neq v_i\}|$, where $\boldsymbol{u} = (u_1, u_2, \ldots, u_n)$, $\boldsymbol{v} = (v_1, v_2, \ldots, v_n)$.

The minimum (Hamming) distance of C is defined by

$$d(C) = \min_{\boldsymbol{u},\boldsymbol{v}\in C, \boldsymbol{u}\neq\boldsymbol{v}} d(\boldsymbol{u},\boldsymbol{v}).$$

The (Hamming) weight $w(\boldsymbol{c})$ of a vector $\boldsymbol{c} = (c_0, c_1, \ldots, c_{n-1})$ is defined by $w(\boldsymbol{c}) := |\{i : c_i \neq 0\}|$, i.e. the number of the nonzero entries of $\boldsymbol{u}$. The fact that for a linear code C the minimum weight $w(C) = \min_{\boldsymbol{c}\in C, \boldsymbol{c}\neq\boldsymbol{0}} w(\boldsymbol{c})$ is equal to the minimum distance significantly reduces the number of computations needed to find the minimum distance of a linear code.

DEFINITION 1.1. A vector subspace C of F_q^n of dimension k and $w(C) = d$ is called an $[n, k, d]_q$ linear code.

1991 *Mathematics Subject Classification.* 94B05, 94B65.

The **Hamming weight enumerator**, $W_C(x,y)$, of a code C is defined by

$$W_C(y) = \sum_{\boldsymbol{u}\in C} y^{w(\boldsymbol{u})} = \sum_i A_i y^i \tag{1.1}$$

where $A_i = |\{\boldsymbol{u} \in C | w(\boldsymbol{u}) = i\}|$ i.e. the number of codewords in C whose weights equal to i.

The smallest non-zero exponent of y with a nonzero coefficient in $W_C(y)$ is equal to the minimum distance of the code.

A linear code C is called a t-error correcting code if $t = \lfloor \frac{d-1}{2} \rfloor$, where $d = d(C)$. Given a code length n and a dimension k, we try to construct codes with largest minimum distance d over the field F_q.

DEFINITION 1.2. The largest possible minimum distance for a linear code of length n and dimension k over the finite field of order q is denoted by $d_q[n,k]$

Let $n = lm$ where $l, m \in \mathbb{N}$, $a \in F_q^\times := F_q - \{0\}$ and

$$\mu_{a,l} : C \to V$$

$$\mu_{a,l}\big((c(0), \dots, c(n-1))\big) = \\ (a \cdot c(\overline{0-l}), \dots, a \cdot c(\overline{(l-1)-l}), c(\overline{l-l}), \dots, c(\overline{n-1-l}))$$

where $\overline{i-l} = i - l \mod n$ for all $0 \le i \le n-1$.

DEFINITION 1.3. A linear code C is called l-quasi-twisted (l-QT) if $\mu_{a,l}(C) = C$.

In words, a constacyclic shift of a codeword by l positions is still a codeword, where a constacyclic shift of a codeword $\big(c(0), \dots, \dots, c(n-1)\big)$ is $\big(ac(n-1), c(0), \dots, \dots, c(n-2)\big)$. Some of the most important classes of codes can be realized as special cases of QT codes. For example the case $a = 1$ gives quasi-cyclic codes, $l = 1$ gives constacyclic codes, $l = 1$ and $a = 1$ yields cyclic codes.

Since a code C is l-QT if and only if it is (l,n)-QT (an immediate generalization of the corresponding fact for QC codes in [**15**]), where (l,n) denotes the greatest common divisor of n and l, we will assume, without loss of generality, that $l|n$, so that $n = ml$ for some integer m. Note that if $(l,n) = 1$, the code is constacyclic.

There has been much research on QC codes due to the facts that they have simpler algebraic structures compared to arbitrary linear codes and they meet Gilbert-Varshamov bound [**9**]. Many new record breaking codes over finite fields of orders 2, 3, 5, 7, 8, and 9 have been constructed which are quasi-cyclic. Most of the work can be found in [**18**], [**19**], [**20**], [**21**], [**12**], [**22**], [**23**], [**10**], [**11**], [**13**], [**14**], [**15**], [**17**], [**16**], [**6**] and [**28**]. Aside from [**14**], [**27**] and [**5**], there has not been quite as much search on QT codes as on QC codes. Since the class of QT codes is larger, it would be no surprise to find "good" linear codes which are QT. That is part of the reason we search over QT codes.

We refer the reader to [**5**] for well-known properties of constacyclic codes, factorizations of the polynomials of the form $x^n - a$ over finite fields, and a BCH-type bound on minimum distance. It is also shown in [**5**] that when $(n, q-1)|i$ then a constacyclic code of length n is equivalent to a cyclic code of the same length over

$GF(q)$, where $a = \alpha^i$ and α is a primitive element of $GF(q)$. In the next section we review structural properties of QT codes, then we conclude the paper by presenting new codes and their generators together with (Hamming) weight enumerators.

2. Structure of 1-Generator QT codes

Let

$$G_0 = \begin{bmatrix} g_0 & g_1 & g_2 & \cdots & g_{m-1} \\ ag_{m-1} & g_0 & g_1 & \cdots & g_{m-2} \\ ag_{m-2} & ag_{m-1} & g_0 & \cdots & g_{m-3} \\ \vdots & \vdots & \vdots & & \vdots \\ ag_1 & ag_2 & ag_3 & \cdots & g_0 \end{bmatrix}_{m\times m}. \tag{2.1}$$

An $(m \times m)$ matrix of the type G_0 is called a twistulant matrix of order m or simply a twistulant matrix.

It is shown in [**8**] that the generator matrices of QC codes can be transformed into blocks of circulant (twistulant with $a = 1$) matrices by suitable permutation of columns. We can adopt a similar proof for QT codes here: Let C be QT code over F_q. Let $\mathbf{c_1}, \mathbf{c_2}, \ldots, \mathbf{c_r}$ be the rows of a generator matrix of C. Form another generator matrix for C by taking all possible $\mu_{a,l}(\mathbf{c_i})$ (l quasi-twisted shifts). Thus we form a $rm \times n$ generator matrix for C. Next permute the columns $C_1, C_2, \ldots, C_n$ of the generator matrix so that they appear in the order

$$C_1, C_{l+1}, \ldots, C_{(m-1)l+1}, C_2, C_{2+l}, \ldots, C_{(m-1)l+2}, \ldots, C_l, C_{2l}, \ldots, C_{ml}.$$

Then, the resulting matrix will be in the blocks of twistulant matrices. Therefore, generator matrices of an r-generator and 1-generator QT codes can be assumed to be in the following forms:

$$\begin{bmatrix} G_{11} & G_{12} & \cdots & G_{1l} \\ G_{21} & G_{22} & \cdots & G_{2l} \\ \vdots & \vdots & & \vdots \\ G_{r1} & G_{r2} & \cdots & G_{rl} \end{bmatrix}_{rm\times n}, \text{ and } \begin{bmatrix} G_1 & G_2 & \cdots & G_l \end{bmatrix}_{m\times n},$$

respectively, where each G_{ij} (or G_k) is a twistulant matrix of the form (1).

Similar to quasi-cyclic case, an l-QT code over F_q of length $n = ml$ can be viewed as an $F_q[x]/(x^m - a)$ submodule of $(F_q[x]/(x^m - a))^l$. Then an r-generator QT code is spanned by r elements of $(F_q[x]/(x^m - a))^l$. In this paper we restrict ourselves to 1-generator QT codes. 1-Generator QC codes and their structural properties have been studied in [**24**] and [**7**], respectively. Recently, in [**1**] the structure of r-generator QC codes has been investigated by use of Gröbner basis.

Let $1 \le i \le l$. For fixed i consider the following i^{th} **restriction map** on an l-QT code C of length $n = ml$:

$$\Pi_i : F_q^n \to F_q^m$$

$$(c(0), c(2), \ldots, c(ml-1)) \to (c((i-1)m), c(1 + (i-1)m), \ldots, c(m-1+(i-1)m)).$$

In view of the structure of QT codes described above, $\Pi_i(C)$ is a constacyclic code for all i. This will yield the following theorem.

THEOREM 2.1. *Let C be a 1-generator l-QT code over F_q of length $n = ml$. Then, a generator $\boldsymbol{g}(\boldsymbol{x}) \in (F_q[x]/\langle x^m - a\rangle)^l$ of C has the following form*

$$\boldsymbol{g}(\boldsymbol{x}) = (f_1(x)g_1(x), f_2(x)g_2(x), \dots, f_l(x)g_l(x)),$$

where $g_i(x)|(x^m - a)$ and $(f_i(x), (x^m - a)/g_i(x)) = 1$ for all $1 \le i \le l$.

PROOF. Since $\Pi_i(C)$ is a constacyclic code for every i we have the result. □

The following is the main theorem which plays an important role in our research. The proof can be found in [**5**].

THEOREM 2.2. [**5**] *Let C be a 1-generating l-QT code of length $n = ml$ with a generator of the form:*

$$\boldsymbol{g}(\boldsymbol{x}) = (g(x)f_1(x), f_2(x)g(x), \dots, f_l(x)g(x)) \tag{2.2}$$

where $g(x)|(x^m - a), g(x), f_i(x) \in F[x]/\langle x^m - a\rangle$, and $(f_i(x), h(x)) = 1$, $h(x) = \frac{x^m - a}{g(x)}$ for all $1 \le i \le l$. Then $l \cdot (d+1) \le d(C)$, where $\{\delta\zeta^i : s \le i \le s + (d-1)\}$ are among the zeros of $g(x)$ for some integers s, d $(d > 0)$ and dimension of C is equal to $n - \deg(g(x))$.

EXAMPLE 2.3. In this example we illustrate construction of some *optimal* constacyclic and QT codes using BCH bound for constacyclic codes and Theorem 2.2 for QT codes. We also use the table [**2**] to check optimal parameters of codes under consideration. Let $q = 5$ and $n = 26$ and consider constacyclic codes of length 26 over F_5 with $a = 3$. We remark that the condition $(n, q-1) \not| i$ implies that it suffices to consider only even lengths over F_5 with $a = 2$ or 3 (to possibly obtain constacyclic codes not equivalent to cyclic ones). We find that $r = 4$ (order of a in $F_5^\times$) and therefore $(x^{26} - 3)|(x^{104} - 1)$. The factorization of $x^{26} - 3$ over F_5 is as follows:

$$\begin{aligned} x^{26} - 3 &= \prod_{i=0}^{25}(x - \delta\zeta^i) = \prod_{i=0}^{25}(x - \delta^{4i+1}) \\ &= (x^4 + 2x^3 + x + 4)(x^4 + 3x^3 + 4x + 4)(x^4 + 4x^3 + 2x^2 + 2x + 4) \\ &\quad (x^4 + 2x^3 + 4x^2 + x + 4)(x^4 + x^3 + 2x^2 + 3x + 4) \\ &\quad (x^4 + 3x^3 + 4x^2 + 4x + 4)(x^2 + 2) \end{aligned}$$

where δ is a primitive 104-th root of 1 and $\zeta = \delta^4$ is a primitive 26-th root of 1 over F_5. The exponents of δ in this factorization are exactly integers of the form $4k+1$ modulo 104 and they are partitioned into the following cyclotomic cosets:
{1,5,21,25}, {9,17,45,85}, {13,65}, {29,41,89,101}, {33,61,69,97}, {37,49,81,93}, and {53,57,73,77}.
Let $h(x)|x^{26} - 3$ be the polynomial of smallest degree having δ among its root (then $h(x) = x^4 + 2x^3 + x + 4$) and let $g(x) = \frac{x^{26}-3}{h(x)}$. Then the zeros of $g(x)$ contain $\delta\zeta^i, 7 \le i \le 25$ and the constacyclic code (with $a = 3$) of length 26 generated by $g(x)$ has dimension 4 and minimum distance at least 20. By the Brouwer's table[**2**] this is the best possible minimum distance of a linear code of length 26 and dimension 4 over F_5. Therefore we obtain an *optimal* code.

Continuing with this example, if we take $g'(x)|x^{26} - 3$ to be the polynomial of smallest degree containing δ^i, $i = 1, 9, 13$ among its roots then $g'(x)$ has degree 10 with $\delta\zeta^i, 0 \le i \le 6$ also among its roots. Therefore, the constacyclic code generated by $g'(x)$ has length 26, dimension 16 and minimum distance at least 8. Again, it turns out that this code is optimal.

Now we consider some QT codes releted to the codes above. By theorem 2.2, a QT code of the form $(g(x), g(x)f(x))$ with $(f(x), h(x)) = 1$ has length 52, dimension 4 and minimum distance at least 40. Checking [**2**], we see that we obtain another optimal code. Note that we can simply take $f(x)$ to be the constant polynomial 1. Similarly, a QT code of the form $(g(x), g(x)f_1(x), g(x)f_2(x))$ with $(f_i(x), h(x)) = 1, i = 1, 2$ has length 78, dimension 4 and minimum distance at least 60. Again, these turn out to be the parameters of an optimal code.

A sufficient condition for QT codes to be equivalent to QC codes is also given in [**5**].

THEOREM 2.4. [**5**]*Let $a = \alpha^i$ where α is a primitive element of F_q. If $(m, q-1)|i$, a QT code of length $n = ml$ over F_q is equivalent to a QC code of length n over F_q.*

3. New Codes And Their Generator Matrices

3.1. The Search Method. We have restricted our search to 1-generator QT codes with generators of the form:

$$(g(x), f_2(x)g(x), \dots, f_l(x)g(x)).$$

In order to refine the search we looked at cyclotomic cosets of appropriate modulus and formed generator polynomials having long strings of consecutive integer powers of ζ among their zeros. After fixing $g(x)$ (hence determining the dimension of the code as well as the block length m), we searched over $f_i(x)$ (by the help of a computer). In most cases $l = 2$ or 3. When $l = 2$, we only search for one $f(x)$ with $\deg(f(x)) < m - \deg(g(x)$. In this case the search is exhaustive over the QT codes with the prescribed block length m and dimension, if the dimension is not too large. For an illustration of the method, we work the following example in details.

Let q=5, m=42 and a=3 and consider constacyclic codes of length 42 with $a = 3$ over F_5. The order of 3 mod 5 is 4 and $x^{42} - 3$ factors over F_5 as

$$x^{42} - 3 = \prod_{i=0}^{41}(x - \delta^{4i+1}).$$

The exponents of δ (a primitive 168-th root of 1) are all the integers of the form $4k+1$ mod 168 which are partitioned into the following cyclotomic cosets mod 168:
{1,5,25,101,121,125}, {9,45,57,69,81,117}, {13,61,65,113,137,157},
{17, 37,41,85,89,109}, {29,53,73,97,145,149}, {33,93,129,141,153,165},
{21,105}, {49,77}, and {133,161}.
Let $h(x)$ be the polynomial corresponding to cyclotomic cosets containing 1,21, and 133 and let

$$g(x) = \frac{x^{42} - 3}{h(x)} = x^{32} + 3x^{30} + 2x^{29} + 3x^{28} + 3x^{26} + x^{25} + 3x^{23} + 4x^{22} + 4x^{21} + x^{20} + 3x^{17} + x^{16} + x^{15} + x^{12} + 3x^{11} + x^{10} + 4x^9 + 2x^7 + 2x^6 + 3x^4 + 4x^3 + 2x^2 + 1.$$

Then $g(x)$ has degree 32 and contains $\delta\zeta^i$, $7 \le i \le 24$ among its roots. Therefore, the constacyclic code of length 42 over F_5 with $a = 3$ generated by $g(x)$ has dimension 10 and minimum distance ≥ 19 and a QT code of the form (g, gf) with $(f, \frac{x^{42}-3}{g}) = 1$ has length 84, dimension 10 and minimum distance at least 38, by Theorem 2.2. Let $f = 4x^8 + 2x^7 + 2x^6 + x^5 + x^4$, then f satisfies the last condition and a computer search shows that the QT code with these generators has, in fact, minimum distance 52, yielding a better code than the previously known $[84, 10, 51]_5$-code [**2**]. The weight enumerator of this code is as follows:

$$0^1 52^{1344} 53^{840} 54^{2352} 55^{6888} 56^{14808} 57^{31080} 58^{51576}$$
$$59^{100632} 60^{171360} 61^{247968} 62^{375648} 63^{532168} 64^{671160} 65^{854448}$$
$$66^{968520} 67^{1052016} 68^{1056720} 69^{986328} 70^{841944} 71^{655200} 72^{462504}$$
$$73^{317352} 74^{185808} 75^{102312} 76^{46368} 77^{19008} 78^{5544} 79^{2688} 80^{672} 81^{336} 84^{32}.$$

3.2. Generators and Weight Enumerators. We conclude by giving the generator matrices and weight enumerators of the new codes. Since a generator matrix of a 1-generator QT code is determined by the first row alone (and the constant a), we only present the first row separating the blocks with a comma.

The codes in the following list improve the minimum distances of linear codes over $GF(5)$ given in [**2**]. These are all QT codes with $a = 3$ or $a = 4$ over $GF(5)$, and the subscript 5 will be dropped in the notation $[n, k, d]_q$.

1. A $[84, 10, 52]$ code:

This is given in the last example.

2. A $[62, 6, 45]$ code, $a = 4$:

$$(1030421204023423030444342100000,$$
$$0114404142342411330211102021023)$$

The weight enumerator of this code is

$0^1 45^{3720} 50^{9424} 55^{2480}$.

3. A $[62, 7, 42]$ code, $a = 4$:

$$(1344300144422441011030231000000,$$
$$0011342042114424401041100344023)$$

The weight enumerator of this code is

$$0^1 42^{744} 43^{1736} 44^{2356} 45^{3720} 46^{6944} 47^{3720} 48^{6820} 49^{9920}$$
$$50^{11780} 51^{6696} 52^{10168} 53^{5704} 54^{4464} 55^{744} 56^{1860} 57^{744} 62^4.$$

4. A $[93, 6, 68]$ code, $a = 4$:

$$(142323010331413011213320410000\,0,\ 000100132131211240232222024202\,4,\ 011424002333101130243433003202\,2)$$

The weight enumerator of this code is

$$0^1 68^{744} 69^{1488} 70^{372} 71^{1116} 72^{1612} 73^{744} 74^{1860} 75^{2108} 76^{1488} 77^{744} 78^{992} 79^{372} 80^{744} 81^{1116} 87^{124}.$$

5. A $[93, 7, 65]_3$ code, $a = 4$:

$$(1441342004112203214040311000000,\ 0014414311333121144343112100311,\ 0134011424242310312341023211133)$$

The weight enumerator of this code is

$$0^1 65^{1116} 66^{1116} 67^{1240} 68^{1984} 69^{3224} 70^{6324} 71^{3100} 72^{4464} 73^{5952} 74^{6200} 75^{14632} 76^{6696} 77^{5952} 78^{4712} 79^{2728} 80^{5828} 81^{1364} 82^{744} 83^{248} 84^{248} 85^{124} 86^{124} 93^4.$$

6. A $[38, 9, 21]$ code, $a = 4$:

(1131002220100000000, 0000121213223334313)

The weight enumerator of this code is

$$0^1 21^{912} 22^{2128} 23^{5472} 24^{15580} 25^{32148} 26^{68096} 27^{113468} 28^{183312} 29^{245860} 30^{311372} 31^{309852} 32^{275880} 33^{196460} 34^{116660} 35^{53504} 36^{18240} 37^{3724} 38^{456}.$$

7. A $[38, 10, 20]$ code, $a = 4$:

(1033234341000000000, 0000111334140224004)

The weight enumerator of this code is

$$0^1 20^{1140} 21^{3572} 22^{10944} 23^{29412} 24^{73720} 25^{158840} 26^{335692} 27^{576992} 28^{914812} 29^{1251340} 30^{1524408} 31^{1563700} 32^{1370584} 33^{985720} 34^{586340} 35^{264936} 36^{93936} 37^{17480} 38^{2056}.$$

8. A $[57, 9, 34]$ code, $a = 4$:

$$(1131002220100000000, 0000124421004233201,$$
$$0001431302133334101)$$

The weight enumerator of this code is

$$0^{1}34^{836}35^{1368}36^{2888}37^{6992}38^{12616}39^{25384}40^{49932}41^{78356}$$
$$42^{116432}43^{166744}44^{213560}45^{246620}46^{263188}47^{241072}48^{206416}$$
$$49^{137636}50^{92568}51^{52516}52^{25004}53^{9424}54^{2584}55^{836}56^{152}.$$

9. A $[57, 10, 33]$ code, $a = 4$:

$$(1033234341000000000, 0000113101200244341,$$
$$0000331440211144121)$$

The weight enumerator of this code is

$$0^{1}33^{1292}34^{3192}35^{7144}36^{13300}37^{31768}38^{66196}39^{132240}40^{232940}$$
$$41^{389804}42^{610204}43^{841548}44^{1046292}45^{1253620}46^{1293140}47^{1186588}$$
$$48^{1014828}49^{734768}50^{464968}51^{260528}52^{114760}53^{48336}54^{13908}55^{3952}56^{228}57^{80}.$$

10. A $[76, 9, 48]$ code, $a = 4$:

$$(1131002220100000000, 0000124410244211000,$$
$$0001141322220314013, 1212423010403214213)$$

The weight enumerator of this code is

$$0^{1}48^{1140}49^{1976}50^{3876}51^{7980}52^{12312}53^{22116}54^{34656}55^{52668}$$
$$56^{80104}57^{108148}58^{150556}59^{187188}60^{214320}61^{227088}62^{218652}$$
$$63^{197448}64^{157244}65^{117040}66^{75164}67^{44536}68^{23864}69^{10336}$$
$$70^{3420}71^{1064}72^{152}73^{76}.$$

11. A $[76, 10, 46]$ code, $a = 4$:

$$(0103323434100000000, 0000011310022201000,$$
$$4000012444111331402, 1023314213140123200)$$

The weight enumerator of this code is

$$0^1 46^{1824} 47^{2432} 48^{5168} 49^{9424} 50^{17252} 51^{34124} 52^{61636} 53^{103968} 54^{169328} 55^{272384} 56^{402116} 57^{559364} 58^{765244} 59^{916712} 60^{1059440} 61^{1133236} 62^{1104812} 63^{979488} 64^{784396} 65^{585276} 66^{381444} 67^{227620} 68^{114380} 69^{48640} 70^{18772} 71^{6156} 72^{836} 73^{152}.$$

12. A $[95, 9, 62]$ code, $a = 4$:

$$(1131002220100000000, 0000124410244211000, 0022101201104241104, 1332410100321123142, 0000121213223334313)$$

The weight enumerator of this code is

$$0^1 62^{1140} 63^{1824} 64^{3496} 65^{5928} 66^{10564} 67^{14288} 68^{27816} 69^{39444} 70^{59052} 71^{80864} 72^{109972} 73^{139536} 74^{165908} 75^{188252} 76^{202464} 77^{195244} 78^{84680} 79^{161044} 80^{128516} 81^{93328} 82^{59508} 83^{40432} 84^{21356} 85^{11400} 86^{4180} 87^{1748} 88^{912} 89^{152} 90^{76}.$$

13. A $[42, 9, 24]$ code, $a = 4$:

(103124242130100000000, 000424123404310023013)

The weight enumerator of this code is

$$0^1 24^{1260} 25^{2520} 26^{6888} 27^{13608} 28^{34332} 29^{62832} 30^{112560} 31^{168504} 32^{227220} 33^{281820} 34^{302820} 35^{280416} 36^{205212} 37^{140280} 38^{73248} 39^{27888} 40^{9828} 41^{1596} 42^{292}.$$

14. A $[39, 8, 23]$ code, $a = 4$:

(4123112013432, 4410232413114, 0103301000000)

The weight enumerator of this code is

$$0^1 23^{416} 24^{2132} 25^{4056} 26^{8372} 27^{14092} 28^{24804} 29^{40040} 30^{53144} 31^{60112} 32^{61620} 33^{52416} 34^{34684} 35^{21684} 36^{9776} 37^{2548} 38^{624} 39^{104}.$$

15. A $[70, 8, 46]$ code, $a = 3$:

$$(24234110000000, 23213402221343, 24201430214241, 20303311304122, 14130120102401)$$

The weight enumerator of this code is

$$0^1 46^{1008} 47^{2520} 48^{3304} 49^{4824} 50^{9744} 51^{15456} 52^{20832}$$
$$53^{31976} 54^{34440} 55^{43008} 56^{45872} 57^{45080} 58^{42280} 59^{32424}$$
$$60^{25200} 61^{15736} 62^{9296} 63^{4480} 64^{2352} 65^{504} 66^{168} 67^{112} 70^{8}.$$

16. A $[84, 8, 56]$ code, $a = 4$:

(410410024123140043031000000, 4422121132004424123432430120,
3011400024324401111422232411)

The weight enumerator of this code is

$$0^1 56^{1568} 57^{896} 58^{2520} 59^{3584} 60^{6720} 61^{8400} 62^{15344}$$
$$63^{21056} 64^{30240} 65^{32256} 66^{39872} 67^{42224} 68^{41664} 69^{37408}$$
$$70^{34968} 71^{23968} 72^{20944} 73^{13216} 74^{7280} 75^{3360} 76^{2072}$$
$$77^{672} 78^{224} 79^{112} 84^{56}.$$

The above results give the following improvements on the lower bounds of minimum distances of codes with certain parameters over GF(5).

THEOREM 3.1.

$52 \leq d_5[84, 10] \leq 59$	$45 \leq d_5[62, 6] \leq 46$	$42 \leq d_5[62, 7] \leq 45$
$68 \leq d_5[93, 6] \leq 71$	$65 \leq d_5[93, 7] \leq 70$	$21 \leq d_5[38, 9] \leq 24$
$20 \leq d_5[38, 10] \leq 24$	$34 \leq d_5[57, 9] \leq 39$	$33 \leq d_5[57, 10] \leq 39$
$48 \leq d_5[76, 9] \leq 55$	$46 \leq d_5[76, 10] \leq 54$	$62 \leq d_5[95, 9] \leq 70$
$24 \leq d_5[42, 9] \leq 27$	$23 \leq d_5[39, 8] \leq 25$	$46 \leq d_5[70, 8] \leq 50$
$56 \leq d_5[84, 8] \leq 62.$		

Before we conclude, we would like to point that by methods such as puncturing or shortening these new codes, we obtain several other improvements on minimum distances.

References

[1] K. Lally and p. Fitzpatrick, *Construction and classification of quasi-cyclic codes*, WCC 99, Workshop on Coding and Cryptography January 11-14, 1999, Paris (France).

[2] A.E. Brouwer, *Linear code bound* (server), Eindhoven University of Technology, The Netherlands, http://www.win.tue.nl/win/math/dw/ voorlincod.html.

[3] ______, *Universal bounds for codes and designs*, Handbook of Coding Theory (V. Pless and W. C. Huffman eds.), vol. 1, Elsevier Science, Amsterdam, 1998, pp. 499-648.

[4] F.J. MacWilliams and N.J.A Sloane, *The Theory Of Error Correcting Codes*, Amsterdam, North-Holland, 1977.

[5] N. Aydin, I. Siap and D. K. Ray-Chaudhuri, *The structure of 1-generator quasi-twisted codes and new linear codes*, (2000), preprint.

[6] I. Siap, N. Aydin and D. K. Ray-Chaudhuri, *New ternary quasi-cyclic codes with better minimum distances*, IEEE Trans. Inform. Theory, to appear.

[7] Jean Conan and Gerald Seguin, *Structural Properties and Enumeration of quasi-cyclic Codes*, AAECC 4, pp. 25-39, (1993).

[8] Koshy Thomas, *Polynomial Approach to Quasi-Cyclic Codes* , Bul. Cal. Math. Soc. **69** (1977), 51–59.

[9] T. Kasami, *A Gilbert-Varshamov bound for quasi-cyclic codes of rate* 1/2, IEEE Trans. Inform. Theory, **20**(1974), 679.

[10] Stafford E. Tavares, V.K. Bhargava and S.G.S. Shiva, *Some rate p/(p+1) quasi-cyclic codes*, IEEE Trans. on Information Theory, **20**(1974), 133–135.

[11] T. Aaron Gulliver, Vijay K. Bhargava, *Some best rate 1/p quasi-cyclic codes over GF(5)*, Information Theory and Applications II, Springer-Verlag Lecture Notes in Computer Science **1133** (1996), 28–40.

[12] T. Aaron Gulliver, *New optimal ternary linear codes of dimension 6*, Ars Combin. **40**(1995), 97–108.

[13] R.N. Daskalov, T. A. Gulliver and E. Metodieva, *New good quasi-cyclic ternary and quaternary linear codes*, IEEE Trans. on Information Theory **43**(1997), 1647–1650.

[14] R.N. Daskalov, T. A. Gulliver and E. Metodieva, *New ternary linear codes*, IEEE Trans. on Information Theory, **45** (1999), 1687–1688.

[15] P. P. Greenough and R. Hill, *Optimal ternary quasi-cyclic codes*, Designs Codes and Cryptography **2**(1992), 81–91.

[16] Zhi Chen, *Six new binary Quasi-Cyclic codes*, IEEE Trans. on Information Theory, **40**(1994), 1666–1667.

[17] T. Aaron Gulliver, Vijay K. Bhargava, *Two new rate 2/p binary quasi-cyclic codes*, IEEE Trans. on Information Theory, **40**(1994), 1667–1668.

[18] T. Aaron Gulliver, Vijay K. Bhargava, *Nine good rate* $(m-1)/pm$ *quasi-cyclic codes*, IEEE Trans. on Information Theory, **38**(1992), 1366–1369.

[19] T. Aaron Gulliver, Vijay K. Bhargava, *Some best rate* $1/p$ *and rate* $(p-1)/p$ *systematic quasi-cyclic codes over GF(3) and GF(4),* IEEE Trans. on Information Theory, **38**(1992),1369–1374.

[20] T. Aaron Gulliver, Patric R.J. Östergard, *Improved bounds for ternary linear codes of dimension 7*, IEEE Trans. on Information Theory,**43**(1997), 1377–1381.

[21] T. Aaron Gulliver, Vijay K. Bhargava, *New good rate* $(m-1)/pm$ *ternary and quaternary quasi-cyclic codes,* Designs Codes and Cryptography, **7**(1996),223-233.

[22] Henk van Tilborg, *On quasi-cyclic codes with rate 1/m*, IEEE Trans. on Information Theory, **24**(1978), no. 2, 628–630.

[23] Viyaj K. Bhargava, Gerald E. Séguin, J.M. Stein, *Some* (mk, k) *cyclic codes in quasi-cyclic form*, IEEE Trans. on Information Theory, **24**(1978), no. 2, 630–632.

[24] G.E. Séguin and G. Drolet, *The Theory Of* 1*-Generator Quasi-Cyclic Codes*, preprint, 1990.

[25] S. Roman, *Field Theory,* Springer-Verlag, 1995.

[26] S. Roman, *Coding and Information Theory,* Springer-Verlag, 1992.

[27] R. Hill and p. p. Greenough, *Optimal quasi-twisted codes*, Proc. Int. Workshop Algebraic and Comb.Coding Theory, Voneshta Voda, Bulgaria, pp. 92–97, June 1992.

[28] R.N. Daskalov, T. Aaron Gulliver, *Bounds on minimum distances for linear codes over GF(5)*, Appl. Algebra Engrg. Comm. Comput., **9**(1999),547–558.

DEPARTMENT OF MATHEMATICS, OHIO STATE UNIVERSITY, COLUMBUS, OHIO 43210
Current address: Department of Mathematics, Ohio State University, Columbus, Ohio 43210
E-mail address: `isiap@math.ohio-state.edu`

DEPARTMENT OF MATHEMATICS, OHIO STATE UNIVERSITY, COLUMBUS, OHIO 43210
E-mail address: `naydin@math.ohio-state.edu`

DEPARTMENT OF MATHEMATICS, OHIO STATE UNIVERSITY, COLUMBUS, OHIO 43210
E-mail address: `dijen@math.ohio-state.edu`

DIMACS Series in Discrete Mathematics
and Theoretical Computer Science
Volume **56**, 2001

Communication Complexity and Orthogonal Polynomials

Ulrich Tamm

ABSTRACT. We shall study the communication complexity of certain functions defined on an association scheme. The communication complexity can be lower bounded by the rank of matrices which are linear combinations of the matrices in the association scheme. These can be simultaneously diagonalized and hence the eigenvalues are linear combinations of the orthogonal polynomials corresponding to the association scheme. The eigenvalues in the association schemes important for Coding Theory can be obtained recursively via a formula discussed by Delsarte in [**9**], which turns out to be very useful in the determination of the communication complexity for several functions under discussion.

1. Introduction

Let $D_0, \ldots, D_n$ be a collection of $\{0,1\}$-matrices forming an association scheme. Since $\sum_{k=0}^{n} D_k$ is the all-one matrix, a linear combination $\sum_{k=0}^{n} c_k D_k$ can be regarded as the matrix $(f(x,y))_{x,y}$ representing the function f defined by $f(x,y) = c_k$ if (x,y) is in relation corresponding to matrix D_k. As an example, consider the Hamming distance d_H arising for the matrices of the Hamming association scheme with the choice of $c_k = k$ for all $k = 0, \ldots, n$. Parameters of functions thus obtained may now be studied exploiting properties of the association scheme.

One such parameter is the communication complexity $C(f)$, which is the number of bits that two persons have to exchange in order to evaluate $f(x,y)$, when initially one person only knows $x \in \mathcal{X}$ and the other person only knows $y \in \mathcal{Y}$. Communication complexity, as introduced by Yao [**28**] for functions $f : \mathcal{X} \times \mathcal{Y} \longrightarrow \mathcal{Z}$ with finite $\mathcal{X}, \mathcal{Y}, \mathcal{Z}$, turned out to be an important topic in computer science, especially in the derivation of bounds for area-time tradeoff in VLSI-layout and for depth of circuits, decision trees and branching programs. We refer to the recent books by Kushilevitz and Nisan [**23**] or Hromkovic [**18**] for further details. Connections between communication complexity and information theory are discussed e.g. by Karp [**20**] or Körner and Orlitsky [**22**].

Communication proceeds according to a predetermined (deterministic) protocol. The two persons exchange bits of information until they both know the result $f(x,y)$. The set of messages a person is allowed to send is required to be prefix-free in order that the second person immediately recognizes the end of the message.

1991 *Mathematics Subject Classification.* Primary 05E35, 05E30; Secondary 68R05, 94A11.

We assume in the following that the persons communicate over a noiseless channel. Note that communication complexity for noisy channels had been studied, e. g., in **[26]** and **[19]**. The model in the latter paper is connected to the theory of identification via channels initiated by Ahlswede and Dueck **[2]**. Closely related to communication complexity is also Orlitsky's model of interactive communication, e. g. **[25]**, where the function is only partially defined, i. e., there are restrictions on the domain.

An upper bound on the communication complexity is always obtained via the following *trivial protocol*. Person 1 sends all the bits of his input x to Person 2 using at most $\lceil \log |\mathcal{X}| \rceil$ bits. Person 2 then knows x and y and hence is able to calculate $f(x, y)$, which he returns to Person 1 using at most $\lceil \log |\mathcal{Z}| \rceil$ bits. The logarithm throughout this paper is always taken to the base 2. Hence (w. l. o. g. $|\mathcal{X}| \leq |\mathcal{Y}|$)

$$C(f) \leq \lceil \log |\mathcal{X}| \rceil + \lceil \log |\mathcal{Z}| \rceil \tag{1.1}$$

Lower bounds for $C(f)$ are obtained from the matrices $M_z(f) = (a_{xy})_{x,y}$ defined by $a_{xy} = \begin{cases} 1 & \text{if } f(x,y) = z \\ 0 & \text{if } f(x,y) \neq z \end{cases}$. In this paper we only need the following rank lower bound

$$C(f) \geq \lceil \log \sum_z \text{rank} M_z(f) \rceil . \tag{1.2}$$

We shall need some basic facts about association schemes, which are represented as a set of matrices $D_0, \ldots, D_n$ with the properties

$$D_i \text{ is symmetric, for all } i$$

$$\sum_{i=0}^{n} D_i \text{ is the all-one matrix}$$

$$D_0 \text{ is the identity matrix}$$

$$D_i D_j = D_j D_i = \sum_{k=0}^{n} c_{ijk} D_k$$

Since the matrices are symmetric and commute with each other they are simultaneously diagonalizable, which will be quite important in the sequel. The eigenvalues of the matrices in an association scheme are orthogonal polynomials evaluated at certain positions. For instance the eigenvalues of the matrix D_k in the Hamming scheme are the Krawtchouk polynomials (where q denotes the alphabet size over which the Hamming distance is defined)

$$K_k(x, n, q) = \sum_{j=0}^{k} (-1)^j \binom{x}{j} \binom{n-x}{k-j} (q-1)^{k-j}$$

evaluated at the integers $x = i, i = 0, \ldots n$ and the eigenvalues of the matrix D_k in the Johnson scheme are the Eberlein polynomials

$$E_k(x, n, l) = \sum_{j=0}^{k} (-1)^j \binom{x}{j} \binom{n-x}{k-j} \binom{l-n-x}{k-j}$$

evaluated at the integers $x = i, i = 0, \dots n$.

In Section 2, we are going to study scenaria, in which the analysis of integral zeros of orthogonal polynomials is the central tool. Functions obtained by choosing $c_k = k$ modulo 2 for the Hamming and Johnson scheme are discussed in Section 3. Here a recurrence formula discussed by Delsarte [**9**] is crucial.

2. Zeros of Orthogonal Polynomials

Let us first consider the case that all the c_k's are different. Then the matrices $M_z(f)$ are just the matrices of the association scheme. In order that all of them have full rank, all their eigenvalues must be different from 0. For the Hamming association scheme the problem is to show that the Krawtchouk polynomials have only very few integral zeros. This way the communication complexity of the binary Hamming distance can be determined up to one bit. For alphabet size $q \geq 3$, it is still an open problem to derive the same result (at least using the above rank lower bound, for $q = 2$ ([**12**]) and $q = 4, 5$ ([**1**]) $C(d_H)$ has been determined up to one bit using a different approach via constant distance code pairs, cf. also [**17, 11, 24, 16**]).

THEOREM 2.1. *([**12, 1**]): The communication complexity of the Hamming distance for alphabet size $q = 2, 4, 5$ can be determined up to one bit, namely*

$$\lceil n \cdot \log q \rceil + \lceil \log(n+1) \rceil - 1 \leq C(d_H) \leq \lceil n \cdot \log q \rceil + \lceil \log(n+1) \rceil \tag{2.1}$$

The upper bound in Theorem 2.1 obviously stems from the trivial protocol, which requires $\lceil n \cdot \log q \rceil$ bits for the transmission of the value $x^n \in \{0, \dots, q-1\}^n$ and $\lceil \log(n+1) \rceil$ bits for the transmission of the result $d_H(x^n, y^n)$.

Observe that if all the eigenvalues of the Hamming association scheme were different from 0, i. e., the Krawtchouk polynomials $K_k(x, n, q)$ do not have integral zeros for all $k = 0, \dots, n$, then

$$\lceil n \cdot \log q + \log(n+1) \rceil \leq C(d_H). \tag{2.2}$$

Integral zeros of Krawtchouk polynomials have been studied thoroughly in [**6, 15, 21**], since they play an important role, e. g. in the study of perfect codes and the invertibility of the Radon transform. Since, because of the ceilings in the upper bound of Theorem 2.1 and the lower bound (2.2), for alphabet size not divisible by 2 the two bounds may differ by one bit, it is not, in general, possible to determine the communication complexity exactly using the rank lower bound. So, the difference of the two bounds by one bit - which is hence the optimal we can prove using this method - is still guaranteed, if there are only very few integral zeros. Krasikov and Litsyn [**21**] conjecture that a Krawtchouk polynomial for $q \geq 3$ can have at most four integral zeros, see also Habsieger's paper [**14**] in this volume.

Another function for which the analysis of zeros of orthogonal polynomials is a crucial tool, is the function f defined by

$$f(x^n, y^n) = \begin{cases} 1 & \text{if } d_H(x^n, y^n) \leq t \\ 0 & \text{if } d_H(x^n, y^n) > t \end{cases} \tag{2.3}$$

yielding the matrices $M_1 = \sum_{k=0}^{t} D_k$, $M_0 = \sum_{k=t+1}^{n} D_k$ Since the matrices D_k, $k = 0, 1, \dots, n$ of the Hamming scheme are simultaneously diagonalizable, the eigenvalues of M_1 are sums of the first t Krawtchouk polynomials $\sum_{k=0}^{t} K_k(i, n, q)$, for $i = 0, \dots, n$. These are just Lloyd polynomials important in the study of perfect codes. The same argumentation obviously extends to any function defined in a similar way as by (2.3) in a metric association scheme, cf. [**7**].

THEOREM 2.2. *Let* D_k, $k = 0, \dots, n$ *be the matrices of a metric association scheme on which a Boolean function* f *is defined (for fixed* t*) by the function matrices* $M_1 = \sum_{k=0}^{t} D_k$, $M_0 = \sum_{k=t+1}^{n} D_k$ *of size* $Q \times Q$. *Then*

$$C(f) = \lceil \log Q \rceil + 1 \tag{2.4}$$

if all the Lloyd polynomials $\Psi_t(i)$*, the sum polynomials (from* 0 *to* t *over the eigenvalues of the association scheme) are different from* 0.

PROOF: Obviously, the right hand side in (2.4) is just the number of bits required by the trivial protocol. On the other hand, it also coincides with the rank lower bound (1.2) if the function matrices M_1 and M_0 have full rank. Because of the simultaneous diagonalizability of the matrices in the association scheme, this is guaranteed if all the Lloyd polynomials are different from 0.

3. Hamming and Johnson Scheme Modulo 2

Now we are going to consider further functions for which $c_k = c_l$ for several pairs $k \neq l$, more exactly $c_k = k \bmod 2$. In this case one might again exploit the simultaneous diagonalizability of the matrices $D_0, \dots, D_n$ of the association scheme in order to study the eigenvalues of the matrices $M_z(f) = \sum_{k \equiv z \bmod 2} D_k$, $z = 0, 1$

These eigenvalues then will be linear combinations of the orthogonal polynomials arising as eigenvalues for the matrices of the association scheme.

One such function is the Hamming distance modulo 2 with matrices $M_0 = \sum_{k \text{ even}} D_k$, $M_1 = \sum_{k \text{ odd}} D_k$ whose eigenvalues

$$e_i(z, n, q) = \sum_{k \equiv z \bmod 2} K_k(i, n, q), \quad i = 0, \dots, n$$

were shown to be all different from 0 (for alphabet size $q > 3$) by the following identity for Krawtchouk polynomials.

THEOREM 3.1. *(*[**27**]*): For* $z = 0, 1$ *it is*

$$\sum_{k \equiv z (\bmod 2)} K_k(i, q, n) = \begin{cases} \frac{1}{2}(q^n + (-1)^z (2-q)^n) & i = 0 \\ (-1)^z 2^{i-1} (2-q)^{n-i+1} & i = 1, \dots, n \end{cases}$$

With Theorem 3.1 it is immediate with the rank lower bound that the communciation complexity of the Hamming distance modulo 2 is that of the trivial protocol.

COROLLARY: *Let* p_n *denote the Hamming distance modulo 2. For alphabet size* $q \geq 3$

$$C(p_n) = \lceil n \cdot \log q \rceil + 1$$

In [**27**] we proved Theorem 3.1 making use of the block structure of its function matrices. We shall now provide a new proof, based on a recurrence for the Krawtchouk polynomials, which was thoroughly studied by Delsarte in [**9**], namely the recursion

$$F(i, k, n) = b^k F(i-1, k, n-1) - b^{k-1} F(i-1, k-1, n-1) \tag{3.1}$$

PROOF OF THEOREM 3.1: Delsarte [9] showed that for the Krawtchouk polynomials recursion (3.1) is fulfilled by choosing $F(i,k,n) = K_k(i,n,q)$ and $b = 1$, so

$$K_k(i,n,q) = K_k(i-1,n-1,q) - K_{k-1}(i-1,n-1,q) \tag{3.2}$$

Now because of the simultaneous diganolizability of the matrices in the Hamming scheme this recursion extends to the eigenvalues $e_i(z,n,q)$, $z = 0,1$, of the function matrices M_z of the parity of the Hamming distance via

$$e_i(z,n,q) = \sum_{k \equiv z \bmod 2} K_k(i,n,q)$$

$$= \sum_{k \equiv z \bmod 2} (K_k(i-1,n-1,q) - K_{k-1}(i-1,n-1,q))$$

$$= \sum_{k \equiv z \bmod 2} K_k(i-1,n-1,q) - \sum_{k \equiv z \bmod 2} K_{k-1}(i-1,n-1,q)$$

$$= \sum_{k \equiv z \bmod 2} K_k(i-1,n-1,q) - \sum_{k \equiv z-1 \bmod 2} K_k(i-1,n-1,q)$$

$$= e_{i-1}(z,n-1,q) - e_{i-1}(z-1 \bmod 2, n-1,q)$$

Since for $i > 0$ it is $e_i(0,n,q) + e_i(1,n,q) = 0$ by simultaneous diagonalizability and the fact that $M_0 + M_1$ is the all-one matrix. But now

$$e_i(0,n,q) = e_{i-1}(0,n-1,q) - e_{i-1}(1,n-1,q)$$

$$= e_{i-1}(0,n-1,q) - e_{i-1}(1,n-1,q) - e_{i-1}(0,n-1,q) + e_{i-1}(0,n-1,q)$$

$$= 2 \cdot e_{i-1}(0,n-1,q) = \ldots = 2^{i-1}\left(e_0(0,n-i+1,q) - e_0(1,n-i+1,q)\right)$$

$$= 2^{i-1} \sum_{k=0}^{n-i+1} (-1)^k K_k(0,n-i+1,q)$$

$$= 2^{i-1} \sum_{k=0}^{n-i+1} (-1)^k \binom{n}{k} (q-1)^k = 2^{i-1}(2-q)^{n-i+1}$$

By the previous considerations, of course $e_i(1,n,q) = -e_i(0,n,q)$ for $i > 1$, for $i = 0$ the identity in Theorem 3.1 can be analyzed by iterating the recursion $K_k(i,n,q) = K_k(i,n-1,q) + (q-1)K_{k-1}(i,n-1,q)$,

This proof has the advantage that it extends to further association schemes, whose eigenvalues obey a similar recursion.

For instance, it is now possible to study the communication complexity of the function g_n defined in an analogous way (as the Hamming distance modulo 2) on the Johnson scheme by its function matrices

$$M_0(g_n) = \sum_{k=0 \bmod 2} D_k, \quad M_1(g_n) = \sum_{k=1 \bmod 2} D_k. \tag{3.3}$$

where the D_k's are now the matrices in the Johnson scheme.

THEOREM 3.2. *The function determined by (3.3) has communication complexity determined by the trivial protocol, if for all $i = 1, \cdots, n$ the Krawtchouk polynomials $K_{n-i+1}(n-i+1, l+2i-2, 2)$ are different from 0.*

PROOF: Following the considerations for the Hamming scheme modulo 2, the eigenvalues are now linear combinations of Eberlein polynomials

$$e_i(z,n,l) = \sum_{k \equiv z \bmod 2} E_k(i,n,l)$$

For the Eberlein polynomials also the recursion (3.1) holds. First observe that the eigenvalues $e_0(z,n,l)$ (and $n \geq 1$) are both positive for $z = 0, 1$ as the sum of positive terms and hence different from 0. In analogy to the considerations for the Krawtchouk polynomials modulo 2 for $i \geq 1$ it is $e_i(0,n,l) + e_i(1,n,l) = 0$. With this

$$e_i(0,n,l) = e_{i-1}(0,n-1,l+2) - e_{i-1}(1,n-1,l+2)$$

$$= e_{i-1}(0,n-1,l+2) - 2e_{i-1}(1,n-1,l+2) + e_{i-1}(0,n-1,l+2)$$

$$= 2 \cdot e_{i-1}(0,n-1,l+2) = \ldots =$$

$$= 2^{i-1}\left(e_0(0,n-i+1,l+2i-2) - e_0(1,n-i+1,l+2i-2)\right)$$

$$= 2^{i-1} \sum_{k=0}^{n-i+1} (-1)^k E_k(0,n-i+1,l+2i-2)$$

$$= 2^{i-1} \sum_{k=0}^{n-i+1} (-1)^k \binom{n-i+1}{k} \binom{(l+2i-2)-(n-i+1)}{k}$$

$$= 2^{i-1} \sum_{k=0}^{n-i+1} (-1)^k \binom{n-i+1}{(n-i+1)-k} \binom{(l+2i-2)-(n-i+1)}{k}$$

$$= (-1)^{n-i+1} 2^{i-1} \sum_{k=0}^{n-i+1} (-1)^k \binom{n-i+1}{k} \binom{(l+2i-2)-(n-i+1)}{(n-i+1)-k}$$

$$= (-1)^{n-i+1} 2^{i-1} K_{n-i+1}(n-i+1, l+2i-2, 2)$$

So the problem now is to determine, when a Krawtchouk polynomial $K_k(k,m,2)$ (the degree and the first variable being the same) can be 0. This is possible for m even and $k = \frac{m}{2}$. We didn't find any other parameter pair (k,m) with this property yet.

A third family of orthogonal polynomials discussed by Delsarte [**9**] obeying the recursion (3.1) are the b-analogues ($b \geq 2$) of the Krawtchouk polynomials

$$\sum_{j=0}^{k} (-1)^j b^{\binom{j}{2}} \binom{i}{j}_b \binom{n-i}{k-j}_b \prod_{t=0}^{k-j-1} (cb^n - b^{i+t})$$

where c is some constant to be specified and $\binom{m}{s}_b = \prod_{j=0}^{s-1} \frac{b^{m-j}-1}{b^{s-j}-1}$ denotes the Gaussian binomial coefficient. In [**9**] are mentioned two special cases. The eigenvalues of the association schemes of bilinear forms over $GF(b)$ have as parameters a prime power b and $c = b^r$ for some nonnegative integer r. The eigenvalues of the association schemes of alternating bilinear forms have as parameters $b = p^2$ the square of a prime p and $c = p$ or $c = \frac{1}{p}$.

Since the Gaussian binomial coefficients $\binom{m}{s}_b$ modulo $b-1$ are just the usual binomial coefficients $\binom{m}{s}$, we can follow the computations for the Hamming and the Johnson scheme modulo 2 if we calculate modulo $b-1$.

For the b-analogues of the Krawtchouk polynomials the analysis is reduced to checking if the following alternating sum times a power of 2 can be 0 modulo $b-1$.

(3.4)
$$\sum_{k=0}^{n}(-1)^k\binom{n}{k}_b \cdot \prod_{t=0}^{k-1}(cb^n - b^t) = \sum_{k=0}^{n}(-1)^k\binom{n}{k}(c-1)^k = (2-c)^n \bmod (b-1)$$

THEOREM 3.3. *Let a function f be defined as under (3.3), where now the D_k's, $k = 0, \dots, n$, are the matrices of an association scheme of bilinear forms over $GF(b)$ or the matrices of an association scheme of alternating bilinear forms. Further let the prime p defining the parameters b and c be odd. Then the trivial protocol is optimal for all such functions if $b-1$ is not a power of 2.*

PROOF: Again for $i = 0$ the eigenvalues are positive as sums of positive terms. As mentioned above, for $i \geq 1$ calculation modulo $b-1$ follows the same lines as in the proof of Theorem 3.1 (for the Hamming scheme modulo 2). We denote the eigenvalues of the function matrix $M_z = \sum_{k \equiv z \bmod 2} D_k$ by $e_i(z,n,c) = \sum_{k\equiv z \bmod 2} F(k,i,n)$, where the $F(k,i,n)$ now are the b-analogues of the Krawtchouk polynomials (observe that we also list the parameter c not important for the recursion in the definition of the eigenvalues but for the final closed formula). Then

$$e_i(z,n,c) = e_{i-1}(z,n-1,c) - e_{i-1}(z-1 \bmod 2, n-1, c) \bmod (b-1)$$

and hence

$$e_i(0,n,c) = e_{i-1}(0,n-1,c) - e_{i-1}(1,n-1,c) \bmod (b-1)$$
$$= 2 \cdot e_{i-1}(0,n-1,c) = \dots =$$
$$2^{i-1}(e_0(0,n-i+1,c) - e_0(1,n-i+1,c)) \bmod (b-1)$$
$$= 2^{i-1}\sum_{k=0}^{n-i+1}(-1)^k K_k(0,n-i+1,c) = 2^{i-1}(2-c)^{n-i+1} \bmod (b-1)$$

by the simultaneous diagonalizability and (3.4). Now, for the association scheme of bilinear forms over $GF(b)$, c is a power of b and hence $2-c$ is congruent 1 modulo $b-1$. For the scheme of alternating bilinear forms the parameters are $c = p$ (or $c = p^{-1} \equiv p \bmod p^2 - 1$) and $b = p^2$, so $p^2 - 1$ is even and $(2-p)^n$ is odd and hence incongruent 0 modulo p^2-1. In both cases $(2-c)^n$ cannot be 0 modulo $b-1$. Since $b-1$ is assumed not to be a power of 2, also 2^{i-1} must be different from 0 modulo $b-1$.

REMARK 3.4. Observe that in our theorem $b = p^r$ is power of an odd prime and we calculate modulo $p^r - 1 = (p-1)\cdot(p^{r-1} + p^{r-2} + \cdots + p + 1)$ for some nonnegative integer r. If additionally r is even, then $p^r - 1 = (p^{r/2}-1)\cdot(p^{r/2}+1)$, which can only be a power of 2 in the case $p = 3, t = 2$, i.e., for the association schemes of alternating bilinear forms, where $b = p^2$, we have to exclude only this case (and, of course, the case $p = 2$). If $r \geq 3$ is odd then $p^{r-1} + p^{r-2} + \cdots + p + 1$ is also odd as the sum of an odd number of odd summands and hence $b - 1$ cannot be a power of 2. So for the association schemes of bilinear forms it remains to study the case $r = 1$, i.e. b is a prime, in which case $b - 1$ can be a power of 2 only if $b = 2^{2^s}$ is a Fermat prime.

4. Concluding Remarks

The communication complexity of further functions defined on association schemes has been studied in [**27**], for instance the Lee distance, the Lee distance modulo 2.

In order to derive a lower bound on the communication complexity of the Hamming distance modulo 3 again it is possible to exploit recursion (3.1), which extends to the eigenvalues

$$e_i(z, n, q) = \sum_{k \equiv z \bmod 3} K_k(i, n, q)$$

of the function matrices $M_z = \sum_{k \equiv z \bmod 3} D_k$, $z = 0, 1, 2$ where the D_k's are the matrices of the Hamming scheme. So it is

$$e_i(z, n, q) = e_{i-1}(z, n-1, q) - e_{i-1}(z - 1 \bmod 3, n-1, q)$$

$$= e_{i-2}(z, n-1, q) - 2e_{i-2}(z - 1 \bmod 3, n-1, q) + e_{i-2}(z - 1 \bmod 3, n-1, q)$$

$$= (-3)e_{i-2}(z - 1 \bmod 3, n-1, q)$$

as long as $i > 2$. So we only have to analyze the values $e_i(z, n, q)$ for $i = 0, 1, 2$. However, this already can be a difficult task. For alphabet size $q \geq 4$ in [**27**] we could determine the communication complexity up to one bit, since the e_i's are not divisible by q in most cases. For $q = 2, 3$ and also for the Johnson scheme modulo 3, where an analogous reduction is possible, binomial identities have to be considered, which seem hard to attack.

References

[1] R. Ahlswede, *On code pairs with specified Hamming distances*, Combinatorics, Eger, 1987, Colloquia Math. Soc. J. Bolyai 52, 1988, 9 - 47.

[2] R. Ahlswede and G. Dueck, *Identification via channels*, IEEE Trans. Inform. Theory 35, 1989, 15-29.

[3] R. Ahlswede, A. El Gamal, and K. F. Pang, *A two-family extremal problem in Hamming space*, Discrete Math. 49, 1984, 1–5.

[4] E. Bannai and T. Ito, *Algebraic Combinatorics I: Association Schemes*, Benjamin-Cummings, 1984.

[5] P. Camion, *Codes and association schemes*, pp. 1441-1566 in V. S. Pless and W. C. Huffman (eds.), *Handbook of Coding Theory*, Volume II, North-Holland, 1998.

[6] L. Chihara and D. Stanton, *Zeros of generalized Krawtchouk polynomials*, J. Approx. Theory 60, 1990, 43-57.

[7] R. Clayton, *Perfect multiple coverings*, pp. 51-64, in D. Ray-Chaudhuri (ed.), *Coding Theory and Design Theory*, Part I Coding Theory, Springer 1990.

[8] P. Delsarte, *An algebraic approach to the association schemes of coding theory*, Philips Res. Rpts. Suppl., No. 10, 1973.

[9] P. Delsarte, *Properties and applications of the recursion* $F(i+1, k+1, n+1) = q^{k+1}F(i, k+1, n) - q^k F(i, k, n)$, SIAM J. Appl. Math. 31, No. 2, 1976, 262-270.

[10] P. Delsarte and V. I. Levenshtein, *Association schemes and coding theory*, IEEE Trans. Inform. Theory, Commemorative Issue, 44, No. 6, 1998, 2477-2504.

[11] P. Delsarte and P. Piret, *An extension of an inequality by Ahlswede, El Gamal and Pang for pairs of binary codes*, Discr. Math. 55, 1985, 313–315.

[12] A. El Gamal and K. F. Pang, *Communication complexity of computing the Hamming distance*, SIAM J. Comp. 15, no 4, 1986, 932–947.

[13] G. D. Godsil, *Algebraic Combinatorics*, Chapman & Hall, 1993.

[14] L. Habsieger, *Integral zeros of Krawtchouk polynomials*, this volume.

[15] L. Habsieger and D. Stanton, *More zeros of Krawtchouk polynomials*, Graphs and Combinatorics 9, 1993, 163-172.

[16] W. Haemers, *Disconnected vertex sets and equidistant code pairs*, Electron. J. Combin. 4, no. 1, 1997, 10 pp.

[17] J. I. Hall and J. H. van Lint, *Constant distance code pairs*, Proc. Kon. Ned. Akad. v. Wet. (A) 88, 1985, 41 - 45.

[18] J. Hromkovic, *Communication complexity and parallel computing*, Springer, 1997.

[19] J. Ja'Ja', *Identification is easier than decoding*, Proc. IEEE Symp. Foundations of Computer Science, 1985, 43-50.

[20] R. M. Karp, *ISIT'98 Plenary Lecture Report: Variations on the theme of 'Twenty Questions'*, IEEE Information Theory Society Newsletter, Vol. 49, No. 1, March 1999, 1-5 and 21-22.

[21] I. Krasikov and S. Litsyn, *On integral zeros of Krawtchouk polynomials*, J. Comb. Theory A 74, 1996, 71-99.

[22] J. Körner and A. Orlitsky, *Zero-error information theory*, IEEE Trans. Inform. Theory, Commemorative Issue, 44, No. 6, 1998, 2207 - 2229.

[23] E. Kushilevitz and N. Nisan, *Communication complexity*, Cambridge University Press, 1997.

[24] J. H. van Lint, *Distance theorems for code pairs*, Combinatorial Mathematics: Proceedings of the Third International Conference (New York, 1985), Ann. New York Acad. Sci. 555, 1989, 421 - 424.

[25] A. Orlitsky, *Worst case interactive communication I: two messages are almost optimal*, IEEE Trans. Inform. Theory 36, 1990, 1111-1126.

[26] L. J. Schulman, *Deterministic coding for interactive communication*, Proc. 25th ACM Symp. Theory of Computing, 1993, 747-756.

[27] U. Tamm, *Communication complexity of sum - type functions invariant under translation*, Inform. and Computation 116, no. 2, 1995, 162 - 173.

[28] A. C. Yao, *Some complexity questions related to distributive computing*, Proc. ACM Symp. Theory of Computing, 1979, 209–213.

Department of Mathematics, University of Bielefeld, P.O. Box 100131, 33501 Bielefeld, Germany

E-mail address: tamm@mathematik.uni-bielefeld.de

DIMACS Series in Discrete Mathematics
and Theoretical Computer Science
Volume **56**, 2001

Estimates of the distance distribution of nonbinary codes, with applications

Alexei Ashikhmin, Alexander Barg, and Simon Litsyn

ABSTRACT. We use the polynomial method to derive upper and lower bounds on the distance distribution of nonbinary codes in the Hamming space. Applications of the bounds include a better asymptotic upper bound on the weight distribution of Goppa codes from maximal curves, a new upper bound on the size of a q-ary code with a given rth generalized weight, improved estimates of the size of secant spaces of algebraic curves over F_q, and remarks on the error exponents for the q-ary symmetric channel.

1. Introduction

Polynomial method is a powerful technique for obtaining bounds on parameters of codes in distance-transitive spaces [**20**], [**6**]. Its applications to bounding distance distribution of codes with various restrictions on their minimum distance d and/or dual distance $d^{\perp}$ are by now well understood [**16**], [**18**], [**3**], [**21**], [**5**]. They include nontrivial asymptotic (upper and lower) estimates of the average number of neighbors of a code vector in a code with a given d. This was first studied in a particular case in a series of works on BCH codes [**23**], [**11**], [**14**], and on a more general level in [**5**]. If, on the other hand, the value of $d^{\perp}$ is fixed, then it is possible to prove that the distance distribution of any code is bounded above by a binomial distribution in some segment of weights around $n/2$ and by some other function outside it [**16**], [**18**], [**5**]. Ramifications of this technique include, in particular, an upper bound on the minimum distance of self-dual binary codes [**17**], [**19**], proved via establishing the presence of a binomial component A_w in the weight distribution for sufficiently low w.

For nonbinary linear codes, and in particular, for codes from algebraic curves, estimates of the weight distribution were derived in [**12**], [**13**]. The present paper aims at improving these results by deriving inequalities, which are tighter in many cases and in addition do not rely on linearity of codes. The general method is presented in Section 2. It is applicable to any linear form of the distance coefficients, for instance, probability of undetected error (see [**3**], [**4**] for details) or probability of error for bounded-distance decoding.

1991 *Mathematics Subject Classification.* 94B65.

Research supported in part by Binational Science Foundation (BSF) under grant 1999099.

Known applications of these estimates include an improved lower bound on the reliability function of the q-ary symmetric channel for large q [**13**]. Another application of the bounds derived below is related to bounds on the size of codes with a given d_r (rth higher weight), and follows the lines of [**7**] where only the case $q = 2$ was studied. We improve the results in [**27**] for $r \geq 2$. In [**28**] the bounds of [**27**] were applied to deriving new estimates of the minimal size of secant spaces of algebraic curves. These results are also improved in the present paper.

We usually omit proofs that are generalized straightforwardly from the binary case.

2. Preliminaries

2.1. Notation. Let $X = F_q^n$ be the Hamming space and $C \subset X$ a code. Let $R = (1/n)\log_q |C|$ be the *rate* of C. Its *distance distribution* is the $(n+1)$-vector $(A_0, A_1, \ldots, A_n)$, where

$$A_i = A_i(C) := (1/|C|)|\{(\mathbf{c}, \mathbf{c}') \in C^2 : \mathrm{dist}(\mathbf{c}, \mathbf{c}') = i\}|.$$

The minimal $d \geq 1$ such that $A_i \neq 0$ is called the *distance* of C. Let $a_\xi = \log_q A_{\xi n}$.

The dual distance distribution is defined as a vector $(A_0^\perp, A_1^\perp, \ldots, A_n^\perp)$, where

$$A_j^\perp = \frac{1}{|C|}\sum_{i=0}^{n} A_i K_j(i), \tag{1}$$

$$\sum_j A_j^\perp = |X|/|C|,$$

where $K_j(i)$ is the Krawtchouk number (see Sect. 2.3 below). We have $A_0^\perp = 1, A_i^\perp \geq 0$. The *dual distance* of C is defined as $d^\perp = \min_{1\leq i\leq n}\{i : A_i^\perp > 0\}$. Let $\delta = d/n, \delta^\perp = d^\perp/n$.

Let

$$\begin{aligned} H_q(x) &:= x\log_q(q-1) - x\log_q x - (1-x)\log_q(1-x), \\ \alpha_0(x) &:= R - 1 + H_q(x), \\ \mu(i) &:= \binom{n}{i}(q-1)^i/q^n. \end{aligned}$$

2.2. The method. Let $g_w(x)$ be a real function defined on $\{0, 1, \ldots, n\}$, where $w, 0 \leq w \leq n$, is an integer parameter. Define a moment function of the distance distribution of C,

$$F_w(C) := \sum_{i=d}^{n} g_w(i)A_i(C).$$

Below we derive bounds on F_w for any code with a given d and/or $d^\perp$. In this case we write $F_w(d, d^\perp)$. The same meaning is ascribed to the distance coefficients $A_w(d, d^\perp)$ and their exponents $a_\xi(\delta, \delta^\perp)$, where we put $i = \xi n$, $d = \delta n$, $d^\perp = \delta^\perp n$. The following proposition is a starting point of our estimates.

PROPOSITION 1. [**5**] *Let $Z_w(x) = \sum_{i=0}^n z_i K_i(x)$ be a polynomial such that*

$$z_i \leq 0 \text{ for } d^\perp \leq i \leq n, \text{ and } Z_w(i) \geq g_w(i) \text{ for } d \leq i \leq n. \tag{2}$$

Then

(3) $$F_w(d,d^\perp) \le |C|z_0 - Z_w(0).$$

Let $Y_w(x) = \sum_{i=0}^n y_i K_i(x)$ *be a polynomial such that*

(4) $$y_i \ge 0 \textit{ for } d^\perp \le i \le n, \textit{ and } Y_w(i) \le g_w(i) \textit{ for } d \le i \le n.$$

Then

(5) $$F_w(d,d^\perp) \ge |C|y_0 - Y_w(0).$$

2.3. Krawtchouk polynomials. This is a family of polynomials $K_k(x) = K_k(q;n,x)$ orthogonal on $\{0,1,\dots,n\}$ with weight equal $\mu(i)$ at $i=0,\dots,n$ and 0 otherwise. We have

$$\langle K_i, K_j\rangle = q^n \mu(i)\delta_{ij}.$$

In particular, $K_0 = 1$. For any polynomial $Z(x) = \sum_{i=0} z_i K_i(x)$ we thus have

$$z_i = \frac{\langle Z, K_i\rangle}{\|K_i\|^2}.$$

The following properties are standard:

(6) $$\frac{K_{k+1}(x)K_k(y) - K_k(x)K_{k+1}(y)}{y-x} = \frac{q\mu(k)}{k+1}\sum_{j=0}^{k}\frac{K_j(x)K_j(y)}{\mu(j)}$$

(7) $$\sum_{i=0}^{n}\binom{n-i}{n-w}K_i(j) = q^w\binom{n-j}{w}$$

(8) $$\mu(i)K_j(i) = \mu(j)K_i(j)$$

(9) $$\sum_{i=0}^{n}K_j(i)K_i(k) = q^n\delta_{jk}$$

(10) $$K_i(x)K_j(x) = \sum_{\sigma=0}^{n}p_{ij}^\sigma K_\sigma(x),$$

where $p_{ij}^\sigma \ge 0$ are the intersection numbers of the Hamming scheme:

$$p_{ij}^\sigma = \sum_{\alpha=0}^{\lfloor\frac{i+j-\sigma}{2}\rfloor}\binom{\sigma}{j-\alpha}\binom{j-\alpha}{\sigma+\alpha-i}\binom{n-\sigma}{\alpha}(q-1)^\alpha(q-2)^{i+j-\sigma-2\alpha}$$

Let $t = \tau n$, then for the minimal zero x_t of K_t we have **[2]** $x_t \to n\gamma(\tau)$, where

$$\gamma(\tau) = \big(q-1-(q-2)\tau - 2\sqrt{(q-1)\tau(1-\tau)}\,\big)/q.$$

Note that γ maps the segment $[0,(q-1)/q]$ onto itself, is monotone decreasing, and that $\gamma\circ\gamma = \mathrm{id}$.

Asymptotics of $K_k(x)$ for $q=2$ outside the oscillatory segment was found in **[10]**; for arbitrary q this method yields the following:

(11) $$n^{-1}\log_q K_{\sigma n}(\xi n) = \phi(\sigma,\xi) + o(1),$$

where

$$\phi(\sigma,\xi) = H_q(\sigma) + \int_0^\xi \log_q\frac{(1-y)\beta + y - q\sigma + \sqrt{((1-y)\beta+y-q\sigma)^2 - 4\beta y(1-y)}}{2\beta(1-y)}\,dy,$$

with $\beta := q-1$. In particular, for $\xi = \gamma(\sigma)$, this gives

$$\phi(\sigma,\gamma(\sigma)) = (1 + H_q(\sigma) - H_q(\xi))/2. \tag{12}$$

3. The Singleton bound

In this section we basically re-derive some known results [**12**], [**13**], with a slight improvement and by a different method (described above). The "Singleton" bound is particularly simple and illustrates well the general principle.

In this section C is a code over an alphabet of size $q \ge 2$. Let $k = \log_q |C|, k^\perp = n-k$. Let us choose $g_w(i)$ as follows:

$$g_w(i) = \binom{n-i}{n-w}, w \ge d.$$

Take

$$Z_w(i) = \binom{n-i}{n-w} - c \cdot \binom{n-i}{n-(d-1)}, \tag{13}$$

where c is a constant, chosen later. Using (7), one obtains

$$z_j = q^{w-n}\binom{n-j}{w} - c \cdot q^{d-1-n}\binom{n-j}{d-1}.$$

Since z_j must be nonpositive when j is greater than or equal to $d^\perp$, it is natural to choose c so that $z_{d^\perp} = 0$, i.e.,

$$c = \begin{cases} q^{w-d+1}\frac{\binom{n-d^\perp}{w}}{\binom{n-d^\perp}{d-1}}, & \text{if } d \le w < n-d^\perp+1, \\ 0, & \text{if } n-d^\perp+1 \le w \le n. \end{cases} \tag{14}$$

The ratio

$$\frac{\binom{n-j}{w}}{\binom{n-j}{d-1}} = \frac{(d-1)!}{w!}(n-j-d+1)\dots(n-j-w+1)$$

is decreasing on j. Hence $z_j \le 0, j = d^\perp, \dots, n$. Further,

$$Z_w(i) = \binom{n-i}{n-w} = g_w(i), i = d, \dots, n.$$

Thus the polynomial $Z_w(i)$ is feasible, and we arrive at the following theorem.

THEOREM 2.

$$F_w(d,d^\perp) \le \binom{n}{w}(q^{k+w-n}-1) + c\binom{n}{d-1}(1-q^{k+d-n-1}), 1 \le d \le w,$$

Now let us assume that only the distance (or dual distance) of a code is known. We start with the case of known d. To guarantee feasibility of polynomial (13), we put $d^\perp = 1$ in (14). Now from Theorem 2 we obtain the following corollary.

COROLLARY 1.

$$F_w(d,1) \le \binom{n}{w}\Big[\frac{(n-w)+q^{d-(n-k+1)}(w-d+1)}{n-d+1}q^{w-d+1} - 1\Big] \tag{15}$$

$$\le \binom{n}{w}(q^{w-d+1}-1), \ d \le w. \tag{16}$$

PROOF. The first inequality follows from Theorem 2 by direct substitution. For the second one note that the coefficient of the term q^{w-d+1} in (15) equals 1 if $d = n - k + 1$ and is less than 1 otherwise. $\square$

The following bound was obtained in [**3**] by an application of the polynomial method.

THEOREM 3. [**3**]

$$F_w(d,1) \geq \min\{0, \binom{n}{w}(q^{w-n+k} - 1)\}, \; d \leq w \leq n.$$

It is well known that for MDS codes this bound is tight. Let us prove this using the results in this section. Consider the following example.

EXAMPLE 1. *Let C be an MDS code, that is $d = n - k + 1$. Then by Corollary 1 we have*

$$F_w(d,1) \leq \binom{n}{w}(q^{w-n+k} - 1), d \leq w \leq n.$$

Comparing this with Theorem 3, we see that this bound on $F_w(C)$ is in fact exact. This gives a triangular system of equations for the distance distribution of MDS codes; solving it, we obtain the distribution itself.

Now let us assume that we know only the dual distance $d^\perp$. For the same reason as in the previous case we assume $d = 1$. To derive a bound on $F_w(0, d^\perp)$ we use the MacWilliams equation. From the proof in [**22**] it follows that

$$F_w(C) = q^{w-k^\perp} F_{n-w}(C^\perp) + \binom{n}{w}(q^{w-k^\perp} - 1).$$

Now using the estimate for $F_w(C^\perp)$ from Corollary 1, we obtain the following.

COROLLARY 2.

$$F_w(1, d^\perp) \leq \binom{n}{w}\left[\frac{w + q^{d^\perp-(k+1)}(n - w - d^\perp + 1)}{n - d^\perp + 1} q^{k-d^\perp+1} - 1\right] \tag{17}$$

$$\leq \binom{n}{w}(q^{k-d^\perp+1} - 1), \quad d^\perp \leq w \leq n - d^\perp. \tag{18}$$

PROOF. To prove the first part, substitute (15) into the MacWilliams equation. For the second part note that the coefficient of the term $q^{k-d^\perp+1}$ in (17) equals 1 if $d^\perp = k + 1$ and is less than 1 otherwise. $\square$

Remark: Estimates (16) and (18) were obtained in [**12**] by linear-algebraic arguments.

4. Asymptotic upper bounds and applications

4.1. Upper bounds. The mean distance distribution of a code C chosen in X with uniform probability is given by

$$\mathcal{A}_i = \mu(i)|C| = \binom{n}{i}(q-1)^i|C|/q^n \quad (1 \leq i \leq n)$$

and it is straightforward to show that there exist codes with

$$A_i \leq n^2\binom{n}{i}(q-1)^i|C|/q^n = q^{n\alpha_0(i/n)+o(n)} \quad (1 \leq i \leq n). \tag{19}$$

Codes with distance distribution bounded above by the right-hand side of (19) have a number of important properties; therefore, the bounds obtained in coding theory are usually compared to this "binomial" distribution. In particular, the following upper bound was derived in [**13**] relying on Corollaries 1 and 2 above[1,2]:

$$a_\xi \le \begin{cases} H_q(\xi) - \frac{q\delta}{q-1}, & \delta \le \xi \le \xi^*, \\ \alpha_0(\xi) + (1-\delta^\perp)(1 - H_q(\frac{\xi}{1-\delta^\perp})), & \xi^* \le \xi \le \frac{(q-1)(1-\delta^\perp)}{q}, \\ \alpha_0(\xi), & \frac{(q-1)(1-\delta^\perp)}{q} \le \xi \le \frac{q-1}{q}, \end{cases} \tag{21}$$

where $\xi^* = (1-\delta^\perp)H_q^{-1}((R - \delta^\perp + q\delta/(q-1))/(1-\delta^\perp))$. This estimate of the distance distribution of nonbinary codes is the best available in the literature.

We will state a number of results that improve upon these estimates. Their proofs are given in the next subsection.

THEOREM 4.

$$a_\xi(\delta, 0) \le H_q(\gamma(\delta)) + H_q(\xi) - 1, \quad \delta \le \xi \le (q-1)/q. \tag{22}$$

Comparing this result with the first bound in (21), we see that the estimates differ by $1 - q\delta/(q-1) - H_q(\gamma(\delta))$; this is the difference between the asymptotic Plotkin and Aaltonen [**2**] bounds on the code rate as a function of code's distance. So whenever the latter bound is better (generally, for small code rates), Theorem 4 is an improvement over (21).

THEOREM 5.

$$a_\xi(0, \delta^\perp) \le \begin{cases} R + H(\frac{\delta^\perp}{2}) - \phi(\delta^\perp/2, \xi), & \delta \le \xi \le \gamma(\frac{\delta^\perp}{2}), \\ \alpha_0(\xi), & \gamma(\frac{\delta^\perp}{2}) \le \xi \le \frac{q-1}{q}. \end{cases} \tag{23}$$

Comparing (23) with the second and third parts in (21), we note that whenever $\gamma(\delta^\perp/2) < (q-1)(1-\delta^\perp)/q$, the last theorem yields a wider range in which the distance distribution is asymptotically at most the binomial distance distribution $\mathcal{A}_i$. This holds true for all $\delta^\perp$ if q is small and for small $\delta^\perp$ for large q. The first bound in Theorem 5 also improves the second bound in (21).

THEOREM 6.

$$a_\xi(0, \delta^\perp) \le R - 1 + 2H_q(\omega^*) - g(\xi, \omega^*), \quad 0 \le \xi \le (q-1)/q, \tag{24}$$

where

$$g(x,y) := (\log_2 q)\big[xH_2(\frac{y-\alpha}{x}) + (y-\alpha)H_2(\frac{x+\alpha-y}{y-\alpha})\big] + (1-x)H_q(\frac{\alpha}{1-x}) + (2y - x - 2\alpha)\log_q(q-2), \tag{25}$$

$y - \frac{x}{2} \le \alpha \le y - x$, *and*

$$\omega^* = \begin{cases} \omega^{**} & \text{if } \gamma(\delta^\perp) \le \omega^{**} \le (q-1)/q, \\ \gamma(\delta^\perp) & \text{if } 0 \le \omega^{**} \le \gamma(\delta^\perp), \end{cases}$$

---

[1] In this formula and similar formulas below we disregard the $o(1)$ terms.

[2] In fact, Corollary 1 implies the bound

$$a_\xi \le H_q(\xi) - \delta. \tag{20}$$

The first of the estimates (21), which is slightly better than this, relies on the Plotkin bound.

where

$$\omega^{**} = \arg\min_{\omega}[2H_q(\omega) - g(\xi, \omega)],$$

Note that function $g(x, y)$ has a unique maximum on α in the segment $y - \frac{x}{2} \le \alpha \le y - x$ attained for α that satisfies the equation

$$(26) \qquad (q-1)(1-x-\alpha)(2y-x-2\alpha) = (q-2)^2\alpha(x+\alpha-y)^2.$$

In the case that $\omega^* = \gamma(\delta^\perp)$, the bound of this theorem actually is $\alpha_0(\xi)$ (we do not prove this). The bound in this theorem improves in many cases the second estimate in (21). For instance, for $q = 4, R = 0.8, \delta^\perp = 0.1$, the value ω^{**} reaches $\gamma(\delta^\perp)$ for $\xi = 0.5531 < (q-1)(1-\delta^\perp)/q = 0.6375$. Thus, (24) gives the "binomial" bound starting from smaller values of ξ; it is also less than the second estimate in (21) for $\xi \le 0.5531$. The qualitative picture is unchanged if we take $\delta^\perp = 0.0559$, the Varshamov-Gilbert distance for $R^\perp = 1 - R$. Finally, for high rates (24) is also better than (23). Directions for further (minor) improvements of Theorems 5 and 6 are outlined in [**5**].

Remark: As noted in the end of Sect. 3, estimates (16) and (18), which led to (21), were derived [**12**], [**13**] by another method. Namely, it is well known that the sum $F_w(C) = \sum_{i=0}^{w} \binom{n-i}{n-w} A_i(C)$ equals the sum over all the w-subsets E of $\{1, 2, \ldots, n\}$ of the size of the subcodes $\mathcal{C}(E)$ of C with $\text{supp}(\mathcal{C}(E)) \subseteq E$. Since the distance of $\mathcal{C}(E)$ is at least that of C, we can use an upper bound on the size of a code of length w and distance d to bound $F_w(C)$ above. In [**13**] the authors use the Singleton and Plotkin bounds (cf. (20) and the first inequality in (21)). They also remark that it is possible to use better bounds on the size of codes, which makes the result "slightly better but the formulae are much more complicated and the calculations are very cumbersome." We concur with this remark. More importantly, we tried a few examples with Aaltonen's bound [**2**] to compare this option with Theorem 4. In all the examples, for small and large d and q Theorem 4 gives a uniformly better estimate.

For linear codes from algebraic curves estimate (21) can be improved [**13**]. To state this result let us assume that C is a q-ary geometric Goppa code constructed from a smooth absolutely irreducible projective curve of genus g and an F_q-rational divisor of degree $a \ge 2g - 1$. Let δn and $\delta^\perp n$ be the *designed* distance and dual distance of C. Note that by the construction, $2R = 1 - \delta + \delta^\perp + o(1)$. Then

$$(27) \quad a_\xi \le \alpha_0(\xi) + (1 - R - \frac{\delta}{2} - \xi \log_q(\sqrt{q}+1)), \quad 0 \le \xi \le \frac{(1-\delta^\perp)(\sqrt{q}-1)}{\sqrt{q}}.$$

This bound is better than (21) for normalized weights in the segment $\delta(\frac{1}{q-1} + \frac{1}{2})/\log_q(\sqrt{q}+1) \le \xi \le \frac{(1-\delta^\perp)(\sqrt{q}-1)}{\sqrt{q}}$. The improvement is due to an application of the Clifford theorem that bounds above the dimension of the space of functions associated with a special divisor.

The bound in Theorem 4 is inferior to (27). However, it can be improved as follows. We begin with an auxiliary result.

LEMMA 7. *Let $C \subset F_q^n$ be a code with distance $d(C) = d$. Then for any fixed i, $1 \le i \le n$, and t, $1 \le t \le n - i$, there exists a code $C(t) \subset F_q^{n-t}$ with distance*

$d(C(t)) = d$ such that

$$A_i(C) \le \frac{\binom{n}{t}}{\binom{n-i}{t}} A_i(C(t)).$$

Let $a_\xi(\delta, 0) \le f(\xi, \delta)$, where $f(\xi, \delta)$ is any upper bound on the distance distribution of codes. Then Lemma 7 implies the inequality

$$a_\xi \le \min_{0 \le \tau \le 1-\xi} H_q(\tau) - (1-\xi) H_q\Big(\frac{\tau}{1-\xi}\Big) + (1-\tau) f\Big(\frac{\xi}{1-\tau}, \frac{\delta}{1-\tau}\Big). \tag{28}$$

As it turns out, the first bound in (21) and bounds (20) and (27) are fixed points of this transformation. However, bound (22) is not.

PROPOSITION 8. *Let $0 \le \delta \le (q-2)^2/q^2$ and $\delta \le \xi \le q(q-1)\delta/(q-2)^2$. Then*

$$a_\xi(\delta, 0) \le H_q(\xi) - \big(\delta q \log_q(q-1)\big)/(q-2).$$

EXAMPLE 2. *Let $q = 49$ and consider a sequence of codes from maximal curves, i.e., those with $R = 1 - \delta - (\sqrt{q} - 1)^{-1}$. For $\delta = 0.6$ we then obtain $R = 0.333$, which is better than the Varshamov-Gilbert bound $1 - H_q(\delta) = 0.324$. For ξ in the neighborhood of δ Proposition 8 improves all the other bounds mentioned in this section.*

4.2. Proofs. A. Theorem 4. The proof is generalized straightforwardly from the binary case [**5**].

B. Theorem 5. Take in Proposition 1

$$Z(x) = K_t^2(x) + K_{t+1}^2(x),$$

where $2t + 2 \le d^\perp - 1$, and $g_w = Z$. Then by (10) $z_j = p_{tt}^j + p_{t+1,t+1}^j$, so

$$z_0 = \mu(t) q^n \frac{n(q-1) - t(q-2) + 1}{t+1} > 0$$

and $z_j = 0$ for $j \ge d^\perp$. So the polynomial $Z(x)$ satisfies the necessary restrictions, and furthermore,

$$A_i \le z_0 |C| / Z(i).$$

We still have to choose $t = \tau n$. If $\gamma(\xi) < \delta^\perp/2$, put $\tau = \gamma(\xi)$. Then taking logarithms and using (12), we obtain the first part of the claim. The condition on ξ in the first part follows by applying γ to both parts of $\gamma(\xi) < \delta^\perp/2$. If $\gamma(\xi) \ge \delta^\perp/2$, take $\tau = \delta^\perp/2$. This gives the second part.

C. Theorem 6. Take in Proposition 1 $g_w(i) = Z_w(i)$ and $Z_w(x) = \sum z_j K_j(x)$, where

$$z_j = \frac{w+1}{q} \frac{(K_w(j) + K_{w+1}(j))^2}{a - j},$$

$w = \omega n$ is chosen later, and a is the smallest root of the numerator. Clearly, $z_j < 0$ for $a < j$, so the restriction on z_j is satisfied as long as $d^\perp > a$, and therefore, also if $d^\perp > x_t$. Note that

$$\begin{aligned} z_0 &= \frac{w+1}{aq} q^{2n} (\mu(w) + \mu(w+1))^2 \\ &= \frac{n(q-1) - w(q-2) + 1}{aq} \binom{n}{w}^2 (q-1)^{2w} \\ &= \exp(2n H_q(\omega) + o(n)). \end{aligned}$$

Let us use (6) and (10) in succession to rewrite z_j as follows:

$$z_j = \frac{\binom{n}{w}}{K_w(a)} \sum_{i=0}^{w} \frac{K_i(a)}{\binom{n}{i}} \sum_{s=0}^{n} (p_{w+1,i}^s + p_{wi}^s) K_s(j).$$

Now use (9) to find, for integer x,

$$Z_w(x) = q^n \frac{\binom{n}{w}}{K_w(a)} \sum_{i=0}^{w} \frac{K_i(a)}{\binom{n}{i}} (p_{w+1,i}^x + p_{wi}^x),$$

and, in particular, $Z_w(0) = q^n \binom{n}{w}(q-1)^w$. Further, $K_i(a) \geq 0$ for $0 \leq i \leq w$, so $A_{\xi n} \leq z_0|C|/Z_w(\xi n)$. We have for $Z_w(x)$ the inequality

$$\begin{aligned} Z_w(x) &\geq q^n \frac{\binom{n}{w}}{K_w(a)} \max_{0 \leq i \leq w} \frac{K_i(a)}{\binom{n}{i}} (p_{w+1,i}^x + p_{wi}^x) \\ &\geq q^n p_{ww}^x. \end{aligned}$$

Now let $x = \xi n$, $w = \omega n$. Then we obtain

$$a_\xi \leq R + 2H_q(\omega) - 1 - n^{-1} \log_q p_{ww}^{\xi n} \quad (\gamma(\omega) < \delta^\perp).$$

Further, taking logarithm of the summation term in the sum for p_{ww}^x we obtain $g(\xi, \omega)$. It remains to optimize on the (hitherto free) parameter ω. We take $\omega = \omega^{**}$ as long as the coefficients z_j of $Z_{\omega n}(x)$ are negative for $j = d^\perp, d^\perp + 1, \ldots, n$. This holds true for large n if $\gamma(\omega^{**}) < \delta^\perp$ (which is the same as $\gamma(\delta^\perp) \leq \omega^{**}$). Otherwise we put $\omega = \gamma(\delta^\perp)$.

D. Lemma 7. Let $C(j, \alpha), 1 \leq j \leq n$, be a subcode of C of size $M(j, \alpha)$ defined as follows:

$$C(j, \alpha) = \{\mathbf{v} \in C : v_j = \alpha\}.$$

It is obvious that for any j

$$\sum_{\alpha \in F_q} M(j, \alpha) = |C|.$$

Let $B_i(j, \alpha)$ be the number of ordered pairs of codewords of $C(j, \alpha)$ at a distance i apart, i.e.,

$$B_i(j, \alpha) = |\{\mathbf{v}, \mathbf{u} \in C(j, \alpha) : \operatorname{dist}(\mathbf{v}, \mathbf{u}) = i\}|.$$

Note that the code $C(j, \alpha)$ has distance distribution $A_i(j, \alpha) = B_i(j, \alpha)/M(j, \alpha)$. It is not difficult to see that

$$\sum_{j=1}^{n} \sum_{\alpha \in F_q} B_i(j, \alpha) = (n - i) A_i |C|.$$

Hence there exists an ℓ such that

$$\sum_{\alpha \in F_q} B_i(\ell, \alpha) \geq \frac{n-i}{n} A_i |C|.$$

Now let $\beta \in F_q$ be such that

$$A_i(\ell, \beta) = \frac{B_i(\ell, \beta)}{M(\ell, \beta)} = \max_{\alpha \in F_q} \frac{B_i(\ell, \alpha)}{M(\ell, \alpha)}.$$

Then

$$\begin{aligned}\frac{n-i}{n}A_i|C| &\le \sum_{\alpha} B_i(\ell,\alpha) \le \sum_{\alpha} M(\ell,\alpha)\frac{B_i(\ell,\beta)}{M(\ell,\beta)}\\ &\le \frac{B_i(\ell,\beta)}{M(\ell,\beta)}|C| = A_i(\ell,\beta)|C|.\end{aligned}$$

Let us shorten C with respect to these ℓ and β, i.e., restrict ourselves to the set of vectors $C(\ell,\beta)$ and project them on $(1,\dots,n)\setminus\ell$. Denote the resulting code by $C(1)$; clearly, it satisfies the assertion of the lemma for $t=1$. The full claim now follows by applying the described procedure recursively t times.

E. Proposition 8. Substitute (22) in (28). Upon simplifying we obtain

$$a_\xi \le H_q(\xi) - 1 + \min_{0\le\tau\le 1-\xi}\big(\tau + (1-\tau)H_q(\gamma(\frac{\delta}{1-\tau}))\big).$$

Note that since (22) is valid for $\xi \le (q-1)/q$, for the shortened code this implies a restriction on τ of the form $\xi \le (1-\tau)(q-1)/q$. The minimum on τ is attained for $\tau = 1-q^2\delta/(q-2)^2$ if $0\le\delta<(q-2)^2/q^2$ and for $\tau=0$ if $(q-2)^2/q^2 \le \delta \le (q-1)/q$ (cf. [**1**]). In the first case the value of the minimum equals $(\delta q\log_q(q-1))/(q-2)$; substituting, we obtain the claimed bound. This bound is valid whenever

$$\xi \le \frac{(1-\tau)(q-1)}{q} = \frac{q^2\delta(q-1)}{q(q-2)^2}.$$

4.3. Applications. The main motivation for the estimate (27) in [**13**] was improving upper bounds on the error probability P_e of decoding for q-ary codes used on the q-ary symmetric channel (qSC) with error probability p. The error probability for a code C is defined as

$$P_e(C,p) = \frac{1}{|C|}\sum_{\mathbf{c}\in C}\Pr\{X\setminus D(\mathbf{c})\}$$

where $D(\mathbf{c})$ is the Voronoi region of $\mathbf{c}$ and the probability is computed under the binomial distribution "centered" at $\mathbf{c}$. Let

$$E(R,p) = \limsup \frac{1}{n}\log_q \min_{C\in X, |C|=q^{Rn}} \frac{1}{P_e(C,p)}.$$

It is known that $E(R,p)>0$ for $R\in(0,1-H_q(p))$, and one of the main problems of asymptotic coding theory is establishing the exact form of this function [**9**]. The best known lower estimate of $E(R,p)$ is obtained by computing $P_e(C,p)$ for a sequence of codes C_n with distance distribution that satisfies (19) where $p(n)$ is a function of polynomial growth (note that these codes asymptotically attain the Varshamov-Gilbert bound $\delta := \delta_0(R) = H_q^{-1}(1-R)$). Without loss of generality we can assume that codes are linear and the transmitted vector is 0. The result (the "random coding exponent") has the form

$$E(R,p) \ge \begin{cases} -\delta_0 \log_q \pi_q(p), & 0\le p\le p_e, \quad \text{(a)}\\ 1-R-\log_q(1+(q-1)\pi_q(p)), & p_e\le p\le p_c, \quad \text{(b)}\\ T_q(\delta_0,p)-1+R, & p_c\le p\le\delta_0. \quad \text{(c)}\end{cases} \tag{29}$$

Here

$$\begin{aligned} p_e &= \frac{(2q - q\delta_0 - 2) - 2\sqrt{(q-1)(q-1-q\delta_0)}}{q^2(1-\delta_0)} \\ p_c &= \frac{\delta_0^2}{q\delta_0^2 + (q-1)(1-2\delta_0)} \\ \pi_q(p) &= p\frac{q-2}{q-1} + 2\sqrt{\frac{p(1-p)}{q-1}} \\ T_q(x,p) &= x\log_q(q-1) - x\log_q p - (1-x)\log_q(1-p). \end{aligned}$$

For codes C_n with distance distribution satisfying (19) and error probability in the channel $p \in (0, p_e)$, the error probability of decoding is determined by decoding errors that result in code vectors of weight $\delta_0 n$. Furthermore, for $p \in (p_e, p_c)$, the typical weight of the code vector in the case of error event is $n(q-1)\pi_q(p)/(1+(q-1)\pi_q(p))$. This value equals $n\delta_0$ for $p = p_e$ and shifts away from it for larger p. For codes on maximal curves and large q there is an interval of code rates in which the minimum distance $\delta = 1 - R - (\sqrt{q}-1)^{-1} > \delta_0$. Together with the bound (27) this was used in [**13**] to improve the lower estimate (29) for low p. Since the estimate of Proposition 8 is better than (27), this yields further sharpening of the bound (29a).

As proved by Vlăduţ (see [**13**]), there exist families of algebraic-geometry codes with relative distance

$$\delta \geq \max(\delta_0(R), 1 - R - (\sqrt{q}-1)^{-1})$$

and weight spectrum that, depending on R, is either at most binomial (19) or bounded above by a function smaller than binomial. This result gives better estimates of $E(R,p)$ than the results discussed in this section. However, Vlăduţ's proof is an existence theorem, while the family of codes on maximal curves discussed here is polynomially constructible.

5. Asymptotic lower bounds and applications

5.1. Lower bounds. The following theorem is generalized directly from [**21**].

THEOREM 9. *Let C be a sequence of codes of asymptotic rate R. Then for every $\beta \in [0, H_q^{-1}(R)]$ there exists a number $\xi \in (0, \gamma(\beta))$ such that*

$$a_\xi(C) \geq R + H_q(\beta) - (2/n)\log_q K_{\beta n}(\xi n).$$

This is proved by taking in (5) $Y(x)$ equal to the Christoffel-Darboux kernel of Krawtchouk polynomials. The exponent of $K_{\beta n}(\xi n)$ is given in (11).

Let us sketch a possible improvement of this bound. As suggested in [**21**] for $q = 2$, one can derive a similar lower bound for the distance distribution of constant weight codes (codes in the Johnson space $J_2^{w,n}$) and then use the Bassalygo-Elias argument to transfer the bound to the Hamming space. The nonbinary case is much harder because $J_q^{w,n}$ is a symmetric space of rank 2; hence the zonal spherical functions are expressed by bivariate polynomials $Q_{rs}(x,y)$. We have [**25**]

$$Q_{rs}(i,j) = \frac{\binom{n}{s}}{\binom{w}{s}} K_s(q-1; w-j, i-j) H_{r-s}(n-s, w-s; j),$$

where the $H_m(n, w; x)$ are the dual Hahn polynomials (the zonal spherical functions for $J_2^{n,w}$). Consequently, applications of the polynomial method in this space are technically more difficult, and the only nontrivial result is that of [**1**].

For two vectors $\mathbf{x}, \mathbf{y} \in J_q^{n,w}$ let

$$n(\mathbf{x}, \mathbf{y}) = |\{i : x_i \neq 0, y_i \neq 0\}|, \quad e(\mathbf{x}, \mathbf{y}) = |\{i : x_i = y_i \neq 0\}|,$$

and let $m = \min(w, n - w)$,

$$\begin{aligned} N &= \{(i,j) \in Z^2, \ 0 \le j \le m, j \le i \le w\}, \quad N^* = N - (0,0), \\ K &= \{(r,s) \in Z^2, 0 \le s \le r \le w, \ r - s \le m\} \\ P(p) &= \{(i,j) \in N : \ i + j > p\}. \end{aligned}$$

Let $G \subset J_q^{n,w}$ be a code. Define the set of numbers

$$A_{ij} = A_{ij}(G) = \frac{1}{|G|}\{(\mathbf{x}, \mathbf{y} \in G : \ e(\mathbf{x}, \mathbf{y}) = w - i, \ n(\mathbf{x}, \mathbf{y}) = w - j\}, \quad (i,j) \in N.$$

We have

$$\sum_{(i,j) \in N} A_{ij} = |G|.$$

Let $\{A_u^{(w)}(G), \ 0 \le u \le 2w\}$ be the distance distribution of G. We have

$$A_u^{(w)}(G) = \sum_{(i,j) \in N, i+j=u} A_{ij}(G).$$

By Delsarte's theory we have the following set of inequalities [**25**]:

$$A_{rs}^{\perp} := \frac{1}{|G|} \sum_{(i,j) \in N} A_{ij} Q_{rs}(i,j) \ge 0, \quad (r,s) \in K, \tag{30}$$

Let use use these inequalities to establish the following lower bound on the distance distribution of C. The proof parallels [**21**].

THEOREM 10. *Let C be a code, $1 \le w \le n, 1 \le p \le \min(2w, n)$. Let $F(x,y) = \sum_{(r,s) \in K} f_{rs} Q_{rs}(x,y)$ be a polynomial such that*

$$\begin{aligned} &(i) \ \ f_{rs} \ge 0 && (r,s) \in K, \\ &(ii\ ) \ F(i,j) \le 0 && (i,j) \in P(p), \\ &(iii) \ F(i,j) \ge 0 && (i,j) \in N^* - P(p), \end{aligned}$$

where $1 \le p \le 2w$. Then there exists a number $u, 1 \le u \le p - 1$, and a pair $(m, \ell) \in N^* - P(p)$ such that

$$A_u(C) \ge \frac{q^n \mu(w)}{w^2 h(u) F(m, \ell)} (f_{00}|C|\mu(w) - F(0,0)).$$

PROOF. Let $C(\mathbf{x}, w) := (C - \mathbf{x}) \cap J_q^{n,w}$. We need the following lemma which extends Lemma 2 in [**21**].

LEMMA 11. *Let $\{A_u\}$ be the distance distribution of a code $C \subset X$. Then*

$$|C| A_u(C) h(u) = \sum_{(i,j) \in N, i+j=u} \ \sum_{\mathbf{x} \in X} |C(\mathbf{x}, w)| A_{ij}(C(\mathbf{x}, w)), \tag{31}$$

where $h(u) = p_{ww}^u$ is the intersection number. Explicitly,

$$h(u) = \sum_{s=0}^{u} \binom{u}{s} \binom{s}{2s - u} \binom{n - u}{w - s} (q-1)^{w-s} (q-2)^{2s-u}.$$

PROOF. Let $\mathbf{a}, \mathbf{b}$ be two codewords in C with $\operatorname{dist}(\mathbf{a}, \mathbf{b}) = u$. Let us count the number of vectors $\mathbf{x} \in X$ such that $\mathbf{a}' = \mathbf{x} - \mathbf{a}$ and $\mathbf{b}' = \mathbf{x} - \mathbf{b}$ are both in $J_q^{n,w}$ and $e(\mathbf{a}', \mathbf{b}') = w - i$ and $n(\mathbf{a}', \mathbf{b}') = w - j$. For a given $\mathbf{x}$ the number of such pairs is $|C(\mathbf{x}, w)| A_{ij}(C(\mathbf{x}, w))$, where $i + j = u$, so we obtain the total of

$$\sum_{(i,j)\in N, i+j=u} \sum_{\mathbf{x}\in X} |C(\mathbf{x}, w)| A_{ij}(C(\mathbf{x}, w))$$

such shifts. So the proof will be complete if we show that for a pair $\mathbf{a}, \mathbf{b}$ there are $h(u)$ vectors $\mathbf{x}$ that take it to a pair $\mathbf{a}', \mathbf{b}'$ with the needed properties. We have $\operatorname{dist}(\mathbf{a}, \mathbf{b}) = u$, $\operatorname{dist}(\mathbf{a}, \mathbf{x}) = \operatorname{dist}(\mathbf{b}, \mathbf{x}) = w$. Hence we are looking for the number of triangles with the sides u, w, w; this is precisely p^u_{ww}. □

Next observe that $\sum_{\mathbf{x}\in X} |C(\mathbf{x}, w)| = q^n \mu(w)|C|$. Then

$$\sum_{\mathbf{x}\in X} |C(\mathbf{x}, w)|^2 \ge q^n \mu^2(w)|C|^2$$

since to minimize the sum of squares of some numbers whose sum is fixed we have to take them all equal to each other. Now we compute

$$\begin{aligned}
q^n \mu^2(w)|C|^2 f_{00} &\le f_{00} \sum_{\mathbf{x}\in X} |C(\mathbf{x}, w)|^2 \\
&= f_{00} \sum_{\mathbf{x}\in X} |C(\mathbf{x}, w)|^2 A^\perp_{00}(C(\mathbf{x}, w)) && (A^\perp_{00}(C(\mathbf{x}, w)) = 1) \\
&\le \sum_{\mathbf{x}\in X} |C(\mathbf{x}, w)|^2 \sum_{(r,s)\in K} f_{rs} A^\perp_{rs}(C(\mathbf{x}, w)) && \text{(by (i) and (30))} \\
&= \sum_{\mathbf{x}\in X} |C(\mathbf{x}, w)| \sum_{(r,s)\in K} f_{rs} \sum_{(i,j)\in N} A_{ij}(C(\mathbf{x}, w)) Q_{rs}(i,j) && \text{(by (30))} \\
&= \sum_{\mathbf{x}\in X} |C(\mathbf{x}, w)| \sum_{(i,j)\in N} A_{ij}(C(\mathbf{x}, w)) F(i,j) \\
&= F(0,0) q^n \mu(w)|C| + \sum_{(i,j)\in N^*} \sum_{\mathbf{x}\in X} |C(\mathbf{x}, w)| A_{ij}(C(\mathbf{x}, w)) F(i,j) \\
&\le F(0,0) \mu(w)|C| + \sum_{(i,j)\in N^* - P(p)} \sum_{\mathbf{x}\in X} |C(\mathbf{x}, w)| A_{ij}(C(\mathbf{x}, w)) F(i,j),
\end{aligned}$$

the last inequality by assumption (ii). Now by (iii) we conclude that there exists a pair $(m, \ell) \in N^* - P$ such that

$$\sum_{\mathbf{x}\in X} |C(\mathbf{x}, w)| A_{m\ell}(C(\mathbf{x}, w)) \ge \frac{q^n \mu(w)|C|(f_{00}\mu(w)|C| - F(0,0))}{w^2 F(m, \ell)}.$$

The proof is completed by using this inequality to bound the right-hand side of (31). □

In particular, it is possible to use the polynomial suggested in [**1**] to improve the estimate of Theorem 9.

5.2. Applications. We describe two related applications of Theorem 9 along the lines of [**7**] and [**28**]. Let $C \subset X$ be a linear code. For a code vector $\mathbf{c}$ let $\operatorname{supp}(\mathbf{c}) = \{e : c_e \ne 0\}$. For a subset $A \subset C$ define its support as

$\operatorname{supp}(A) = \cup_{\mathbf{a}\in A} \operatorname{supp}(\mathbf{a})$. The rth generalized weight of C [**29**] is defined as the minimal support of the r-dimensional linear subcode of C. Let

$$\delta_r(R) = \limsup_{n\to\infty} \frac{d_r(C_n)}{n},$$

where the limit is calculated over all sequences of codes of growing length and rate at least R. It is known that $\delta_r(R) : [0,1] \to [0, 1-q^{-r}]$. The following upper bound on $\delta_r(R)$ (the Bassalygo–Elias bound) was obtained in [**8**]:

$$\delta_r(R) \le \delta_r^{\mathrm{el}}(R) := 1 - (\lambda(R))^{r+1}(q-1)^{-r} - (1-\lambda(R))^{r+1}$$

where $\lambda(x) := H_q^{-1}(1-x)$. The following recursive estimate of $\delta_r(R)$ was obtained in [**27**]:

$$\delta_r(R) \le \min_{1\le s\le r-1} \Big(\delta_s^{\mathrm{up}}(R) + (1-\delta_s^{\mathrm{up}}(R))\delta_{r-s}^{\mathrm{up}}\big(\frac{R}{1-\delta_s^{\mathrm{up}}(R)}\big)\Big), \tag{32}$$

where $\delta_i^{\mathrm{up}}(R)$ is any asymptotic upper bound on $\delta_i(R)$. This bound is valid whenever the r.h.s. is defined. Together the last two inequalities give the best upper bound on δ_r currently known for $q \ge 3$. Theorem 9 can be used to improve this result for $r = 2$ (see [**7**] for the binary case).

PROPOSITION 12. *Let $\delta^{\mathrm{up}}(R,\xi)$ be any upper bound on the relative distance of codes of rate R and constant weight ξn. Then*

$$\delta_2(R) \le \min_{0\le\beta\le H_q^{-1}(R)} \max_{0\le\xi\le\gamma(\beta)} \big(\frac{2}{q}\xi + \frac{q-1}{q}\delta^{\mathrm{up}}(R',\xi)\big),$$

where $R' = R + H_q(\beta) - (2/n)\log_q K_{\beta n}(\xi n)$.

PROOF. Let C be a linear code of rate R. By Theorem 9 for every $\beta \in (0, H_q^{-1}(R))$ there exists a constant weight code of size $q^{R'n}$ and weight ξn, $0 < \xi < \gamma(\beta)$. Let $\mathbf{x}, \mathbf{y}$ be two noncollinear vectors from C and $\langle \mathbf{x}, \mathbf{y}\rangle$ their linear span. Let $T = \operatorname{supp}(\mathbf{x}) \cap \operatorname{supp}(\mathbf{y})$, $t = |T|$, and $s = |\{i \in T : x_i = y_i\}|$. Since C is linear, we can assume that $0 \le s \le t/(q-1)$. Now we compute

$$\begin{aligned}\operatorname{dist}(\mathbf{x},\mathbf{y}) &= 2\xi n - t - s \ge 2\xi n - t - \frac{t}{q-1}\\ &= \frac{q}{q-1}(2\xi n - t) - \frac{2}{q-1}\xi n,\end{aligned}$$

which implies the claimed inequality for $|\operatorname{supp}(\langle \mathbf{x},\mathbf{y}\rangle)| = 2\xi n - t$. □

The result of this proposition can be used in (32) to improve upper bounds for $r \ge 3$, though the complete optimization in (32) soon becomes unmanageable [**7**].

EXAMPLE 3. *Let $q = 3, r = 2, R = 0.19$. Let us compare the bounds on $\delta_2(R)$ in (32) and Proposition 12, taking in (32) $s = 1$. Take $\beta = H_q^{-1}(R) = 0.0434$. We still have to choose an upper bound on $\delta_1(R)$ to use in (32) and an upper bound $\delta^{\mathrm{up}}(R,\xi)$. For instance, bounding $\delta_1(R)$ by $\delta_1^{\mathrm{el}}(R)$, we obtain in (32) $\delta_2(R) \le 0.6899$. The best known upper bounds on $\delta_1(R)$ and $\delta(R,\xi)$) were obtained in* [**1**]. *We have $\delta_1(R) \le 0.4600$; using this in* (32), *we obtain $\delta_2(R) \le 0.6878$. Prop. 12 gives a much better estimate $\delta_2(R) \le 0.6173$. The lower VG bound for this rate is* 0.5231.

The second application is related to a more geometric view of higher weights. Let C be a k-dimensional linear code with the generator matrix $A = [\mathbf{a}_1, \dots, \mathbf{a}_n]$, where $\mathbf{a}_i$ denotes the ith column. C is a linear mapping takes $\mathbf{m} \in F_q^k$ to $\mathbf{m}A \in F_q^n$.

If moreover we assume that A contains no all-zero columns, then the $\mathbf{a}_i$'s can be viewed as a (multi)set of points in PF_q^{k-1}, so that $\mathbf{m} \in (PF_q^{k-1})^*$. This view of linear codes was suggested in [**24**]. Hence we have

$$d(C) = n - \max\big(\#(\mathbf{a}_1, \dots, \mathbf{a}_n) \cap H : \ H \text{ a hyperplane in } PF_q^{k-1}\big).$$

This definition is extended naturally to cover higher weights [**26**]:

$$d_r(C) = n - \max\big(\#(\mathbf{a}_1, \dots, \mathbf{a}_n) \cap H : \ H \text{ a space of codim } r \text{ in } PF_q^{k-1}\big)$$

Now let X be a curve over F_q (projective smooth absolutely irreducible) with $\geq n$ F_q-rational points and let H be a projective subspace of PF_q^{k-1} of dimension $\ell = k-1-r$. If $\#H \cap X = m$, then H is called an m-secant ℓ-plane of X. Size of secant spaces is important for the study of geometry of curves over finite fields [**28**]. As observed in that paper, an upper bound on d_r of the form $d_r \leq f_r(n,k)$ implies the existence of an $(n - f_r(n,k))$-secant ℓ-plane for X. Hence Proposition 12 improves the lower asymptotic bound on the size of F_q-rational secant planes.

Finally, lower bounds on the distance distribution were used in [**21**] to improve upper bounds on $E(R,p)$ for the binary symmetric channel. Similar results are possible for the qSC; we will not elaborate on this.

6. Further applications

Let us outline some further results that are possible along the lines of this paper and related works. In Theorems 4–9 we were only interested in the asymptotic bounds on the distance distribution. It is possible to formulate these bounds for specific codes of any given length, taking into account further properties of the codes. As shown in [**5**], this enables one sometimes to improve general results for specific codes. Apart from this, one can use Proposition 1 to derive bounds on other linear forms of the distance coefficients of the code. Among the most well-known examples is the probability of undetected error for a code with the distance distribution $\{A_i, 0 \leq i \leq n\}$ used over a q-ary symmetric channel with crossover probability p, which equals

$$P_{ue}(C,p) = \sum_{i=1}^{n} A_i p^i (q-1)^{-i} (1-p)^{n-i}.$$

Putting $F_w(C) = P_{ue}(C,p)$ in Proposition 1 we can derive lower bounds on this probability (related results were obtained in [**4**], pt.II). Similar estimates are also possible for the error probability of bounded distance decoding as long as the decoding spheres are disjoint.

References

[1] M. Aaltonen, *A new upper bound on nonbinary block codes*, Discrete Mathematics **83** (1990), no. 2-3, 139–160.

[2] M. J. Aaltonen, *Linear programming bounds for tree codes*, IEEE Trans. Inform. Theory **25** (1977), 85–90.

[3] A. Ashikhmin and A. Barg, *Binomial moments of the distance distribution: Bounds and applications*, IEEE Trans. Inform. Theory **45** (1999), no. 2, 438–452.

[4] A. Ashikhmin, A. Barg, E. Knill, and S. Litsyn, *Quantum error detection, I-II*, IEEE Trans. Inform. Theory **46** (2000), no. 3, 778–800.

[5] A. Ashikhmin, A. Barg, and S. Litsyn, *Estimates of the distance distribution of codes and designs*, IEEE Trans. Inform. Theory, to appear.

[6] ———, *Polynomial method in coding and information theory*, Lanl e-print math.CO/9910175, 1999.
[7] ———, *New bounds on generalized weights*, IEEE Trans. Inform. Theory **45** (1999), no. 4, 1258–1263.
[8] G. Cohen, S. Litsyn, and G Zémor, *Upper bounds on generalized distances*, IEEE Trans. Inform. Theory **40** (1994), 2090–2092.
[9] R. G. Gallager, *Information theory and reliable communication*, John Wiley & Sons, New York e.a., 1968.
[10] G. Kalai and N. Linial, *On the distance distribution of codes*, IEEE Trans. Inform. Theory **41** (1995), no. 5, 1467–1472.
[11] T. Kasami, T. Fujiwara, and S. Lin, *An approximation to the weight distribution of binary linear codes*, IEEE Trans. Inform. Theory **31** (1985), 769–780.
[12] G. L. Katsman and M. A. Tsfasman, *Spectra of algebraic-geometric codes*, Problemy Peredachi Informatsii **23** (1987), no. 4, 19–34, (In Russian).
[13] G. L. Katsman, M. A. Tsfasman, and S. G. Vladuţ, *Spectra of linear codes and error probability of decoding*, Coding theory and algebraic geometry (Luminy, 1991), Springer, Berlin, 1992, pp. 82–98.
[14] O. Keren and S. Litsyn, *More on the distance distribution of BCH codes*, IEEE Trans. Inform. Theory **45** (1999), no. 1, 251–255.
[15] I. Krasikov, *Nonnegative quadratic forms and bounds on orthogonal polynomials*, (1999), preprint.
[16] I. Krasikov and S. Litsyn, *Estimates for the range of binomiality in codes' spectra*, IEEE Trans. Inform. Theory **43** (1997), no. 3, 987–991.
[17] ———, *Linear programming bounds for doubly-even self-dual codes*, IEEE Trans. Inform. Theory **43** (1997), no. 4, 1238–1244.
[18] ———, *Bounds on spectra of codes with known dual distance*, Des. Codes Cryptogr. **13** (1998), no. 3, 285–297.
[19] ———, *An improved bound on the minimum distance of doubly-even codes*, IEEE Trans. Inform. Theory **46** (2000), no. 1, 274–278.
[20] V. I. Levenshtein, *Universal bounds for codes and designs*, Handbook of Coding Theory (V. Pless and W. C. Huffman, eds.), vol. 1, Elsevier Science, Amsterdam, 1998, pp. 499–648.
[21] S. Litsyn, *New upper bounds on error exponents*, IEEE Trans. Inform. Theory **45** (1999), no. 2, 385–398.
[22] F. J. MacWilliams, *A theorem in the distribution of weights in a systematic code*, Bell Syst. Techn. Journ. **42** (1963), 79–94.
[23] V. M. Sidelnikov, *The spectrum of weights of binary Bose-Chaudhuri-Hocquenghem codes*, Problemy Peredači Informacii **7** (1971), no. 1, 14–22.
[24] D. Slepian, *A class of binary signalling alphabets*, Bell Syst. Techn. Journ. **35** (1956), no. 1, 203–234.
[25] H. Tarnanen, M. J. Aaltonen, and J.-M. Goethals, *On the nonbinary Johnson scheme*, **6** (1985), 279–285.
[26] M. A. Tsfasman and S. G. Vlăduţ, *Geometric approach to higher weights*, IEEE Trans. Inform. Theory **41** (1995), 1564–1588.
[27] S. G. Vlăduţ, *An upper bound for generalized Hamming weights*, Arithmetic, Geometry and Coding Theory (R. Pellikaan and S. G. Vlăduţ, eds.), W. de Gruyter, Berlin, 1996, pp. 263–267.
[28] ———, *Secant spaces and Clifford's theorem over finite fields*, Finite Fields and Applications **4** (1998), 101–111.
[29] V. Wei, *Generalized Hamming weights for linear codes*, IEEE Trans. Inform. Theory **37** (1991), no. 5, 1412–1418.

Bell Labs, Lucent Technologies, 600 Mountain Ave., Rm. 2C-180, Murray Hill, NJ 07974
E-mail address: aea@research.bell-labs.com

Bell Labs, Lucent Technologies, 600 Mountain Ave., Rm. 2C-375, Murray Hill, NJ 07974
E-mail address: abarg@research.bell-labs.com

EE-Systems, Tel Aviv University, 69987 Ramat Aviv , Israel
E-mail address: litsyn@eng.tau.ac.il

Selected Titles in This Series

(Continued from the front of this publication)

26 **David S. Johnson and Michael A. Trick, Editors,** Cliques, Coloring, and Satisfiability: Second DIMACS Implementation Challenge

25 **Gilbert Baumslag, David Epstein, Robert Gilman, Hamish Short, and Charles Sims, Editors,** Geometric and Computational Perspectives on Infinite Groups

24 **Louis J. Billera, Curtis Greene, Rodica Simion, and Richard P. Stanley, Editors,** Formal Power Series and Algebraic Combinatorics/Séries Formelles et Combinatoire Algébrique, 1994

23 **Panos M. Pardalos, David I. Shalloway, and Guoliang Xue, Editors,** Global Minimization of Nonconvex Energy Functions: Molecular Conformation and Protein Folding

22 **Panos M. Pardalos, Mauricio G. C. Resende, and K. G. Ramakrishnan, Editors,** Parallel Processing of Discrete Optimization Problems

21 **D. Frank Hsu, Arnold L. Rosenberg, and Dominique Sotteau, Editors,** Interconnection Networks and Mapping and Scheduling Parallel Computations

20 **William Cook, László Lovász, and Paul Seymour, Editors,** Combinatorial Optimization

19 **Ingemar J. Cox, Pierre Hansen, and Bela Julesz, Editors,** Partitioning Data Sets

18 **Guy E. Blelloch, K. Mani Chandy, and Suresh Jagannathan, Editors,** Specification of Parallel Algorithms

17 **Eric Sven Ristad, Editor,** Language Computations

16 **Panos M. Pardalos and Henry Wolkowicz, Editors,** Quadratic Assignment and Related Problems

15 **Nathaniel Dean and Gregory E. Shannon, Editors,** Computational Support for Discrete Mathematics

14 **Robert Calderbank, G. David Forney, Jr., and Nader Moayeri, Editors,** Coding and Quantization: DIMACS/IEEE Workshop

13 **Jin-Yi Cai, Editor,** Advances in Computational Complexity Theory

12 **David S. Johnson and Catherine C. McGeoch, Editors,** Network Flows and Matching: First DIMACS Implementation Challenge

11 **Larry Finkelstein and William M. Kantor, Editors,** Groups and Computation

10 **Joel Friedman, Editor,** Expanding Graphs

9 **William T. Trotter, Editor,** Planar Graphs

8 **Simon Gindikin, Editor,** Mathematical Methods of Analysis of Biopolymer Sequences

7 **Lyle A. McGeoch and Daniel D. Sleator, Editors,** On-Line Algorithms

6 **Jacob E. Goodman, Richard Pollack, and William Steiger, Editors,** Discrete and Computational Geometry: Papers from the DIMACS Special Year

5 **Frank Hwang, Fred Roberts, and Clyde Monma, Editors,** Reliability of Computer and Communication Networks

4 **Peter Gritzmann and Bernd Sturmfels, Editors,** Applied Geometry and Discrete Mathematics, The Victor Klee Festschrift

3 **E. M. Clarke and R. P. Kurshan, Editors,** Computer-Aided Verification '90

2 **Joan Feigenbaum and Michael Merritt, Editors,** Distributed Computing and Cryptography

1 **William Cook and Paul D. Seymour, Editors,** Polyhedral Combinatorics